ADDISON-WESLEY
GENERAL MATHEMATICS

MERVIN L. KEEDY

MARVIN L. BITTINGER

STANLEY A. SMITH

PAUL A. ANDERSON

ADDISON-WESLEY PUBLISHING COMPANY

Menlo Park, California • Reading, Massachusetts
Don Mills, Ontario • Wokingham, England • Amsterdam
Sydney • Singapore • Tokyo • Madrid • Bogotá
Santiago • San Juan

ILLUSTRATIONS

Teresa Camozzi: 17, 25, 29, 45, 49, 55, 57, 74, 77, 83, 85, 99, 103, 105, 109, 111, 113, 117, 136, 137, 141, 145, 153, 155, 169, 171, 175, 177, 181, 183, 184, 185, 197, 201, 203, 209, 224, 228, 229, 236, 253, 257, 265, 287, 291, 293, 295, 297, 299, 351, 369, 381, 403, 405, 407, 409, 423, 427, 429, 437

Bill Shields: Cover and chapter openers.

All other illustrations by Addison-Wesley staff.

PHOTOGRAPHS

Elihu Blotnick*: 280, 324

George B. Fry III*: 30, 58, 80, 81, 86, 87, 88, 114, 122, 123, 158, 179, 186, 204, 212, 214, 228, 236, 238, 260, 268, 269, 288, 302, 316, 330, 342, 352, 376, 383, 384, 424, 434, 438

Historical Pictures Service, Inc.: 121

© Leo Touchet/Van Cleve Photography: 350, 412

*Photographs provided expressly for the publisher.

ACKNOWLEDGMENTS

p. 236 left: from Recreational Equipment, Inc. The Co-op Spring 1979 Catalogue, © 1979 by Recreational Equipment, Inc. Reprinted by permission of R.E.I. Co-op, Seattle, Washington.

ISBN 0-201-20647-1

BCDEFGHIJK-VH-8876

AUTHORS

MERVIN L. KEEDY is Professor of Mathematics at Purdue University. He received his Ph.D. degree at the University of Nebraska, and formerly taught at the University of Maryland. He has also taught mathematics and science in junior and senior high schools. He is the author of many books on mathematics, and he is co-author with Marvin L. Bittinger and Stanley A. Smith of *Algebra One, Algebra Two,* and *Algebra Two and Trigonometry* (Addison-Wesley, 1982).

MARVIN L. BITTINGER is Professor of Mathematics Education at Indiana University-Purdue University at Indianapolis. He earned his Ph.D. degree at Purdue University. He is the author of several books and is co-author with Mervin L. Keedy and Stanley A. Smith of *Algebra One, Algebra Two,* and *Algebra Two and Trigonometry* (Addison-Wesley, 1982).

STANLEY A. SMITH is Coordinator, Office of Mathematics (K–12), for Baltimore County Public Schools, Maryland. He has taught junior high school mathematics and science and senior high school mathematics. He earned his M.A. degree at the University of Maryland. He is co-author with Mervin L. Keedy and Marvin L. Bittinger of *Algebra One, Algebra Two,* and *Algebra Two and Trigonometry* (Addison-Wesley, 1982).

PAUL A. ANDERSON is a teacher in the Clark County School District, Las Vegas, Nevada. He earned his M.Ed. degree at the University of Nevada. He is a former principal and instructor of mathematics education at the University of Nevada, Las Vegas. He is a co-author of several K–8 mathematics series.

SPECIAL THANKS

To Marilyn Ohriner, Las Vegas, Nevada, for her research and preparation of the Application and Career lessons.

To Ray Chayo, Norte Del Rio High School, Sacramento, California, for his work on the *Making Practice Fun* supplement.

To all the administrators and teachers who aided in reviewing this text:

Wendell L. Dain
Mathematics Teacher
Earl Wooster High School
Reno, Nevada

Philip P. Halloran
Supervisor of Mathematics
Springfield Public Schools
Springfield, Massachusetts

Geraldine L. Thompson
Mathematics Teacher
William Penn High School
New Castle, Delaware

Dyanne B. Dandridge
Instructional Services Coordinator
Chicago Public Schools
Chicago, Illinois

Peggy Pase
Mathematics Teacher
Trevor G. Browne High School
Phoenix, Arizona

Curtis B. Wolf
Mathematics Department Chairman
Golden Ring Junior High School
Baltimore, Maryland

Bill R. Fisher
Mathematics Teacher
Arlington High School
Indianapolis, Indiana

Georgia Strecker
Mathematics Program Supervisor
Berkeley School District
Berkeley, California

AN INTRODUCTION TO THE TEXT

This text has been designed to help you learn and use mathematics skills.

Here are some of the special features that should make learning easy and effective for you.

1. Full-Color Art and Photography
2. Flowchart Presentation
3. Examples and Try This Exercises
4. Lots of Practice
5. Frequent Problems and Applications
6. Complete Testing Program
7. Career and Calculator Strands

BEFORE THE LESSONS

Each chapter begins with a Pre-Test and a Getting Started activity.

First take the Pre-Test to see which skills you have already mastered.

Then enjoy Getting Started practicing mathematics skills. The whole class or a small group can work together in the activity.

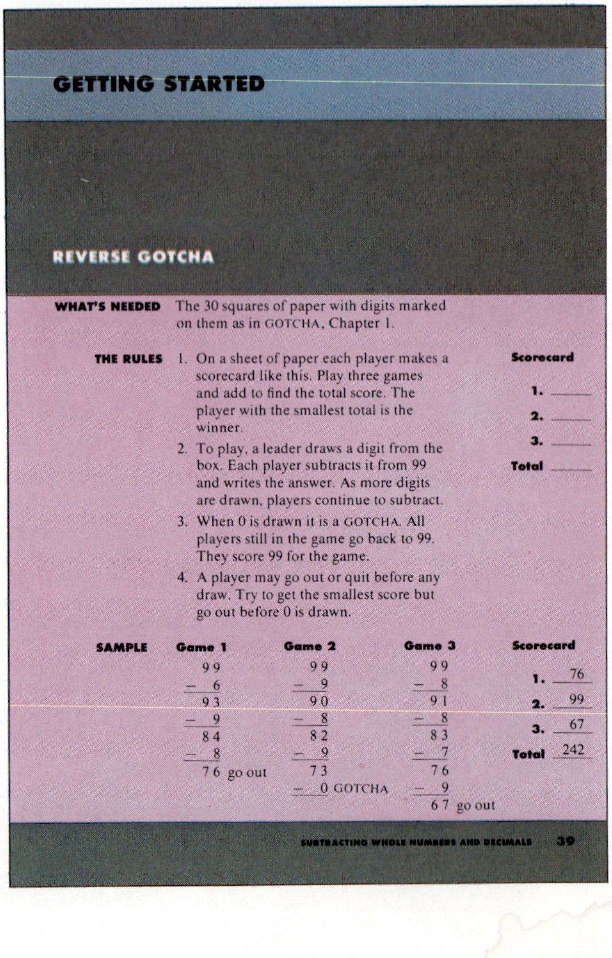

GETTING STARTED

REVERSE GOTCHA

WHAT'S NEEDED The 30 squares of paper with digits marked on them as in GOTCHA, Chapter 1.

THE RULES

1. On a sheet of paper each player makes a scorecard like this. Play three games and add to find the total score. The player with the smallest total is the winner.
2. To play, a leader draws a digit from the box. Each player subtracts it from 99 and writes the answer. As more digits are drawn, players continue to subtract.
3. When 0 is drawn it is a GOTCHA. All players still in the game go back to 99. They score 99 for the game.
4. A player may go out or quit before any draw. Try to get the smallest score but go out before 0 is drawn.

Scorecard
1. ____
2. ____
3. ____
Total ____

SAMPLE

Game 1	Game 2	Game 3
99	99	99
− 6	− 9	− 8
93	90	91
− 9	− 8	− 8
84	82	83
− 8	− 9	− 7
76 go out	73	76
	− 0 GOTCHA	− 9
		67 go out

Scorecard
1. 76
2. 99
3. 67
Total 242

SUBTRACTING WHOLE NUMBERS AND DECIMALS 39

LESSONS

The lessons are presented in an easy-to-read style. First, follow the Flowchart steps. Then work through some Examples. Finally, do the TRY THIS exercises to see if you have learned the skill.

Now you are ready to Practice the skills by completing the exercises and Problem Corner applications which follow the lessons.

If you need More Practice, do the exercises on the indicated pages.

You can Review and Test your skills with exercises at the end of each chapter.

ZEROS IN SUBTRACTING

Subtract. 700 − 268

We must rename to subtract. → Rename a special way. → Subtract.

$$\begin{array}{r} 700 \\ -\,268 \\ \hline \end{array} \qquad \begin{array}{r} \overset{6\,9\,10}{\not{7}\not{0}\not{0}} \\ -\,268 \\ \hline \end{array} \quad 700 = 69 \text{ tens} + 10 \qquad \begin{array}{r} \overset{6\,9\,10}{\not{7}\not{0}\not{0}} \\ -\,268 \\ \hline 432 \end{array}$$

EXAMPLE 1. Subtract. 803 − 379

$$\begin{array}{r} \overset{7\,9\,13}{\not{8}\not{0}\not{3}} \\ -\,379 \\ \hline 424 \end{array}$$

Rename. 803 = 79 tens + 13
Subtract.

EXAMPLE 2. Subtract. 7000 − 3497

$$\begin{array}{r} \overset{6\,9\,9\,10}{\not{7}\not{0}\not{0}\not{0}} \\ -\,3497 \\ \hline 3503 \end{array}$$

Rename. 7000 = 699 tens + 10
Subtract.

TRY THIS Subtract.

1. 600 − 423	**2.** 8004 − 375	**3.** $ 50.03 − 29.47	**4.** $ 30.00 − 6.37

EXAMPLE 3. Subtract. 54,007 − 12,478

$$\begin{array}{r} 5\overset{3\,9\,9\,17}{\not{4},\not{0}\not{0}\not{7}} \\ -\,12,478 \\ \hline 41,529 \end{array}$$

To rename, go to the first non-zero digit.
4007 = 399 tens + 17

TRY THIS Subtract.

5. 36,001 − 15,275	**6.** 70,040 − 19,281	**7.** $ 500.17 − 192.88	**8.** $ 700.09 − 367.82

44 CHAPTER 2

PRACTICE

Subtract.

1. 700 − 341	**2.** 804 − 25	**3.** 300 − 6	**4.** 605 − 197	**5.** 507 − 19
6. 207 − 198	**7.** 100 − 73	**8.** 500 − 371	**9.** 603 − 214	**10.** 400 − 282
11. 7000 − 1234	**12.** 8002 − 3246	**13.** 9000 − 1725	**14.** 5004 − 2119	**15.** 7008 − 1999
16. 4001 − 974	**17.** 3000 − 326	**18.** 2000 − 7	**19.** 1001 − 2	**20.** 8005 − 27
21. 76,005 − 15,738	**22.** 76,050 − 15,783	**23.** 9007 − 2433	**24.** 50,050 − 25,597	**25.** 60,004 − 18,795
26. 42,004 − 23,541	**27.** 7800 − 2974	**28.** 44,000 − 378	**29.** 76,004 − 6,005	**30.** 90,000 − 823
31. $ 4.00 − 1.73	**32.** $ 7.04 − 2.95	**33.** $ 3.00 − 0.37	**34.** $ 8.02 − 0.03	**35.** $ 6.06 − 1.77
36. $ 10.00 − 1.75	**37.** $ 50.05 − 27.68	**38.** $ 38.00 − 17.59	**39.** $ 80.54 − 17.73	**40.** 60.02 − 14.55

PROBLEM CORNER

1. You buy a record for $ 7.98. You give the clerk a ten-dollar bill. How much change should you receive?
2. You buy two tapes for $ 14.75. You give the clerk a twenty-dollar bill. How much change should you receive?

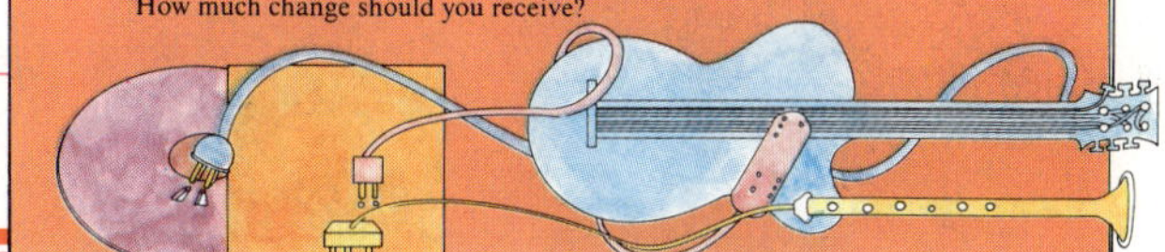

More Practice, page 62 SUBTRACTING WHOLE NUMBERS AND DECIMALS 45

SOLVING PROBLEMS AND CALCULATOR ACTIVITIES

Solving Problems is an important skill you use every day. There are several lessons to help you develop and practice this ability. Master the steps to become a good problem solver.

The Calculator activities will help you practice and sharpen your skills. Discover interesting number patterns and relationships by using a calculator. Some of the activities can be performed by the whole class or a small group.

SOLVING PROBLEMS

When solving a problem, we sometimes need to decide whether to do a mental estimate, a paper and pencil calculation, or use a calculator. For each of the following problem situations, tell whether you would use a calculator, do a paper and pencil calculation, or make a mental estimate.

SITUATION 1 You are at the ballgame. Do you have enough money to buy two hot dogs and a bag of peanuts?

SITUATION 2 You are completing your income tax return.

SITUATION 3 You are deciding about how much it will cost to buy furniture for a bedroom.

SITUATION 4 You are planning a vacation. From a map you want to estimate the approximate number of kilometers you will drive.

SITUATION 5 You are balancing your checkbook.

SITUATION 6 You are in a pastry shop with \$6. You are to purchase a dozen pastries for a party. Pastries sell for \$1.20/dozen, but there is a sign telling that there is a 10% discount for the purchase of 5 dozen or more. Should you take advantage of the discount and buy 5 dozen?

280 CHAPTER 9

CALCULATOR CORNER

WIPE OUT

Players two
Materials two calculators, paper and pencils

SAMPLE Chin enters a five-digit whole number, such as 59,762, on the display. Kristen enters a five-digit whole number, such as 48,829, on the display.

They exchange calculators. Each player must now attempt to "wipe out" the number and get 0 on the display.

To do so they may add, subtract, multiply, or divide by any one or two-digit whole number or decimal except 0.

Kristen		Chin	
Play 1	59,762 −62 59,700	**Play 1**	48,829 −99 48,730
Play 2	×0.01 597	**Play 2**	−30 48,700
Play 3	−97 500	**Play 3**	×0.01 487
Play 4	×0.1 50	**Play 4**	−87 400
Play 5	−50 0	**Play 5**	×0.01 4
		Play 6	−4 0

Kristen's score is 5. Chin's score is 6.

Kristen wins the game, because she reaches 0 in fewer plays.

244 CHAPTER 8

APPLICATIONS AND CAREERS

Learn how skills are used in real-life situations in the Applications. There are two lessons in each chapter in which you can apply mathematics to problems you encounter every day. Discover why you need to master mathematics skills.

You will also see how skills are applied in various Careers. Each chapter contains a Career which focuses on some of the mathematics used in that job. You will learn about the Career while practicing many important skills.

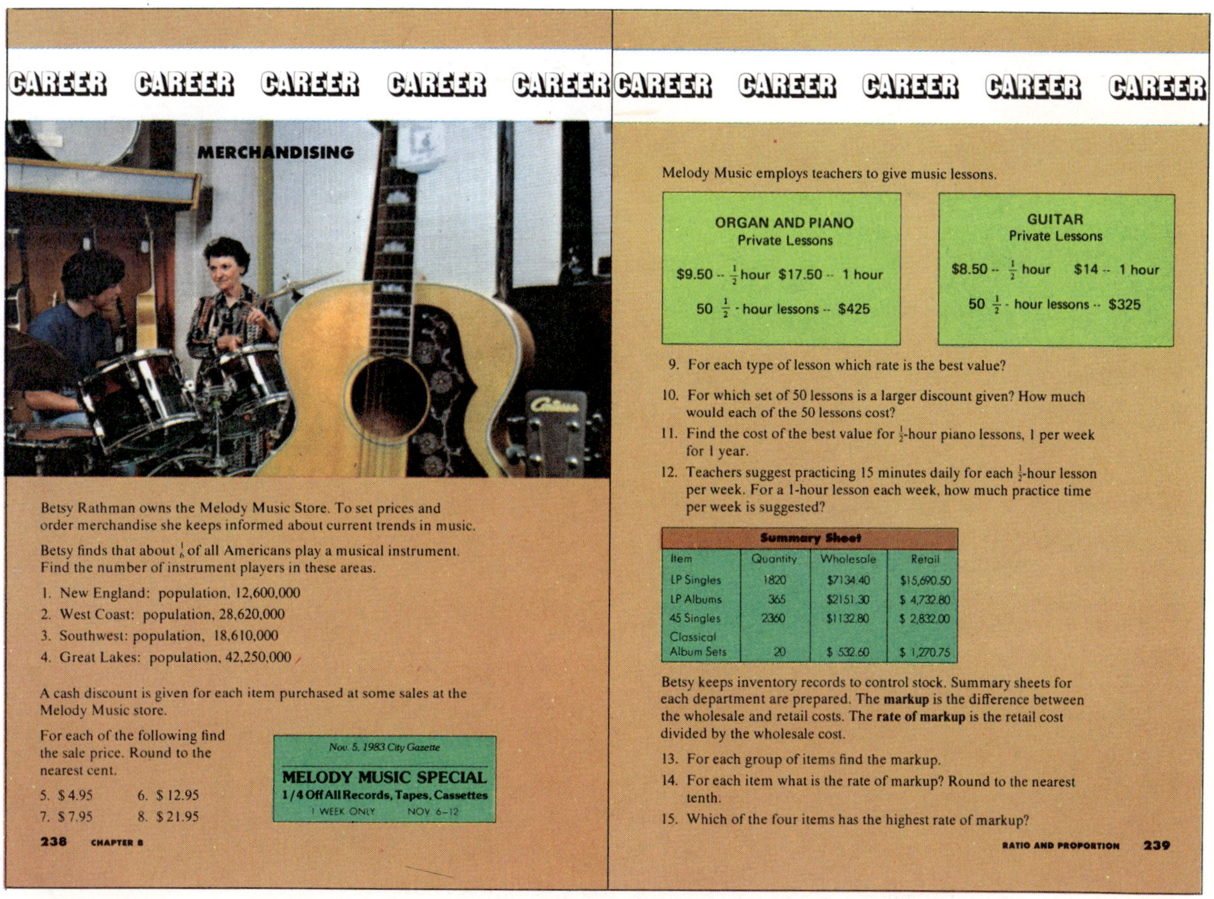

CAREER CAREER CAREER CAREER CAREER CAREER CAREER CAREER CAREER CAREER

MERCHANDISING

Betsy Rathman owns the Melody Music Store. To set prices and order merchandise she keeps informed about current trends in music.

Betsy finds that about $\frac{1}{6}$ of all Americans play a musical instrument. Find the number of instrument players in these areas.

1. New England: population, 12,600,000
2. West Coast: population, 28,620,000
3. Southwest: population, 18,610,000
4. Great Lakes: population, 42,250,000

A cash discount is given for each item purchased at some sales at the Melody Music store.

For each of the following find the sale price. Round to the nearest cent.

5. $4.95
6. $12.95
7. $7.95
8. $21.95

Nov. 5, 1983 City Gazette
MELODY MUSIC SPECIAL
1/4 Off All Records, Tapes, Cassettes
1 WEEK ONLY NOV. 6–12

238 CHAPTER 8

Melody Music employs teachers to give music lessons.

ORGAN AND PIANO
Private Lessons
$9.50 -- $\frac{1}{2}$ hour $17.50 -- 1 hour
50 $\frac{1}{2}$-hour lessons -- $425

GUITAR
Private Lessons
$8.50 -- $\frac{1}{2}$ hour $14 -- 1 hour
50 $\frac{1}{2}$-hour lessons -- $325

9. For each type of lesson which rate is the best value?
10. For which set of 50 lessons is a larger discount given? How much would each of the 50 lessons cost?
11. Find the cost of the best value for $\frac{1}{2}$-hour piano lessons, 1 per week for 1 year.
12. Teachers suggest practicing 15 minutes daily for each $\frac{1}{2}$-hour lesson per week. For a 1-hour lesson each week, how much practice time per week is suggested?

Summary Sheet			
Item	Quantity	Wholesale	Retail
LP Singles	1820	$7134.40	$15,690.50
LP Albums	365	$2151.30	$ 4,732.80
45 Singles	2360	$1132.80	$ 2,832.00
Classical Album Sets	20	$ 532.60	$ 1,270.75

Betsy keeps inventory records to control stock. Summary sheets for each department are prepared. The **markup** is the difference between the wholesale and retail costs. The **rate of markup** is the retail cost divided by the wholesale cost.

13. For each group of items find the markup.
14. For each item what is the rate of markup? Round to the nearest tenth.
15. Which of the four items has the highest rate of markup?

RATIO AND PROPORTION 239

TABLE OF CONTENTS

CHAPTER 1

CHAPTER 2

CHAPTER 3

CHAPTER 4

Dividing Whole Numbers and Decimals

CHAPTER 5

Fractions

CHAPTER 6

CHAPTER 7

CHAPTER 8

CHAPTER 9

Percent

CHAPTER 10

Measurement

CHAPTER 11

Perimeter and Area

CHAPTER 12

CHAPTER 13

CHAPTER 14

CHAPTER 15

APPENDIX

APPLICATIONS AND CAREERS CONTENTS

APPLICATIONS

CAREERS

READY FOR CHAPTER 1

Write standard notation.

1. six thousand, three hundred forty-five

2. five thousand, seventy-one

3. 724 thousand, 607

4. 41 million, 600 thousand

Write money notation.

5. fifty-five dollars and thirty-four cents

6. seventy dollars and seven cents

Write decimal notation.

7. 13 and 424 thousandths

8. 5 and 24 ten-thousandths

Add.

9. $\begin{array}{r} 849 \\ +\ 276 \\ \hline \end{array}$

10. $\begin{array}{r} 8749 \\ +\ 9828 \\ \hline \end{array}$

11. $\begin{array}{r} 498 \\ 756 \\ 49 \\ +\ \ \ 8 \\ \hline \end{array}$

12. $\begin{array}{r} \$\ 7.49 \\ 6.24 \\ +\ 0.29 \\ \hline \end{array}$

13. $\begin{array}{r} 6.9 \\ +\ 7.8 \\ \hline \end{array}$

14. $\begin{array}{r} 14.248 \\ +\ \ 9.876 \\ \hline \end{array}$

15. $\begin{array}{r} 3.98 \\ 1.8 \\ +\ 7.349 \\ \hline \end{array}$

16. $\begin{array}{r} 19.748 \\ 3.98 \\ 7.8 \\ +\ \ 0.72 \\ \hline \end{array}$

Estimate the sums.

17. $\begin{array}{r} 948 \\ 658 \\ +\ 850 \\ \hline \end{array}$

18. $\begin{array}{r} 3486 \\ 2785 \\ +\ \ 320 \\ \hline \end{array}$

19. $\begin{array}{r} \$\ 7.98 \\ 6.78 \\ +\ 3.64 \\ \hline \end{array}$

20. $\begin{array}{r} 3.176 \\ 4.92 \\ +\ 3.8 \\ \hline \end{array}$

Application

21. Write a check from Jason Sanders to Acme Fuel Company in the amount of $ 276.79. Date it October 12, 1981.

22. Find the net deposit.

		Dollars	Cents
Cash	Currency	86	00
	Coin	18	75
Checks	12-24	8	76
	1-107	148	16
Total			
Less Cash Received			-0-
Net Deposit			

1

ADDING WHOLE NUMBERS AND DECIMALS

GETTING STARTED

GOTCHA

WHAT'S NEEDED 30 squares of paper or cardboard about this size

Write 0 on three of them, 1 on three of them, 2 on three of them, and so on up to 9. Put them in a box and mix well.

THE RULES

1. On a sheet of paper each player makes a scorecard like this. Play three games and add to find the total score. The player with the largest total wins.
2. To play, a leader draws digits from the box, one at a time. All players write them down. They are to be added as they are drawn to get the total for the game.
3. When 0 is drawn, it is a GOTCHA. Each player still in the game gets a 0 for that game.
4. A player can quit or go out at any time. Try to get a big score but go out before 0 is drawn.

Scorecard

1. ______

2. ______

3. ______

Total ______

SAMPLE

Game 1	Game 2	Game 3
4	5	9
+ 6	+ 1	+ 8
10	6	17
+ 8	+ 8	+ 9
18	14	26
+ 9	+ 0 GOTCHA	+ 4
27 go out		30
		+ 8
		38 go out

Scorecard

1. 27

2. 0

3. 38

Total 65

USING NUMBERS

We use numbers in many ways.

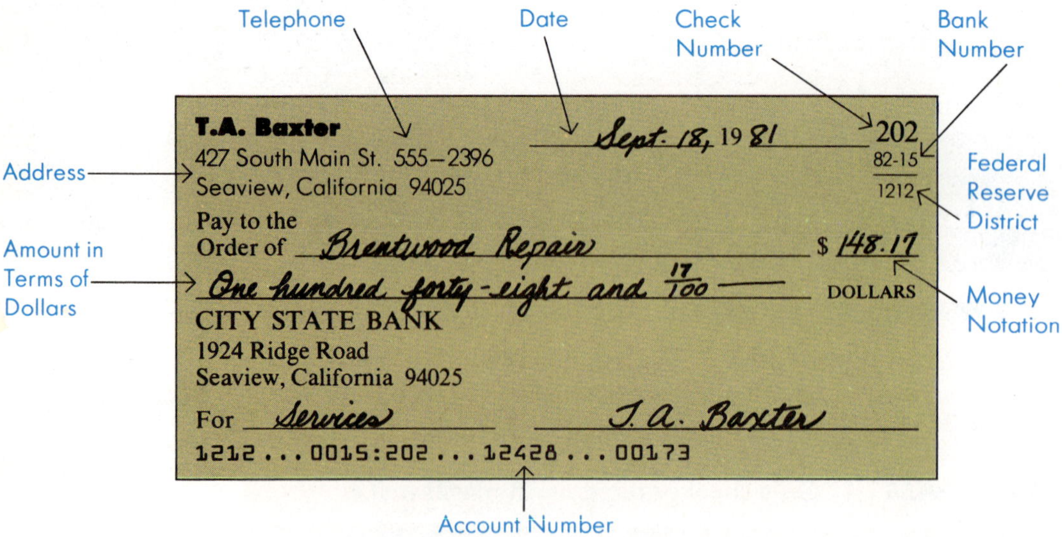

EXAMPLE 1. Write August 14, 1979, in three ways.

We may also write *14 August 79.*

8–14–79	month-day-year	
79–8–14	year-month-day	*(used on some forms)*
14–8–79	day-month-year	*(used by government agencies)*

TRY THIS

Write these dates in three ways.

1. October 21, 1936 **2.** December 25, 1981 **3.** May 7, 1978

EXAMPLES. Tell a likely use.

2. 452–367–2184 Telephone Number *(area code–number)*
3. 312–44–0786 Social Security Number
4. 89102 Zip Code

TRY THIS

Tell a likely use.

4. 512–62–1407 **5.** 46035 **6.** 801–457–0826

To write time notation we use colons to separate hours, minutes, and seconds.

EXAMPLES. Write time notation.

5. 10 hours and 7 minutes — 10:07
6. 9 minutes and 29 seconds — 9:29
7. 15 hours, 8 minutes, and 47 seconds — 15:08:47
8. 12 minutes after 6 o'clock — 6:12

TRY THIS

Write time notation.

7. 28 minutes and 9 seconds
8. 7 hours
9. 16 minutes after 7 o'clock
10. 14 minutes
11. 11 hours and 15 minutes
12. 2 seconds
13. 9 hours, 9 minutes and 9 seconds
14. 14 seconds

PRACTICE

Write these dates in three ways.

1. January 15, 1980
2. February 7, 1979
3. March 21, 1982
4. May 2, 1981
5. December 7, 1941
6. July 4, 1963

Tell a likely use.

7. 317–463–6129
8. 46032
9. 427–86–5104
10. 60651
11. 736–81–4104
12. 21107
13. 415–854–3000
14. 453-54-2016

Write time notation.

15. 7 hours and 45 minutes
16. 11 minutes and 30 seconds
17. 30 minutes after 3 o'clock
18. 5 hours, 18 minutes, and 4 seconds
19. 9 hours and 9 minutes
20. 15 minutes after 1 o'clock
21. 10 hours, 50 minutes and 10 seconds
22. 40 minutes and 40 seconds

PLACE VALUE TO THOUSANDS

To name numbers we use place value.

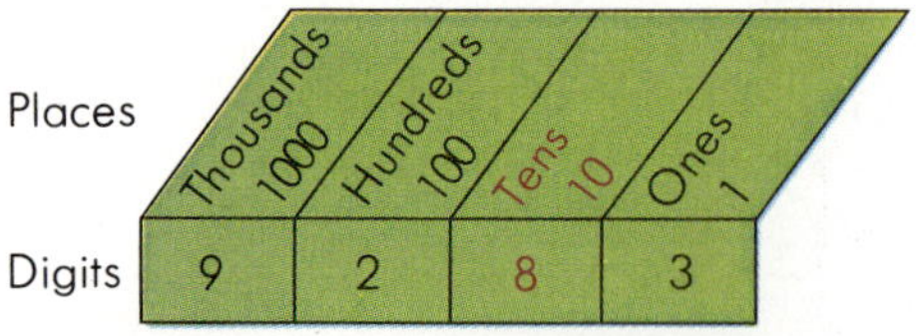

The 8 is in the tens' place. It means 8 × 10 (digit times place), so its value is 80.

EXAMPLE 1. Find the place and value of the underlined digit.

$8\underline{2}79$ The 2 is in the hundreds' place. Its value is 200.

TRY THIS

Find the place and value of the underlined digit.

1. $37\underline{4}1$ **2.** $\underline{5}286$ **3.** $401\underline{5}$ **4.** $\underline{2}536$ **5.** $7\underline{8}61$

To write word names we place a comma after thousands and a hyphen between tens and ones.

EXAMPLES. Write word names.

2. 9283 nine thousand, two hundred eighty-three
3. 580 five hundred eighty
4. 508 five hundred eight

TRY THIS

Write word names.

6. 98 **7.** 376 **8.** 5287 **9.** 6006 **10.** 4160

EXAMPLES. Write standard notation.

5. eight thousand, three hundred forty-nine 8349
6. five hundred five 505
7. seven thousand, seventy 7070

TRY THIS

Write standard notation.

11. three thousand, six hundred fifty-one

12. sixty-two

13. five thousand, eighty-eight

14. eight hundred

15. three thousand, three

16. two thousand

PRACTICE

Find the place and value of the underlined digit.

1. $\underline{4}286$ **2.** $9\underline{8}74$ **3.** $398\underline{5}$ **4.** $2\underline{7}05$ **5.** $173\underline{6}$

6. $6\underline{2}94$ **7.** $\underline{4}876$ **8.** $\underline{9}000$ **9.** $40\underline{7}8$ **10.** $268\underline{4}$

Write word names.

11. 68 **12.** 86 **13.** 41 **14.** 90 **15.** 73 **16.** 89

17. 374 **18.** 276 **19.** 470 **20.** 407 **21.** 593 **22.** 742

23. 3948 **24.** 2789 **25.** 5780 **26.** 6074 **27.** 9871 **28.** 4680

Write standard notation.

29. nine hundred sixty-three

30. seven hundred forty-two

31. one hundred twelve

32. six hundred six

33. four thousand, four hundred forty-four

34. eight thousand, one hundred ninety-three

35. one thousand, five hundred seventy

36. nine thousand, six hundred five

37. eight thousand, nine hundred

38. seven thousand, ninety

39. six thousand, nine

40. five thousand

41. eight hundred thirty-seven

42. three hundred sixteen

43. two thousand, eleven

44. one thousand, sixty-two

More Practice, page 34

WRITING MONEY NOTATION

Money Notation	**Word Name**
73¢ or $ 0.73	seventy-three cents
$ 18 or $ 18.00	eighteen dollars
$ 18.73	eighteen dollars and seventy-three cents
$ 307.90	three hundred seven dollars and ninety cents

We write *and* for the decimal point to separate dollars and cents.

EXAMPLES. Write word names.

1. $ 0.04 four cents **2.** $ 6.00 six dollars
3. $ 9.05 nine dollars and five cents

TRY THIS Write word names.

1. $ 0.01 **2.** $ 17.46 **3.** $ 236.00 **4.** $ 620.18

EXAMPLES. Write money notation.

4. twenty-five cents 25¢ or $ 0.25
5. twenty-five dollars $ 25 or $ 25.00
6. five dollars and eighteen cents $ 5.18
7. ninety dollars and nine cents $ 90.09

TRY THIS Write money notation.

5. thirty cents
6. fifteen dollars and eleven cents
7. thirty dollars
8. six cents
9. one dollar and seven cents
10. eight dollars

For checks we write money notation in terms of dollars. Cents are written as hundredths of a dollar.

EXAMPLES. Write in terms of dollars.

8. $ 73.56 seventy-three and $\frac{56}{100}$ dollars
9. $ 901.07 nine hundred one and $\frac{07}{100}$ dollars
10. $ 7.00 seven and $\frac{00}{100}$ dollars

TRY THIS Write in terms of dollars.

11. $ 9.89 **12.** $ 56.04 **13.** $ 174.00 **14.** $ 16.10

PRACTICE

Write word names.

1. 79¢ **2.** $ 0.87 **3.** $ 0.07 **4.** 8¢ **5.** 30¢

6. $ 19.00 **7.** $ 42 **8.** $ 738 **9.** $ 404.00 **10.** $ 68

11. $ 26.37 **12.** $ 9.84 **13.** $ 38.98 **14.** $ 47.50 **15.** $ 1.67

16. $ 506.05 **17.** $ 870.70 **18.** $ 100.01 **19.** $ 390.09 **20.** $ 350.84

Write money notation.

21. four cents

22. thirteen cents

23. sixty-five cents

24. ninety cents

25. three dollars

26. twenty-one dollars

27. five dollars and forty-two cents

28. two dollars and thirteen cents

29. fifty-three dollars and fifteen cents

30. seventy-five dollars and eleven cents

31. fifty dollars and fifty cents

32. five dollars and five cents

33. sixty dollars and six cents

34. one hundred dollars and one cent

Write in terms of dollars.

35. $ 8.78 **36.** $ 4.25 **37.** $ 39.56 **38.** $ 19.75

39. $ 326.48 **40.** $ 427.86 **41.** $ 125.87 **42.** $ 999.99

43. $ 7.04 **44.** $ 16.09 **45.** $ 407.09 **46.** $ 328.05

47. $ 5.00 **48.** $ 0.95 **49.** $ 10.00 **50.** $ 0.08

More Practice, page 34

APPLICATION

WRITING CHECKS

Most payments are made by checks. Checks are used for safety, time saving and record keeping.

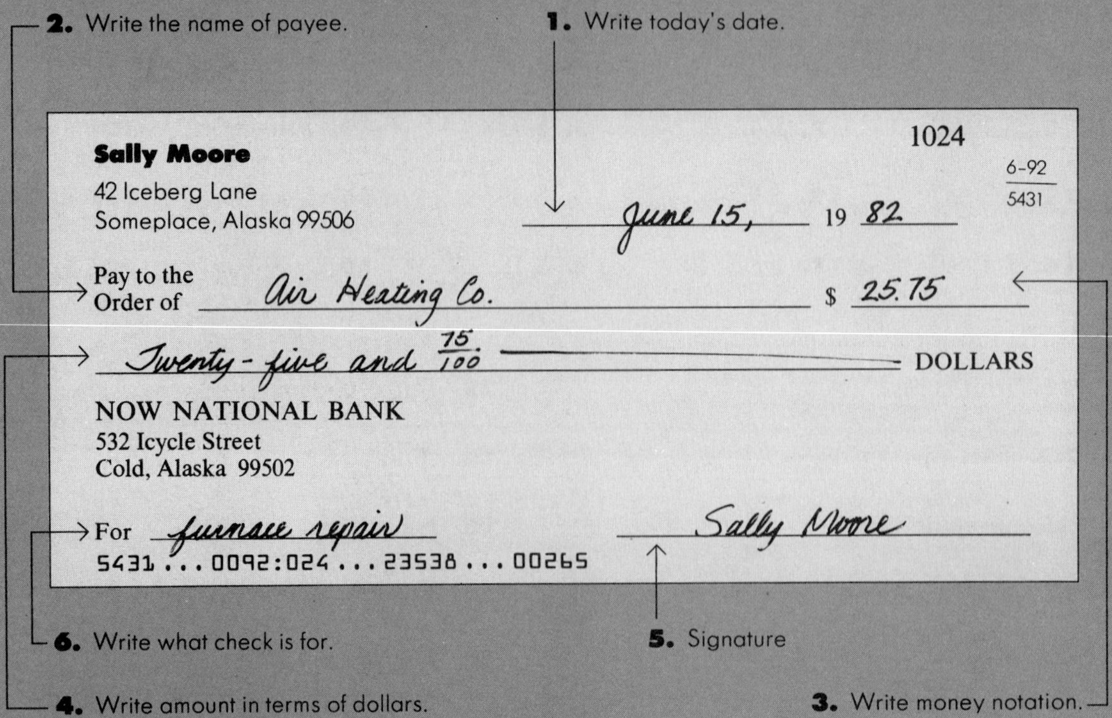

CONSUMER HINTS

1. Use the correct date. Do not date ahead.
2. The payee is the person to whom the money is paid.
3. How much in money notation? Start at the left.
4. How much in terms of dollars? Fill up the space.
5. Sign the same way each time.
6. This is a reminder of what the check was for.

Make copies of checks like this.

19 ______

Pay to the Order of ______ $ ______

______ DOLLARS

For ______ ______

Write these checks. Use the given information.

1. Lucy Lee made a house payment of $ 258.54 to Western Mortgage Company on September 9, 1984.

2. On October 25, 1981 Kazuo Sata wrote a $ 38.24 check to Anderson Food Services for groceries.

3. Ace Catering Service sent Alice Washington a bill for $ 89.00 for a party. Alice paid the bill on December 18, 1980.

4. Paulo Lopez bought a bike from Joe's Cycle Shop on February 19, 1982. It cost $ 158.45. He paid by check.

Write these checks. Use your own name and today's date. Choose the payee and the amount.

5. A restaurant for dinner for two.

6. A friend for payment of a loan.

7. A radio shop for stereo equipment.

8. A sporting goods store for camping gear.

9. A record store for two albums.

10. A clothing store for a sweater.

11. A camera store for having two rolls of film developed.

12. A magazine for a yearly subscription.

PLACE VALUE TO BILLIONS

Periods	Billions			Millions			Thousands			Ones		
Places	Hundred-billions 100,000,000,000	Ten-billions 10,000,000,000	Billions 1,000,000,000	Hundred-millions 100,000,000	Ten-millions 10,000,000	Millions 1,000,000	Hundred-thousands 100,000	Ten-thousands 10,000	Thousands 1000	Hundreds 100	Tens 10	Ones 1
Digits	8	9	0,	1	2	3,	5	6	7,	4	8	9

The 5 is in the hundred-thousands' place. It means 5 × 100,000 (digit times place), so its value is 500,000.

EXAMPLE 1. Find the place and value of the underlined digit.

276,4$\underline{8}$9,124,019 The 8 is in the ten-millions' place.
Its value is 80,000,000.

TRY THIS

Find the place and value of the underlined digit.

1. 1,4$\underline{7}$8,269 **2.** $\underline{7}$65,078,947 **3.** $\underline{8}$76,491,024,000

For large numbers we often use short word names.

EXAMPLES. Write short word names.

2. 123,456,789 123 million, 456 thousand, 789
3. 40,350,027,080 40 billion, 350 million, 27 thousand, 80

TRY THIS

Write short word names.

4. 496,287,104 **5.** 380,087,006,100 **6.** 52,890,003,167

EXAMPLES. Write standard notation.

4. 76 million, 280 thousand, 400 76,280,400
5. 900 billion 900,000,000,000
6. 9 million, 8 thousand 9,008,000

TRY THIS

Write standard notation.

7. 986 million, 280 thousand, 186 **8.** 4 billion **9.** 20 million, 20
10. 300 billion, 200 million, 100 thousand

PRACTICE

Find the place and value of the underlined digit.

1. $\underline{3}7{,}486$ **2.** $4\underline{2}5{,}807$ **3.** $1\underline{9}{,}875$ **4.** $\underline{2}6{,}108$

5. $3\underline{6}{,}498{,}248$ **6.** $\underline{7}{,}864{,}297$ **7.** $\underline{3}68{,}270{,}027$ **8.** $\underline{5}{,}734{,}802$

9. $\underline{9}{,}846{,}701{,}425$ **10.** $\underline{8}9{,}678{,}130{,}031$ **11.** $42\underline{9}{,}878{,}000{,}000$ **12.** $5\underline{1}{,}438{,}921{,}035$

13. $\underline{7}84{,}965$ **14.** $\underline{8}6{,}000{,}000$ **15.** $\underline{8}6{,}000{,}000{,}000$ **16.** $692{,}170{,}\underline{3}85$

Write short word names.

17. 37,826 **18.** 428,874 **19.** 5687 **20.** 173,840

21. 80,000 **22.** 984,126,481 **23.** 82,476,125 **24.** 64,200

25. 9,009,009 **26.** 80,800,008 **27.** 9,008,007,006 **28.** 780,000

29. 73,816,479,100 **30.** 728,000,000 **31.** 7,000,000,007 **32.** 135,000,000

Write standard notation.

33. 28 thousand, 514 **34.** 348 thousand, 704

35. 36 thousand, 15 **36.** 999 thousand

37. 7 million **38.** 28 million

39. 18 million, 238 thousand, 375 **40.** 326 million, 400 thousand, 18

41. 5 million, 5 thousand, 5 **42.** 18 million, 18 thousand

43. 1 billion **44.** 38 billion

More Practice, page 34

ADDING WHOLE NUMBERS

Add. 378 + 519

Line up vertically. → Add the ones. → Add the tens. → Add the hundreds.

$$\begin{array}{r} 378 \\ +\,519 \\ \hline \end{array} \qquad \begin{array}{r} \overset{1}{7}8 \\ +\,519 \\ \hline 7 \end{array} \qquad \begin{array}{r} 3\overset{1}{7}8 \\ +\,519 \\ \hline 97 \end{array} \qquad \begin{array}{r} 3\overset{1}{7}8 \\ +\,519 \\ \hline 897 \end{array}$$

EXAMPLES. Add.

1. $\begin{array}{r} 316 \\ +\;\;43 \\ \hline 359 \end{array}$ **2.** $\begin{array}{r} \overset{1}{4}\overset{1}{8}9 \\ +\,637 \\ \hline 1126 \end{array}$

TRY THIS

Add.

1. 476 + 13
2. 878 + 239 + 18
3. 91 + 9 + 476

4. $\begin{array}{r} 708 \\ +\,897 \\ \hline \end{array}$ **5.** $\begin{array}{r} 348 \\ 16 \\ +\;\;\;7 \\ \hline \end{array}$ **6.** $\begin{array}{r} 28 \\ 113 \\ +\;\;99 \\ \hline \end{array}$

EXAMPLE 3. Add. \$6.58 + \$0.97 + \$0.06.

$$\begin{array}{r} \$\,\overset{1}{6}.\overset{2}{5}8 \\ 0.97 \\ +\,0.06 \\ \hline \$7.61 \end{array}$$

Add as with whole numbers.

Write money notation in the answer.

TRY THIS

Add.

7. \$4.96 + \$0.86 + \$0.05 **8.** \$0.76 + \$0.09 + \$0.87

9. $\begin{array}{r} \$\,4.93 \\ 0.29 \\ +\,6.34 \\ \hline \end{array}$ **10.** $\begin{array}{r} \$\,9.48 \\ 7.04 \\ +\,0.09 \\ \hline \end{array}$ **11.** $\begin{array}{r} \$\,0.49 \\ 0.32 \\ +\,1.78 \\ \hline \end{array}$ **12.** $\begin{array}{r} \$\,7.64 \\ 0.08 \\ +\,9.37 \\ \hline \end{array}$

PRACTICE

Add.

1. 14 + 72
2. 13 + 46
3. 81 + 18
4. 36 + 33
5. 14 + 42
6. 238 + 611
7. 721 + 275
8. 976 + 22
9. 80 + 914
10. 7 + 720
11. 98 + 76
12. 12 + 99
13. 76 + 8
14. 47 + 9
15. 66 + 77
16. 347 + 286
17. 255 + 55
18. 975 + 87
19. 498 + 74
20. 787 + 9
21. 731 + 979 + 486
22. 483 + 56 + 809
23. 829 + 331 + 97
24. 648 + 926 + 9
25. 999 + 99 + 9
26. 498 + 786 + 594 + 687
27. 678 + 43 + 295 + 9
28. 79 + 98 + 98 + 42
29. 876 + 429 + 15 + 4
30. 987 + 36 + 8 + 7
31. $8.76 + 4.29
32. $8.07 + 9.99
33. $9.47 + 3.86
34. $7.49 + 0.87
35. $7.93 + 0.08
36. $0.76 + 0.94 + 0.07
37. $7.95 + 0.08 + 5.42
38. $8.94 + 0.75 + 9.27
39. $1.23 + 4.56 + 0.10
40. $9.87 + 6.54 + 0.46

PROBLEM CORNER

Use the bar graph.

1. How many cars were parked on the third floor?
2. On which floor were the most cars parked?
3. The top three floors were reserved for all-day parking. How many cars parked all day?

Cars Parked in a City Garage

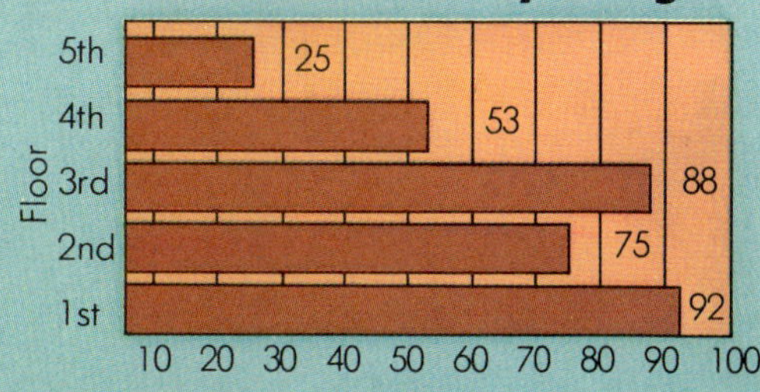

More Practice, page 34

ADDING LARGER NUMBERS

Add. 6297 + 8059

Line up vertically. → Add each column. Start at the right.

$$\begin{array}{r} 6297 \\ +\ 8059 \\ \hline \end{array} \qquad \begin{array}{r} {\scriptstyle 1\,1\ \ } \\ 6297 \\ +\ 8059 \\ \hline 14{,}356 \end{array}$$

EXAMPLE 1. Add. 897,348 + 37,496

$$\begin{array}{r} {\scriptstyle 1\,1\ \ 1\,1\ \ } \\ 897{,}348 \\ +\ \ 37{,}496 \\ \hline 934{,}844 \end{array}$$

Line up.

Add.

TRY THIS

Add.

1. 13,247 + 9129

2. 29,100 + 684

3. $\begin{array}{r} 427{,}986 \\ +\ 794{,}129 \\ \hline \end{array}$

4. $\begin{array}{r} \$\ 764.97 \\ +\ 274.86 \\ \hline \end{array}$

5. $\begin{array}{r} \$\ 298.67 \\ +\ 702.33 \\ \hline \end{array}$

EXAMPLE 2. Add. 734,198 + 98,374 + 9,479

$$\begin{array}{r} {\scriptstyle 1\,2\,1\,2\,2\ } \\ 734{,}198 \\ 98{,}374 \\ +\ \ \ 9{,}479 \\ \hline 842{,}051 \end{array}$$

Line up.

Add.

TRY THIS

Add.

6. 978,419 + 376,673 + 42,999 + 1248

7. $\begin{array}{r} \$\ 427.91 \\ 987.46 \\ +\ \ 94.87 \\ \hline \end{array}$

8. $\begin{array}{r} 398{,}276 \\ 672{,}893 \\ 728{,}936 \\ +\ \ 92{,}487 \\ \hline \end{array}$

9. $\begin{array}{r} 465{,}201 \\ 9{,}455 \\ 68{,}139 \\ +\ 559{,}246 \\ \hline \end{array}$

PRACTICE

Add.

1. 7486 + 9281	**2.** 4236 + 6143	**3.** 7281 + 3409	**4.** 9876 + 1234
5. 46,287 + 94,789	**6.** 72,287 + 63,943	**7.** 74,198 + 82,798	**8.** 63,078 + 92,937
9. 498,276 + 987,409	**10.** 763,297 + 429,846	**11.** 943,217 + 629,805	**12.** 348,296 + 296,348
13. 73,846 + 9,017	**14.** 498,766 + 9,487	**15.** 368,499 + 386	**16.** 369,421 + 86,493
17. 14,798 29,287 + 98,075	**18.** 424,688 327,941 + 327,198	**19.** 8294 3789 + 6081	**20.** 728,495 62,481 + 9,987
21. 9824 7698 5247 + 4098	**22.** 18,476 29,487 38,265 + 47,154	**23.** 268,497 862,794 321,543 + 629,980	**24.** 798,476 420,895 15,746 + 9,285
25. $ 85.98 + 37.68	**26.** $ 246.97 + 579.23	**27.** $ 17.47 + 9.88	**28.** $ 258.17 + 9.49

29. $ 16.49 + $ 87.68 + $ 15.45

30. $ 37.48 + $ 26.49 + $ 5.97 + $ 0.74

PROBLEM CORNER

Watch for extra information.

1. Mrs. James bought a dress for $ 19.95, marked down from $ 30.00. She also bought a scarf for $ 3.29. How much did she spend in all?

2. There were 29,348 adults and 15,478 children at the game. There were seats for 54,345 people. How many people were at the game?

More Practice, page 35

APPLICATION

MAKING DEPOSIT SLIPS

When you deposit money in your checking account you must fill out a deposit slip.

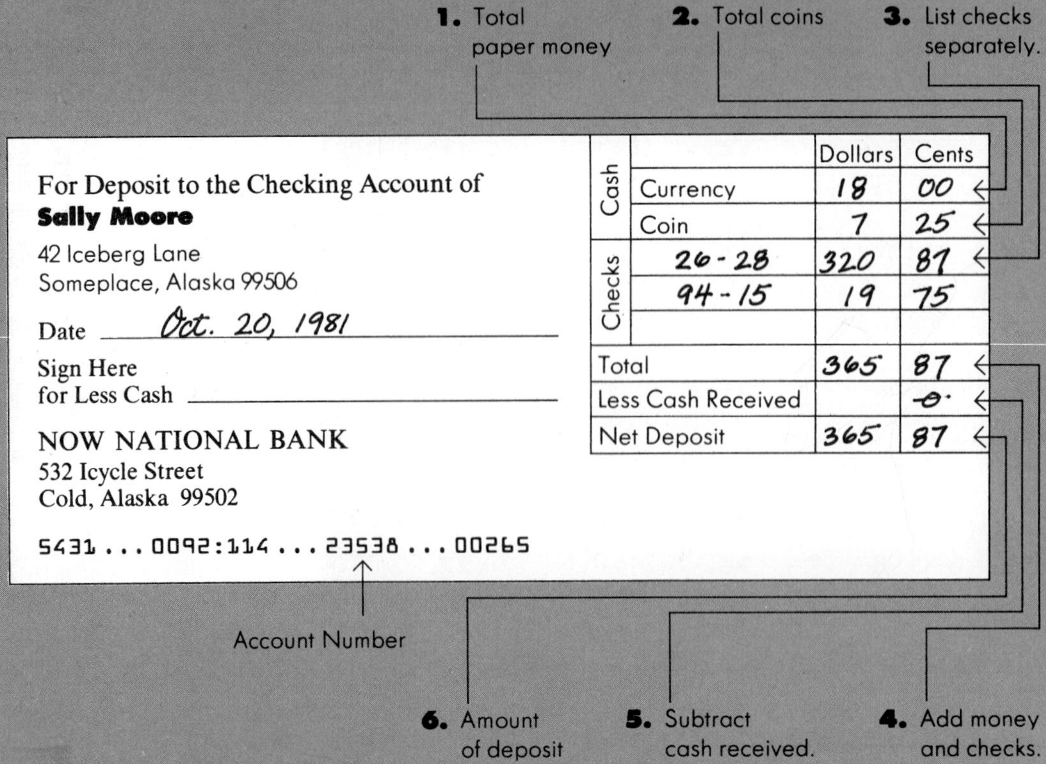

CONSUMER HINTS

1. Arrange bills from greatest on top to least.
2. Put coins in paper rolls, if you have a large number of coins.
3. Endorse each check in ink. List them separately by bank number and amount.
4. Add currency, coins and checks.
5. Subtract the amount you want back. Sign your name to indicate you received the money.
6. Record the amount actually deposited.

Find the net deposits.

1.

		Dollars	Cents
Cash	Currency	12	00
	Coin	7	45
Checks	21-17	18	24
	16-12	7	98
Total			
Less Cash Received			
Net Deposit			

2.

		Dollars	Cents
Cash	Currency	5	00
	Coin		50
Checks	21-87	8	75
Total			
Less Cash Received			
Net Deposit			

3.

		Dollars	Cents
Cash	Currency		
	Coin		
Checks	19-29	7	49
	18-26	14	98
Total			
Less Cash Received			
Net Deposit			

4.

		Dollars	Cents
Cash	Currency	8	00
	Coin		
Checks	21-86	195	00
Total			
Less Cash Received			
Net Deposit			

5.

		Dollars	Cents
Cash	Currency	28	00
	Coin	18	95
Checks			
Total			
Less Cash Received			
Net Deposit			

6.

		Dollars	Cents
Cash	Currency	10	00
	Coin	7	86
Checks	59-18	185	45
	62-52	30	00
	28-21	156	00
Total			
Less Cash Received			
Net Deposit			

Make copies of deposit slips like the ones above. Write these deposits.

7. 1 ten-dollar bill, 2 five-dollar bills, 3 one-dollar bills, 3 half-dollar coins, 1 quarter, 2 dimes, 1 nickel, check 94-18 for $ 19.50, check 25-16 for $ 75.00.

8. 2 twenty-dollar bills, 3 ten-dollar bills, 7 one-dollar bills, 1 two-dollar roll of nickels, 2 ten-dollar rolls of dimes, check 46-3 for $ 17.85.

9. 10 half-dollars, 8 nickels, check 21-20 for $ 127.05, check 24-20 for $ 84.46.

10. 2 one-hundred-dollar bills, 3 fifty-dollar bills, 5 ten-dollar bills, check 92-18 for $ 375.00, check 52-12 for $ 795.95.

DECIMAL PLACE VALUE

The decimal point separates the whole number from the part less than 1.

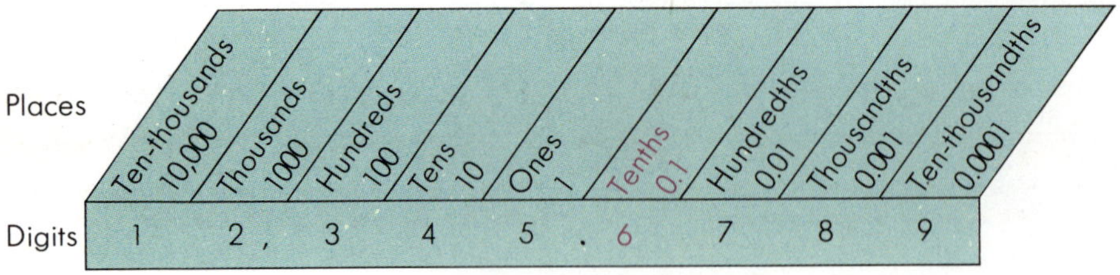

The 6 is in the tenths' place. It means 6 × 0.1 (digit times place), so its value is 0.6.

EXAMPLE 1. Find the place and value of the underlined digit.

7.04$\underline{8}$2 The 8 is in the thousandths' place. Its value is 0.008.

TRY THIS

Find the place and value of the underlined digit.

1. 73.$\underline{4}$294 **2.** 6.7$\underline{9}$6 **3.** 927.48$\underline{6}$3 **4.** 65.034$\underline{5}$

To write short word names we write *and* for the decimal point. We use the name of the last decimal place.

EXAMPLES. Write short word names.

2. 6.5 6 and 5 tenths
3. 6.54 6 and 54 hundredths
4. 6.05 6 and 5 hundredths
5. 6.543 6 and 543 thousandths
6. 6.1002 6 and 1002 ten-thousandths
7. 0.0007 7 ten-thousandths

TRY THIS

Write short word names.

5. 7.3 **6.** 27.386 **7.** 7.0492 **8.** 75.08

EXAMPLES. Write decimal notation.

8. 9 and 6 tenths	9.6	
9. 7 and 42 hundredths	7.42	
10. 87 thousandths	0.087	*Write 0's to fill the places.*
11. 5 and 126 ten-thousandths	5.0126	

TRY THIS Write decimal notation.

9. 6 hundredths **10.** 47 and 58 thousandths **11.** 18 hundredths

PRACTICE

Find the place and value of the underlined digit.

1. $9.\underline{4}872$	**2.** $47.\underline{4}6$	**3.** $0.\underline{5}98$	**4.** $376.\underline{5}$	**5.** $1.4\underline{8}$
6. $46.9\underline{7}24$	**7.** $5.07\underline{6}$	**8.** $31.907\underline{5}$	**9.** $286.\underline{4}98$	**10.** $27.08\underline{2}$
11. $94.704\underline{9}$	**12.** $8.0\underline{1}24$	**13.** $0.00\underline{6}$	**14.** $3.123\underline{4}$	**15.** $5.00\underline{7}$

Write short word names.

16. 3.7	**17.** 49.8	**18.** 73.28	**19.** 9.326	**20.** 4.98
21. 17.4295	**22.** 95.0875	**23.** 0.38	**24.** 1.141	**25.** 3.46
26. 295.6	**27.** 36.9853	**28.** 8.7	**29.** 2.74	**30.** 27.402
31. 3.095	**32.** 0.785	**33.** 7.0007	**34.** 7000.7	**35.** 60.004

Write decimal notation.

36. 7 and 15 hundredths	**37.** 95 and 7 tenths	**38.** 16 and 2 tenths
39. 16 and 7 hundredths	**40.** 3 and 3 thousandths	**41.** 1 and 3 hundredths
42. 26 and 275 thousandths	**43.** 1 ten-thousandth	**44.** 280 and 4 tenths
45. 8 and 27 hundredths	**46.** 5 tenths	**47.** 78 and 5 thousandths
48. 784 thousandths	**49.** 19 and 48 thousandths	**50.** 4 hundredths

More Practice, page 35

ADDING DECIMALS

Add. 21.668 + 5.491

Line up the decimal points.	→ Add as with whole numbers	→ Place the decimal point in the sum.
21.668	21.668	21.668
+ 5.491	+ 5.491	+ 5.491
	27159	27.159

EXAMPLE 1. Add. 18.9 + 9.376

```
   1 1
   18.900     Line up the decimal points.
 +  9.376     Write 0's to fill the places.
   28.276     Add.
```

TRY THIS

Add.

1. 18.76 + 19.9 **2.** 14.7 + 9.435

3. 9.786 + 0.895

4. 165.97 + 36.075 **5.** 0.765 + 0.89 **6.** 26.3 + 0.985

EXAMPLE 2. Add. 9.7 + 0.89 + 16.756 + 8.6

```
   2 2 1
    9.700     Write 0's to fill the places.
    0.890
   16.756     Add.
 +  8.600
   35.946     Place the decimal point in the sum.
```

TRY THIS

Add.

7. 4.3 + 0.876 + 21.42 **8.** 0.96 + 0.096 + 9.6

9. 4.95, 0.879, 19.8, + 0.007 **10.** 0.15, 0.19, 0.875, + 0.919 **11.** 5.64, 0.7, 0.831, + 16.92

PRACTICE

Add.

1. 19.6 + 7.9 **2.** 27.9 + 16.1 **3.** 37.3 + 19.8 **4.** 0.9 + 8.8

5. 47.98 + 16.84 **6.** 2.87 + 9.03 **7.** 6.87 + 9.9 **8.** 3.76 + 0.9

9. $\begin{array}{r} 7.398 \\ +\ 0.475 \\ \hline \end{array}$ **10.** $\begin{array}{r} 37.485 \\ +\ 18.08 \\ \hline \end{array}$ **11.** $\begin{array}{r} 2.786 \\ +\ 0.9 \\ \hline \end{array}$ **12.** $\begin{array}{r} 63.284 \\ +\ 19.986 \\ \hline \end{array}$

13. $\begin{array}{r} 8.736 \\ +\ 9.49 \\ \hline \end{array}$ **14.** $\begin{array}{r} 36.892 \\ +\ 18.08 \\ \hline \end{array}$ **15.** $\begin{array}{r} 0.7 \\ +\ 0.6 \\ \hline \end{array}$ **16.** $\begin{array}{r} 297.8 \\ +\ 96.88 \\ \hline \end{array}$

17. $\begin{array}{r} 13.767 \\ 4.382 \\ +\ 7.689 \\ \hline \end{array}$ **18.** $\begin{array}{r} 3.4 \\ 9.6 \\ +\ 7.3 \\ \hline \end{array}$ **19.** $\begin{array}{r} 7.86 \\ 4.08 \\ +\ 0.9 \\ \hline \end{array}$ **20.** $\begin{array}{r} 13.621 \\ 9.78 \\ +\ 0.5 \\ \hline \end{array}$

21. $\begin{array}{r} 0.78 \\ 6.94 \\ 8.72 \\ +\ 4.81 \\ \hline \end{array}$ **22.** $\begin{array}{r} 16.284 \\ 9.786 \\ 4.395 \\ +\ 5.007 \\ \hline \end{array}$ **23.** $\begin{array}{r} 9.8 \\ 19.4 \\ 0.98 \\ +\ 5.9 \\ \hline \end{array}$ **24.** $\begin{array}{r} 874.1 \\ 87.41 \\ 8.741 \\ +\ 0.874 \\ \hline \end{array}$

25. $\begin{array}{r} 7.9 \\ 6.84 \\ 14.3 \\ +\ 8.9 \\ \hline \end{array}$ **26.** $\begin{array}{r} 16.9 \\ 3.18 \\ 0.8 \\ +\ 0.9 \\ \hline \end{array}$ **27.** $\begin{array}{r} 0.68 \\ 6.8 \\ 3.9 \\ +\ 0.395 \\ \hline \end{array}$ **28.** $\begin{array}{r} 4.875 \\ 0.129 \\ 97.6 \\ +\ 0.976 \\ \hline \end{array}$

PROBLEM CORNER

Use the chart.

1. Which city usually gets more rain in January? June?

2. How much rain does Los Angeles usually get in the winter? summer?

3. Does Albany get more rain in December or July?

Average Expected Rainfall in Centimeters

	Winter			Summer		
	Dec	Jan	Feb	Jun	Jul	Aug
Los Angeles, Cal.	6.07	6.4	5.89	0.08	0.03	0.05
Albany, N.Y.	7.44	5.59	5.36	7.62	7.92	7.29

More Practice, page 35

SOLVING PROBLEMS

When solving problems, keep these steps in mind.

1. UNDERSTAND What must I find?
What information is given?
Do I have enough information?

2. PLAN Have I solved a problem like this before?
Should I restate the problem?
What operations are needed?
Is an approximate answer sufficient?
Should I use a calculator?

3. CARRY OUT Did I calculate correctly?
Did I check my calculations?

4. LOOK BACK Is my answer reasonable?
Does the answer fit the problem?

EXAMPLE. Mr. and Mrs. Capan plan to rent an apartment. The basic rent is $ 310 with a $ 24 maintenance fee. A $ 55 security deposit is required. What is the total payment for the first month?

UNDERSTAND I must find the total.
I have enough information.

PLAN I must add to find the total.

CARRY OUT

$$\begin{array}{r} \$\ 310 \\ 24 \\ +\ 55 \\ \hline \$\ 389 \end{array}$$

It checks.

LOOK BACK The answer is reasonable.
It makes sense in the problem.

TRY THIS

Terry, a college student, rents an efficiency apartment for $ 145 a month. There is a charge of $ 23 for electricity and gas. Find the total monthly payment.

PRACTICE

1. Makato and Yin Lee bought a new house. The closing costs were:

Title Fee	$ 143.90
State Stamps	54.50
Transfer Tax	304.00

   Find the total closing costs.
2. Phyllis paid auto insurance costs of $ 115 for bodily injury liability, $ 72 for property damage, and $ 118 for collision. What was her total annual premium?

3. The Emperor Jahangir of India died in 1627. He owned 931,500 carats of emeralds, 376,600 carats of rubies, 279,450 carats of diamonds, and 186,300 carats of jade. How many carats of gems did he own?
4. Recent records show there are 4,600,000 men in the ABC (American Bowling Congress). There are 4,034,000 women in the WIBC (Women's International Bowling Congress). How many bowlers are in these two organizations?
5. Bob Hart bought an old car for $ 325.85. The materials to restore and repair it cost $ 527.47. What was his cash investment?
6. Cecilia Ortiz bought a pair of shoes for $ 36.50, a dress for $ 47.17, and a purse for $ 21.83. How much did she spend in all?

ROUNDING NUMBERS

EXAMPLE 1. Round to the nearest thousand. 4 3 6,8 2 7

Find the place to be rounded. The digit to the right is 5 or more.

4 3 6,8 2 7

Round up. Change to 0's.

4 3 7,0 0 0

EXAMPLE 2. Round to the nearest hundred-thousand. 4 3 6,8 2 7

Find the place to be rounded. The digit to the right is less than 5.

4 3 6,8 2 7

Keep the digit the same. Change to 0's.

4 0 0,0 0 0

TRY THIS

Round to the nearest hundred.

1. 486 **2.** 6713 **3.** 14,354 **4.** 256,662

EXAMPLE 3. Round to the nearest tenth. 7.3 7 2

Find the place to be rounded. The digit to the right is 5 or more.

7.3 7 2

Round up. Drop the digits to the right.

7.4

EXAMPLE 4. Round to the nearest whole number. 2 8.3 9 1

Find the ones' place. The digit to the right is less than 5.

2 8.3 9 1

Keep the digit the same. Drop the digits to the right.

2 8

EXAMPLE 5. Round to the nearest dollar. $ 1 2 9.5 0

Dollar. The place to the right is 5 or more.

$ 1 2 9.5 0

Round up. Drop the digits to the right.

$ 1 3 0

TRY THIS

Round to the nearest hundredth.

5. 7.372 **6.** 9.4891 **7.** 10.3356

Round to the nearest whole number.

8. 8.7 **9.** 342.3259 **10.** 29.731

Round to the nearest dollar.

11. $3.75 **12.** $41.28 **13.** $65.43

PRACTICE

Round to the nearest ten.

1. 48 **2.** 859 **3.** 596 **4.** 1412 **5.** 26

Round to the nearest hundred.

6. 563 **7.** 215 **8.** 32,850 **9.** 43,119 **10.** 4235

Round to the nearest thousand.

11. 5482 **12.** 5582 **13.** 428,750 **14.** 870,991 **15.** 62,489

Round to the nearest tenth.

16. 8.71 **17.** 0.85 **18.** 14.98 **19.** 3.7099 **20.** 7.86

Round to the nearest hundredth.

21. 8.765 **22.** 0.423 **23.** 17.396 **24.** 0.0075 **25.** 8.065

Round to the nearest whole number.

26. 7.3 **27.** 8.9 **28.** 91.476 **29.** 15.72 **30.** 43.7

Round to the nearest dollar.

31. $ 8.88 **32.** $ 17.45 **33.** $ 29.51 **34.** 225.50 **35.** $ 37.95

36. $53.29 **37.** $ 124.32 **38.** $ 37.67 **39.** $ 1.15 **40.** $2.04

ESTIMATING SUMS

Estimate the sum. 436 + 297 + 548

Line up vertically. → Round each number. → Add.

$$\begin{array}{r} 436 \\ 297 \\ +548 \\ \hline \end{array} \qquad \begin{array}{r} 400 \\ 300 \\ +500 \\ \hline \end{array} \qquad \begin{array}{r} 400 \\ 300 \\ +500 \\ \hline 1200 \end{array}$$

We usually round to the highest place in the smallest number.

EXAMPLES. Estimate the sums.

1. $\begin{array}{r} 426 \\ 31 \\ +\ 85 \\ \hline \end{array} \qquad \begin{array}{r} 430 \\ 30 \\ +\ 90 \\ \hline 550 \end{array}$

2. $\begin{array}{r} 4296 \\ 3500 \\ +7301 \\ \hline \end{array} \qquad \begin{array}{r} 4000 \\ 4000 \\ +7000 \\ \hline 15{,}000 \end{array}$

3. $\begin{array}{r} 17.8 \\ 3.14 \\ +\ 6.359 \\ \hline \end{array} \qquad \begin{array}{r} 18 \\ 3 \\ +\ 6 \\ \hline 27 \end{array}$

Round each number to the nearest whole number. Add.

4. $\begin{array}{r} \$29.15 \\ 32.75 \\ +\ 5.98 \\ \hline \end{array} \qquad \begin{array}{r} \$29 \\ 33 \\ +\ 6 \\ \hline \$68 \end{array}$

Round each amount to the nearest dollar. Add.

TRY THIS

Estimate the sums.

1. $\begin{array}{r} 78 \\ 25 \\ +34 \\ \hline \end{array}$ **2.** $\begin{array}{r} 4568 \\ 379 \\ +\ 615 \\ \hline \end{array}$ **3.** $\begin{array}{r} 3427 \\ 148 \\ +\ 261 \\ \hline \end{array}$

4. $\begin{array}{r} 36.781 \\ 9.31 \\ +\ 6.8 \\ \hline \end{array}$ **5.** $\begin{array}{r} \$36.49 \\ 5.50 \\ +\ 7.11 \\ \hline \end{array}$ **6.** $\begin{array}{r} \$67.51 \\ 3.29 \\ +\ 4.44 \\ \hline \end{array}$

PRACTICE

Estimate the sums.

1. 27 + 34	**2.** 39 + 28	**3.** 748 + 258	**4.** 698 + 307
5. 9286 + 6148	**6.** 4736 + 8367	**7.** 3,748,976 + 3,487,296	**8.** 4,286,987 + 8,628,146
9. 74 28 + 78	**10.** 348 296 755 + 508	**11.** 2748 3476 4985 + 6297	**12.** 3,287,496 5,349,891 + 6,548,196
13. 871 364 + 58	**14.** 725 75 + 86	**15.** 4297 385 + 115	**16.** 7284 6813 + 743
17. 8.7 + 6.5	**18.** 8.743 + 3.289	**19.** 15.87 + 7.3	**20.** 36.784 + 45.39
21. 3.9 4.8 + 8.9	**22.** 3.76 5.98 + 7.42	**23.** 1.742 3.9 + 6.1	**24.** 15.8 9.87 + 3.147
25. $ 7.75 + 2.98	**26.** $ 8.37 + 9.65	**27.** $ 16.19 + 24.78	**28.** $ 38.79 + 73.86
29. $ 9.15 6.25 + 8.60	**30.** $ 3.49 2.78 + 3.85	**31.** $ 24.73 9.68 + 5.45	**32.** $ 28.79 38.44 + 8.50

PROBLEM CORNER

1. Estimate the cost of a skirt and two blouses.
2. You have $ 10. Can you buy a shirt and two ties? Estimate to find the answer.

CAREER CAREER CAREER CAREER CAREER

AGRICULTURAL PILOT

Agricultural pilots are sometimes called cropdusters. They dust crops to kill insects. They also spray to kill weeds, apply fertilizer, and so on. The job is only for skillful pilots. Most of the flying is close to the ground.

Jason and Sherry Sanders own "J & S Agricultural Flying Service." One of their big expenses is airplane fuel. Their three gas storage tanks are refilled each month. They use a chart to record monthly fuel deliveries.

1. How much gas was delivered to tank #2 in May?
2. How much gas was put in the three tanks in March?
3. How much gas was put in the three tanks in April?
4. In which month was more gas used, May or June?

Monthly Fuel Deliveries—1983 (in liters)

	Mar 30	Apr 30	May 30	Jun 30
Tank #1	1138	2210	3344	3461
Tank #2	2161	2131	2876	2827
Tank #3	1294	1158	3851	3798
Total	4593			

J & S Service can be hired to check irrigation ditches. They fly low to spot any problems with the ditches. This is faster than checking the ditches by truck. This map shows some check routes. The distances are given in miles.

5. How long is Route #2?
6. Which is longer, Route #1 or Route #3?
7. One plane flew all three routes. Estimate the distance flown.

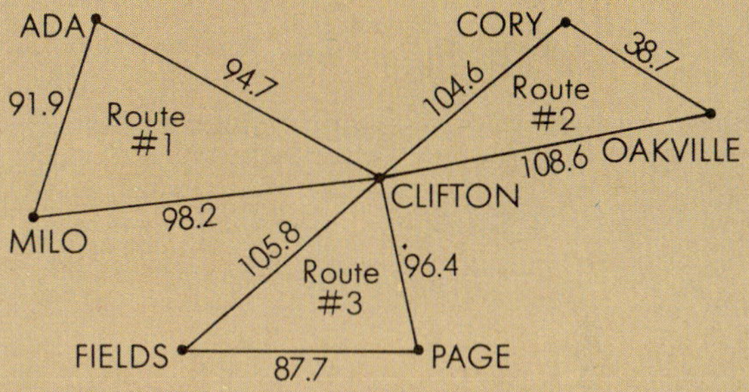

Ed Bluespruce is a farmer. He wants to have his fields sprayed to prevent insects. He compares estimates for two spraying methods.

J & S Estimate	
Time	3 hours
Chemicals	$165
Pilot and Plane	180
Assistant	30
Total	

Estimate of Ground Spraying	
Time	15 hours
Sprayer Rental	$ 75
Truck Costs	35
Labor	75
Chemicals	140
Total	

8. Which method takes less time?

9. Which method costs less?

On Aug. 7, 1981, J & S Service had $352.48 in a checking account. Jason wanted to buy the following items from Chemical Supply Corp.

1 barrel	P2-7	$ 92.53
1 barrel	TB804	$ 118.18
100 liters	filler fluid	$ 51.95
50 kg	pest powder	$ 40.50

10. Estimate Jason's cost.
11. Can Jason pay by check before making an account deposit?
12. Calculate Jason's exact cost.
13. Using the information above, copy and complete a check like the one on page 11.

CHAPTER 1 REVIEW

Write standard notation. p. 6

1. nine thousand, four hundred seventy-one
2. eight thousand, six
3. 320 thousand, 720
4. 800 million, 8 thousand

Write money notation. p. 8

5. eighty-six dollars and fifteen cents
6. forty dollars and five cents

Write decimal notation. p. 20

7. 18 and 14 hundredths
8. 7 and 6 thousandths

Add. pp. 14–22

9. 376 + 984

10. 9824 + 6958

11. 732 + 599 + 57 + 6

12. \$8.95 + 6.87 + 0.78

13. 482,239 + 58,804

14. 5.4 + 8.7

15. 17.291 + 8.789

16. 4.72 + 3.9 + 1.426

17. 18.984 + 7.3 + 8.47 + 0.987

18. 0.69 + 12.34 + 0.765 + 28.8

Estimate the sums. p. 28

19. 2725 + 3895 + 4350

20. 428 + 67 + 41

21. \$8.75 + 2.30 + 1.98

22. 4.2 + 8.75 + 3.199

23. \$6.65 + 8.37

Application pp. 10, 18

24. Write a check from Charles Jones to J and S Agricultural Service in the amount of \$725.75. Date it December 14, 1980.

25. Find the net deposit.

		Dollars	Cents
Cash	Currency	25	00
	Coin	6	10
Checks	16-28	37	48
	11-32	128	45
	17-94	5	00
Total			
Less Cash Received			-0-
Net Deposit			

CHAPTER 1 TEST

Write standard notation.

1. two thousand, six hundred fifty-eight

2. nine thousand, ninety

3. 386 thousand, 317

4. 40 million, 400 thousand

Write money notation.

5. seventy-four dollars and fifty cents

6. sixteen dollars and seven cents

Write decimal notation.

7. 6 and 8 tenths

8. 29 and 123 ten-thousandths

Add.

9. $429 + 738$

10. $4298 + 8764$

11. $862 + 291 + 87 + 9$

12. $\$7.49 + 8.50 + 0.73$

13. $190{,}987 + 69{,}253$

14. $9.72 + 8.69$

15. $3.429 + 0.984$

16. $5.762 + 8.7 + 0.94$

17. $37.382 + 7.48 + 9.9 + 8.739$

18. $0.42 + 18.75 + 0.318 + 33.9$

Estimate the sums.

19. $498 + 878 + 212$

20. $7384 + 921 + 650$

21. $\$9.34 + 6.50 + 7.75$

22. $7.6 + 8.34 + 6.149$

23. $\$5.77 + 3.42$

Application

24. Write a check to Ace Chemical Company from Sherie Sanders in the amount of \$ 78.50. Date it March 22, 1983.

25. Find the net deposit.

		Dollars	Cents
Cash	Currency	5	00
	Coin	8	75
Checks	19-25	18	00
	8-406	10	00
	4-15	135	68
Total			
Less Cash Received			-0-
Net Deposit			

MORE PRACTICE

Write standard notation. Use after page 7.

1. six hundred forty-three
2. five hundred ninety-eight
3. one thousand, six hundred ten
4. eight thousand, eighty
5. three thousand, five hundred
6. nine thousand, nine
7. twelve thousand, ninety-nine
8. ten thousand, two hundred eight
9. eighty thousand, one hundred eleven
10. twenty-six thousand, four hundred nineteen

Write money notation. Use after page 9.

11. eight dollars and ninety-five cents
12. seventeen dollars
13. eighty-one dollars and fifty cents
14. thirty dollars and six cents
15. one hundred twenty dollars and twenty cents
16. seventy-seven dollars and seven cents
17. five dollars and eighty-nine cents
18. sixty-five dollars
19. seventeen dollars and ninety-eight cents
20. two hundred dollars and sixty cents

Write standard notation. Use after page 13.

21. 112 thousand, 600
22. 800 thousand, 4
23. 26 million, 455 thousand
24. 11 million

Add. Use after page 15.

25. $74 + 96$

26. $49 + 5$

27. $746 + 298$

28. $348 + 67$

29. $348 + 267 + 98$

30. $948 + 76 + 18 + 7$

31. $\$4.78 + 2.97 + 6.49$

32. $\$8.47 + 0.49 + 6.40 + 0.86$

Add. Use after page 17.

1. $\begin{array}{r} 7284 \\ +\ 3796 \\ \hline \end{array}$ **2.** $\begin{array}{r} 14,781 \\ +\ 39,499 \\ \hline \end{array}$ **3.** $\begin{array}{r} 347,821 \\ +\ 981,294 \\ \hline \end{array}$ **4.** $\begin{array}{r} 349,278 \\ +\ 97,146 \\ \hline \end{array}$ **5.** $\begin{array}{r} 609,183 \\ +\ 497,256 \\ \hline \end{array}$

6. $\begin{array}{r} 7269 \\ 3784 \\ 6985 \\ +\ 3784 \\ \hline \end{array}$ **7.** $\begin{array}{r} 47,264 \\ 69,209 \\ +\ 15,875 \\ \hline \end{array}$ **8.** $\begin{array}{r} 349,287 \\ 190,180 \\ 762,498 \\ +\ 875,287 \\ \hline \end{array}$ **9.** $\begin{array}{r} \$\ 746.84 \\ 198.78 \\ +\ 58.74 \\ \hline \end{array}$ **10.** $\begin{array}{r} \$\ 486.22 \\ 176.49 \\ +\ 89.23 \\ \hline \end{array}$

Write decimal notation. Use after page 21.

11. 1 and 6 tenths **12.** 12 and 24 hundredths **13.** 4 thousandths

14. 6 and 7 hundredths **15.** 10 and 326 thousandths **16.** 9 and 1 tenth

17. 4 and 28 thousandths **18.** 3 and 4248 ten-thousandths **19.** 16 and 3 thousandths

20. 86 thousandths **21.** 2 and 5 ten-thousandths **22.** 9 tenths

23. 9 hundredths **24.** 9 ten-thousandths **25.** 10 and 23 hundredths

Add. Use after page 23.

26. $\begin{array}{r} 8.7 \\ +\ 9.2 \\ \hline \end{array}$ **27.** $\begin{array}{r} 6.4 \\ +\ 8.7 \\ \hline \end{array}$ **28.** $\begin{array}{r} 3.78 \\ +\ 9.65 \\ \hline \end{array}$ **29.** $\begin{array}{r} 0.918 \\ +\ 6.942 \\ \hline \end{array}$ **30.** $\begin{array}{r} 14.675 \\ +\ 0.351 \\ \hline \end{array}$

31. $\begin{array}{r} 4.98 \\ 3.7 \\ +\ 5.876 \\ \hline \end{array}$ **32.** $\begin{array}{r} 9.7 \\ 6.82 \\ +\ 9.487 \\ \hline \end{array}$ **33.** $\begin{array}{r} 7.328 \\ 0.09 \\ 1.4268 \\ +\ 0.6 \\ \hline \end{array}$ **34.** $\begin{array}{r} 4.298 \\ 5.29 \\ 6.2 \\ +\ 7.0 \\ \hline \end{array}$ **35.** $\begin{array}{r} 3.48 \\ 0.569 \\ 28.3 \\ +\ 5.42 \\ \hline \end{array}$

36. $5.62 + 0.039 + 37.8$ **37.** $0.2 + 29.6 + 7.12 + 0.594$ **38.** $12.7 + 6.3$

Estimate the sums. Use after page 29.

39. $\begin{array}{r} 86 \\ 94 \\ +\ 38 \\ \hline \end{array}$ **40.** $\begin{array}{r} 376 \\ 298 \\ +\ 815 \\ \hline \end{array}$ **41.** $\begin{array}{r} 324 \\ 58 \\ +\ 63 \\ \hline \end{array}$ **42.** $\begin{array}{r} 4286 \\ 750 \\ +\ 310 \\ \hline \end{array}$ **43.** $\begin{array}{r} 7259 \\ 861 \\ +\ 248 \\ \hline \end{array}$

44. $\begin{array}{r} \$\ 8.76 \\ +\ 9.48 \\ \hline \end{array}$ **45.** $\begin{array}{r} \$\ 3.98 \\ 2.47 \\ 6.50 \\ +\ 1.25 \\ \hline \end{array}$ **46.** $\begin{array}{r} 1.78 \\ +\ 3.25 \\ \hline \end{array}$ **47.** $\begin{array}{r} 4.298 \\ 7.9 \\ 6.48 \\ +\ 7.007 \\ \hline \end{array}$ **48.** $\begin{array}{r} 25.81 \\ 6.44 \\ 0.79 \\ +\ 1.08 \\ \hline \end{array}$

CALCULATOR CORNER

ADDITION ESTIMATION

Players two
Materials two calculators, paper and pencils

SAMPLE

1. Nick asks Devon to estimate a sum, such as 422 + 576 + 182. Devon asks Nick to estimate a sum, such as 386 + 94 + 214.
2. Devon and Nick write down their problems. They estimate and add.
3. Devon and Nick calculate the exact answers to their problems.
4. They then calculate the differences between their estimates and the exact answers.

Devon

422	400
576	600
+ 182	200
	1200
estimate	1200
exact answer	− 1180
	20

Nick

386	390
94	90
+ 214	210
	690
exact answer	694
estimate	− 690
	4

Since Nick's estimate was closer to the exact answer, he gets 1 point.

Play continues until a player gets 5 points.

READY FOR CHAPTER 2

Subtract.

1. 85 − 23	**2.** 72 − 39	**3.** 421 − 368	**4.** \$ 6.48 − 2.89
5. 6287 − 1938	**6.** 62,486 − 19,197	**7.** 427,685 − 388,747	**8.** \$ 238.46 − 95.88
9. 400 − 146	**10.** 5001 − 3247	**11.** 75,003 − 19,437	**12.** \$ 10.00 − 2.45
13. 7.43 − 4.68	**14.** 3.786 − 1.9	**15.** 6.3 − 1.428	**16.** 8 − 7.34

Compare. Use =, >, or <.

17. 0.2 and 0.24	**18.** 0.7 and 0.07	**19.** 0.87 and 0.870
20. 0.2 and 0.200	**21.** 0.3 and 0.29	**22.** 0.367 and 0.37
23. 0.62 and 0.620	**24.** 0.03 and 0.003	**25.** 0.465 and 0.473

Estimate the differences.

26. 471 − 182	**27.** 6349 − 2481	**28.** 758 − 86	**29.** 9.1 − 3.986

Application

Find the balance carried forward.

30.

Date *July 5,* 19*82* *\$72.46* 291
To *Foreign Auto Repair*
For *tune-up*

	Dollars	Cents
Balance Forward	348	26
Amount of Deposit		-0-
Total		
Amount of Check		
Balance Carried Forward		

31.

Date *Aug. 16,* 19*82* *\$28.50* 325
To *Long's Market*
For *groceries*

	Dollars	Cents
Balance Forward	268	74
Amount of Deposit	198	70
Total		
Amount of Check		
Balance Carried Forward		

2
SUBTRACTING WHOLE NUMBERS AND DECIMALS

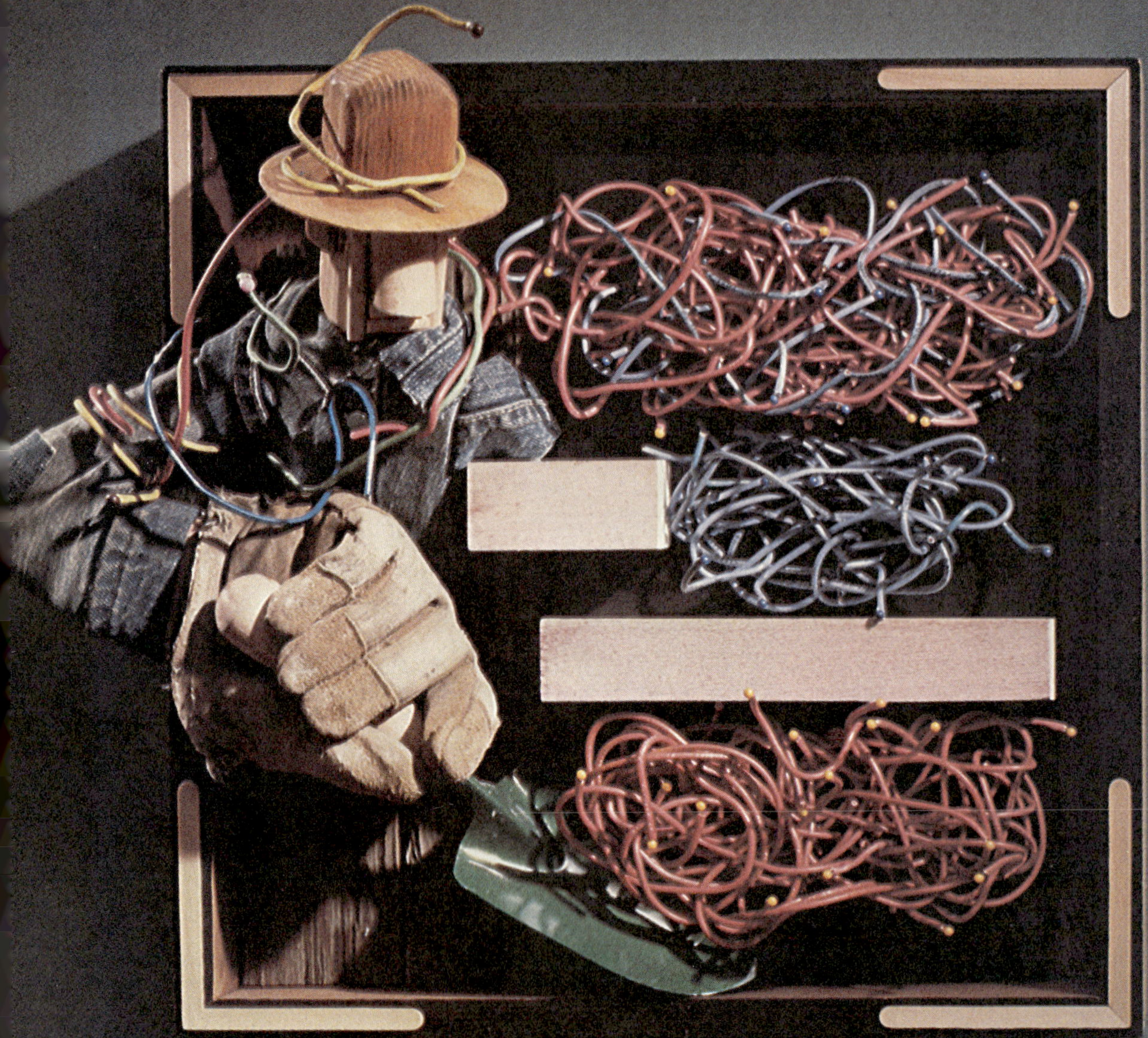

GETTING STARTED

REVERSE GOTCHA

WHAT'S NEEDED The 30 squares of paper with digits marked on them as in GOTCHA, Chapter 1.

THE RULES

1. On a sheet of paper each player makes a scorecard like this. Play three games and add to find the total score. The player with the smallest total is the winner.
2. To play, a leader draws a digit from the box. Each player subtracts it from 99 and writes the answer. As more digits are drawn, players continue to subtract.
3. When 0 is drawn it is a GOTCHA. All players still in the game go back to 99. They score 99 for the game.
4. A player may go out or quit before any draw. Try to get the smallest score but go out before 0 is drawn.

Scorecard

1.	______
2.	______
3.	______
Total	______

SAMPLE

Game 1	Game 2	Game 3
99	99	99
− 6	− 9	− 8
93	90	91
− 9	− 8	− 8
84	82	83
− 8	− 9	− 7
76 go out	73	76
	− 0 GOTCHA	− 9
		67 go out

Scorecard

1.	76
2.	99
3.	67
Total	242

SUBTRACTING WHOLE NUMBERS

Subtract. 486 − 157

Rename. → Subtract the ones. → Subtract the tens. → Subtract the hundreds.

$$\begin{array}{r} 4\overset{7}{\cancel{8}}\overset{16}{\cancel{6}} \\ -\ 157 \\ \hline \end{array} \qquad \begin{array}{r} 4\overset{7}{\cancel{8}}\overset{16}{\cancel{6}} \\ -\ 157 \\ \hline 9 \end{array} \qquad \begin{array}{r} 4\overset{7}{\cancel{8}}\overset{16}{\cancel{6}} \\ -\ 157 \\ \hline 29 \end{array} \qquad \begin{array}{r} 4\overset{7}{\cancel{8}}\overset{16}{\cancel{6}} \\ -\ 157 \\ \hline 329 \end{array}$$

EXAMPLE 1. Subtract. 759 − 327

$$\begin{array}{r} 759 \\ -\ 327 \\ \hline 432 \end{array}$$

Subtract the ones.
Subtract the tens.
Subtract the hundreds.

EXAMPLE 2. Subtract. 942 − 148

$$\begin{array}{r} \overset{8}{\cancel{9}}\overset{13}{\cancel{4}}\overset{12}{\cancel{2}} \\ -\ 148 \\ \hline 794 \end{array}$$

Rename. Subtract the ones.
Rename. Subtract the tens.
Subtract the hundreds.

TRY THIS

Subtract.

1. $\begin{array}{r} 78 \\ -\ 24 \\ \hline \end{array}$ **2.** $\begin{array}{r} 876 \\ -\ 45 \\ \hline \end{array}$ **3.** $\begin{array}{r} 78 \\ -\ 39 \\ \hline \end{array}$ **4.** $\begin{array}{r} 673 \\ -\ 385 \\ \hline \end{array}$

EXAMPLE 3. Subtract. \$8.48 − \$3.29

$$\begin{array}{r} \$\ 8.\overset{3}{\cancel{4}}\overset{18}{\cancel{8}} \\ -\ 3.29 \\ \hline \$\ 5.19 \end{array}$$

Subtract as with whole numbers.
Write money notation in the answer.

TRY THIS

Subtract.

5. $\begin{array}{r} \$\ 7.67 \\ -\ 1.31 \\ \hline \end{array}$ **6.** $\begin{array}{r} \$\ 1.19 \\ -\ 0.98 \\ \hline \end{array}$ **7.** $\begin{array}{r} \$\ 7.64 \\ -\ 1.89 \\ \hline \end{array}$ **8.** $\begin{array}{r} \$\ 8.62 \\ -\ 5.91 \\ \hline \end{array}$

PRACTICE

Subtract.

1. 95 − 31	**2.** 76 − 25	**3.** 86 − 40	**4.** 78 − 45	**5.** 47 − 25
6. 426 − 123	**7.** 837 − 501	**8.** 298 − 236	**9.** 425 − 113	**10.** 296 − 183
11. 75 − 38	**12.** 86 − 37	**13.** 42 − 17	**14.** 72 − 43	**15.** 52 − 18
16. 725 − 118	**17.** 477 − 269	**18.** 914 − 382	**19.** 328 − 148	**20.** 828 − 264
21. 721 − 389	**22.** 424 − 398	**23.** 724 − 377	**24.** 711 − 222	**25.** 431 − 298
26. 83 − 8	**27.** 424 − 19	**28.** 613 − 38	**29.** 386 − 8	**30.** 255 − 26
31. \$ 0.86 − 0.19	**32.** \$ 7.62 − 3.47	**33.** \$ 8.71 − 1.90	**34.** \$ 9.47 − 9.38	**35.** \$ 6.79 − 0.99
36. \$ 4.27 − 1.88	**37.** \$ 3.76 − 0.87	**38.** \$ 6.92 − 5.93	**39.** \$ 6.50 − 3.75	**40.** 5.54 − 3.56

PROBLEM CORNER

Use the chart.

1. What is the yearly electricity cost of a tube type color TV?

2. Which is the least expensive TV to watch per year?

3. How much more does it cost per year to watch a tube type color TV than a solid state color TV?

Yearly Electricity Cost For TV	
Black and White (tube type)	\$19.71
Black and White (solid state)	2.74
Color (tube type)	35.59
Color (solid state)	18.07

More Practice, page 62

SUBTRACTING LARGER NUMBERS

Subtract. 8148 − 2529

Rename. Subtract the ones. → Subtract the tens. → Rename. Subtract the hundreds. → Subtract the thousands.

$$\begin{array}{r} \scriptstyle 3\ 18 \\ 81\not{4}\not{8} \\ -\ 2529 \\ \hline 9 \end{array} \qquad \begin{array}{r} \scriptstyle 3\ 18 \\ 81\not{4}\not{8} \\ -\ 2529 \\ \hline 19 \end{array} \qquad \begin{array}{r} \scriptstyle 7\ 11\ 3\ 18 \\ \not{8}\not{1}\not{4}\not{8} \\ -\ 2529 \\ \hline 619 \end{array} \qquad \begin{array}{r} \scriptstyle 7\ 11\ 3\ 18 \\ \not{8}\not{1}\not{4}\not{8} \\ -\ 2529 \\ \hline 5619 \end{array}$$

EXAMPLE 1. Subtract. 73,206 − 38,543

$$\begin{array}{r} \scriptstyle 6\ 12\ 11\ 10 \\ \not{7}\not{3},\not{2}\not{0}6 \\ -\ 38,543 \\ \hline 34,663 \end{array}$$

Subtract the ones.
Rename. Subtract the tens.
Rename. Subtract the hundreds.
Rename. Subtract the thousands.
Subtract the ten-thousands.

TRY THIS

Subtract.

1. $\begin{array}{r} 4873 \\ -\ 2379 \\ \hline \end{array}$ **2.** $\begin{array}{r} 28,497 \\ -\ 9,598 \\ \hline \end{array}$ **3.** $\begin{array}{r} 346,291 \\ -\ 253,834 \\ \hline \end{array}$

EXAMPLE 2. Subtract. \$794.61 − \$345.87

$$\begin{array}{r} \scriptstyle 8\ 13\ 15\ 11 \\ \$\,7\not{9}\not{4}.\not{6}\not{1} \\ -\ 345.87 \\ \hline \$\,448.74 \end{array}$$

Subtract as with whole numbers.
Write money notation in the answer.

TRY THIS

Subtract.

4. $\begin{array}{r} \$\,47.28 \\ -\ 26.42 \\ \hline \end{array}$ **5.** $\begin{array}{r} \$\,4276.94 \\ -\ 3148.75 \\ \hline \end{array}$ **6.** $\begin{array}{r} \$\,7385.65 \\ -\ 948.79 \\ \hline \end{array}$

PRACTICE

Subtract.

1. 8472 − 5380
2. 5298 − 3876
3. 9412 − 7308
4. 1476 − 925
5. 1525 − 776
6. 3641 − 1278
7. 4371 − 1719
8. 7526 − 1958
9. 36,297 − 18,429
10. 73,248 − 28,076
11. 42,573 − 19,861
12. 44,444 − 27,705
13. 72,486 − 29,549
14. 34,937 − 17,679
15. 52,648 − 37,959
16. 68,860 − 9,493
17. 498,276 − 379,818
18. 769,287 − 589,549
19. 315,264 − 276,581
20. 728,651 − 354,974
21. 864,590 − 127,498
22. 726,627 − 398,893
23. 345,215 − 68,758
24. 421,742 − 8,756
25. $ 79.48 − 27.39
26. $ 48.36 − 19.28
27. $ 68.14 − 39.25
28. $ 81.46 − 7.98
29. $ 182.87 − 94.38
30. $ 361.27 − 284.82
31. $ 432.87 − 76.98
32. $ 153.82 − 77.77

PROBLEM CORNER

1. Which is the largest island on Earth?
2. How much larger is New Guinea than Borneo?
3. The area of Texas is 692,405 km². How much larger is Borneo than Texas?

Three Largest Islands on Earth	
Greenland	2,175,600 km²
New Guinea	821,030 km²
Borneo	744,366 km²

More Practice, page 62

ZEROS IN SUBTRACTING

Subtract. 700 − 268

We must rename to subtract. → Rename a special way. → Subtract.

$$\begin{array}{r} 700 \\ -\ 268 \\ \hline \end{array} \qquad \begin{array}{r} {\scriptstyle 6\ 9\ 10} \\ \not{7}\not{0}\not{0} \\ -\ 268 \\ \hline \end{array} \qquad 700 = 69 \text{ tens} + 10 \qquad \begin{array}{r} {\scriptstyle 6\ 9\ 10} \\ \not{7}\not{0}\not{0} \\ -\ 268 \\ \hline 432 \end{array}$$

EXAMPLE 1. Subtract. 803 − 379

$$\begin{array}{r} {\scriptstyle 7\ 9\ 13} \\ \not{8}\not{0}\not{3} \\ -\ 379 \\ \hline 424 \end{array}$$

Rename. 803 = 79 tens + 13
Subtract.

EXAMPLE 2. Subtract. 7000 − 3497

$$\begin{array}{r} {\scriptstyle 6\ 9\ 9\ 10} \\ \not{7}\not{0}\not{0}\not{0} \\ -\ 3497 \\ \hline 3503 \end{array}$$

Rename. 7000 = 699 tens + 10
Subtract.

TRY THIS

Subtract.

1. 600 − 423
2. 8004 − 375
3. \$ 50.03 − 29.47
4. \$ 30.00 − 6.37

EXAMPLE 3. Subtract. 54,007 − 12,478

$$\begin{array}{r} {\scriptstyle 3\ 9\ 9\ 17} \\ 5\,\not{4}{,}\not{0}\not{0}\not{7} \\ -\ 12{,}478 \\ \hline 41{,}529 \end{array}$$

To rename, go to the first non-zero digit.
4007 = 399 tens + 17

TRY THIS

Subtract.

5. 36,001 − 15,275
6. 70,040 − 19,281
7. \$ 500.17 − 192.88
8. \$ 700.09 − 367.82

PRACTICE

Subtract.

1. 700 − 341	**2.** 804 − 25	**3.** 300 − 6	**4.** 605 − 197	**5.** 507 − 19
6. 207 − 198	**7.** 100 − 73	**8.** 500 − 371	**9.** 603 − 214	**10.** 400 − 282
11. 7000 − 1234	**12.** 8002 − 3246	**13.** 9000 − 1725	**14.** 5004 − 2119	**15.** 7008 − 1999
16. 4001 − 974	**17.** 3000 − 326	**18.** 2000 − 7	**19.** 1001 − 2	**20.** 8005 − 27
21. 76,005 − 15,738	**22.** 76,050 − 15,783	**23.** 9007 − 2433	**24.** 50,050 − 25,597	**25.** 60,004 − 18,795
26. 42,004 − 23,541	**27.** 7800 − 2974	**28.** 44,000 − 378	**29.** 76,004 − 6,005	**30.** 90,000 − 823
31. $ 4.00 − 1.73	**32.** $ 7.04 − 2.95	**33.** $ 3.00 − 0.37	**34.** $ 8.02 − 0.03	**35.** $ 6.06 − 1.77
36. $ 10.00 − 1.75	**37.** $ 50.05 − 27.68	**38.** $ 38.00 − 17.59	**39.** $ 80.54 − 17.73	**40.** 60.02 − 14.55

PROBLEM CORNER

1. You buy a record for $ 7.98. You give the clerk a ten-dollar bill. How much change should you receive?

2. You buy two tapes for $ 14.75. You give the clerk a twenty-dollar bill. How much change should you receive?

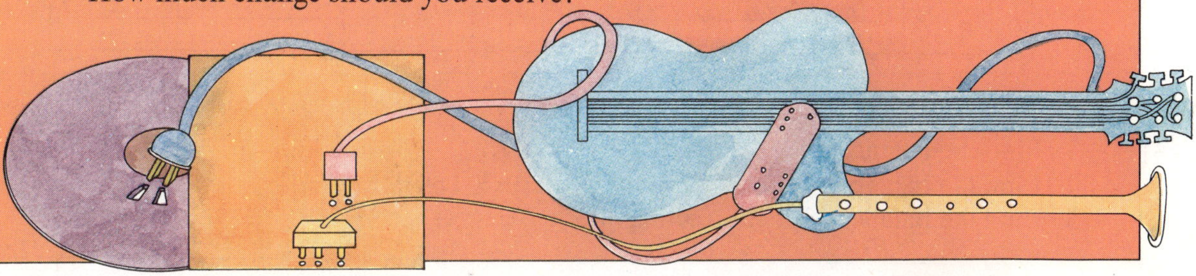

More Practice, page 62

APPLICATION

KEEPING A RUNNING BALANCE

One method of recording checks is the check stub attached to the check.

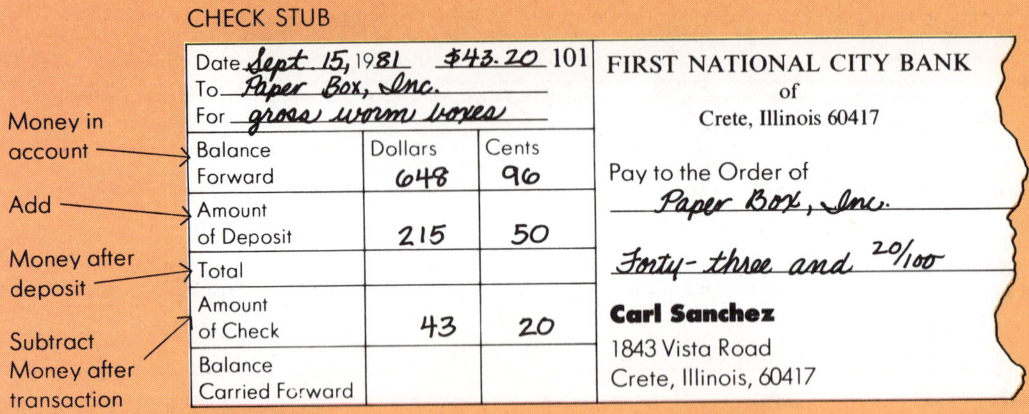

Use the check stub above.

1. Find the account total by adding.
2. Find the balance forward by subtracting.

Checks can also be recorded in a check register. It is in a booklet separate from your checks.

This part shows a running balance.

CHECK REGISTER

Check Number	Date	Checks issued to or description of deposit	Amount of check		✓	Amount of deposit		Balance Forward		
									$648	96
101	9-15	To Paper Box, Inc.	43	20				Check or deposit	43	20
		For gross worm boxes						Balance		
		To Deposit				215	50	Check or deposit	215	50
		For receipts 9-14-80						Balance		
103	9-16	To City Electric Co.	72	95				Check or deposit	72	95
		For electricity through 9-12						Balance		

Use the register.

3. Find the balance after subtracting check 101.
4. Find the balance after adding the deposit on 9-15-80.
5. Find the balance after subtracting check 103.

Copy and complete the right-hand portion of each check stub.

6. Date Oct. 1, 1981 $24.95 135
To Sun Lighting
For Lamp

Balance Forward	Dollars $487	Cents 56
Amount of Deposit		-0-
Total	487	56
Amount of Check	24	95
Balance Carried Forward		

Balance Carried Forward to Next Stub

7. Date Oct. 2, 1981 $16.85 136
To Joe's Market
For Food

Balance Forward	Dollars	Cents
Amount of Deposit		-0-
Total		
Amount of Check	$16	85
Balance Carried Forward		

8. Date Jan. 12, 1982 $18.75 258
To John Burke
For Birthday gift

Balance Forward	Dollars $392	Cents 15
Amount of Deposit	65	00
Total		
Amount of Check	18	75
Balance Carried Forward		

Balance Carried Forward to Next Stub

9. Date Jan. 14, 1982 $106.11 259
To IRS
For Tax payment

Balance Forward	Dollars	Cents
Amount of Deposit	$10	95
Total		
Amount of Check		
Balance Carried Forward		

10. Copy and complete the right-hand portion of this check register.

Check Number	Date	Checks issued to or description of deposit	Amount of check		✓	Amount of deposit		Balance Forward	$398	59
273	2-15	To Sun Fashions For white skirt	16	77				Check or deposit Balance		
	2-15	To Deposit For				175	80	Check or deposit Balance		
274	2-17	To Ready Start For battery	68	25				Check or deposit Balance		
275	2-18	To Altos Apts. For March rent payment	295	00				Check or deposit Balance		
	2-22	To Deposit For				175	80	Check or deposit Balance		
276	2-23	To Northern Gas Co. For gas bill 2-19	31	18				Check or deposit Balance		

COMPARING DECIMALS

In some decimals writing or dropping zeros shows that the decimals are equal.

EXAMPLE 1. Write a decimal in hundredths equal to 0.5.

$0.5 = 0.50$ *Write one 0.*

EXAMPLE 2. Write a decimal in tenths equal to 0.300.

$0.300 = 0.3$ *Drop two 0's.*

TRY THIS

Write each decimal in hundredths.

1. 0.2 **2.** 0.700 **3.** 0.90000 **4.** 0.06000

We can compare decimals using $=$ (is equal to), $>$ (is greater than), or $<$ (is less than).

Compare. 0.8 and 0.73

Write 0's to make the number of decimal places the same. → Compare the decimal parts. → Write the comparison.

$0.8 = 0.80$
$0.73 = 0.73$

$80 > 73$

$0.8 > 0.73$

EXAMPLE 3. Compare. 0.54 and 0.6

$0.54 = 0.54$ *Write 0's to make the number of decimal*
$0.6 = 0.60$ *places the same.*
$54 < 60$ *Compare the decimal parts.*
$0.54 < 0.6$ *Write the decimals.*

TRY THIS

Compare. Use $=$, $>$ or $<$.

5. 0.86 and 0.91
6. 0.75 and 0.57
7. 0.6 and 0.529
8. 0.60 and 0.600
9. 0.7 and 0.700
10. 0.439 and 0.5

PRACTICE

Write each decimal in tenths.

1. 0.80 **2.** 0.800 **3.** 0.8000 **4.** 0.70 **5.** 0.700

6. 0.7000 **7.** 0.300 **8.** 0.60 **9.** 0.10 **10.** 0.500

Write each decimal in hundredths.

11. 0.1 **12.** 0.2 **13.** 0.8 **14.** 0.9 **15.** 0.4

16. 0.300 **17.** 0.4000 **18.** 0.150 **19.** 0.3800 **20.** 0.080

Write each decimal in thousandths.

21. 0.1 **22.** 0.20 **23.** 0.3 **24.** 0.40 **25.** 0.8

26. 0.32 **27.** 0.48 **28.** 0.5000 **29.** 0.9000 **30.** 0.0600

Compare. Use =, >, or <.

31. 0.3 and 0.300 **32.** 0.4 and 0.3 **33.** 0.72 and 0.7 **34.** 0.02 and 0.2

35. 0.319 and 0.391 **36.** 0.319 and 0.32 **37.** 0.45 and 0.450 **38.** 0.567 and 0.576

39. 0.4 and 0.04 **40.** 0.007 and 0.07 **41.** 0.310 and 0.31 **42.** 0.26 and 0.28

43. 0.8 and 0.8000 **44.** 0.9 and 0.09 **45.** 0.08 and 0.008 **46.** 0.012 and 0.0120

47. 0.07 and 0.7 **48.** 0.914 and 0.92 **49.** 0.5 and 0.500 **50.** 0.857 and 0.85

51. 0.421 and 0.42 **52.** 0.628 and 0.625 **53.** 0.24 and 0.27 **54.** 0.030 and 0.03

PROBLEM CORNER

1. Ana's fish weighed 0.73 kilograms. Luis's fish weighed 0.712 kilograms. Whose fish weighed more?

2. Cory ran 6.51 kilometers. Devon ran 6.53 kilometers. Who ran further?

SUBTRACTING DECIMALS

Subtract. 28.35 − 16.57

Line up the decimal points. Subtract the hundredths. → Subtract the tenths. → Subtract the whole numbers. → Place the decimal point in the answer.

$$\begin{array}{r} {\scriptstyle 2\ 15} \\ 28.\not{3}\not{5} \\ -\ 16.57 \\ \hline 8 \end{array} \qquad \begin{array}{r} {\scriptstyle 7\ 12\ 15} \\ 2\not{8}.\not{3}\not{5} \\ -\ 16.57 \\ \hline 78 \end{array} \qquad \begin{array}{r} {\scriptstyle 7\ 12\ 15} \\ 2\not{8}.\not{3}\not{5} \\ -\ 16.57 \\ \hline 1178 \end{array} \qquad \begin{array}{r} 28.35 \\ -\ 16.57 \\ \hline 11.78 \end{array}$$

Sometimes we write 0's to fill the places.

EXAMPLES. Subtract.

1. 7.48 − 3.7 **2.** 9.3 − 3.754

$$\begin{array}{r} {\scriptstyle 6\ 14} \\ \not{7}.\not{4}8 \\ -\ 3.70 \\ \hline 3.78 \end{array} \qquad \begin{array}{r} {\scriptstyle 8\ 12\ 9\ 10} \\ \not{9}.\not{3}\not{0}\not{0} \\ -\ 3.754 \\ \hline 5.546 \end{array}$$

Line up the decimal points.
Subtract.

TRY THIS

Subtract.

1. 3.71 − 2.16 **2.** 6.3 − 1.48
3. 15.287 − 7.6

4. $\begin{array}{r} 8.7 \\ -\ 4.24 \\ \hline \end{array}$ **5.** $\begin{array}{r} 5.67 \\ -\ 2.8 \\ \hline \end{array}$ **6.** $\begin{array}{r} 13.5 \\ -\ 7.482 \\ \hline \end{array}$

When one number is a whole number, we write a decimal point and 0's.

EXAMPLES. Subtract.

3. 6.743 − 2 **4.** 8 − 3.62

$$\begin{array}{r} 6.743 \\ -\ 2.000 \\ \hline 4.743 \end{array} \qquad \begin{array}{r} {\scriptstyle 7\ 9\ 10} \\ \not{8}.\not{0}\not{0} \\ -\ 3.62 \\ \hline 4.38 \end{array}$$

Line up.
Subtract.
Place the decimal point in the answer.

TRY THIS

Subtract.

7. 7.48 − 3

8. 9 − 3.7

9. $\begin{array}{r} 14 \\ -\ 6.21 \\ \hline \end{array}$

10. $\begin{array}{r} 19.32 \\ -\ 7 \\ \hline \end{array}$

11. $\begin{array}{r} 6 \\ 3.714 \\ \hline \end{array}$

PRACTICE

Subtract.

1. $\begin{array}{r} 7.3 \\ -\ 3.9 \\ \hline \end{array}$ **2.** $\begin{array}{r} 5.25 \\ -\ 1.89 \\ \hline \end{array}$ **3.** $\begin{array}{r} 13.71 \\ -\ 4.52 \\ \hline \end{array}$ **4.** $\begin{array}{r} 9.386 \\ -\ 3.739 \\ \hline \end{array}$ **5.** $\begin{array}{r} 6.74 \\ -\ 3.81 \\ \hline \end{array}$

6. 9.36 − 4.7 **7.** 7.8 − 3.1 **8.** 12.84 − 6.9 **9.** 9.04 − 3.2 **10.** 17.61 − 7.8

11. $\begin{array}{r} 8.76 \\ -\ 3.8 \\ \hline \end{array}$ **12.** $\begin{array}{r} 5.51 \\ -\ 2.7 \\ \hline \end{array}$ **13.** $\begin{array}{r} 3.743 \\ -\ 1.92 \\ \hline \end{array}$ **14.** $\begin{array}{r} 4.765 \\ -\ 3.9 \\ \hline \end{array}$ **15.** $\begin{array}{r} 9.38 \\ -\ 2.5 \\ \hline \end{array}$

16. $\begin{array}{r} 4.26 \\ -\ 3.9 \\ \hline \end{array}$ **17.** $\begin{array}{r} 9.11 \\ -\ 5.8 \\ \hline \end{array}$ **18.** $\begin{array}{r} 4.27 \\ -\ 1.835 \\ \hline \end{array}$ **19.** $\begin{array}{r} 6.7 \\ -\ 4.381 \\ \hline \end{array}$ **20.** $\begin{array}{r} 7.2 \\ -\ 3.84 \\ \hline \end{array}$

21. $\begin{array}{r} 9.3 \\ -\ 7 \\ \hline \end{array}$ **22.** $\begin{array}{r} 4.2 \\ -\ 3 \\ \hline \end{array}$ **23.** $\begin{array}{r} 6.75 \\ -\ 3 \\ \hline \end{array}$ **24.** $\begin{array}{r} 8.764 \\ -\ 5 \\ \hline \end{array}$ **25.** $\begin{array}{r} 6.98 \\ -\ 5 \\ \hline \end{array}$

26. $\begin{array}{r} 6 \\ -\ 1.8 \\ \hline \end{array}$ **27.** $\begin{array}{r} 9 \\ -\ 7.34 \\ \hline \end{array}$ **28.** $\begin{array}{r} 8 \\ -\ 3.27 \\ \hline \end{array}$ **29.** $\begin{array}{r} 7 \\ -\ 3.843 \\ \hline \end{array}$ **30.** $\begin{array}{r} 5 \\ -\ 1.25 \\ \hline \end{array}$

31. 7.9 − 3 **32.** 6 − 4.5 **33.** 15.1 − 8 **34.** 9 − 4.6 **35.** 11 − 6.7

PROBLEM CORNER

1. Which is the heaviest book; the lightest book?
2. How much heavier is the math book than the English book?
3. How much lighter is the math book than the dictionary?
4. How much heavier is the English book than the notebook?

Weighty Knowledge	
Math Book	0.72 kg
English Book	0.513 kg
Note Book	0.098 kg
Dictionary	1.174 kg

APPLICATION

RECONCILING MONTHLY STATEMENTS

Each month you receive a bank statement. Seeing if it agrees with your records is called *reconciling.*

FIRST NATIONAL CITY BANK
of
Crete, Illinois 60417

Account #13-05762-2173-94

page 1 of 1 Enclosures-4

Stella Davis
1843 Vista Road
Crete, Illinois 60417

Account Summary and Detail

For period ending 02-03-81

513.81	Beginning balance this period
428.39	Plus deposits/credits
394.26	Less checks/debits
3.25	Less service charges/fees
544.69	Ending balance this period

Non-check transactions

Amount	Date	Description
275.19	01-01-81	deposit
153.20	10-14-81	deposit

Check transactions

Amount	Check	Date
97.18	100	01-05-81
35.70	101	01-15-81
172.15	103	01-19-81
89.23	105	01-26-81

Daily Balance

Amount	Date	+/−
789.00	01-01-81	+
691.82	01-05-81	-
845.02	01-14-81	+
809.32	01-15-81	-
637.17	01-19-81	-
547.94	01-26-81	-
544.69	02-03-81	-

1. What was the balance in Stella Davis's account on 02-03-81?
2. What was the service charge?
3. Which checks have not been deducted?

CONSUMER HINTS

1. Reconcile the account immediately upon receiving the statement.
2. Sort and match the checks to your stubs or check register.

The bank provides a form for reconciling your account. Stella Davis completes this form after receiving her statement.

Checks not shown on statement				
Number	Dollars	Cents	Bank balance shown on statement	$ 631.42
117	48	19	Add deposits not shown on statement.	$ 89.16
120	56	00	TOTAL	$ ______
			Subtract checks not shown on statement.	$ ______
Total			BALANCE (Should agree with check stub or register)	$ ______

Stella's next check stub shows:

Date ______ 19__		______ 121
To ______		
For ______		
Balance Forward	Dollars 616	Cents 39
Amount of Deposit		
Total		
Amount of Check		
Balance Carried Forward		

4. What is the total after deposits are added?
5. What is the total of checks not shown on statement?
6. Subtract checks not shown on statement.
7. There was no service charge. Is the account reconciled?

The following is information from bank statements. Do the two records agree? Complete copies of the bank form above to help you decide.

	Bank Balance	Deposits Not Shown By Bank	Total Checks Not Shown On Statement	Service Charge	Your Balance
8.	$ 173.95	$187.15	$ 58.15	0	$ 302.95
9.	$ 315.00	0	$128.25	$2.25	$ 184.50
10.	$ 463.19	$ 92.00	$106.22	0	$ 458.97
11.	$ 36.21	$126.14	0	$1.75	$ 160.60
12.	$1627.58	$743.29	$615.98	0	$1745.98
13.	$ 598.74	$315.85	$130.67	$2.25	$ 781.67

ESTIMATING DIFFERENCES

Estimate the difference. 874 − 386

Line up. → Round the numbers. → Subtract.

$$\begin{array}{r} 874 \\ -\ 386 \\ \hline \end{array} \qquad \begin{array}{r} 900 \\ -\ 400 \\ \hline \end{array} \qquad \begin{array}{r} 900 \\ -\ 400 \\ \hline 500 \end{array}$$

EXAMPLES. Estimate the differences.

1. 7286 − 3500

$$\begin{array}{r} 7286 \\ -\ 3500 \\ \hline \end{array} \qquad \begin{array}{r} 7000 \\ -\ 4000 \\ \hline 3000 \end{array}$$

2. 876 − 57

$$\begin{array}{r} 876 \\ -\ \ 57 \\ \hline \end{array} \qquad \begin{array}{r} 880 \\ -\ \ 60 \\ \hline 820 \end{array}$$

3. 25.3 − 9.476

$$\begin{array}{r} 25.3 \\ -\ \ 9.476 \\ \hline \end{array} \qquad \begin{array}{r} 25 \\ -\ \ 9 \\ \hline 16 \end{array}$$

Round to the nearest whole number.
Subtract.

4. \$78.98 − \$6.89

$$\begin{array}{r} \$\,78.98 \\ -\ \ 6.89 \\ \hline \end{array} \qquad \begin{array}{r} \$\,79 \\ -\ \ 7 \\ \hline \$\,72 \end{array}$$

Round to the nearest dollar.
Subtract.

TRY THIS

Estimate the differences.

1. $\begin{array}{r} 87 \\ -\ 38 \\ \hline \end{array}$ **2.** $\begin{array}{r} 4428 \\ -\ \ 735 \\ \hline \end{array}$ **3.** $\begin{array}{r} 6235 \\ -\ \ 490 \\ \hline \end{array}$

4. $\begin{array}{r} 627 \\ -\ \ 43 \\ \hline \end{array}$ **5.** $\begin{array}{r} 8.14 \\ -\ 3.2 \\ \hline \end{array}$ **6.** $\begin{array}{r} 762 \\ -\ 439 \\ \hline \end{array}$

7. $\begin{array}{r} \$\,12.85 \\ -\ \ 3.90 \\ \hline \end{array}$ **8.** $\begin{array}{r} 12.6 \\ -\ \ 8.1 \\ \hline \end{array}$ **9.** $\begin{array}{r} \$\,19.27 \\ -\ \ 8.61 \\ \hline \end{array}$

PRACTICE

Estimate the differences.

1. 83 − 24	**2.** 78 − 39	**3.** 69 − 31	**4.** 74 − 15	**5.** 62 − 34
6. 428 − 139	**7.** 672 − 386	**8.** 811 − 298	**9.** 689 − 203	**10.** 575 − 448
11. 4276 − 1398	**12.** 5984 − 1190	**13.** 8743 − 6819	**14.** 8612 − 2500	**15.** 6504 − 2953
16. 378 − 49	**17.** 265 − 36	**18.** 731 − 88	**19.** 349 − 85	**20.** 426 − 36
21. 2744 − 349	**22.** 6273 − 580	**23.** 3811 − 299	**24.** 7313 − 827	**25.** 1460 − 542
26. 9.382 − 6.45	**27.** 8.194 − 3.9	**28.** 3.156 − 2.2	**29.** 7.1 − 3.985	**30.** 6.610 − 2.7
31. 12.75 − 3.864	**32.** 14.382 − 8.4	**33.** 16.482 − 9.39	**34.** 17.3 − 8.498	**35.** 14.9 − 7.821
36. $ 7.98 − 2.99	**37.** $ 9.15 − 3.20	**38.** $ 7.53 − 2.65	**39.** $ 9.07 − 1.98	**40.** $ 5.40 − 2.56
41. $ 36.47 − 8.49	**42.** $ 74.36 − 38.40	**43.** $ 35.15 − 8.89	**44.** $ 63.95 − 40.16	**45.** $ 14.67 − 10.39

PROBLEM CORNER

1. You pay for a record with a $ 10 bill. Estimate how much change you should receive.
2. You pay for a tape with a $ 20 bill. Estimate how much change you should receive.

SOLVING PROBLEMS

Many real life problems do not have enough information for an immediate solution.

EXAMPLE 1. What is the cost for two new snow tires?

UNDERSTAND I must find the total cost.
I do not know the cost of one tire or the tax.

PLAN Find an ad showing prices and tax.
I will add when the price is known.

CARRY OUT The ad shows tires for $ 32 each, plus $ 2.18 tax.

```
$32.00
 32.00
  2.18
  2.18
------
$68.36
```

LOOK BACK The answer is reasonable.

Sometimes there is too much information. We need to decide what to use.

EXAMPLE 2. Mr. Weimer bought a chain saw with a 17 in. blade and a 3.7 cubic inch motor chain saw for $ 199.99. He had to pay $ 10 sales tax. What was the total bill?

UNDERSTAND I must find the total cost.
The blade length and engine size are extra information.
I know the cost and the tax.

PLAN I will add.
I do not need a calculator.

CARRY OUT

```
  $ 199.99
+    10.00
----------
  $ 209.99
```

LOOK BACK The answer is reasonable.

PRACTICE

Make up or find missing information. Solve each problem.

1. Jeri ordered a pair of nylon jogging shoes and two pairs of striped knee highs. What will the items cost?

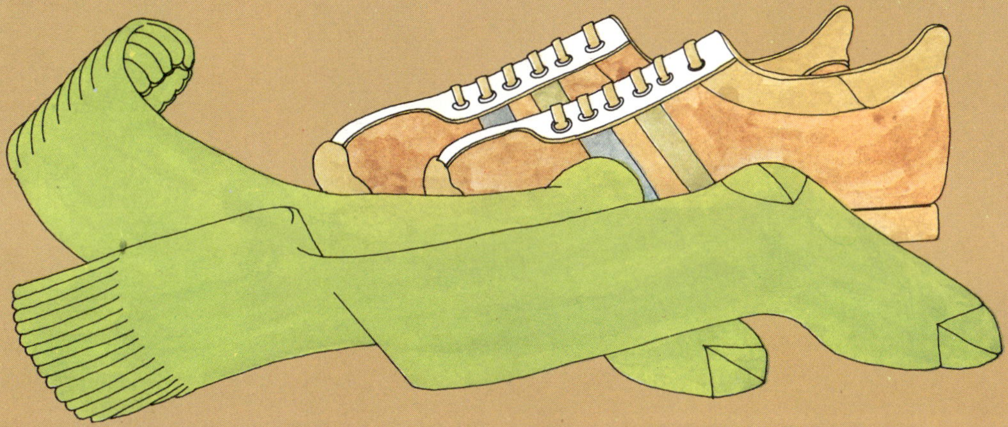

2. Bill Engelmeyer bought 100 shares of utility stock. What was the total cost?
3. Rhode Island is the smallest state; it has an area of 1214 square miles. Alaska is the largest state. How much larger is Alaska than Rhode Island?
4. Mrs. Bruder's phone bill was $ 18.50, and her gas bill was $ 21.20. Her electric bill was the highest. How much did she spend for phone and utilities?

Decide what information is extra. Solve each problem.

5. Kris King bought a 32-gallon trash container with a 4-year warranty. The trash container was on sale for $ 13.99. The regular price was $ 19.85. How much was saved?
6. In 1965 there were about 127 million dollars in two-dollar bills in circulation in the 48 contiguous states. By 1975 there were $ 135,000,000 in two-dollar bills in circulation. How much more was in circulation in 1975 than in 1965?
7. Recently, Yale University library spent $ 2,298,882 for new books and $ 6,676,271 for salaries. Stanford University spent $ 2,427,272 for books and $ 5,799,732 for salaries. Which university spent more for books? How much more?

CAREER CAREER CAREER CAREER CAREER

WORM FARMING

Carl Sanchez owns the Blue Box Worm Farm. He began his business while he was in high school.

Worms are used for bait and in gardening. People say the African Night Crawler is the best worm for fishing. The Red Worm is used by gardeners to improve the soil.

1. Last June the Blue Box Worm Farm sold a total of 225,000 worms. How many African Night Crawlers were sold?
2. A large bait company bought 75,500 worms. How many worms were sold to other customers?

Worm Sales—June

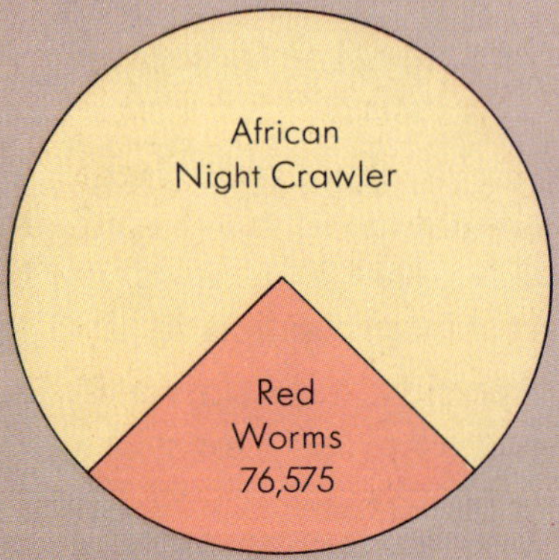

CAREER CAREER CAREER CAREER CAREER

The best worms are handpicked and sold individually. The Blue Box Worm Farm hires pickers to handpick worms.

3. How many worms did Rhea pick?
4. Who picked the least?
5. How many more worms did Mike pick than Sal?
6. How many worms were picked in all?

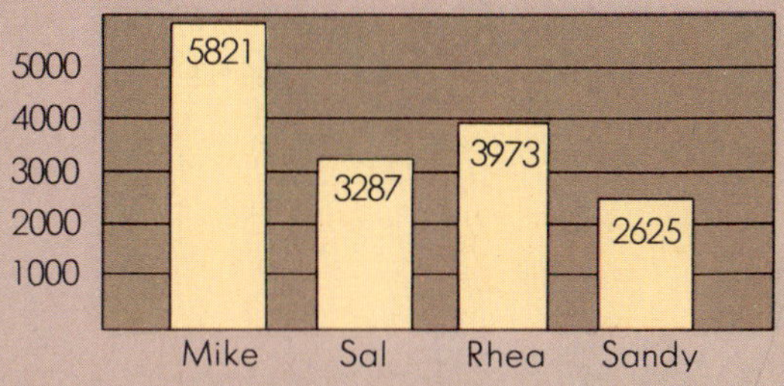

Worms must be fed properly to be fat, healthy, and saleable. The Blue Box Farm uses this recipe.

7. How many kilograms of food do they mix in one batch?
8. How much more almond dust do they use than walnut meal?
9. How much more rolled oats do they use than brown sugar?

Worm Food Recipe: 1 batch

100 kg dry manure
74 kg almond dust
62.5 kg walnut meal
9.5 kg rolled oats
8.75 kg brown sugar

Mix well. Sprinkle over worm bed.

Worms eat less on cool days than on warm days.

10. How much was eaten on Sunday?
11. Which was the coolest day of the week?
12. Which was the warmest day of the week?
13. It rained 1.5 centimeters on Friday. How much less was eaten on Friday than Saturday?
14. How much more was eaten on Thursday than on Monday?

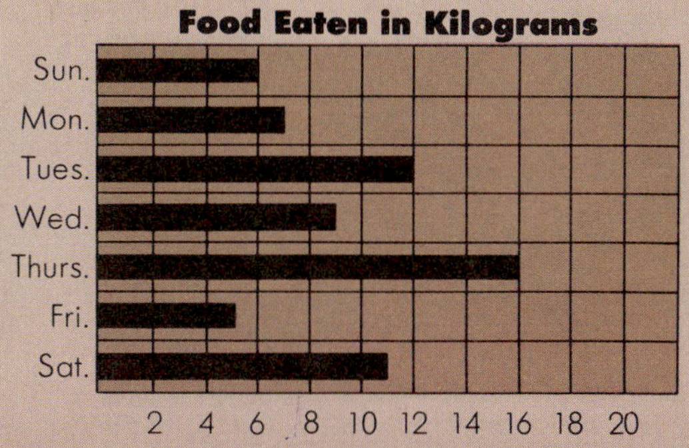

CHAPTER 2 REVIEW

Subtract. pp. 40–50

1.	746 − 125	**2.**	86 − 37	**3.**	713 − 385	**4.**	\$ 4.15 − 3.86
5.	8246 − 3948	**6.**	71,387 − 24,198	**7.**	726,414 − 345,547	**8.**	\$ 486.79 − 87.80
9.	500 − 148	**10.**	8004 − 5628	**11.**	37,000 − 19,756	**12.**	\$ 30.05 − 27.86
13.	9.7 − 3.8	**14.**	4.86 − 1.9	**15.**	7.2 − 3.65	**16.**	5. − 1.75

Compare. Use =, >, or <. p. 48

17. 0.8 and 0.08 **18.** 0.004 and 0.4 **19.** 0.5 and 0.50

20. 0.700 and 0.7 **21.** 0.16 and 0.163 **22.** 0.27 and 0.268

23. 0.03 and 0.030 **24.** 0.07 and 0.16 **25.** 0.259 and 0.248

Estimate the differences. p. 54

26.	85 − 36	**27.**	4500 − 1684	**28.**	5482 − 675	**29.**	6.7 − 3.84	**30.**	\$ 5.68 − 1.04

Application

Find the balance carried forward. p. 46.

31.

Date Nov. 18, 19 82 \$25.00 473 To Sports, Inc. For magazine subscription		
Balance Forward	Dollars 219	Cents 15
Amount of Deposit		0
Total	219	15
Amount of Check	25	00
Balance Carried Forward	194	15

32.

Date Dec. 24, 19 81 \$36.75 519 To Eaton's For shoes		
Balance Forward	Dollars 736	Cents 58
Amount of Deposit	75	00
Total	811	58
Amount of Check	36	75
Balance Carried Forward	774	83

CHAPTER 2 TEST

Subtract.

1.	84 − 33	**2.**	92 − 28	**3.**	821 − 657	**4.**	\$ 5.34 − 3.76
5.	9487 − 3891	**6.**	82,491 − 28,735	**7.**	487,241 − 398,655	**8.**	\$ 728.41 − 649.58
9.	900 − 261	**10.**	4006 − 1288	**11.**	63,005 − 18,247	**12.**	\$ 90.00 − 16.43
13.	3.82 − 1.94	**14.**	7.21 − 3.4	**15.**	9.21 − 3.786	**16.**	4 − 3.281

Compare. Use =, >, or <.

17. 0.6 and 0.06 **18.** 0.27 and 0.271 **19.** 0.08 and 0.080

20. 0.12 and 0.21 **21.** 0.504 and 0.54 **22.** 0.600 and 0.6

23. 0.03 and 0.030 **24.** 0.07 and 0.7 **25.** 0.847 and 0.816

Estimate the differences.

26.	432 − 141	**27.**	7876 − 2904	**28.**	756 − 87	**29.**	9.68 − 3.7

Application

Find the balance carried forward.

30.

Date Jan. 25, 19 81 \$76.25 265
To Maxwell's
For tires

	Dollars	Cents
Balance Forward	287	00
Amount of Deposit		−0−
Total		
Amount of Check		
Balance Carried Forward		

31.

Date Oct. 21, 19 84 \$16.84 787
To Sound, Inc.
For records

	Dollars	Cents
Balance Forward	325	16
Amount of Deposit	276	15
Total		
Amount of Check		
Balance Carried Forward		

MORE PRACTICE

Subtract. Use after page 41.

1. 87 − 69	**2.** 47 − 38	**3.** 60 − 47	**4.** 91 − 36
5. 346 − 278	**6.** 831 − 129	**7.** 243 − 69	**8.** 715 − 47
9. \$ 0.76 − 0.47	**10.** \$ 6.48 − 1.94	**11.** \$ 8.75 − 2.98	**12.** \$ 4.53 − 0.68

Subtract. Use after page 43.

13. 4287 − 3196	**14.** 8421 − 6789	**15.** 6426 − 1238	**16.** 7134 − 2929
17. 73,197 − 28,088	**18.** 61,217 − 19,073	**19.** 46,810 − 38,134	**20.** 93,241 − 8,645
21. 691,286 − 424,878	**22.** 876,215 − 198,378	**23.** 921,486 − 67,498	**24.** 175,034 − 68,318
25. \$ 63.40 − 19.65	**26.** \$ 476.15 − 138.20	**27.** \$ 5286.15 − 3874.66	**28.** \$ 2410.25 − 976.43

Subtract. Use after page 45.

29. 600 − 143	**30.** 801 − 645	**31.** 4000 − 1286	**32.** 5003 − 647
33. 42,000 − 12,876	**34.** 43,008 − 9,429	**35.** 60,080 − 12,345	**36.** 7006 − 1809
37. 80,000 − 34,261	**38.** 900,008 − 6,789	**39.** 86,040 − 19,428	**40.** 31,000 − 2,499
41. \$ 20.00 − 16.57	**42.** \$ 30.00 − 21.83	**43.** \$ 400.50 − 399.86	**44.** \$ 100.00 − 62.86

Compare. Use $=$, $>$, or $<$. Use after page 49.

1. 0.7 and 0.70 **2.** 0.3 and 0.03 **3.** 0.04 and 0.4

4. 0.12 and 0.120 **5.** 0.6 and 0.59 **6.** 0.10 and 0.1

7. 0.08 and 0.079 **8.** 0.9 and 0.90 **9.** 0.012 and 0.021

10. 0.460 and 0.46 **11.** 0.104 and 0.12 **12.** 0.3 and 0.34

13. 0.68 and 0.612 **14.** 0.740 and 0.704 **15.** 0.4 and 0.400

16. 0.20 and 0.200 **17.** 0.539 and 0.568 **18.** 0.27 and 0.264

Subtract. Use after page 51.

19. $\begin{array}{r} 8.6 \\ -\ 1.9 \\ \hline \end{array}$ **20.** $\begin{array}{r} 7.38 \\ -\ 1.29 \\ \hline \end{array}$ **21.** $\begin{array}{r} 6.071 \\ -\ 4.268 \\ \hline \end{array}$ **22.** $\begin{array}{r} 4.37 \\ -\ 0.19 \\ \hline \end{array}$

23. $\begin{array}{r} 9.87 \\ -\ 1.9 \\ \hline \end{array}$ **24.** $\begin{array}{r} 6.742 \\ -\ 3.87 \\ \hline \end{array}$ **25.** $\begin{array}{r} 6.123 \\ -\ 3.9 \\ \hline \end{array}$ **26.** $\begin{array}{r} 9.26 \\ -\ 2.7 \\ \hline \end{array}$

27. $\begin{array}{r} 8.7 \\ -\ 1.93 \\ \hline \end{array}$ **28.** $\begin{array}{r} 6.74 \\ -\ 0.986 \\ \hline \end{array}$ **29.** $\begin{array}{r} 4.3 \\ -\ 2.87 \\ \hline \end{array}$ **30.** $\begin{array}{r} 10.6 \\ -\ 7.93 \\ \hline \end{array}$

31. $5.78 - 3.9$ **32.** $14.3 - 6.58$ **33.** $5.5 - 1.79$ **34.** $16.471 - 8.62$

35. $\begin{array}{r} 18.427 \\ -\ 9 \\ \hline \end{array}$ **36.** $\begin{array}{r} 26.32 \\ -\ 7 \\ \hline \end{array}$ **37.** $\begin{array}{r} 4 \\ -\ 1.123 \\ \hline \end{array}$ **38.** $\begin{array}{r} 6 \\ -\ 0.84 \\ \hline \end{array}$

Estimate the differences. Use after page 55.

39. $\begin{array}{r} 93 \\ -\ 34 \\ \hline \end{array}$ **40.** $\begin{array}{r} 685 \\ -\ 290 \\ \hline \end{array}$ **41.** $\begin{array}{r} 448 \\ -\ 149 \\ \hline \end{array}$ **42.** $\begin{array}{r} 6285 \\ -\ 3396 \\ \hline \end{array}$

43. $\begin{array}{r} 643 \\ -\ 78 \\ \hline \end{array}$ **44.** $\begin{array}{r} 8764 \\ -\ 385 \\ \hline \end{array}$ **45.** $\begin{array}{r} 4785 \\ -\ 67 \\ \hline \end{array}$ **46.** $\begin{array}{r} 7684 \\ -\ 498 \\ \hline \end{array}$

47. $\begin{array}{r} 4.7 \\ -\ 3.87 \\ \hline \end{array}$ **48.** $\begin{array}{r} 6.482 \\ -\ 1.39 \\ \hline \end{array}$ **49.** $\begin{array}{r} 15.3 \\ -\ 6.1 \\ \hline \end{array}$ **50.** $\begin{array}{r} 27.6 \\ -\ 19.4 \\ \hline \end{array}$

51. $\begin{array}{r} \$\,8.78 \\ -\ 1.98 \\ \hline \end{array}$ **52.** $\begin{array}{r} \$\,6.14 \\ -\ 3.15 \\ \hline \end{array}$ **53.** $\begin{array}{r} \$\,40.43 \\ -\ 28.89 \\ \hline \end{array}$ **54.** $\begin{array}{r} \$\,17.51 \\ -\ 9.49 \\ \hline \end{array}$

CALCULATOR CORNER

ADDITION–SUBTRACTION "IN BETWEEN"

Players two

Materials two calculators, paper and pencils

SAMPLE Juan gives Carlos a calculator with a number, such as 174, on the display. Juan tells Carlos that by adding a number to 174, he must get a number on the display which is between 285 and 295.

Carlos gives Juan a calculator with a number, such as 475, on the display. Carlos tells Juan that by subtracting a number from 475, he must get a number on the display which is between 345 and 355.

ROUND 1 **Carlos** Carlos writes down 174 and the interval 285–295.

Play 1 He adds 90 to 174. He records 90.
Result: 264—too small

Play 2 He adds 100 to 174. He records 100.
Result: 274—too small

Play 3 He adds 110 to 174. He records 110.
Result: 284—too small

Play 4 He adds 115 to 174. He records 115.
Result: 289–OK

It took Carlos 4 plays, so his score is 4.

Juan Juan writes down 475 and the interval 345–355.

Play 1 Juan subtracts 100 from 475. He records 100.
Result: 375—too large

Play 2 Juan subtracts 120 from 475. He records 120.
Result: 355—too large

Play 3 Juan subtracts 125 from 475. He records 125.
Result: 350–OK

It took Juan 3 plays, so his score is 3.

In the next round Carlos subtracts, and Juan adds.

Play continues for 6 rounds. The player with the lowest score wins.

READY FOR CHAPTER 3

Multiply.

1. 41×6 **2.** 87×4 **3.** 8394×5 **4.** $\$9.78 \times 7$

5. 85×30 **6.** 73×26 **7.** 2745×89 **8.** $\$5.38 \times 64$

9. 827×500 **10.** 3827×567 **11.** 498×703 **12.** $\$9.42 \times 358$

13. 80×70 **14.** 900×30 **15.** 4000×60 **16.** 8000×700

17. 9.6×8 **18.** 3.42×0.7 **19.** 0.6×0.3 **20.** 0.04×0.02

21. 10×8.7 **22.** 100×9.4 **23.** 0.1×36 **24.** 0.001×1.2

Estimate.

25. 82×9 **26.** 73×68 **27.** 694×72 **28.** 4215×876

29. $\$7.14 \times 39$ **30.** $\$27.98 \times 41$ **31.** 8.94×6 **32.** 3.1427×8.81

Application

33. Is the change counted correctly?
Charge: \$2.85. Paid with: \$5.00.

Clerk gives	Clerk says
Item	\$2.85
1 nickel	2.90
1 dime	3.00
1 dollar	4.00
1 dollar	5.00

34. Complete the list to give change.
Charge: \$7.75. Paid with: \$10.00.
Machine computes \$2.25 change.

Clerk gives	Clerk says

3
MULTIPLYING WHOLE NUMBERS AND DECIMALS

GETTING STARTED

MULTIPLICATION GOTCHA

WHAT'S NEEDED The squares of paper with digits marked on them as used in GOTCHA, Chapter 1. Use only 10 squares, one of each digit.

THE RULES

1. On a sheet of paper each player makes a scorecard like this. Play three games and add to find the total score. The player with the highest total score wins.
2. To play, a leader draws digits from the box, one at a time. All players write them down. They are to be multiplied as they are drawn to get the total for the game.
3. When 0 is drawn, it is a GOTCHA. Each player still in the game gets a 0 for that game.
4. A player may go out or quit before any draw. Try to get the biggest score but go out before 0 is drawn.

Scorecard

1. ______

2. ______

3. ______

Total ______

SAMPLE

Game 1	Game 2	Game 3
7	9	7
× 3	× 8	× 9
21	72	63
× 9	× 6	× 8
189 go out	432	504
	× 0	× 5
	0 GOTCHA	2520 go out

Scoreboard

1. 189

2. 0

3. 2520

Total 2709

MULTIPLYING BY 1-DIGIT NUMBERS

Multiply. 714×6

Multiply the ones. → Multiply the tens. Add the extra tens. → Multiply the hundreds.

$$\begin{array}{r} \overset{2}{7\,1}\,4 \\ \times \quad 6 \\ \hline 4 \end{array} \qquad \begin{array}{r} \overset{2}{7\,1}\,4 \\ \times \quad 6 \\ \hline 8\,4 \end{array} \qquad \begin{array}{r} \overset{2}{7\,1}\,4 \\ \times \quad 6 \\ \hline 4\,2\,8\,4 \end{array}$$

EXAMPLE 1. Multiply. 52×3

$$\begin{array}{r} 5\,2 \\ \times \quad 3 \\ \hline 1\,5\,6 \end{array}$$

Multiply the ones.
Multiply the tens.

EXAMPLE 2. Multiply. 784×8

$$\begin{array}{r} \overset{6\;3}{7\,8\,4} \\ \times \quad 8 \\ \hline 6\,2\,7\,2 \end{array}$$

Multiply the ones.
Multiply the tens.
Multiply the hundreds.

TRY THIS

1. 643×2 **2.** 39×4 **3.** 8015×9 **4.** 3172×6

EXAMPLE 3. Multiply. $\$3.73 \times 5$

$$\begin{array}{r} \$\ \overset{3\;1}{3.7\,3} \\ \times \quad 5 \\ \hline \$\,1\,8.6\,5 \end{array}$$

Multiply as with whole numbers.
Write money notation in the answer.

TRY THIS

Multiply.

5. $\$0.71 \times 8$ **6.** $\$7.58 \times 4$ **7.** $\$235.19 \times 7$ **8.** $\$136.05 \times 5$

PRACTICE

Multiply.

1. 74×2 **2.** 32×3 **3.** 71×6 **4.** 90×8

5. 731×3 **6.** 824×2 **7.** 121×4 **8.** 811×5

9. 78×6 **10.** 42×8 **11.** 65×7 **12.** 83×9

13. 374×2 **14.** 876×3 **15.** 347×5 **16.** 986×4

17. 3287×2 **18.** 9827×4 **19.** 3270×6 **20.** 8947×8

21. $12{,}497 \times 3$ **22.** $24{,}685 \times 5$ **23.** $73{,}496 \times 7$ **24.** $32{,}083 \times 9$

25. $425{,}681 \times 8$ **26.** $542{,}614 \times 6$ **27.** $320{,}521 \times 4$ **28.** $681{,}299 \times 7$

29. $\$0.87 \times 2$ **30.** $\$3.76 \times 4$ **31.** $\$18.78 \times 8$ **32.** $\$316.74 \times 5$

PROBLEM CORNER

A $ 50.00 United States Savings Bond costs $ 37.50 to purchase.
It must earn interest for 5 years before its value is $ 50.00.
Joseph Shemanski celebrated becoming a U.S. citizen.
He bought seven $ 50.00 bonds.

1. How much did Joseph pay for the bonds?

2. How much will they be worth in 5 years?

3. How much interest will they earn in 5 years?

MULTIPLYING BY 2-DIGIT NUMBERS

Multiply. 36 × 43

Multiply by the ones' digit.	→	Write 0 in the ones' place. Multiply by the tens' digit.	→	Add.

$$\begin{array}{r} 36 \\ \times\, 43 \\ \hline 108 \end{array} \qquad \begin{array}{r} 36 \\ \times\, 43 \\ \hline 108 \\ 1440 \\ \hline \end{array} \qquad \begin{array}{r} 36 \\ \times\, 43 \\ \hline 108 \\ 1440 \\ \hline 1548 \end{array}$$

EXAMPLE 1. Multiply. 84 × 60

$$\begin{array}{r} 84 \\ \times\, 60 \\ \hline 5040 \end{array}$$

Write 0 in the ones' place.
Multiply by the tens' digit.

EXAMPLE 2. Multiply. 246 × 38

$$\begin{array}{r} 246 \\ \times\ \ 38 \\ \hline 1968 \\ 7380 \\ \hline 9348 \end{array}$$

Multiply by the ones' digit.
Write 0 in the ones' place.
Multiply by the tens' digit.
Add.

TRY THIS

Multiply.

1. $\begin{array}{r} 892 \\ \times\ 70 \\ \hline \end{array}$ **2.** $\begin{array}{r} 98 \\ \times\, 27 \\ \hline \end{array}$ **3.** $\begin{array}{r} 4764 \\ \times\ \ 12 \\ \hline \end{array}$ **4.** $\begin{array}{r} 2653 \\ \times\ \ 36 \\ \hline \end{array}$

EXAMPLE 3. Multiply. \$7.48 × 23

$$\begin{array}{r} \$\,7.48 \\ \times\ \ 23 \\ \hline 2244 \\ 14960 \\ \hline \$\,172.04 \end{array}$$

Multiply as with whole numbers.

Write money notation in the answer.

TRY THIS

Multiply.

5. $\begin{array}{r} \$\,0.73 \\ \times\ \ 12 \\ \hline \end{array}$	**6.** $\begin{array}{r} \$\,6.25 \\ \times\ \ 34 \\ \hline \end{array}$	**7.** $\begin{array}{r} \$\,56.79 \\ \times\ \ 56 \\ \hline \end{array}$	**8.** $\begin{array}{r} \$\,30.06 \\ \times\ \ 98 \\ \hline \end{array}$

PRACTICE

Multiply.

1. $\begin{array}{r} 73 \\ \times 20 \\ \hline \end{array}$	**2.** $\begin{array}{r} 84 \\ \times 40 \\ \hline \end{array}$	**3.** $\begin{array}{r} 76 \\ \times 60 \\ \hline \end{array}$	**4.** $\begin{array}{r} 73 \\ \times 80 \\ \hline \end{array}$	**5.** $\begin{array}{r} 25 \\ \times 30 \\ \hline \end{array}$
6. $\begin{array}{r} 249 \\ \times\ \ 30 \\ \hline \end{array}$	**7.** $\begin{array}{r} 648 \\ \times\ \ 50 \\ \hline \end{array}$	**8.** $\begin{array}{r} 240 \\ \times\ \ 70 \\ \hline \end{array}$	**9.** $\begin{array}{r} 704 \\ \times\ \ 90 \\ \hline \end{array}$	**10.** $\begin{array}{r} 120 \\ \times\ \ 40 \\ \hline \end{array}$
11. $\begin{array}{r} 3789 \\ \times\ \ 10 \\ \hline \end{array}$	**12.** $\begin{array}{r} 1827 \\ \times\ \ 80 \\ \hline \end{array}$	**13.** $\begin{array}{r} 4286 \\ \times\ \ 20 \\ \hline \end{array}$	**14.** $\begin{array}{r} 3219 \\ \times\ \ 40 \\ \hline \end{array}$	**15.** $\begin{array}{r} 1904 \\ \times\ \ 50 \\ \hline \end{array}$
16. $\begin{array}{r} 46 \\ \times 24 \\ \hline \end{array}$	**17.** $\begin{array}{r} 38 \\ \times 42 \\ \hline \end{array}$	**18.** $\begin{array}{r} 75 \\ \times 15 \\ \hline \end{array}$	**19.** $\begin{array}{r} 19 \\ \times 48 \\ \hline \end{array}$	**20.** $\begin{array}{r} 69 \\ \times 31 \\ \hline \end{array}$
21. $\begin{array}{r} 728 \\ \times\ \ 31 \\ \hline \end{array}$	**22.** $\begin{array}{r} 436 \\ \times\ \ 13 \\ \hline \end{array}$	**23.** $\begin{array}{r} 806 \\ \times\ \ 44 \\ \hline \end{array}$	**24.** $\begin{array}{r} 980 \\ \times\ \ 16 \\ \hline \end{array}$	**25.** $\begin{array}{r} 242 \\ \times\ \ 57 \\ \hline \end{array}$
26. $\begin{array}{r} 5284 \\ \times\ \ 33 \\ \hline \end{array}$	**27.** $\begin{array}{r} 8047 \\ \times\ \ 27 \\ \hline \end{array}$	**28.** $\begin{array}{r} 7297 \\ \times\ \ 64 \\ \hline \end{array}$	**29.** $\begin{array}{r} 9040 \\ \times\ \ 87 \\ \hline \end{array}$	**30.** $\begin{array}{r} 6113 \\ \times\ \ 22 \\ \hline \end{array}$
31. $\begin{array}{r} \$\,7.25 \\ \times\ \ 30 \\ \hline \end{array}$	**32.** $\begin{array}{r} \$\,8.76 \\ \times\ \ 60 \\ \hline \end{array}$	**33.** $\begin{array}{r} \$\,4.29 \\ \times\ \ 50 \\ \hline \end{array}$	**34.** $\begin{array}{r} \$\,7.08 \\ \times\ \ 40 \\ \hline \end{array}$	**35.** $\begin{array}{r} \$\,4.50 \\ \times\ \ 90 \\ \hline \end{array}$
36. $\begin{array}{r} \$\,0.98 \\ \times\ \ 21 \\ \hline \end{array}$	**37.** $\begin{array}{r} \$\,4.27 \\ \times\ \ 43 \\ \hline \end{array}$	**38.** $\begin{array}{r} \$\,17.38 \\ \times\ \ 54 \\ \hline \end{array}$	**39.** $\begin{array}{r} \$\,46.50 \\ \times\ \ 75 \\ \hline \end{array}$	**40.** $\begin{array}{r} \$\,32.98 \\ \times\ \ 36 \\ \hline \end{array}$

PROBLEM CORNER

In a leap-frog marathon fourteen students covered 400 laps around a track in 23 hours and 11 minutes. They averaged 42 leaps each lap. How many times did the students leap in the marathon?

More Practice, page 92

MULTIPLYING BY 3-DIGIT NUMBERS

Multiply. 478 × 293

Multiply by the ones' digit. → Write 0. Multiply by the tens' digit. → Write 00. Multiply by the hundreds' digit. → Add.

$$\begin{array}{r} 478 \\ \times\, 293 \\ \hline 1434 \end{array} \qquad \begin{array}{r} 478 \\ \times\, 293 \\ \hline 1434 \\ 43020 \end{array} \qquad \begin{array}{r} 478 \\ \times\, 293 \\ \hline 1434 \\ 43020 \\ 95600 \\ \hline \end{array} \qquad \begin{array}{r} 478 \\ \times\, 293 \\ \hline 1434 \\ 43020 \\ 95600 \\ \hline 140{,}054 \end{array}$$

EXAMPLE 1. Multiply. 246 × 300

$$\begin{array}{r} 246 \\ \times\, 300 \\ \hline 73{,}800 \end{array}$$

Write 00.
Multiply by the hundreds' digit.

EXAMPLE 2. Multiply. 724 × 631

$$\begin{array}{rl} 631 & \\ \times\, 724 & \\ \hline 2524 & \textit{Multiply by the ones' digit.} \\ 12620 & \textit{Write 0. Multiply by the tens' digit.} \\ 441700 & \textit{Write 00. Multiply by the hundreds' digit.} \\ \hline 456{,}844 & \textit{Add.} \end{array}$$

TRY THIS

1. $\begin{array}{r} 437 \\ \times\, 200 \\ \hline \end{array}$ **2.** $\begin{array}{r} 345 \\ \times\, 234 \\ \hline \end{array}$ **3.** $\begin{array}{r} \$\,4.75 \\ \times\, 321 \\ \hline \end{array}$ **4.** $\begin{array}{r} \$\,2.65 \\ \times\, 483 \\ \hline \end{array}$

EXAMPLE 3. Multiply. 302 × 456

$$\begin{array}{rl} 456 & \textit{Multiply by the ones' digit.} \\ \times\, 302 & \textit{We can leave out multiplying by 0.} \\ \hline 912 & \textit{Write 00.} \\ 136800 & \textit{Multiply by the hundreds' digit.} \\ \hline 137{,}712 & \textit{Add.} \end{array}$$

TRY THIS

Multiply.

5.	647 × 405	**6.**	647 × 450	**7.**	\$ 19.62 × 303	**8.**	\$ 17.39 × 420

PRACTICE

Multiply.

1.	948 × 200	**2.**	763 × 900	**3.**	6334 × 400	**4.**	3879 × 500	**5.**	5062 × 300
6.	\$ 7.68 × 600	**7.**	\$ 3.97 × 400	**8.**	\$ 72.49 × 200	**9.**	\$ 87.61 × 800	**10.**	\$ 14.73 × 300
11.	325 × 523	**12.**	761 × 427	**13.**	7159 × 741	**14.**	5828 × 764	**15.**	6302 × 428
16.	\$ 3.91 × 428	**17.**	\$ 8.39 × 182	**18.**	\$ 24.27 × 724	**19.**	\$ 77.77 × 777	**20.**	\$ 25.19 × 243
21.	624 × 404	**22.**	624 × 440	**23.**	1398 × 670	**24.**	4896 × 409	**25.**	2187 × 353
26.	\$ 7.65 × 870	**27.**	\$ 8.24 × 703	**28.**	\$ 19.05 × 490	**29.**	\$ 77.30 × 307	**30.**	\$ 68.75 × 450

PROBLEM CORNER

Jocelyn Sims is the manager of the school cafeteria. She computes the average daily sales for the 180-day school year.

Average Daily Sales	
Whole Milk	185
Chocolate Milk	118
Orange Juice	150

1. How many whole milks are sold? chocolate milks?

2. How many orange juices are sold?

3. How many students attend the school?

More Practice, page 92

APPLICATION

COUNTING MONEY

Evaluating a pile of bills and coins requires careful counting.

THE A, B, C's OF MONEY COUNTING

Always Sort coins and bills into groups of like denominations.

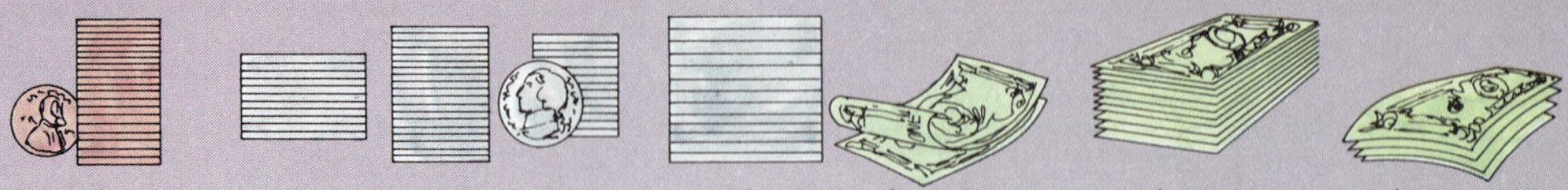

Be Sure to Count the amounts in two ways.

1. Count the number of coins and the number of bills. Then multiply.

3 × $5 = $15 12 × $10 = $120 30 × $0.05 = $1.50

2. Count the total value of the coins and the bills by adding.

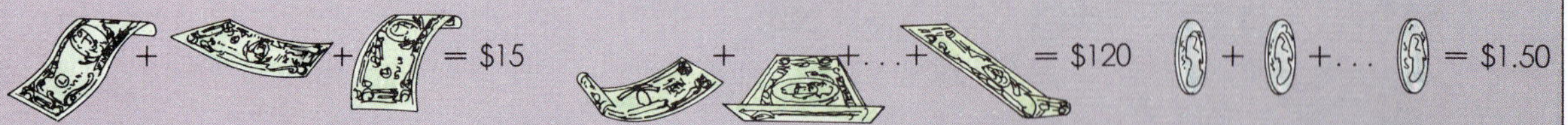

Carefully Add groups together to find the total.

This money has been sorted.

How much money is there:

1. in $ 20 bills? **2.** in $ 10 bills? **3.** in $ 5 bills? **4.** in $ 1 bills?

5. in quarters? **6.** in nickels? **7.** in dimes? **8.** in pennies?

9. What is the total in bills?

10. What is the total in coin?

11. What is the total amount of money?

Mr. Stewart manages a restaurant. The small bills and coins he uses to start each day are called the daily bank.

	Bills			**Coins**				
Denomination	$10	$ 5	$ 1	50¢	25¢	10¢	5¢	1¢
Number of Coins or Bills	2	7	17	20	40	50	40	10

12. How much of the daily bank is in bills?

13. How much of the daily bank is in coins?

14. How much is Mr. Stewart's daily bank?

Mr. Stewart needs to balance his cash and receipts. He totals the cash, checks and charges. Then he subtracts the daily bank. This tells the total amount of daily sales which should match the cash register sales.

Cash Drawer			
Bills	Coins	Checks	Charges
2 – $100	13 – 50¢	$28.44	$25.20
2 – $ 50	23 – 25¢	$14.38	$22.09
16 – $ 20	48 – 10¢	$17.96	
8 – $ 10	15 – 5¢		
11 – $ 5	28 – 1¢		

Cash Register Tape

$796.25 Total Sales

15. How much money is in bills?

16. How much money is in coins?

17. What is the total in checks?

18. What is the total charges in charges?

19. What is the total in the cash drawer?

20. Subtract the amount in the daily bank.

21. Does Mr. Stewart's count agree with the cash register tape?

SPECIAL PRODUCTS AND ESTIMATING

Multiply. 800×70

$$\begin{array}{rl} 800 & \textit{two 0's} \\ \times\ \ 70 & \textit{one 0} \\ \hline 56{,}000 & \textit{three 0's} \end{array}$$

1. Count the 0's in the factors.
2. Write the total number of 0's in the answer.
3. Multiply the non-zero digits.

Estimate. 795×68

$$\begin{array}{r} 795 \\ \times\ \ 68 \\ \hline \end{array} \longrightarrow \begin{array}{r} 800 \\ \times\ \ 70 \\ \hline 56{,}000 \end{array}$$

To estimate products:

1. Round the factors.
2. Multiply the rounded factors.

EXAMPLE 1. Multiply. 500×6

$$\begin{array}{r} 500 \\ \times\ \ \ 6 \\ \hline 3000 \end{array}$$

Count two 0's. Write two 0's in the answer.
Multiply the non-zero digits. $6 \times 5 = 30$

EXAMPLE 2. Multiply. 5000×600

$$\begin{array}{r} 5000 \\ \times\ \ 600 \\ \hline 3{,}000{,}000 \end{array}$$

Count five 0's. Write five 0's in the answer.
Multiply the non-zero digits. $6 \times 5 = 30$

TRY THIS

Multiply.

1. 7000×9
2. 70×10
3. 8000×50
4. 900×80

EXAMPLES. Estimate.

3. $$\begin{array}{r} 486 \\ \times\ \ 6 \\ \hline \end{array} \longrightarrow \begin{array}{r} 500 \\ \times\ \ 6 \\ \hline 3000 \end{array}$$

4. $$\begin{array}{r} 3984 \\ \times\ \ 246 \\ \hline \end{array} \longrightarrow \begin{array}{r} 4000 \\ \times\ \ 200 \\ \hline 800{,}000 \end{array}$$

TRY THIS

Estimate.

5. 46×8
6. 784×65
7. 5204×395
8. 655×439

PRACTICE

Multiply.

1. 80 × 9	**2.** 700 × 6	**3.** 5000 × 4	**4.** 1000 × 9	**5.** 6000 × 6
6. 50 × 70	**7.** 90 × 80	**8.** 900 × 90	**9.** 300 × 80	**10.** 400 × 50
11. 700 × 100	**12.** 800 × 200	**13.** 7000 × 100	**14.** 3000 × 700	**15.** 6000 × 600

Estimate.

16. 76 × 8	**17.** 420 × 9	**18.** 910 × 6	**19.** 498 × 5	**20.** 309 × 8
21. 89 × 88	**22.** 41 × 22	**23.** 590 × 38	**24.** 695 × 18	**25.** 427 × 26
26. 888 × 888	**27.** 421 × 395	**28.** 7284 × 387	**29.** 6980 × 785	**30.** 5638 × 528

PROBLEM CORNER

1. The average attendance at a basketball game was 8198. The team played 42 games at home and 42 games away. Estimate the total number who attended the basketball games.

2. Find the exact number who attended the basketball games.

MORE ON ESTIMATING PRODUCTS

With Money

Estimate. $8.98 × 79

$8.98 × 79	$9 × 80 = $720	*Round to the nearest dollar. Round as usual. Multiply.*

With Decimals

Estimate. 26.148 × 8.9

26.148 × 8.9	26 × 9	30 × 9 = 270	*Round each decimal to the nearest whole number. Round as usual. Multiply.*

EXAMPLES. Estimate.

1. $6.15 × 7 → $6 × 7 = $42

2. $2.75 × 399 → $3 × 400 = $1200

3. $78.85 × 42 → $79 × 40 → $80 × 40 = $3200 *Round to the nearest dollar. Round as usual. Multiply.*

TRY THIS

Estimate.

1. $3.98 × 6
2. $8.13 × 61
3. $28.80 × 5
4. $37.75 × 78

EXAMPLES. Estimate.

4. 42.93 × 17.8 → 43 × 18 → 40 × 20 = 800

5. 3.126 × 6.82 → 3 × 7 = 21

TRY THIS

Estimate.

5.	6.	7.	8.
5.2176 × 68	5.2176 × 3.9	37.245 × 51.17	42.894 × 39.3

PRACTICE

Estimate.

1. \$ 8.95 × 4	2. \$ 3.15 × 6	3. \$ 7.75 × 32	4. \$ 5.87 × 81	5. \$ 4.37 × 58
6. \$ 2.85 × 412	7. \$ 7.89 × 878	8. \$ 5.12 × 210	9. \$ 8.88 × 398	10. \$ 2.65 × 455
11. \$ 41.80 × 5	12. \$ 67.50 × 8	13. \$ 39.95 × 7	14. \$ 86.10 × 3	15. \$ 78.23 × 2
16. \$ 48.95 × 59	17. \$ 73.15 × 43	18. \$ 61.12 × 54	19. \$ 88.95 × 21	20. \$ 94.03 × 34
21. 7.8 × 4	22. 8.13 × 3	23. 1.9147 × 31	24. 4.519 × 44	25. 8.625 × 17
26. 7.9 × 6.1	27. 3.75 × 5.2	28. 3.1416 × 8.79	29. 6.437 × 5.628	30. 2.279 × 3.642
31. 49.1 × 67.3	32. 78.37 × 61.3	33. 41.8 × 78.9	34. 56.248 × 31.98	35. 48.153 × 24.72

PROBLEM CORNER

Arrow Construction Company pays unskilled workers $4.88 per hour. Estimate what a worker makes during:

1. one 8-hour shift.
2. an 8-hour shift for a 6-day week.
3. an 8-hour shift for a 6-day week for 48 weeks per year.

More Practice, page 93

APPLICATION

RECEIVING AND GIVING CHANGE

Change is usually given in the fewest number of coins and bills. There are two common methods of counting change.

METHOD A The change is counted out from the purchase price, smallest unit to largest.

A charge of $ 0.38 is paid with a $ 1 bill.

Clerk gives	Clerk says
1 penny	$ 0.39
1 penny	0.40
1 dime	0.50
1 half-dollar	1.00

METHOD B The cash register computes the change. Then it is counted out largest unit to smallest.

A charge of $ 0.38 is paid with a $ 1 bill. The register computes $ 0.62 change.

Clerk gives	Clerk says
1 half-dollar	$ 0.50
1 dime	0.60
1 penny	0.61
1 penny	0.62

PRACTICE

Was the change counted correctly?

1. Charge: $7.95 Paid with: $10

Clerk gives	Clerk says
1 nickel	$ 8.00
1 dollar	9.00
1 dollar	10.00

2. Charge: $3.48 Paid with: $5

Clerk gives	Clerk says
1 penny	$3.49
1 half-dollar	3.50

3. Charge: $8.15 Paid with: $10

Clerk gives	Clerk says
1 quarter	$ 8.25
1 quarter	8.50
1 half-dollar	9.00
1 dollar	10.00

4. Charge: $4.35 Paid with: $20

Clerk gives	Clerk says
1 nickel	$ 4.40
1 dime	4.50
1 half-dollar	5.00
1 $5 bill	10.00
1 $10 bill	20.00

5. Charge: $16.29 Paid with: $20
Machine computes $3.71 change.

Clerk gives	Clerk says
1 dollar	$1.00
1 dollar	2.00
1 half-dollar	2.50
1 dime	2.60
1 dime	2.70
1 penny	2.71

6. Charge: $8.38 Paid with: $10
Machine computes $1.62 change.

Clerk gives	Clerk says
1 dollar	$1.00
1 half-dollar	1.50
1 dime	1.60
1 penny	1.61
1 penny	1.62

List the change that should be given. Use words and money notation.

7. Charge: $8.25 Paid with: $10
8. Charge: $13.75 Paid with: $20
9. Charge: $3.68 Paid with: $5
10. Charge: $2.64 Paid with: $5
11. Charge: $8.25 Paid with: $10
Machine computes $1.75 change.
12. Charge: $12.15 Paid with: $20
Machine computes $7.85 change.
13. Charge: $17.76 Paid with: $20
Machine computes $2.24 change.
14. Charge: $3.17 Paid with: $5
Machine computes $1.83 change.
15. Charge: $13.19 Paid with: $20
Machine computes $6.81 change.
16. Charge: $6.32 Paid with: $10
Machine computes $3.68 change.

MULTIPLYING DECIMALS

Multiply. 6.58×7.4

Multiply as with whole numbers. → Count the number of decimal places in both factors. → Place the decimal point in the answer. The answer has the same number of decimal places as the two factors.

$$\begin{array}{r} 6.58 \\ \times\ \ 7.4 \\ \hline 2632 \\ 46060 \\ \hline 48692 \end{array} \qquad \begin{array}{r} 6.58 \\ \times\ \ 7.4 \\ \hline 2632 \\ 46060 \\ \hline 48692 \end{array}\Big\}\ \textit{3 decimal places} \qquad \begin{array}{r} 6.58 \\ \times\ \ 7.4 \\ \hline 2632 \\ 46060 \\ \hline 48.692 \end{array}\ \begin{array}{l} \} \textit{ 3 places} \\ \\ \\ \\ \textit{3 places} \end{array}$$

EXAMPLES. Multiply.

1.
$$\begin{array}{r} 6.384 \\ \times\ \ \ \ 7 \\ \hline 44.688 \end{array} \quad \begin{array}{l} \} \textit{ 3 places} \\ \\ \textit{3 places} \end{array}$$

2.
$$\begin{array}{r} 6.784 \\ \times\ \ 8.1 \\ \hline 6784 \\ 542720 \\ \hline 54.9504 \end{array} \quad \begin{array}{l} \} \textit{ 4 places} \\ \\ \\ \\ \textit{4 places} \end{array}$$

3.
$$\begin{array}{r} 3.9 \\ \times\ 0.007 \\ \hline 0.0273 \end{array} \quad \begin{array}{l} \} \textit{ 4 places} \\ \\ \textit{4 places} \end{array}$$

Multiply.
Count the places.
Write 0's in the answer as needed.

TRY THIS

Multiply.

1. 3.8×6 **2.** 7.74×0.9 **3.** 374×0.24 **4.** 8.72×9

5. 0.008×3 **6.** 0.07×0.7 **7.** 0.03×0.003 **8.** 0.409×0.505

PRACTICE

Multiply.

1. 8.6×7	**2.** 47×0.9	**3.** 0.84×7	**4.** 7.3×0.6	**5.** 4.2×0.6
6. 63×0.04	**7.** 7.8×0.09	**8.** 87×0.006	**9.** 8.7×0.06	**10.** 5.3×0.07
11. 32.6×16	**12.** 7.28×5.4	**13.** 0.984×3.3	**14.** 7.489×8.2	**15.** 1.962×3.8
16. 374×0.4	**17.** 569×0.05	**18.** 749×0.008	**19.** 8749×0.003	**20.** 2468×0.006
21. 43.7×0.8	**22.** 43.7×0.08	**23.** 4.37×0.8	**24.** 4.37×0.08	**25.** 0.437×0.8
26. 0.7×0.7	**27.** 0.9×0.4	**28.** 0.08×8	**29.** 0.06×7	**30.** 0.05×9
31. 0.08×0.8	**32.** 0.2×0.2	**33.** 0.3×0.1	**34.** 0.01×7	**35.** 0.02×3
36. 0.06×0.06	**37.** 0.04×0.03	**38.** 0.004×2	**39.** 0.007×1	**40.** 0.008×0.02
41. 0.005×0.05	**42.** 0.007×0.1	**43.** 0.006×1	**44.** 0.003×0.002	**45.** 0.009×0.008

PROBLEM CORNER

Myra drinks an average of 5.48 cups of tea per day. How many cups of tea does she drink in a week? in a month (30 days)?

More Practice, page 93

OTHER SPECIAL PRODUCTS

When we multiply by 10, 100, or 1000, there is a shorter method.

Multiply. 100×743.684

Count the number of 0's. → Move the decimal point to the right that number of places.

There are two. $100 \times 743.684 = 74{,}368.4$

EXAMPLES. Multiply.

1. $10 \times 7.98 = 79.8$

2. $100 \times 4.37 = 437$

3. $1000 \times 16.500 = 16{,}500$

TRY THIS

Multiply.

1. 100×6.3

2. 100×13.586

3. 1000×9.678

4. 1000×8.6093

5. 10×8.64

6. 10×6.542

When we multiply by 0.1, 0.01, or 0.001, there is also a shorter method.

Multiply. 0.01×743.68

Count the number of decimal places. → Move the decimal point to the left that number of places.

There are two. $0.01 \times 743.68 = 7.4368$

EXAMPLES. Multiply.

4. $0.1 \times 52.9 = 5.29$

5. $0.01 \times 21 = 0.21$

6. $0.001 \times 004.83 = 0.00483$

TRY THIS

Multiply.

7. 0.01 × 4.3 **8.** 0.1 × 0.87 **9.** 0.001 × 7.34

PRACTICE

Multiply.

1. 10 × 6.74	**2.** 10 × 0.76	**3.** 10 × 9.785
4. 100 × 7.31	**5.** 100 × 7.842	**6.** 100 × 24.843
7. 1000 × 7.6421	**8.** 1000 × 97.694	**9.** 1000 × 82.984
10. 100 × 7.4	**11.** 100 × 19.4	**12.** 100 × 0.7
13. 1000 × 9.6	**14.** 1000 × 8.7	**15.** 1000 × 0.3
16. 1000 × 9.13	**17.** 1000 × 19.4	**18.** 1000 × 0.7
19. 0.1 × 18.5	**20.** 0.1 × 36.27	**21.** 0.1 × 8.7
22. 0.01 × 17.6	**23.** 0.01 × 297.4	**24.** 0.01 × 17.98
25. 0.001 × 4748.2	**26.** 0.001 × 6284.15	**27.** 0.001 × 428.7
28. 0.01 × 1.76	**29.** 0.01 × 7.4	**30.** 0.01 × 0.5
31. 0.001 × 13.4	**32.** 0.001 × 28.5	**33.** 0.001 × 19.7

PROBLEM CORNER

1. There are 100 centimeters in a meter. How many centimeters are there in 8.762 meters?

2. There are 0.001 kilograms in a gram. How many kilograms are there in 6842 grams?

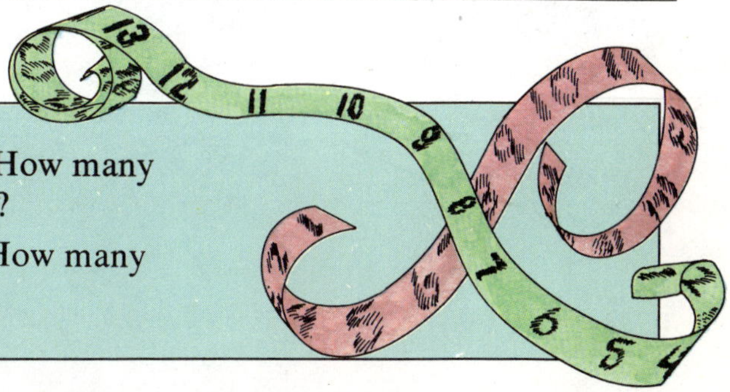

More Practice, page 93

SOLVING PROBLEMS

Sometimes solutions to problems require more than one step.

EXAMPLE. Finn sold 6 trays of popcorn at the stadium. Each tray contained 15 bags of popcorn. The cost of a bag of popcorn was 75¢. How much did Finn receive for the popcorn?

UNDERSTAND I must find the total for all the popcorn.
I have enough information.

PLAN I have solved problems like this before.
I must do two multiplications.
I do not need a calculator.

CARRY OUT

$$\begin{array}{r} 15 \\ \times\ 6 \\ \hline 90 \end{array} \text{ bags of popcorn} \qquad \begin{array}{r} \$\,0.75 \\ \times\ 90 \\ \hline \$\,67.50 \end{array}$$

Finn received $ 67.50 for the popcorn.

LOOK BACK This answer is reasonable.
The answer fits the problem.

When problems require more than one step, we complete each step, one at a time.

TRY THIS Katy bought 3 records at $ 5.95 and 2 records at $ 6.25. What was the total cost for the records?

PRACTICE

1. Freda sold 19 push-up packs of flowers at the school fair. Each pack contained 12 flowers. The cost of a flower was $ 0.08. How much did Freda receive for the flowers?

2. Rico bought 12 cans of soda at $ 0.40 and 6 bags of chips at $ 0.65 for the party. What was the total cost?

3. Simon earns $ 4.50 per hour for working Monday through Friday. He earns $ 6.75 per hour for weekend work. Last week he worked 8 hours a day on Monday, Wednesday, and Saturday. He worked 6 hours on both Tuesday and Sunday. How much did he earn last week?

4. Mull Reef, a famous race horse, broke his leg. His fans sent him 30 get well cards a day during the four weeks of his recuperation. How many cards did he receive?

5. Sandy bought 2 cups of plain yogurt at $ 0.55, 3 cups of strawberry yogurt at $ 0.65, and one cup of boysenberry yogurt at $ 0.69. What was the total cost?

6. The Booster Club bought 6 cartons of peanuts at $ 3.60 per carton. Each carton contained 72 bags of peanuts. The club sold the peanuts for 20¢ a bag. What was their gross profit after selling all the peanuts?

7. Kim Su runs 8 km on Monday, Wednesday, and Friday and 16 km on Tuesday, Thursday, and Saturday. How many kilometers does she run per week?

CAREER CAREER CAREER CAREER CAREER

FAST FOOD RESTAURANT MANAGER

Ana Moreno manages the Apple Cellar Deli. She hires the part-time workers and orders the supplies.

Ana Moreno hires part-time workers for $4.25 per hour. Find the weekly salaries of these part-time workers.

1. Mark works 24 hours per week.
2. Patty works 8 hours each Saturday and Sunday.
3. Ben works 3 days per week; 6 hours per day.
4. Marcia works Monday through Friday for 3 hours each day.
5. Juan works 4 hours on weekdays and 8 hours on Saturday.

Ana finds the total cost of supplies ordered for the deli. Find the total cost for each item. Find the total cost of all items.

	Item	Amount	Cost Per Unit	Total Cost
6.	Bread	167 loaves	$0.46	
7.	Cheese	15 kg	3.88	
8.	Lettuce	22 heads	0.49	
9.	Tuna	36 cans	0.89	
10.	Paper Cups	6 cases	3.07	
11.	Mayonnaise	12 jars	1.26	
				Total

PROFIT = SELLING PRICE − COST

Figuring cost and profit is a major part of Ana Moreno's job.

Cost per Deli Sandwich	
Bread	$0.03
Meat	0.28
Cheese	0.12
Lettuce	0.05
Mayonnaise	0.14

MENU	
Deli Sandwich	$1.49
Tuna Sandwich	0.99
Vegetarian Delight	1.09
Low-Cal Special	0.59
Drinks:	
Small	0.29
Large	0.49

12. What does it cost to make a Deli sandwich?
13. What is the profit per Deli sandwich?
14. Last week 1079 Deli sandwiches were sold. What was the profit?
15. Ana checked this order ticket. What mistakes did she find?

Apple Cellar Deli Order Form

Quantity	Item	Cost	Total
3	Deli Sandwich	$1.49	$4.47
4	Vegetarian Delight	1.09	5.36
2	Tuna Sandwich	.99	1.98
9	Drinks, large	.29	2.61
		Sub-Total	14.32
		Tax	.59
		Pay This Total Amount	$14.81

CHAPTER 3 REVIEW

Multiply. pp. 68–84

1. 82 × 4	**2.** 96 × 8	**3.** 429 × 7	**4.** \$ 48.26 × 5
5. 48 × 60	**6.** 386 × 24	**7.** 7285 × 83	**8.** \$ 19.48 × 17
9. 849 × 400	**10.** 859 × 648	**11.** 3982 × 560	**12.** \$ 74.98 × 603
13. 70 × 60	**14.** 800 × 20	**15.** 5000 × 60	**16.** 9000 × 400
17. 8.74 × 6	**18.** 4.92 × 6.3	**19.** 0.5 × 0.6	**20.** 0.02 × .3
21. 100 × 8.3	**22.** 1000 × 6.76	**23.** 0.01 × 92	**24.** 0.001 × 3.17
25. 10 × 0.43	**26.** 100 × 0.05	**27.** 0.1 × 9.3	**28.** 0.001 × 32.4

Estimate. pp. 76–78

29. 89 × 6	**30.** 47 × 61	**31.** 742 × 891	**32.** 6500 × 412
33. \$ 8.74 × 4	**34.** \$ 37.68 × 59	**35.** 9.12 × 32	**36.** 4.291 × 8.76

Application p. 80

37. Is the change counted correctly?
Charge: \$ 13.80. Paid with: \$ 20.00.

Clerk gives	Clerk says
Item	\$ 13.80
1 quarter	14.00
1 dollar	15.00
5 dollars	20.00

38. Complete the list to give change.
Charge: \$ 2.35. Paid with: \$ 5.00
Machine computes \$ 2.65 change.

Clerk gives	Clerk says

CHAPTER 3 TEST

Multiply.

1. 72×3 **2.** 48×6 **3.** 743×7 **4.** $\$\,92.46 \times 5$

5. 72×60 **6.** 429×81 **7.** 7984×68 **8.** $\$\,49.50 \times 43$

9. 812×900 **10.** 629×493 **11.** 6284×507 **12.** $\$\,42.85 \times 150$

13. 40×30 **14.** 700×10 **15.** 6000×80 **16.** 4000×900

17. 9.23×7 **18.** 8.76×5.3 **19.** 0.7×0.7 **20.** 0.02×0.03

21. 10×8.3 **22.** 1000×9.67 **23.** 0.1×8.6 **24.** 0.01×36.5

Estimate.

25. 72×8 **26.** 56×65 **27.** 525×89 **28.** 3994×313

29. $\$\,6.15 \times 9$ **30.** $\$\,8.79 \times 38$ **31.** 6.874×19 **32.** 5.199×6.91

Application

33. Is the change counted correctly?
Charge: \$ 4.39. Paid with: \$ 10.00.

Clerk gives	Clerk says
Item	\$ 4.39
1 penny	4.40
1 dime	4.50
1 half-dollar	5.00
5 dollars	10.00

34. Complete the list to give change.
Charge: \$ 8.73. Paid with: \$ 20.00
Machine computes \$ 11.27 change.

Clerk gives	Clerk says

MORE PRACTICE

Multiply. Use after page 69.

1. 86×3 2. 97×4 3. 54×5 4. 32×6

5. 148×9 6. 259×8 7. 4360×7 8. $\$62.78 \times 4$

Multiply. Use after page 71.

9. 64×30 10. 87×50 11. 762×40 12. 8246×70

13. 64×34 14. 87×76 15. 43×18 16. 387×49

17. 425×86 18. 619×57 19. 7324×63 20. 7463×58

21. $\$5.98 \times 12$ 22. $\$0.67 \times 24$ 23. $\$87.96 \times 74$ 24. $\$63.87 \times 93$

Multiply. Use after page 73.

25. 678×600 26. 974×300 27. 891×500 28. 680×800

29. 968×432 30. 427×628 31. 649×427 32. $\$4.98 \times 123$

33. 874×603 34. 428×640 35. 798×507 36. 640×640

Multiply. Use after page 77.

37. 30×40 38. 60×80 39. 400×40 40. 600×90

41. 300×600 42. 8000×20 43. 4000×600 44. 7000×6000

Estimate. Use after page 77.

1. 48×7 **2.** 64×8 **3.** 426×9 **4.** 7287×8

5. 65×75 **6.** 87×23 **7.** 429×86 **8.** 788×55

9. 428×679 **10.** 478×310 **11.** 8294×625 **12.** 6942×7180

Estimate. Use after page 79.

13. $\$4.98 \times 8$ **14.** $\$38.74 \times 5$ **15.** $\$4.78 \times 38$ **16.** $\$59.48 \times 69$

17. 374×6.8 **18.** 2.9×7 **19.** 11.83×6 **20.** 5.2×37

21. 8.74×6.9 **22.** 4.316×6.125 **23.** 87.3×31.9 **24.** 33.5×18.4

Multiply. Use after page 83.

25. 9.7×8 **26.** 6.42×7 **27.** 3.768×12 **28.** 224×8.77

29. 6.9×2.7 **30.** 5.4×3.8 **31.** 1.63×8.2 **32.** 4.17×0.09

33. 9.8×3.4 **34.** 2.76×3.8 **35.** 8.14×92.3 **36.** 6.483×0.4

37. 7.6×5.8 **38.** 2.42×1.9 **39.** 17.3×0.6 **40.** 24.8×0.03

41. 0.7×0.6 **42.** 0.08×0.8 **43.** 0.01×0.4 **44.** 0.07×0.02

Multiply. Use after page 85.

45. 10×7.6 **46.** 10×8.74 **47.** 100×9.68 **48.** 1000×3.94

49. 0.1×86 **50.** 0.1×9.7 **51.** 0.01×94.3 **52.** 0.001×87

CALCULATOR CORNER

MULTIPLICATION ESTIMATION

Players two
Materials two calculators, paper and pencils

SAMPLE

1. Mark asks Erin to estimate a product, such as 527 × 37.
   Erin asks Mark to estimate a product, such as 462 × 52.
2. Mark and Erin write down their problems. They estimate and multiply.
3. Mark and Erin calculate the exact answers to their problems.
4. They then calculate the differences between their estimates and the exact answers.

Erin

$$\begin{array}{r} 527 \\ \times\ 37 \\ \hline \end{array} \qquad \begin{array}{r} 500 \\ \times\ 40 \\ \hline 20{,}000 \end{array}$$

estimate	20,000
exact answer	− 19,499
	501

Mark

$$\begin{array}{r} 462 \\ \times\ 52 \\ \hline \end{array} \qquad \begin{array}{r} 460 \\ \times\ 50 \\ \hline 23{,}000 \end{array}$$

exact answer	24,024
estimate	− 23,000
	1,024

Since Erin's estimate was closer to the exact answer, she gets 1 point.

Play continues until a player gets 5 points.

READY FOR CHAPTER 4

Divide.

1. $30 \div 8$ **2.** $84 \div 7$ **3.** $407 \div 6$ **4.** $184 \div 9$

5. $5\overline{)920}$ **6.** $7\overline{)45{,}721}$ **7.** $8\overline{)6412}$ **8.** $4\overline{)\$\,19.53}$

9. $42\overline{)93}$ **10.** $33\overline{)1519}$ **11.** $28\overline{)14{,}571}$ **12.** $56\overline{)\$\,209.64}$

13. $276\overline{)2484}$ **14.** $136\overline{)3138}$ **15.** $834\overline{)60{,}894}$ **16.** $498\overline{)29{,}888}$

17. $6\overline{)7.314}$ **18.** $42\overline{)365.4}$ **19.** $7\overline{)0.049}$ **20.** $8\overline{)3}$

21. $2.7\overline{)9.18}$ **22.** $0.04\overline{)8.76}$ **23.** $0.3\overline{)9}$ **24.** $0.14\overline{)7}$

25. $800 \div 10$ **26.** $732.4 \div 100$ **27.** $46{,}249.4 \div 1000$ **28.** $6 \div 10$

Estimate.

29. $6\overline{)4981}$ **30.** $78\overline{)2359}$ **31.** $7\overline{)\$\,34.72}$ **32.** $9.3\overline{)46.176}$

Find the quotient to the nearest hundredth.

33. $7\overline{)9}$ **34.** $9\overline{)7}$ **35.** $6.4\overline{)9.3}$ **36.** $0.03\overline{)2}$

Application

Period Ending	Hours	Rate	Earnings		Gross Earnings	Deductions				Net Pay
			Amount Earned Regular Rate	Overtime and Other		FICA	Federal Withholding Tax	State Withholding Tax	Others	
7-9-81	30	4.50		33.00		10.30	24.50	3.70	40.00	

Statement of Earnings and Deductions for Employee's Record—Detach Before Cashing Check

37. Find the gross earnings.

38. Find the net pay.

4
DIVIDING WHOLE NUMBERS AND DECIMALS

GETTING STARTED

FISHING FOR FACTORS

WHAT'S NEEDED The 30 squares of paper with digits marked on them as used in GOTCHA, Chapter 1. Take out the 0's.

THE RULES

1. On a sheet of paper each player makes a 5 × 5 grid. Fill in the grid with any 25 numbers.
2. A leader draws two digits from the box, multiplies them and says the product.

   Example: The leader draws 3 and 4 and says 12.
3. Each player crosses out all numbers that are factors of the product just found.

   Example: The factors of 12 are 1, 2, 3, 4, 6 and 12. Each player crosses out a 1, a 2, a 3, a 4, a 6, and a 12.
4. The leader continues to draw digits two at a time. The first player to cross out all 25 numbers is the winner.

~~1~~	1	1	1	~~2~~
2	2	~~3~~	3	4
4	5	5	~~6~~	6
7	8	9	10	~~12~~
15	25	32	48	56

1-DIGIT DIVISORS

Divide and check. 98 ÷ 5

How many 5's are in 9? Try 1. Multiply. Subtract. → Bring down the 8. How many 5's are in 48? Try 9. Multiply. Subtract. → Check.

$$\begin{array}{r} 1 \\ 5\overline{)98} \\ \underline{5} \\ 4 \end{array} \qquad \begin{array}{r} 19 \text{ r } 3 \\ 5\overline{)98} \\ \underline{5}\downarrow \\ 48 \\ \underline{45} \\ 3 \end{array} \qquad \begin{array}{rl} 19 & \leftarrow \text{quotient} \\ \times\ \ 5 & \leftarrow \text{divisor} \\ \hline 95 & \\ +\ \ 3 & \leftarrow \text{remainder} \\ \hline 98 & \leftarrow \text{dividend} \end{array}$$

EXAMPLE 1. Divide and check. 61 ÷ 8

$$\begin{array}{r} 7 \text{ r } 5 \\ 8\overline{)61} \\ \underline{56} \\ 5 \end{array}$$

How many 8's are in 6? None.
How many 8's are in 61? Try 7.
Multiply. Subtract.

Check.

$$\begin{array}{r} 7 \\ \times\ \ 8 \\ \hline 56 \\ +\ \ 5 \\ \hline 61 \end{array}$$

EXAMPLE 2. Divide. 87 ÷ 7

$$\begin{array}{r} 12 \text{ r } 3 \\ 7\overline{)87} \\ \underline{7} \\ 17 \\ \underline{14} \\ 3 \end{array}$$

How many 7's are in 8? Try 1.
Multiply. Subtract. Bring down the 7.
How many 7's are in 17? Try 2.
Multiply. Subtract.

EXAMPLE 3. Divide. 87 ÷ 8

$$\begin{array}{r} 10 \text{ r } 7 \\ 8\overline{)87} \\ \underline{8} \\ 07 \\ \underline{0} \\ 7 \end{array}$$

How many 8's are in 8? Try 1.
How many 8's are in 7? Write 0 in the quotient.
Multiply. Subtract.

TRY THIS

Divide and check.

1. $5\overline{)31}$ 2. $7\overline{)42}$ 3. $9\overline{)60}$ 4. $8\overline{)75}$

5. $2\overline{)93}$ 6. $5\overline{)193}$ 7. $9\overline{)728}$ 8. $6\overline{)379}$

PRACTICE

Divide.

1. $6\overline{)38}$ 2. $7\overline{)65}$ 3. $3\overline{)25}$ 4. $2\overline{)16}$ 5. $9\overline{)88}$

6. $4\overline{)31}$ 7. $5\overline{)42}$ 8. $8\overline{)42}$ 9. $6\overline{)42}$ 10. $9\overline{)42}$

11. $3\overline{)74}$ 12. $2\overline{)74}$ 13. $4\overline{)74}$ 14. $5\overline{)74}$ 15. $6\overline{)74}$

16. $6\overline{)87}$ 17. $7\overline{)87}$ 18. $3\overline{)87}$ 19. $4\overline{)87}$ 20. $5\overline{)87}$

21. $2\overline{)196}$ 22. $3\overline{)196}$ 23. $4\overline{)196}$ 24. $8\overline{)196}$ 25. $7\overline{)196}$

26. $6\overline{)374}$ 27. $7\overline{)429}$ 28. $9\overline{)849}$ 29. $7\overline{)123}$ 30. $4\overline{)249}$

31. $2\overline{)47}$ 32. $8\overline{)749}$ 33. $4\overline{)98}$ 34. $6\overline{)349}$ 35. $3\overline{)261}$

36. $4\overline{)81}$ 37. $7\overline{)76}$ 38. $4\overline{)43}$ 39. $8\overline{)81}$ 40. $2\overline{)61}$

41. $7\overline{)143}$ 42. $9\overline{)456}$ 43. $6\overline{)425}$ 44. $5\overline{)403}$ 45. $8\overline{)645}$

PROBLEM CORNER

1. There are 90 bottles in cartons. There are 6 bottles in each carton. How many cartons are there?
2. There are 75 people sitting in rows in the class. If there are 5 people in each row, how many rows are there?

More Practice, page 126

DIVIDING LARGER NUMBERS

Divide. 748 ÷ 3

How many 3's are in 7? Try 2. → How many 3's are in 14? Try 4. → How many 3's are in 28? Try 9.

$$\begin{array}{r} 2 \\ 3\overline{)748} \\ \underline{6} \\ 1 \end{array} \qquad \begin{array}{r} 24 \\ 3\overline{)748} \\ \underline{6}\downarrow \\ 14 \\ \underline{12} \\ 2 \end{array} \qquad \begin{array}{r} 249 \text{ r } 1 \\ 3\overline{)748} \\ \underline{6} \\ 14 \\ \underline{12}\downarrow \\ 28 \\ \underline{27} \\ 1 \end{array}$$

EXAMPLE 1. Divide. 4278 ÷ 5

$$\begin{array}{r} 855 \text{ r } 3 \\ 5\overline{)4278} \\ \underline{40} \\ 27 \\ \underline{25} \\ 28 \\ \underline{25} \\ 3 \end{array}$$

How many 5's are in 4? None.
How many 5's are in 42? Try 8.
How many 5's are in 27? Try 5.
How many 5's are in 28? Try 5.

EXAMPLE 2. Divide. 7258 ÷ 9

$$\begin{array}{r} 806 \text{ r } 4 \\ 9\overline{)7258} \\ \underline{72} \\ 05 \\ \underline{0} \\ 58 \\ \underline{54} \\ 4 \end{array}$$

How many 9's are in 7? None.
How many 9's are in 72? Try 8.
How many 9's are in 5? Write 0.
How many 9's are in 58? Try 6.

TRY THIS

Divide.

1. $6\overline{)947}$ **2.** $2\overline{)1840}$ **3.** $6\overline{)24,183}$ **4.** $5\overline{)27,043}$

EXAMPLE 3. Divide. $\$18.45 \div 5$

$$\begin{array}{r} \$\ \ 3.69 \\ 5\overline{)\$18.45} \\ \underline{15} \\ 34 \\ \underline{30} \\ 45 \\ \underline{45} \end{array}$$

Divide as with whole numbers. Write money notation for the quotient and remainder.

TRY THIS

Divide.

5. $2\overline{)\$87.46}$ **6.** $8\overline{)\$748.72}$ **7.** $3\overline{)\$18.12}$ **8.** $6\overline{)\$412.02}$

PRACTICE

Divide.

1. $798 \div 3$	**2.** $493 \div 4$	**3.** $584 \div 2$	**4.** $327 \div 5$
5. $824 \div 6$	**6.** $930 \div 5$	**7.** $947 \div 8$	**8.** $628 \div 4$
9. $2\overline{)1587}$	**10.** $7\overline{)3647}$	**11.** $5\overline{)4830}$	**12.** $6\overline{)4896}$
13. $3\overline{)9486}$	**14.** $4\overline{)9484}$	**15.** $2\overline{)7307}$	**16.** $5\overline{)8734}$
17. $8\overline{)17,498}$	**18.** $5\overline{)26,489}$	**19.** $6\overline{)40,294}$	**20.** $9\overline{)65,287}$
21. $7\overline{)709}$	**22.** $6\overline{)3665}$	**23.** $6\overline{)1234}$	**24.** $5\overline{)3650}$
25. $9\overline{)7203}$	**26.** $4\overline{)16,080}$	**27.** $6\overline{)1805}$	**28.** $5\overline{)40,198}$
29. $2\overline{)\$5.98}$	**30.** $5\overline{)\$18.47}$	**31.** $3\overline{)\$36.49}$	**32.** $5\overline{)\$48.75}$
33. $2\overline{)\$1.78}$	**34.** $7\overline{)\$14.09}$	**35.** $5\overline{)\$16.00}$	**36.** $9\overline{)\$297.50}$
37. $8\overline{)\$25.92}$	**38.** $6\overline{)\$53.74}$	**39.** $4\overline{)\$95.12}$	**40.** $7\overline{)\$306.22}$

More Practice, page 126

2-DIGIT DIVISORS

Divide. $792 \div 21$

How many 21's are in 7? None.	→	How many 21's are in 79? Think: How many 2's are in 7? Try 3. $3 \times 21 = 63$ Subtract.	→	How many 21's are in 162? Think: How many 2's are in 16? Try 8. $8 \times 21 = 168$. Too large. Try 7. $7 \times 21 = 147$ Subtract.

```
21)792
```

```
      3
21)792
   63
   16
```

```
     37 r 15
21)792
   63↓
   162
   147
    15
```

EXAMPLE 1. Divide. $187 \div 25$

```
     7 r 12
25)187
   175
    12
```

How many 25's are in 1? None.
How many 25's are in 18? None.
How many 25's are in 187?
Think: How many 2's are in 18?
Try 9. $9 \times 25 = 225$. Too large.
Try 8. $8 \times 25 = 200$. Still too large.
Try 7. $7 \times 25 = 175$. Subtract.

EXAMPLE 2. Divide. $458 \div 24$

```
    19 r 2
24)458
   24
   218
   216
     2
```

How many 24's are in 4? None.
How many 24's are in 45?
Think: How many 2's are in 4?
Try 2. $2 \times 24 = 48$. Too large.
Try 1. $1 \times 24 = 24$. Subtract.
How many 24's are in 218?
Think: How many 2's are in 21?
9 is the largest number to try.
$9 \times 24 = 216$. Subtract.

TRY THIS

Divide.

1. $22\overline{)78}$ 2. $27\overline{)80}$ 3. $21\overline{)798}$ 4. $45\overline{)562}$

5. $68\overline{)1427}$ 6. $56\overline{)48,832}$ 7. $72\overline{)\$96.48}$ 8. $32\overline{)\$67.84}$

PRACTICE

Divide.

1. $96 \div 31$ 2. $88 \div 42$ 3. $648 \div 80$ 4. $400 \div 63$

5. $310 \div 53$ 6. $200 \div 47$ 7. $90 \div 23$ 8. $81 \div 42$

9. $27\overline{)348}$ 10. $31\overline{)738}$ 11. $54\overline{)892}$ 12. $87\overline{)986}$

13. $43\overline{)1246}$ 14. $73\overline{)2654}$ 15. $92\overline{)3176}$ 16. $67\overline{)3825}$

17. $31\overline{)5624}$ 18. $43\overline{)7396}$ 19. $71\overline{)9407}$ 20. $52\overline{)7000}$

21. $63\overline{)13,545}$ 22. $47\overline{)32,641}$ 23. $84\overline{)62,412}$ 24. $72\overline{)41,641}$

25. $23\overline{)9223}$ 26. $32\overline{)2249}$ 27. $44\overline{)26,620}$ 28. $56\overline{)1690}$

29. $62\overline{)33,480}$ 30. $75\overline{)22,565}$ 31. $87\overline{)60,900}$ 32. $94\overline{)34,800}$

33. $31\overline{)\$19.46}$ 34. $52\overline{)\$102.96}$ 35. $74\overline{)\$173.16}$ 36. $45\overline{)\$55.50}$

37. $12\overline{)\$89.76}$ 38. $21\overline{)\$22.89}$ 39. $33\overline{)\$237.60}$ 40. $54\overline{)\$432.07}$

PROBLEM CORNER

1. Jake has 500 dollars in 20-dollar bills. How many 20-dollar bills does he have?
2. There are 16 ounces in a pound. How many pounds are there in 528 ounces?

More Practice, page 126

3-DIGIT DIVISORS

Divide. 19,674 ÷ 438

How many 438's are in 1967? Think: How many 4's are in 19? Try 4. $4 \times 438 = 1752$ Subtract.	→	How many 438's are in 2154? Think: How many 4's are in 21? Try 5. $5 \times 438 = 2190$. Too large. Try 4. $4 \times 438 = 1752$ Subtract.

```
        4                    44 r 402
438)19,674           438)19,674
    1752                 1752↓
     215                  2154
                          1752
                           402
```

EXAMPLE 1. Divide. 2,357 ÷ 386

```
       6 r 41
386)2357
    2316
      41
```

How many 386's are in 2357?
Think: How many 3's are in 23?
Try 7. $7 \times 386 = 2702$. Too large.
Try 6. $6 \times 386 = 2316$. Subtract.

EXAMPLE 2. Divide. 42,497 ÷ 728

```
        58 r 273
728)42,497
    3640
     6097
     5824
      273
```

How many 728's are in 4249?
Think: How many 7's are in 42?
Try 6. $6 \times 728 = 4368$. Too large.
Try 5. $5 \times 728 = 3640$. Subtract.
How many 728's are in 6097?
Think: How many 7's are in 60?
Try 8. $8 \times 728 = 5824$. Subtract.

TRY THIS

Divide.

1. 221)895 **2.** 187)748 **3.** 826)7248

4. 827)9479 **5.** 741)63,874 **6.** 568)17,040

PRACTICE

Divide.

1. 846 ÷ 326	**2.** 890 ÷ 212	**3.** 270 ÷ 125	**4.** 983 ÷ 640
5. 956 ÷ 407	**6.** 548 ÷ 113	**7.** 650 ÷ 386	**8.** 907 ÷ 287
9. $368\overline{)2944}$	**10.** $524\overline{)3700}$	**11.** $750\overline{)4500}$	**12.** $941\overline{)3774}$
13. $815\overline{)4095}$	**14.** $688\overline{)6292}$	**15.** $425\overline{)3000}$	**16.** $250\overline{)2000}$
17. $146\overline{)3289}$	**18.** $395\overline{)5545}$	**19.** $684\overline{)8308}$	**20.** $136\overline{)4624}$
21. $248\overline{)9425}$	**22.** $310\overline{)9641}$	**23.** $405\overline{)9750}$	**24.** $190\overline{)8360}$
25. $525\overline{)45,150}$	**26.** $736\overline{)31,000}$	**27.** $999\overline{)65,934}$	**28.** $888\overline{)78,144}$
29. $333\overline{)30,700}$	**30.** $140\overline{)10,640}$	**31.** $291\overline{)13,100}$	**32.** $407\overline{)19,129}$
33. $874\overline{)43,700}$	**34.** $736\overline{)30,000}$	**35.** $381\overline{)4000}$	**36.** $459\overline{)9200}$
37. $187\overline{)17,000}$	**38.** $512\overline{)40,960}$	**39.** $245\overline{)17,250}$	**40.** $650\overline{)39,000}$

PROBLEM CORNER

1. There are 144 items in a gross. How many gross are there in 1152 items?

2. A plane averaged 328 miles per hour. It traveled 4592 miles. How many hours did the plane travel?

More Practice, page 126

APPLICATION

RATES AND WAGES

One way to be paid is by the length of time worked. The time period can be the hour, the week, the month, or even the year.

The minimum wage law was first passed in 1938. This law is based on an hourly rate. The rate in 1938 was 25¢ per hour. Compare this to some recent rates in the graph.

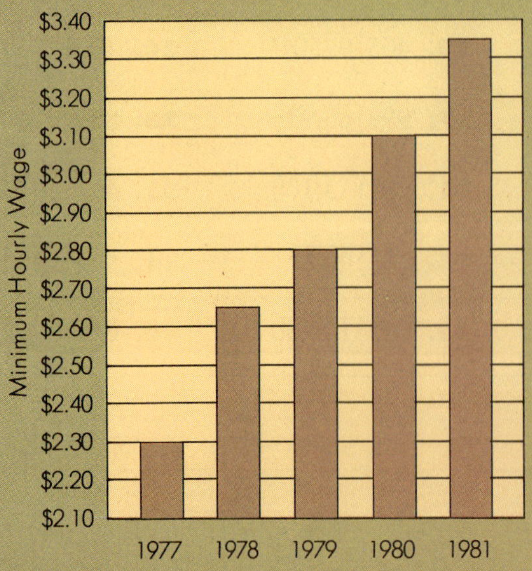

1. Find the minimum wage for 1980.
2. Find the minimum wage for 1981.
3. Find the daily minimum wage for an eight-hour day in:
   a. 1977 b. 1979 c. 1981 d. 1938
4. Find the weekly minimum wage for a forty-hour week in:
   a. 1977 b. 1979 c. 1981 d. 1938
5. A store clerk earned $ 10 for 40 hours at the minimum wage. What year was it?
6. A secretary earned $ 18.40 for 8 hours at the minimum wage. What year was it?

A pay stub is usually found attached to a paycheck.

Period Ending	Hours	Rate	Earnings		Gross Earnings	Deductions				Net Pay
			Amount Earned Regular Rate	Overtime and Other		FICA	Federal Withholding Tax	State Withholding Tax	Others	
3-4-82	40	4.50	180.00	68.00	248.00	14.72	35.18	9.67	Insurance 21.20	161.23

Statement of Earnings and Deductions for Employee's Record—Detach Before Cashing Check

↑ Hours Worked (under Hours)
↑ Hourly Wage (under Rate)
↑ Overtime, Travel Allowance, Meal Allowance (under Overtime and Other)
↑ Social Security (under FICA)
↑ Pay After Deductions (under Net Pay)

7. How much was earned at the regular rate?
8. What were the overtime and other earnings?
9. What were the gross earnings?
10. What was the net pay?

Period Ending	Hours	Rate	Earnings		Gross Earnings	Deductions				Net Pay
			Amount Earned Regular Rate	Overtime and Other		FICA	Federal Withholding Tax	State Withholding Tax	Others	
4-7-82	48	4.00		73.00		15.10	37.15	9.88	Insurance 22.50	

Statement of Earnings and Deductions for Employee's Record—Detach Before Cashing Check

11. Find the amount earned at the regular rate. (Multiply hours by rate.)
12. Find the gross earnings. (Add overtime to amount earned at regular rate.)
13. Find the total deductions.
14. Find the net pay. (Subtract deductions from gross earnings.)

Period Ending	Hours	Rate	Earnings		Gross Earnings	Deductions				Net Pay
			Amount Earned Regular Rate	Overtime and Other		FICA	Federal Withholding Tax	State Withholding Tax	Others	
8-4-82	40	7.25		56.75		25.50	47.75	18.50	Ins. 23.00 Sav. 50.00	

Statement of Earnings and Deductions for Employee's Record—Detach Before Cashing Check

15. Find the gross earnings.
16. Find the net pay.

DIVIDING A DECIMAL BY A WHOLE NUMBER

Divide. 17.421 ÷ 3

Place the decimal point in the quotient. → Divide as with whole numbers.

```
                   5.807
3)17.421         3)17.421
                   15
                    24
                    24
                     02
                      0
                      21
                      21
                       0
```

EXAMPLE 1. Divide. 171.36 ÷ 36

```
     4.76
36)171.36
   144
    273
    252
     216
     216
```

Place the decimal point in the quotient.
Divide as with whole numbers.

TRY THIS

Divide.

1. 2)8.6 **2.** 99)227.7 **3.** 21)70.14 **4.** 37)122.84

EXAMPLES. Divide. 0.154 ÷ 2

```
      0.077
2.  2)0.154
        0
        15
        14
         14
         14
```

0's must be placed in the quotient.

Divide. 15 ÷ 2

```
      7.5
3.  2)15.0
      14
       10
       10
```

TRY THIS

Divide.

5. $5\overline{)2.5}$ **6.** $9\overline{)0.18}$ **7.** $8\overline{)0.016}$

9. $5\overline{)12}$ **10.** $6\overline{)3.9}$ **11.** $8\overline{)3}$

PRACTICE

Divide.

1. $9.3 \div 3$ **2.** $20.8 \div 4$ **3.** $31.44 \div 6$ **4.** $7.38 \div 2$

5. $9.144 \div 8$ **6.** $16.731 \div 9$ **7.** $9.828 \div 4$ **8.** $12.123 \div 3$

9. $23\overline{)89.7}$ **10.** $12\overline{)104.4}$ **11.** $42\overline{)176.4}$ **12.** $68\overline{)360.4}$

13. $25\overline{)206.00}$ **14.** $82\overline{)140.22}$ **15.** $79\overline{)255.96}$ **16.** $93\overline{)516.15}$

17. $3\overline{)1.5}$ **18.** $4\overline{)3.6}$ **19.** $8\overline{)6.4}$ **20.** $6\overline{)5.4}$

21. $2\overline{)1.04}$ **22.** $6\overline{)0.36}$ **23.** $7\overline{)1.33}$ **24.** $5\overline{)0.35}$

25. $8\overline{)1.128}$ **26.** $9\overline{)0.018}$ **27.** $4\overline{)0.076}$ **28.** $3\overline{)1.212}$

29. $4\overline{)6}$ **30.** $6\overline{)15}$ **31.** $5\overline{)11}$ **32.** $2\overline{)1}$

33. $4\overline{)7}$ **34.** $8\overline{)10}$ **35.** $4\overline{)1.4}$ **36.** $6\overline{)4.5}$

37. $8\overline{)1}$ **38.** $8\overline{)5}$ **39.** $8\overline{)7}$ **40.** $2\overline{)1.87}$

PROBLEM CORNER

1. James had 15.3 meters of rope. He wants to cut it into 3 equal pieces. How long should each piece be?

2. Six people are to work 9 pages of problems. How many pages should each do to share the work equally?

CIAL QUOTIENTS

To divide by 10, 100, or 1000, we can use a short method.

Divide. 7 0 0 0 ÷ 1 0 0

Count the 0's in the divisor. → Move the decimal point to the left that same number of places. → Drop the 0's after the decimal point.

7 0 0 0 ÷ 1 0 0   7 0 0 0. ÷ 1 0 0   =   7 0

EXAMPLES. Divide.

1. 5 0 0 ÷ 1 0 = 5 0 0. ÷ 1 0 = 5 0

2. 7 3 2 ÷ 1 0 0 = 7 3 2. ÷ 1 0 0 = 7.3 2

TRY THIS

Divide.

1. 600 ÷ 100   **2.** 30,000 ÷ 1000   **3.** 52 ÷ 10

EXAMPLES. Divide.

3. 4 3.5 ÷ 1 0 = 4.3 5   **4.** 5 2 7.4 5 ÷ 1 0 0 = 5.2 7 4 5

We sometimes need to write 0's.

EXAMPLES. Divide.

5. 8.7 ÷ 1 0 0 = 0 8.7 ÷ 1 0 0 = 0.0 8 7

6. 9 ÷ 1 0 0 0 = 0 0 9. ÷ 1 0 0 0 = 0.0 0 9

TRY THIS

Divide.

4. 68.74 ÷ 10   **5.** 49.6 ÷ 10
6. 6 ÷ 100   **7.** 17.5 ÷ 100
8. 0.7 ÷ 1000   **9.** 2.08 ÷ 1000

PRACTICE

Divide.

1. $80 \div 10$ **2.** $700 \div 10$ **3.** $8000 \div 10$

4. $38 \div 10$ **5.** $4748 \div 10$ **6.** $386 \div 10$

7. $600 \div 100$ **8.** $4000 \div 100$ **9.** $900 \div 100$

10. $326 \div 100$ **11.** $1348 \div 100$ **12.** $248 \div 100$

13. $8000 \div 1000$ **14.** $22{,}000 \div 1000$ **15.** $40{,}000 \div 1000$

16. $3278 \div 1000$ **17.** $4219 \div 1000$ **18.** $73{,}489 \div 1000$

19. $82.5 \div 10$ **20.** $37.86 \div 10$ **21.** $672.5 \div 10$

22. $276.3 \div 100$ **23.** $427.86 \div 100$ **24.** $425.4 \div 100$

25. $8276.4 \div 1000$ **26.** $9387.6 \div 1000$ **27.** $18{,}427.3 \div 1000$

28. $9.3 \div 10$ **29.** $0.9 \div 10$ **30.** $6 \div 10$

31. $8 \div 10$ **32.** $4.7 \div 10$ **33.** $0.3 \div 10$

34. $3.4 \div 100$ **35.** $0.1 \div 100$ **36.** $2 \div 100$

37. $3 \div 100$ **38.** $3.8 \div 100$ **39.** $0.5 \div 100$

40. $4 \div 1000$ **41.** $12 \div 1000$ **42.** $86 \div 1000$

PROBLEM CORNER

1. There are 10 millimeters in a centimeter. How many centimeters are there in 376 millimeters?

2. There are 1000 kilograms in a metric ton. How many metric tons are there in 4236 kilograms?

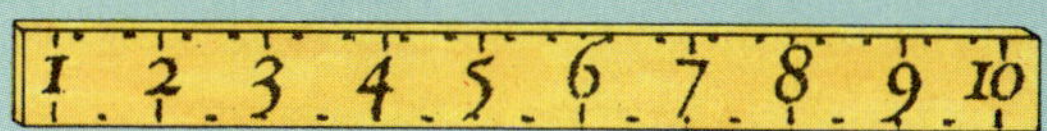

More Practice, page 127

ESTIMATING QUOTIENTS

1-digit divisor: Estimate. 376 ÷ 6

Think: Is 6 × 6 or 6 × 7 closer to 37? Use 6. → Write 0's in the rest of the quotient.

$$\begin{array}{r}6\\6\overline{)376}\end{array} \qquad \begin{array}{r}60\\6\overline{)376}\end{array}$$

EXAMPLE 1. Estimate. 3974 ÷ 5

$$\begin{array}{r}800\\5\overline{)3974}\end{array}$$

Think: Is 5 × 7 or 5 × 8 closer to 39? Use 8. Write 0's in the rest of the quotient.

EXAMPLE 2. Estimate. \$ 18.78 ÷ 4

$$4\overline{)\$18.78} \longrightarrow 4\overline{)\$19} \longrightarrow \begin{array}{r}\$5\\4\overline{)\$19}\end{array}$$

Round to the nearest dollar. 5 × 4 is closest to 19.

TRY THIS

Estimate.

1. $4\overline{)131}$ **2.** $7\overline{)4128}$ **3.** $6\overline{)\$41.47}$ **4.** $5\overline{)\$49.30}$

2-digit divisor: Estimate. 3764 ÷ 63

Write the division. → Round to the highest place of the smaller number. → Divide the divisor and dividend by 10. → Is 6 × 6 or 6 × 7 closer to 37? Use 6. → Write 0's in the rest of the quotient.

$$63\overline{)3764} \longrightarrow 60\overline{)3760} \longrightarrow 6\not{0}\overline{)376\not{0}} \longrightarrow \begin{array}{r}6\\6\overline{)376}\end{array} \longrightarrow \begin{array}{r}60\\6\overline{)376}\end{array}$$

EXAMPLE 3. Estimate. 5481 ÷ 69

$$69\overline{)5481} \longrightarrow 70\overline{)5480} \longrightarrow \begin{array}{r}8\\7\overline{)548}\end{array} \longrightarrow \begin{array}{r}80\\7\overline{)548}\end{array}$$

EXAMPLE 4. Estimate. $29.2 \div 3.89$

$3.89\overline{)29.2}$ $\quad 4\overline{)29}$ $\quad \begin{array}{r} 7 \\ 4\overline{)29} \end{array}$

Round to the nearest whole numbers.
7×4 is closest to 29.

TRY THIS

Estimate.

5. $59\overline{)248}$ **6.** $42\overline{)1931}$ **7.** $9.12\overline{)44.129}$ **8.** $7.75\overline{)33.41}$

PRACTICE

Estimate.

1. $8\overline{)576}$ **2.** $4\overline{)312}$ **3.** $5\overline{)398}$ **4.** $6\overline{)435}$

5. $4\overline{)2519}$ **6.** $7\overline{)3486}$ **7.** $9\overline{)8091}$ **8.** $2\overline{)1234}$

9. $4\overline{)31{,}497}$ **10.** $8\overline{)65{,}497}$ **11.** $5\overline{)44{,}987}$ **12.** $3\overline{)22{,}015}$

13. $9\overline{)\$36.70}$ **14.** $6\overline{)\$41.80}$ **15.** $3\overline{)\$20.80}$ **16.** $7\overline{)\$40.74}$

17. $6.2\overline{)18.13}$ **18.** $8.78\overline{)26.1}$ **19.** $3.124\overline{)16.01}$ **20.** $7.8\overline{)41.8174}$

21. $9.1\overline{)26.7}$ **22.** $4.63\overline{)37.2}$ **23.** $5.4\overline{)59.93}$ **24.** $6.6\overline{)41.17}$

25. $72\overline{)369}$ **26.** $83\overline{)631}$ **27.** $79\overline{)547}$ **28.** $38\overline{)178}$

29. $37\overline{)2756}$ **30.** $93\overline{)4632}$ **31.** $42\overline{)1943}$ **32.** $49\overline{)3721}$

PROBLEM CORNER

1. Six records cost $ 41.98. Estimate the cost of one record.
2. A car traveled 710 kilometers in 8 hours. Estimate how many kilometers it traveled in one hour.
3. Vito bought 3.3 kilograms of fish for $ 11.37. Estimate how much the fish cost per kilogram.

More Practice, page 127

APPLICATION

JOB AND FLAT RATES

Some income is based on the number of items handled or produced.

1. Sylvia delivered circulars. She was paid 3¢ per circular. Find how much she earned each day.

Day	Monday	Tuesday	Wednesday	Thursday	Friday
No. of Circulars	252	176	295	93	184

2. Sharon charges $ 3.50 to mow a lawn. How many lawns does she have to mow to earn:
   a. $ 63 b. $ 45.50 c. $ 87.50 d. $ 42
3. Some wallpaper hangers charge by the roll. Some charge by the hour. Who earned the most?

   Mr. Morales hung 49 rolls in 7 hours. He charged $ 2.30 per roll.

   Mr. Yoshida hung 49 rolls in 7 hours. He charged $ 16.00 per hour.

The Quality Maintenance Service Co. pays employees flat rates for cleaning apartment units. Complete the following weekly payroll register by finding each employee's total pay.

	Employee	No. of Units Cleaned						Rate per Unit
		M	T	W	Th	F	S	
4.	H. H. Elliott	10	13	8	15	7	9	$4.00
5.	R. Salla	9	10	14	8	8	10	$3.50
6.	T. M. Gill	11	12	12	6	18	6	$4.50
7.	S. Blumenthal	14	5	12	9	12	11	$3.25
8.	B. Vantucci	6	9	14	9	13	14	$4.75
9.	N. Love	11	12	9	11	14	16	$5.00

10. What was the total payroll?
11. Suggest some reasons why the rate per unit varies.

Burt Taylor makes armatures for the Electro Corporation. Use the graph to find the number of armatures Burt made each month from July to December.

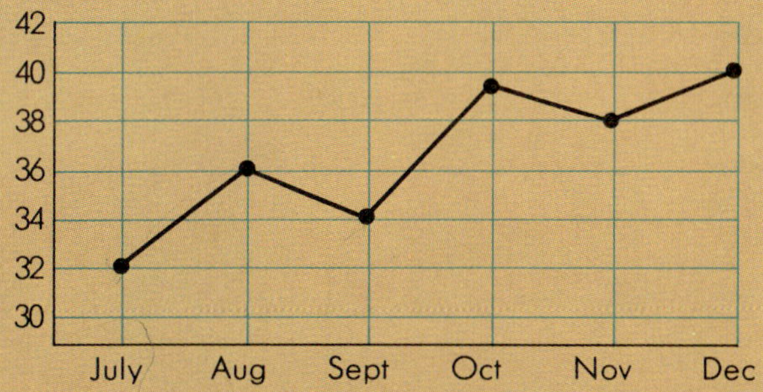

Burt's Monthly Production

12. Burt worked 22 days in July and 19 days in December. Find the average number of armatures produced each day during these months. Round to the nearest tenth.
13. Burt earned a flat rate of $ 21.65 per armature. Find his earnings for August and November.
14. How many more armatures did Burt produce in October than in September?

DIVIDING A DECIMAL BY A DECIMAL

Divide. 8.73 ÷ 0.9

Multiply the divisor by 10 to make a whole number. → Also multiply the dividend by 10. → Divide.

```
0.9)8.73        0.9)8.73          9.7
                              0.9)8.73
                                  81
                                   63
                                   63
```

EXAMPLE 1. Divide. 22.63 ÷ 0.31

```
        73.
0.31)22.63
     217
      93
      93
```

Multiply the divisor by 100 to make a whole number.
Also multiply the dividend by 100.
Divide.

TRY THIS

Divide.

1. 0.3)7.44 **2.** 0.23)9.43 **3.** 0.002)8.642 **4.** 0.52)14.508

EXAMPLE 2. Divide. 9.8 ÷ 0.002

```
       4900.
0.002)9.800
      8
      18
      18
       00
        0
        00
         0
```

Multiply the divisor and the dividend by 1000.
Write needed 0's.
Divide.

EXAMPLE 3. Divide. 8 ÷ 0.02

```
     400.
0.02)8.00
```

Multiply the divisor and the dividend by 100.
Write needed 0's.

TRY THIS

Divide.

5. $0.08\overline{)6.4}$ **6.** $0.012\overline{)7.2}$ **7.** $0.71\overline{)355}$ **8.** $0.45\overline{)9}$

PRACTICE

1. $4.5 \div 0.3$ **2.** $67.4 \div 0.2$ **3.** $82.17 \div 0.9$ **4.** $4.26 \div 0.6$

5. $8.4 \div 1.2$ **6.** $220.1 \div 3.1$ **7.** $78.3 \div 8.7$ **8.** $4.32 \div 3.6$

9. $0.04\overline{)1.68}$ **10.** $0.01\overline{)7.692}$ **11.** $0.02\overline{)41.28}$ **12.** $0.05\overline{)0.845}$

13. $0.12\overline{)8.40}$ **14.** $0.25\overline{)2.25}$ **15.** $0.36\overline{)2.88}$ **16.** $0.93\overline{)5.673}$

17. $0.001\overline{)7.426}$ **18.** $0.002\overline{)7.426}$ **19.** $0.003\overline{)0.933}$ **20.** $0.004\overline{)3.2848}$

21. $0.005\overline{)8.7420}$ **22.** $0.012\overline{)0.072}$ **23.** $0.123\overline{)0.984}$ **24.** $0.423\overline{)2.538}$

25. $0.05\overline{)1.5}$ **26.** $0.13\overline{)5.2}$ **27.** $0.34\overline{)27.2}$ **28.** $0.08\overline{)17.6}$

29. $0.004\overline{)1.6}$ **30.** $0.003\overline{)6.36}$ **31.** $0.012\overline{)1.2}$ **32.** $0.041\overline{)0.82}$

33. $1.2\overline{)6}$ **34.** $3.4\overline{)68}$ **35.** $0.12\overline{)6}$ **36.** $0.25\overline{)5}$

PROBLEM CORNER

1. Mr. Harris has 6 liters of honey. He wants to put it in jars that hold 0.25 liters each. How many jars does he need?

2. If your heart beats once every 0.8 seconds, how many times will it beat in a minute?

More Practice, page 127

ROUNDING QUOTIENTS

Find the quotient to the nearest tenth. 5.3 ÷ 0.7

Divide to 1 more place than to be rounded.	→	Round to the given place.

```
       7.5 7
  0.7)5.3 0 0
      4 9
        4 0
        3 5
          5 0
          4 9
            1
```

$7.57 \approx 7.6$

≈ means is approximately equal to.

EXAMPLE 1. Find the quotient to the nearest hundredth. 7 ÷ 0.9

```
       7.7 7 7 ≈ 7.7 8
  0.9)7.0 0 0 0
      6 3
        7 0
        6 3
          7 0
          6 3
            7 0
            6 3
              7
```

Divide to thousandths.
Round to hundredths.

EXAMPLE 2. Find the quotient to the nearest thousandth. 1 ÷ 3

```
    0.3 3 3 3 ≈ 0.333
  3)1.0 0 0 0
      9
      1 0
        9
        1 0
          9
          1 0
            9
            1
```

Divide to ten-thousandths.
Round to thousandths.

TRY THIS

Find each quotient to the nearest tenth.

1. $8\overline{)6}$ **2.** $0.43\overline{)10.1}$ **3.** $4.3\overline{)8.9}$

Find each quotient to the nearest hundredth.

4. $0.3\overline{)2}$ **5.** $0.14\overline{)7.1}$ **6.** $0.07\overline{)3.6}$

Find each quotient to the nearest thousandth.

7. $7\overline{)8}$ **8.** $0.7\overline{)9.3}$ **9.** $0.06\overline{)9.2}$

PRACTICE

Find each quotient to the nearest tenth.

1. $3\overline{)4}$ **2.** $4\overline{)5}$ **3.** $7\overline{)6}$ **4.** $8\overline{)7}$ **5.** $9\overline{)5}$

6. $3.2\overline{)8.7}$ **7.** $4.4\overline{)9.3}$ **8.** $0.07\overline{)11}$ **9.** $0.09\overline{)8}$ **10.** $0.07\overline{)19}$

Find each quotient to the nearest hundredth.

11. $9\overline{)8}$ **12.** $11\overline{)7}$ **13.** $9\overline{)11}$ **14.** $3\overline{)11}$ **15.** $7\overline{)13}$

16. $0.03\overline{)0.1}$ **17.** $0.03\overline{)2}$ **18.** $9.2\overline{)39.6}$ **19.** $1.7\overline{)8.2}$ **20.** $2.9\overline{)3.7}$

Find each quotient to the nearest thousandth.

21. $3\overline{)10}$ **22.** $7\overline{)10}$ **23.** $7\overline{)3}$ **24.** $9\overline{)1}$ **25.** $3\overline{)2}$

26. $0.3\overline{)1}$ **27.** $0.7\overline{)2}$ **28.** $0.9\overline{)4}$ **29.** $0.9\overline{)7}$ **30.** $0.7\overline{)11}$

PROBLEM CORNER

1. If you do 15 push ups in 11 seconds, how many push ups is that per second? Find the answer to the nearest hundredth.

2. If you do 20 sit ups in 11 seconds, how many sit ups is that per second? Find the answer to the nearest tenth.

More Practice, page 127

SOLVING PROBLEMS

Approximate answers are often all that is needed for a problem.

EXAMPLE. An accounting firm paid the following annual salaries to its office employees.

Stenographer	$ 11,018
Typist	$ 9,276
File Clerk	$ 6,621
Secretary	$ 14,435

About how much money should be budgeted for office employees salaries?

For budget purposes, an approximate answer is sufficient. Round each salary and add to get an approximate total.

Stenographer	$ 11,000	
Typist	9,000	
File Clerk	7,000	
Secretary	+ 14,000	
	$ 41,000	Approximate Total

TRY THIS The accounting firm pays its chief accountant $ 39,895 annually. The two junior accountants are paid $ 18,115 annually. About how much should be budgeted for accountants' salaries?

PRACTICE

1. Fred Wise bought 600 shares of stock. The annual dividend on the stock was $ 2.10. About how much income can Fred expect yearly from the dividends?

2. A population study showed that a large city had 577,642 people living in an area of just 47 square miles. About how many people live in each square mile?

3. In 1936, a 37-year-old man walked from Calcutta to Constantinople and back. He made the 5589 mile trip in just 59 days. About how many miles did he average per day?
4. One of the first automobile races took place in Chicago in 1895. The winner covered the 54.36 mile course in seven hours and 17 minutes. Estimate the average speed.

5. The eruption of the volcano on Krakatoa in 1883 sent a tidal wave halfway around the world. The wave reached heights of 120 feet. It traveled 5450 miles to the South African coast in 12 hours. Estimate its average speed.
6. Alaska has an area of 586,412 square miles. Rhode Island has an area of 1214 square miles. About how many times larger is Alaska than Rhode Island?
7. The distance from San Francisco, California to New York City is 3045 miles. About how many hours would it take to drive from San Francisco to New York City at an average speed of 55 miles per hour?

Terry Willis owns the Quality Maintenance Service. He provides janitorial and maintenance service for factories, offices, apartments, and homes.

Terry's business employs many people. People are employed to clean, supervise, and run his office. Their pay depends on the type of job they do.

1. Terry Willis draws a yearly salary of $ 22,500. How much is this monthly?
2. Job supervisors earn $ 15,000 per year. What is their monthly salary?
3. With overtime pay, full-time skilled employees may make $ 8800 annually. They are usually given 52 weeks pay for 50 weeks work. What is their weekly salary?
4. Shirley King is paid at the rate of $ 2.80 per hour. Her overtime rate is 1.5 times the regular rate. During a busy season, Shirley works a full 40-hour week plus 12 hours overtime. What does she earn that week?

CAREER CAREER CAREER CAREER CAREER

Quality Maintenance uses a sliding scale for the cost of window washing. The more windows washed, the less the cost per window.

Number of Windows	Cost
1–20	$ 60.00
21–30	$ 85.00
31–50	$132.50
51–75	$180.00
76–100	$205.00

How much does it cost per window for the following jobs?

5. 20 windows
6. 22 windows
7. 29 windows
8. 37 windows
9. 53 windows
10. 72 windows
11. 93 windows
12. 15 windows

Terry Willis studied the graph of the gross monthly income.

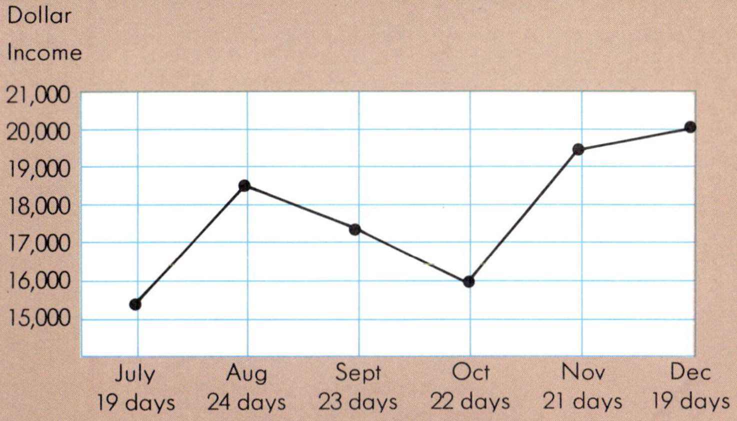

Gross Monthly Income

13. In which month was the income lowest? highest?

Terry computed the daily gross income by considering the number of work days per month.

14. To the nearest dollar, what was the daily gross income per month?
15. Estimate the average daily gross income for the six months.

CHAPTER 4 REVIEW

Divide. p. 98

1. $41 \div 7$ **2.** $72 \div 3$ **3.** $713 \div 8$ **4.** $243 \div 6$

5. $4\overline{)732}$ **6.** $5\overline{)36{,}498}$ **7.** $9\overline{)4511}$ **8.** $7\overline{)\$31.44}$

9. $34\overline{)87}$ **10.** $82\overline{)2641}$ **11.** $65\overline{)52{,}455}$ **12.** $43\overline{)\$355.64}$

13. $512\overline{)4096}$ **14.** $147\overline{)4716}$ **15.** $947\overline{)82{,}389}$ **16.** $747\overline{)29{,}993}$

17. $4\overline{)8.712}$ **18.** $56\overline{)268.8}$ **19.** $3\overline{)0.012}$ **20.** $5\overline{)4}$

21. $3.6\overline{)34.92}$ **22.** $0.05\overline{)3.75}$ **23.** $0.2\overline{)8}$ **24.** $0.25\overline{)5}$

25. $700 \div 100$ **26.** $48.7 \div 10$ **27.** $7348.2 \div 1000$ **28.** $9 \div 100$

Estimate. p. 112

29. $7\overline{)3480}$ **30.** $47\overline{)2436}$ **31.** $8\overline{)\$47.78}$ **32.** $3.12\overline{)13.721}$

33. $6\overline{)3642}$ **34.** $38\overline{)7899}$ **35.** $9\overline{)\$53.65}$ **36.** $3.71\overline{)11.83}$

Find the quotient to the nearest tenth. p. 118

37. $4\overline{)7}$ **38.** $7\overline{)4}$ **39.** $4.7\overline{)8.9}$ **40.** $0.07\overline{)4}$

41. $6\overline{)5}$ **42.** $7\overline{)9}$ **43.** $2.3\overline{)9.6}$ **44.** $0.09\overline{)5}$

Application p. 106

Period Ending	Hours	Rate	Earnings		Gross Earnings	Deductions				Net Pay
			Amount Earned Regular Rate	Overtime and Other		FICA	Federal Withholding Tax	State Withholding Tax	Others	
11-4-80	48	5.00		13.50		15.54	44.00	9.60	19.60	

Statement of Earnings and Deductions for Employee's Record—Detach Before Cashing Check

45. Find the gross earnings.

46. Find the net pay.

CHAPTER 4 TEST

Divide.

1. $35 \div 8$ **2.** $72 \div 4$ **3.** $473 \div 9$ **4.** $284 \div 7$

5. $5\overline{)620}$ **6.** $6\overline{)37{,}492}$ **7.** $3\overline{)1514}$ **8.** $8\overline{)\$25.83}$

9. $45\overline{)96}$ **10.** $93\overline{)1302}$ **11.** $76\overline{)30{,}786}$ **12.** $54\overline{)\$511.38}$

13. $623\overline{)4463}$ **14.** $258\overline{)5934}$ **15.** $369\overline{)35{,}893}$ **16.** $858\overline{)51{,}480}$

17. $5\overline{)7.315}$ **18.** $67\overline{)288.1}$ **19.** $4\overline{)0.024}$ **20.** $4\overline{)3}$

21. $4.7\overline{)40.42}$ **22.** $0.06\overline{)3.12}$ **23.** $0.3\overline{)9}$ **24.** $0.35\overline{)7}$

25. $8000 \div 100$ **26.** $538.3 \div 10$ **27.** $4298.7 \div 1000$ **28.** $3 \div 10$

Estimate.

29. $8\overline{)3142}$ **30.** $58\overline{)3547}$ **31.** $9\overline{)\$62.50}$ **32.** $4.9\overline{)24.73}$

33. $5\overline{)2630}$ **34.** $27\overline{)6234}$ **35.** $8\overline{)\$33.40}$ **36.** $5.8\overline{)41.84}$

Find the quotient to the nearest hundredth.

37. $3\overline{)8}$ **38.** $8\overline{)3}$ **39.** $5.8\overline{)7.3}$ **40.** $0.09\overline{)5}$

41. $3\overline{)1}$ **42.** $7\overline{)8}$ **43.** $3.6\overline{)4.7}$ **44.** $0.07\overline{)5}$

Application

Period Ending	Hours	Rate	Earnings: Amount Earned Regular Rate	Earnings: Overtime and Other	Gross Earnings	Deductions: FICA	Deductions: Federal Withholding Tax	Deductions: State Withholding Tax	Deductions: Others	Net Pay
9-21-82	40	6.50		40.25		18.41	57.50	14.30	29.10	

Statement of Earnings and Deductions for Employee's Record—Detach Before Cashing Check

45. Find the gross earnings.

46. Find the net pay.

MORE PRACTICE

Divide. Use after page 99.

1. $60 \div 8$ **2.** $52 \div 7$ **3.** $15 \div 6$ **4.** $80 \div 9$

5. $2\overline{)87}$ **6.** $6\overline{)96}$ **7.** $4\overline{)174}$ **8.** $6\overline{)438}$

9. $7\overline{)74}$ **10.** $8\overline{)643}$ **11.** $4\overline{)160}$ **12.** $5\overline{)353}$

Divide. Use after page 101.

13. $897 \div 6$ **14.** $873 \div 3$ **15.** $4398 \div 7$ **16.** $3920 \div 5$

17. $6\overline{)7352}$ **18.** $4\overline{)9287}$ **19.** $7\overline{)68{,}749}$ **20.** $5\overline{)47{,}845}$

21. $3\overline{)313}$ **22.** $7\overline{)4256}$ **23.** $4\overline{)24{,}024}$ **24.** $8\overline{)64{,}053}$

25. $9\overline{)8209}$ **26.** $8\overline{)9832}$ **27.** $6\overline{)4490}$ **28.** $4\overline{)33{,}907}$

Divide. Use after page 103.

29. $70 \div 33$ **30.** $90 \div 68$ **31.** $440 \div 87$ **32.** $400 \div 56$

33. $24\overline{)300}$ **34.** $83\overline{)913}$ **35.** $74\overline{)1702}$ **36.** $67\overline{)3179}$

37. $61\overline{)6832}$ **38.** $48\overline{)6336}$ **39.** $78\overline{)47{,}156}$ **40.** $19\overline{)14{,}825}$

41. $17\overline{)2079}$ **42.** $24\overline{)9816}$ **43.** $36\overline{)83{,}495}$ **44.** $59\overline{)11{,}992}$

Divide. Use after page 105.

45. $846 \div 731$ **46.** $4992 \div 624$ **47.** $2292 \div 326$ **48.** $5200 \div 850$

49. $912\overline{)11{,}967}$ **50.** $444\overline{)24{,}879}$ **51.** $399\overline{)16{,}758}$ **52.** $479\overline{)28{,}439}$

53. $248\overline{)17{,}460}$ **54.** $821\overline{)24{,}630}$ **55.** $596\overline{)53{,}765}$ **56.** $199\overline{)16{,}120}$

57. $123\overline{)4786}$ **58.** $487\overline{)60{,}529}$ **59.** $706\overline{)35{,}490}$ **60.** $263\overline{)19{,}248}$

61. $504\overline{)8964}$ **62.** $613\overline{)35{,}560}$ **63.** $345\overline{)29{,}824}$ **64.** $759\overline{)80{,}468}$

65. $699\overline{)49{,}629}$ **66.** $936\overline{)75{,}438}$ **67.** $288\overline{)47{,}735}$ **68.** $177\overline{)14{,}356}$

Divide. Use after page 109.

1. $26.88 \div 7$ **2.** $46.4 \div 8$ **3.** $14.73 \div 3$ **4.** $2.490 \div 6$

5. $82\overline{)787.2}$ **6.** $31\overline{)107.88}$ **7.** $43\overline{)73.1}$ **8.** $62\overline{)7.626}$

9. $6\overline{)0.18}$ **10.** $4\overline{)0.008}$ **11.** $8\overline{)4}$ **12.** $10\overline{)2}$

Divide. Use after page 111.

13. $46.4 \div 10$ **14.** $3.8 \div 10$ **15.** $6 \div 10$ **16.** $0.9 \div 10$

17. $7.8 \div 100$ **18.** $3.76 \div 100$ **19.** $4 \div 100$ **20.** $71 \div 100$

21. $9.7 \div 1000$ **22.** $8.64 \div 1000$ **23.** $87 \div 1000$ **24.** $3 \div 1000$

Estimate. Use after page 113.

25. $212 \div 4$ **26.** $6214 \div 7$ **27.** $31{,}814 \div 5$ **28.** $15{,}749 \div 8$

29. $73\overline{)849}$ **30.** $41\overline{)3349}$ **31.** $79\overline{)33{,}124}$ **32.** $67\overline{)36{,}841}$

33. $4.3\overline{)21.8}$ **34.** $8.17\overline{)42.722}$ **35.** $6.2\overline{)23.916}$ **36.** $3.914\overline{)25.72}$

Divide. Use after page 117.

37. $264.6 \div 6.3$ **38.** $34.08 \div 7.1$ **39.** $2.448 \div 7.2$ **40.** $3.43 \div 4.9$

41. $0.03\overline{)2.943}$ **42.** $7.3\overline{)0.438}$ **43.** $0.28\overline{)95.2}$ **44.** $0.004\overline{)0.7}$

45. $0.3\overline{)9}$ **46.** $0.7\overline{)14}$ **47.** $0.02\overline{)14}$ **48.** $0.08\overline{)64}$

49. $0.24\overline{)48}$ **50.** $0.12\overline{)6}$ **51.** $0.12\overline{)24}$ **52.** $0.013\overline{)52}$

Find each quotient to the nearest tenth. Use after page 119.

53. $7\overline{)9}$ **54.** $4\overline{)13}$ **55.** $8\overline{)11}$ **56.** $3\overline{)4}$

57. $4\overline{)3}$ **58.** $7\overline{)1}$ **59.** $9\overline{)7}$ **60.** $7\overline{)6}$

61. $3.2\overline{)7.3}$ **62.** $9.6\overline{)20.15}$ **63.** $0.3\overline{)13}$ **64.** $0.09\overline{)4}$

65. $2.8\overline{)16}$ **66.** $0.9\overline{)8.25}$ **67.** $7.1\overline{)25.8}$ **68.** $0.24\overline{)16}$

69. $0.8\overline{)29}$ **70.** $5.8\overline{)22}$ **71.** $0.52\overline{)9}$ **72.** $8.3\overline{)34.67}$

SKILLS REVIEW

Add. Refer to page 14.

1. 49 + 76 **2.** 94 + 36 **3.** 38 + 26 + 54 **4.** 428 + 62

5.
```
   37
   49
   63
 + 47
```
6.
```
  728
  479
  280
+ 146
```
7.
```
$  4.17
  13.29
+  6.54
```
8.
```
$  5.87
  28.61
   2.99
+  0.73
```

Add. Refer to page 16.

9.
```
  9428
+ 6759
```
10.
```
  7284
+ 6985
```
11.
```
  37,487
+ 28,643
```
12.
```
  328,412
+ 149,386
```
13.
```
  72,497
  65,388
+  9,246
```
14.
```
  329,421
  658,187
+  86,294
```
15.
```
  37,285
  49,784
  65,118
+ 97,091
```
16.
```
  427,286
  698,487
  407,759
+  65,287
```

Add. Refer to page 22.

17. 8.7 + 3.9 **18.** 7.32 + 9.48

19. 8.746 + 0.943 **20.** 8.69 + 6.738

21.
```
  7.63
  9.87
+ 1.4
```
22.
```
  9.72
  8.746
+ 8.9
```
23.
```
  9.87
  3.86
  1.98
+ 0.7
```
24.
```
  6.4
  4.295
  3.87
+ 9.9
```

25. 2.3 + 0.8 + 19.46 **26.** 12.4 + 6.78 + 0.39 + 0.068

Estimate the sums. Refer to page 28.

27.
```
  49
+ 87
```
28.
```
  3297
+ 8746
```
29.
```
  4293
+  415
```
30.
```
  276
   49
+  62
```
31.
```
  3.738
+ 1.9
```
32.
```
  4.25
+ 6.704
```
33.
```
  4.92
  6.104
+ 8.6
```
34.
```
  9.04
  0.921
+ 8.1
```

Subtract. Refer to page 40.

1. 97 − 28	**2.** 64 − 37	**3.** 80 − 16	**4.** 42 − 37
5. 824 − 673	**6.** 920 − 637	**7.** 824 − 197	**8.** 835 − 444
9. 725 − 68	**10.** 479 − 85	**11.** 741 − 8	**12.** 627 − 9

Subtract. Refer to page 42.

13. 9487 − 2688	**14.** 3241 − 1276	**15.** 3498 − 2719	**16.** 7348 − 2619
17. 28,741 − 19,198	**18.** 36,287 − 27,499	**19.** 326,145 − 198,087	**20.** 628,324 − 597,698

Subtract. Refer to page 44.

21. 500 − 146	**22.** 701 − 387	**23.** 5000 − 3298	**24.** 8002 − 498
25. 26,000 − 19,487	**26.** 5007 − 2681	**27.** 8040 − 1246	**28.** 30,091 − 28,135

Subtract. Refer to page 50.

29. 3.4 − 1.7	**30.** 8.75 − 4.69	**31.** 9.348 − 6.889	**32.** 8.764 − 3.871
33. 4.7 − 3.98	**34.** 8.15 − 6.9	**35.** 9.781 − 4.99	**36.** 9.214 − 6.5
37. 4.4 − 3.76	**38.** 8.74 − 3.987	**39.** 8.7 − 6.812	**40.** 9 − 3.47

Estimate the differences. Refer to page 54.

41. 781 − 93	**42.** 872 − 181	**43.** 3281 − 878	**44.** 3984 − 1299
45. 7.4 − 2.349	**46.** 3.76 − 1.879	**47.** 8.941 − 3.199	**48.** 4.187 − 1.99

Multiply. Refer to page 68.

1. $\begin{array}{r} 89 \\ \times\ 7 \\ \hline \end{array}$	**2.** $\begin{array}{r} 46 \\ \times\ 5 \\ \hline \end{array}$	**3.** $\begin{array}{r} 898 \\ \times\ 4 \\ \hline \end{array}$	**4.** $\begin{array}{r} \$\,7.49 \\ \times\ 7 \\ \hline \end{array}$
5. $\begin{array}{r} 4287 \\ \times\ 9 \\ \hline \end{array}$	**6.** $\begin{array}{r} 7307 \\ \times\ 8 \\ \hline \end{array}$	**7.** $\begin{array}{r} 8291 \\ \times\ 7 \\ \hline \end{array}$	**8.** $\begin{array}{r} 6489 \\ \times\ 5 \\ \hline \end{array}$

Multiply. Refer to page 70.

9. $\begin{array}{r} 56 \\ \times\ 47 \\ \hline \end{array}$	**10.** $\begin{array}{r} 88 \\ \times\ 24 \\ \hline \end{array}$	**11.** $\begin{array}{r} 64 \\ \times\ 28 \\ \hline \end{array}$	**12.** $\begin{array}{r} 48 \\ \times\ 75 \\ \hline \end{array}$
13. $\begin{array}{r} 927 \\ \times\ 42 \\ \hline \end{array}$	**14.** $\begin{array}{r} 729 \\ \times\ 67 \\ \hline \end{array}$	**15.** $\begin{array}{r} \$\,8.40 \\ \times\ 98 \\ \hline \end{array}$	**16.** $\begin{array}{r} \$\,8.26 \\ \times\ 54 \\ \hline \end{array}$
17. $\begin{array}{r} 3784 \\ \times\ 86 \\ \hline \end{array}$	**18.** $\begin{array}{r} 4281 \\ \times\ 81 \\ \hline \end{array}$	**19.** $\begin{array}{r} 7285 \\ \times\ 18 \\ \hline \end{array}$	**20.** $\begin{array}{r} 4784 \\ \times\ 36 \\ \hline \end{array}$

Multiply. Refer to page 72.

21. $\begin{array}{r} 798 \\ \times\ 642 \\ \hline \end{array}$	**22.** $\begin{array}{r} 786 \\ \times\ 424 \\ \hline \end{array}$	**23.** $\begin{array}{r} 8749 \\ \times\ 183 \\ \hline \end{array}$	**24.** $\begin{array}{r} \$\,92.84 \\ \times\ 366 \\ \hline \end{array}$
25. $\begin{array}{r} 376 \\ \times\ 404 \\ \hline \end{array}$	**26.** $\begin{array}{r} 928 \\ \times\ 480 \\ \hline \end{array}$	**27.** $\begin{array}{r} 1376 \\ \times\ 604 \\ \hline \end{array}$	**28.** $\begin{array}{r} \$\,42.98 \\ \times\ 730 \\ \hline \end{array}$

Multiply. Refer to page 76.

29. $\begin{array}{r} 70 \\ \times\ 80 \\ \hline \end{array}$	**30.** $\begin{array}{r} 60 \\ \times\ 40 \\ \hline \end{array}$	**31.** $\begin{array}{r} 30 \\ \times\ 70 \\ \hline \end{array}$	**32.** $\begin{array}{r} 900 \\ \times\ 60 \\ \hline \end{array}$
33. $\begin{array}{r} 300 \\ \times\ 700 \\ \hline \end{array}$	**34.** $\begin{array}{r} 9000 \\ \times\ 80 \\ \hline \end{array}$	**35.** $\begin{array}{r} 8000 \\ \times\ 300 \\ \hline \end{array}$	**36.** $\begin{array}{r} 4000 \\ \times\ 6000 \\ \hline \end{array}$

Multiply. Refer to page 82.

37. $\begin{array}{r} 93.6 \\ \times\ 7 \\ \hline \end{array}$	**38.** $\begin{array}{r} 89.43 \\ \times\ 6 \\ \hline \end{array}$	**39.** $\begin{array}{r} 4.287 \\ \times\ 35 \\ \hline \end{array}$	**40.** $\begin{array}{r} 6.34 \\ \times\ 12 \\ \hline \end{array}$
41. $\begin{array}{r} 8.7 \\ \times\ 3.1 \\ \hline \end{array}$	**42.** $\begin{array}{r} 9.82 \\ \times\ 3.3 \\ \hline \end{array}$	**43.** $\begin{array}{r} 87.1 \\ \times\ 0.32 \\ \hline \end{array}$	**44.** $\begin{array}{r} 9.62 \\ \times\ 8.78 \\ \hline \end{array}$
45. 0.3×0.3	**46.** 0.7×0.06	**47.** 0.1×0.08	**48.** 0.03×0.2

Multiply. Refer to page 84.

1. 10×7.6 **2.** 10×9.87 **3.** 10×8.746 **4.** 10×3.9

5. 100×3.7 **6.** 100×4.28 **7.** 1000×3.7 **8.** 1000×9.42

9. 0.1×7.4 **10.** 0.1×6 **11.** 0.1×9.28 **12.** 0.1×0.4

13. 0.01×8.7 **14.** 0.01×8.73 **15.** 0.001×8.76 **16.** 0.001×8.5

Estimate the products. Refer to page 76.

17. $\begin{array}{r} 87 \\ \times \ 9 \\ \hline \end{array}$ **18.** $\begin{array}{r} 79 \\ \times 48 \\ \hline \end{array}$ **19.** $\begin{array}{r} 386 \\ \times \ 41 \\ \hline \end{array}$ **20.** $\begin{array}{r} 925 \\ \times 648 \\ \hline \end{array}$

21. $\begin{array}{r} 827 \\ \times 650 \\ \hline \end{array}$ **22.** $\begin{array}{r} 3698 \\ \times \ 879 \\ \hline \end{array}$ **23.** $\begin{array}{r} 4219 \\ \times \ 327 \\ \hline \end{array}$ **24.** $\begin{array}{r} 7414 \\ \times 8785 \\ \hline \end{array}$

Estimate the products. Refer to page 78.

25. $\begin{array}{r} 6.82 \\ \times \ 7 \\ \hline \end{array}$ **26.** $\begin{array}{r} 8.75 \\ \times 6.24 \\ \hline \end{array}$ **27.** $\begin{array}{r} 1.947 \\ \times \ 8.7 \\ \hline \end{array}$ **28.** $\begin{array}{r} 3.682 \\ \times \ 4.79 \\ \hline \end{array}$

29. $\begin{array}{r} \$\,6.85 \\ \times \ 7 \\ \hline \end{array}$ **30.** $\begin{array}{r} \$\,9.15 \\ \times \ 4 \\ \hline \end{array}$ **31.** $\begin{array}{r} \$\,38.76 \\ \times \ 29 \\ \hline \end{array}$ **32.** $\begin{array}{r} \$\,42.16 \\ \times \ 58 \\ \hline \end{array}$

Divide. Refer to page 98.

33. $36 \div 7$ **34.** $71 \div 3$ **35.** $92 \div 4$ **36.** $186 \div 7$

37. $5\overline{)328}$ **38.** $4\overline{)321}$ **39.** $6\overline{)62}$ **40.** $7\overline{)352}$

Divide. Refer to page 100.

41. $876 \div 4$ **42.** $947 \div 5$ **43.** $3724 \div 8$ **44.** $4914 \div 6$

45. $7\overline{)8241}$ **46.** $6\overline{)9487}$ **47.** $8\overline{)35{,}487}$ **48.** $5\overline{)36{,}491}$

49. $8\overline{)4012}$ **50.** $6\overline{)3685}$ **51.** $7\overline{)42{,}017}$ **52.** $9\overline{)87{,}319}$

Divide. Refer to page 102.

53. $89 \div 22$ **54.** $98 \div 47$ **55.** $228 \div 65$ **56.** $897 \div 74$

57. $62\overline{)3241}$ **58.** $63\overline{)9487}$ **59.** $87\overline{)52{,}431}$ **60.** $18\overline{)11{,}294}$

Divide. Refer to page 104.

1. $879 \div 624$ **2.** $4149 \div 827$ **3.** $1374 \div 225$ **4.** $20,930 \div 645$

5. $497\overline{)23,860}$ **6.** $429\overline{)34,420}$ **7.** $576\overline{)52,000}$ **8.** $610\overline{)24,626}$

Divide. Refer to page 108.

9. $9.45 \div 5$ **10.** $9.6 \div 6$ **11.** $14.72 \div 8$ **12.** $1.323 \div 3$

13. $71\overline{)610.6}$ **14.** $84\overline{)261.24}$ **15.** $34\overline{)214.2}$ **16.** $51\overline{)16.983}$

17. $4\overline{)0.12}$ **18.** $7\overline{)0.007}$ **19.** $6\overline{)3}$ **20.** $5\overline{)2}$

Divide. Refer to page 110.

21. $78.2 \div 10$ **22.** $9.7 \div 10$ **23.** $3 \div 10$ **24.** $0.4 \div 10$

25. $4.7 \div 100$ **26.** $5.27 \div 100$ **27.** $8 \div 100$ **28.** $42 \div 100$

29. $6.3 \div 1000$ **30.** $7.42 \div 1000$ **31.** $94 \div 1000$ **32.** $8 \div 1000$

Estimate. Refer to page 112.

33. $8\overline{)650}$ **34.** $9\overline{)7147}$ **35.** $6\overline{)41,821}$ **36.** $9\overline{)19,247}$

37. $42\overline{)948}$ **38.** $73\overline{)6241}$ **39.** $87\overline{)33,421}$ **40.** $81\overline{)40,628}$

41. $3.7\overline{)16.5}$ **42.** $9.32\overline{)27.491}$ **43.** $3.3\overline{)23.84}$ **44.** $4.376\overline{)27.98}$

Divide. Refer to page 116.

45. $98.7 \div 4.7$ **46.** $24.57 \div 6.3$ **47.** $1.025 \div 4.1$ **48.** $7.76 \div 9.7$

49. $0.2\overline{)6}$ **50.** $0.4\overline{)16}$ **51.** $0.03\overline{)15}$ **52.** $0.09\overline{)81}$

53. $0.36\overline{)72}$ **54.** $0.25\overline{)6}$ **55.** $0.15\overline{)6}$ **56.** $0.025\overline{)4}$

Find each quotient to the nearest tenth. Refer to page 118.

57. $4\overline{)9}$ **58.** $7\overline{)12}$ **59.** $6\overline{)7}$ **60.** $8\overline{)9}$

61. $7\overline{)4}$ **62.** $9\overline{)1}$ **63.** $7\overline{)3}$ **64.** $6\overline{)5}$

65. $5.1\overline{)6.7}$ **66.** $8.7\overline{)19.42}$ **67.** $0.6\overline{)11}$ **68.** $0.03\overline{)4}$

READY FOR CHAPTER 5

1. What part is shaded?

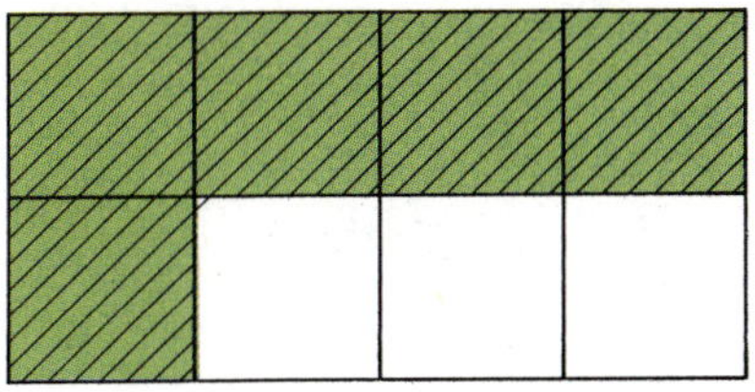

2. What part is shaded?

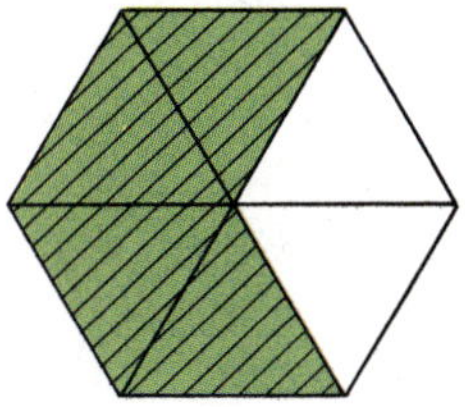

Divide, if possible.

3. $\frac{26}{1}$ **4.** $\frac{26}{0}$ **5.** $\frac{12}{12}$ **6.** $\frac{0}{16}$

Are the fractions equal?

7. $\frac{3}{4}, \frac{6}{8}$ = **8.** $\frac{5}{4}, \frac{9}{7}$

Compare.

9. $\frac{11}{20}$ and $\frac{2}{3}$ **10.** $\frac{2}{3}$ and $\frac{9}{14}$

Convert to a fraction.

11. 1.11 **12.** 0.005 **13.** $3\frac{1}{2}$ **14.** $9\frac{7}{8}$

Convert to a decimal.

15. $\frac{25}{16}$ **16.** $\frac{7}{8}$ **17.** $\frac{947}{100}$ **18.** $\frac{7813}{10,000}$

Convert to a decimal. Round to the nearest thousandth, when necessary.

19. $\frac{49}{12}$ **20.** $\frac{5}{6}$ **21.** $4\frac{3}{10}$ **22.** $677\frac{4}{5}$

Convert to a mixed number.

23. $\frac{9}{2}$ **24.** $\frac{74}{9}$ **25.** 2.8 **26.** 112.73

Application

27. Last week Miguel played his guitar for 4 hr, 45 min. Rita played her guitar for 2 hr, 55 min. How much longer did Miguel play than Rita?

28. When it is 8 PM in Los Angeles, what time is it in New York?

5
FRACTIONS

GETTING STARTED

FRACTION EQUO

WHAT'S NEEDED The 30 squares of paper with digits marked on them as used in GOTCHA, Chapter 1.

THE RULES

1. On a sheet of paper each player writes this:

   $\frac{[\ ]}{[\ ]} = \frac{[\ ]}{[\ ]} = \frac{[\ ]}{[\ ]}$

2. To play, a leader draws digits from the box, one at a time.
3. Each player decides whether or not to use the digit, but once the digit has been chosen it must be kept. The digit can be written in any space. Once chosen, the digit can be moved from space to space.
4. The first player to get three equal fractions wins.

Sample

Jacky: $\frac{[6]}{[9]} = \frac{[2]}{[3]} = \frac{[4]}{[6]}$ Yes, a winner!

Fred: $\frac{[3]}{[5]} = \frac{[6]}{[\ ]} = \frac{[2]}{[\ ]}$ No.

Jacky wins because she has three equal fractions.

USING FRACTIONS

We use fractions to describe some of the objects in a set.

EXAMPLE 1. What part of the set of tools are hammers?

2 out of 7 tools are hammers. We write this as $\frac{2}{7}$.

TRY THIS

1. What part of the set of tools above are wrenches?
2. What part of the set of tools are pliers?

We call $\frac{2}{7}$ a fraction. $\frac{\text{Number of objects considered}}{\text{Number of objects in the set}}$ $\frac{2}{7}$

The top number, 2, is the numerator.
The bottom number, 7, is the denominator.

We also use a fraction to describe part of an object which has been divided into equal parts.

EXAMPLE 2. What part is shaded?

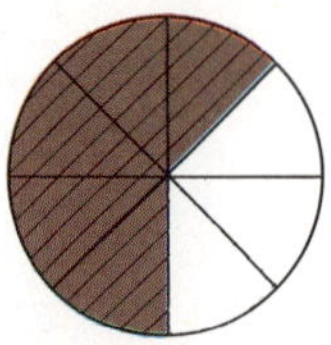

5 of the 8 equal parts are shaded.

We say $\frac{5}{8}$ is shaded.

TRY THIS

What part is shaded?

3. 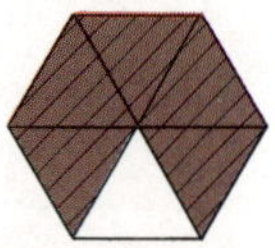

4. 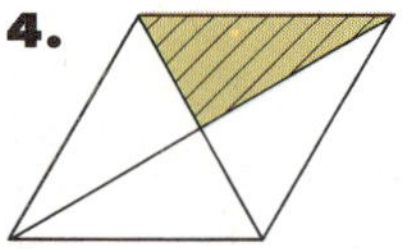

5. 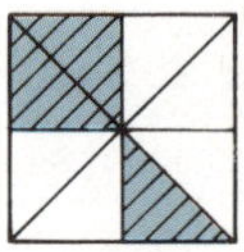

PRACTICE

1. What part are women?

2. What part are men?

What part of each set is shaded?

3.

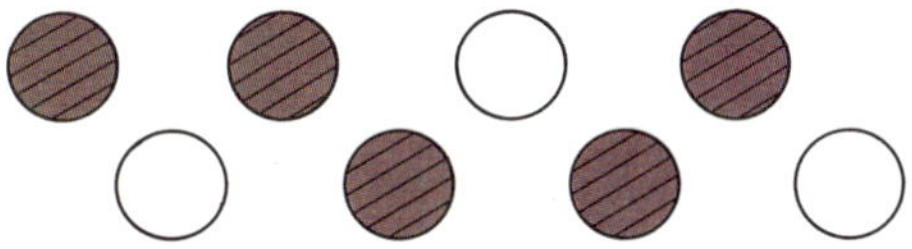

4.

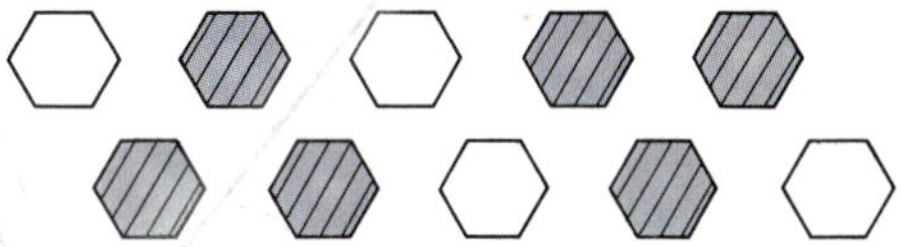

5.

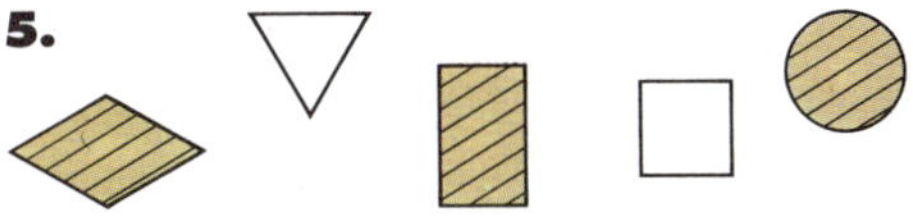

6.

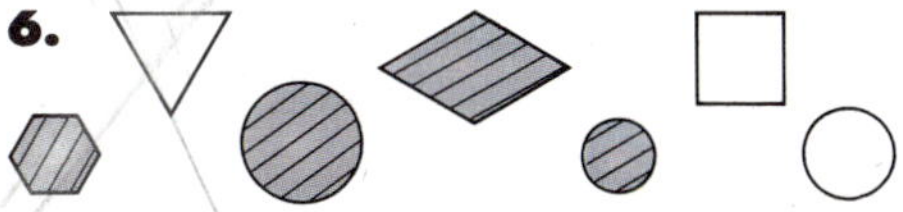

What part is shaded?

7.

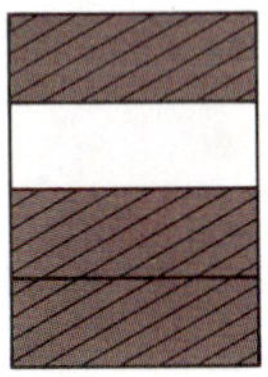

8.

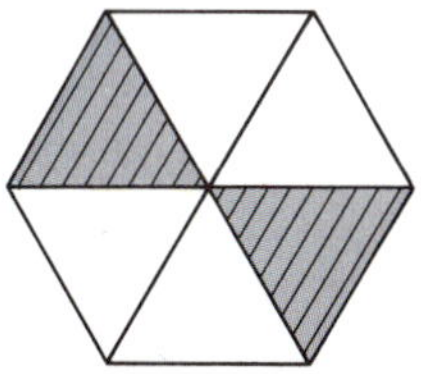

9.

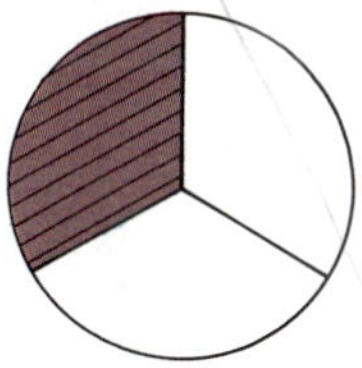

10.

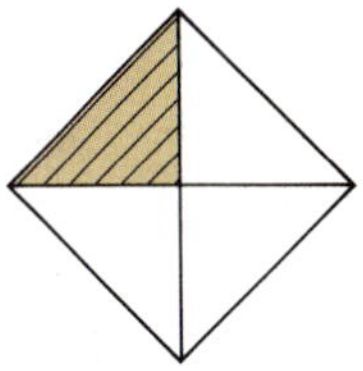

Identify the numerator and denominator of each fraction.

11. $\frac{11}{20}$ **12.** $\frac{14}{25}$ **13.** $\frac{19}{20}$ **14.** $\frac{18}{25}$ **15.** $\frac{51}{100}$ **16.** $\frac{71}{100}$

PROBLEM CORNER

The surface of the Earth is 3 parts water and 1 part land.

1. What fraction of the Earth is water?

2. What fraction on the Earth is land?

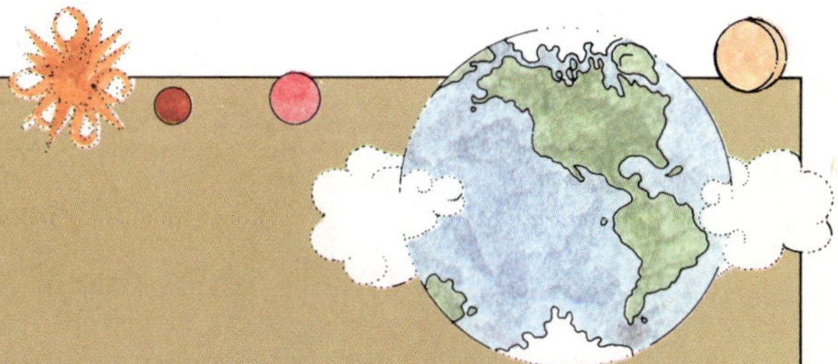

FRACTIONS MEAN DIVISION

We know that $\frac{8}{8} = 1$ because $8 \div 8 = 1$.

Any nonzero number divided by itself is 1.

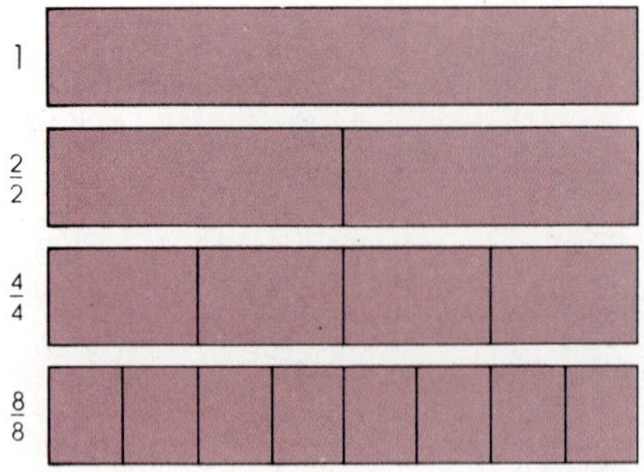

EXAMPLES. Divide.

1. $\frac{6}{6} = 1$ **2.** $\frac{9}{9} = 1$ **3.** $\frac{27}{27} = 1$

We know that $\frac{0}{4} = 0$ because $0 \div 4 = 0$.

0 divided by any nonzero number is 0.

EXAMPLES. Divide.

4. $\frac{0}{3} = 0$ **5.** $\frac{0}{7} = 0$ **6.** $\frac{0}{36} = 0$

TRY THIS

Divide.

1. $\frac{5}{5}$ **2.** $\frac{42}{42}$ **3.** $\frac{0}{8}$ **4.** $\frac{0}{10}$ **5.** $\frac{16}{16}$ **6.** $\frac{0}{59}$

We know that $\frac{8}{1} = 8$ because $8 \div 1 = 8$.

Any number divided by 1 is that number.

EXAMPLES. Divide.

7. $\frac{2}{1} = 2$ **8.** $\frac{6}{1} = 6$ **9.** $\frac{24}{1} = 24$

Division by 0 is not possible.

EXAMPLES. Divide, if possible.

10. $\frac{14}{0}$ Not possible **11.** $\frac{0}{6} = 0$ **12.** $\frac{0}{0}$ Not possible

TRY THIS

Divide, if possible.

7. $\frac{9}{1}$ **8.** $\frac{36}{1}$ **9.** $\frac{14}{1}$ **10.** $\frac{99}{1}$

11. $\frac{0}{12}$ **12.** $\frac{19}{0}$ **13.** $\frac{23}{0}$ **14.** $\frac{0}{4}$

15. $\frac{0}{47}$ **16.** $\frac{24}{0}$ **17.** $\frac{0}{67}$ **18.** $\frac{89}{0}$

PRACTICE

Divide, if possible.

1. $\frac{8}{8}$ **2.** $\frac{10}{10}$ **3.** $\frac{9}{9}$ **4.** $\frac{12}{12}$ **5.** $\frac{7}{7}$ **6.** $\frac{15}{15}$

7. $\frac{6}{1}$ **8.** $\frac{9}{1}$ **9.** $\frac{8}{1}$ **10.** $\frac{7}{1}$ **11.** $\frac{2}{1}$ **12.** $\frac{5}{1}$

13. $\frac{0}{6}$ **14.** $\frac{0}{8}$ **15.** $\frac{0}{5}$ **16.** $\frac{0}{9}$ **17.** $\frac{0}{1}$ **18.** $\frac{0}{3}$

19. $\frac{24}{24}$ **20.** $\frac{24}{1}$ **21.** $\frac{0}{24}$ **22.** $\frac{14}{14}$ **23.** $\frac{14}{1}$ **24.** $\frac{0}{14}$

25. $\frac{0}{22}$ **26.** $\frac{22}{22}$ **27.** $\frac{22}{1}$ **28.** $\frac{0}{15}$ **29.** $\frac{15}{15}$ **30.** $\frac{15}{1}$

31. $\frac{16}{1}$ **32.** $\frac{0}{16}$ **33.** $\frac{18}{18}$ **34.** $\frac{21}{1}$ **35.** $\frac{42}{42}$ **36.** $\frac{42}{1}$

37. $\frac{0}{0}$ **38.** $\frac{19}{0}$ **39.** $\frac{18}{0}$ **40.** $\frac{42}{0}$ **41.** $\frac{1}{0}$ **42.** $\frac{99}{0}$

43. $\frac{13}{13}$ **44.** $\frac{23}{1}$ **45.** $\frac{25}{25}$ **46.** $\frac{0}{42}$ **47.** $\frac{24}{24}$ **48.** $\frac{24}{1}$

49. $\frac{124}{1}$ **50.** $\frac{0}{256}$ **51.** $\frac{0}{357}$ **52.** $\frac{562}{562}$ **53.** $\frac{688}{1}$ **54.** $\frac{975}{975}$

55. $\frac{0}{120}$ **56.** $\frac{164}{164}$ **57.** $\frac{1253}{1253}$ **58.** $\frac{789}{1}$ **59.** $\frac{0}{342}$ **60.** $\frac{2651}{2651}$

61. $\frac{45}{0}$ **62.** $\frac{0}{0}$ **63.** $\frac{32}{32}$ **64.** $\frac{56}{0}$ **65.** $\frac{39}{0}$ **66.** $\frac{762}{762}$

67. $\frac{0}{0}$ **68.** $\frac{13}{1}$ **69.** $\frac{28}{0}$ **70.** $\frac{45}{0}$ **71.** $\frac{49}{1}$ **72.** $\frac{2984}{1}$

More Practice, page 162

EQUALITY

Compare $\frac{2}{4}$ and $\frac{3}{6}$.

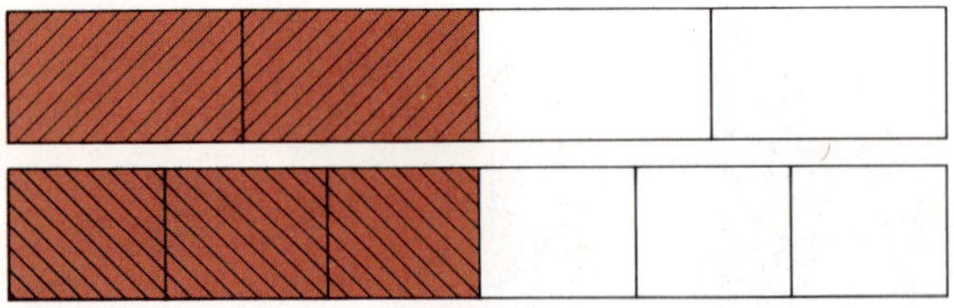

We see that $\frac{2}{4} = \frac{3}{6}$.

Tell whether the fractions are equal. $\frac{2}{4}, \frac{3}{6}$

$\frac{2}{4} \quad \frac{3}{6}$	Write the fractions horizontally.
$2 \times 6 \qquad 4 \times 3$	Find the cross products.
$12 = 12$	Compare.
$\frac{2}{4} = \frac{3}{6}$	Write the same symbol between the fractions.

We call 2×6 and 4×3 cross products.

EXAMPLES. Tell whether the fractions are equal.

1. $\frac{6}{10} \quad \frac{3}{5}$

$6 \times 5 \qquad 10 \times 3$

$30 = 30$

$\frac{6}{10} = \frac{3}{5}$

2. $\frac{6}{7} \quad \frac{7}{8}$

$6 \times 8 \qquad 7 \times 7$

$48 \neq 49$

$\frac{6}{7} \neq \frac{7}{8}$

$\neq$ means "is not equal to."

TRY THIS

Tell whether the fractions are equal.

1. $\frac{2}{6}, \frac{3}{9}$ **2.** $\frac{3}{4}, \frac{5}{7}$ **3.** $\frac{8}{10}, \frac{4}{5}$ **4.** $\frac{12}{4}, \frac{6}{2}$

PRACTICE

Tell whether the fractions are equal.

1. $\frac{3}{4}, \frac{9}{12}$	**2.** $\frac{2}{4}, \frac{4}{6}$	**3.** $\frac{4}{5}, \frac{2}{3}$	**4.** $\frac{4}{8}, \frac{3}{6}$	**5.** $\frac{5}{10}, \frac{3}{6}$
6. $\frac{1}{2}, \frac{2}{4}$	**7.** $\frac{1}{3}, \frac{2}{6}$	**8.** $\frac{1}{5}, \frac{2}{9}$	**9.** $\frac{1}{4}, \frac{3}{12}$	**10.** $\frac{3}{4}, \frac{9}{12}$
11. $\frac{0}{3}, \frac{0}{4}$	**12.** $\frac{0}{1}, \frac{0}{3}$	**13.** $\frac{0}{5}, \frac{0}{7}$	**14.** $\frac{0}{8}, \frac{0}{7}$	**15.** $\frac{0}{9}, \frac{0}{2}$
16. $\frac{4}{5}, \frac{2}{3}$	**17.** $\frac{6}{8}, \frac{9}{12}$	**18.** $\frac{3}{8}, \frac{6}{16}$	**19.** $\frac{2}{6}, \frac{6}{18}$	**20.** $\frac{5}{7}, \frac{4}{8}$
21. $\frac{1}{2}, \frac{1}{3}$	**22.** $\frac{1}{3}, \frac{1}{4}$	**23.** $\frac{1}{4}, \frac{1}{2}$	**24.** $\frac{1}{5}, \frac{1}{3}$	**25.** $\frac{1}{9}, \frac{1}{3}$
26. $\frac{2}{2}, \frac{5}{5}$	**27.** $\frac{3}{3}, \frac{4}{4}$	**28.** $\frac{5}{5}, \frac{3}{3}$	**29.** $\frac{4}{4}, \frac{2}{2}$	**30.** $\frac{6}{6}, \frac{7}{7}$
31. $\frac{2}{5}, \frac{3}{7}$	**32.** $\frac{3}{5}, \frac{5}{10}$	**33.** $\frac{4}{7}, \frac{5}{11}$	**34.** $\frac{12}{16}, \frac{3}{4}$	**35.** $\frac{2}{5}, \frac{4}{10}$
36. $\frac{3}{2}, \frac{9}{6}$	**37.** $\frac{16}{14}, \frac{8}{7}$	**38.** $\frac{9}{3}, \frac{3}{1}$	**39.** $\frac{12}{9}, \frac{8}{6}$	**40.** $\frac{18}{9}, \frac{2}{1}$
41. $\frac{3}{9}, \frac{1}{2}$	**42.** $\frac{5}{2}, \frac{17}{7}$	**43.** $\frac{4}{6}, \frac{17}{7}$	**44.** $\frac{3}{2}, \frac{5}{2}$	**45.** $\frac{6}{5}, \frac{5}{6}$
46. $\frac{3}{10}, \frac{30}{100}$	**47.** $\frac{70}{100}, \frac{700}{1000}$	**48.** $\frac{5}{10}, \frac{520}{1000}$	**49.** $\frac{49}{100}, \frac{50}{1000}$	**50.** $\frac{80}{800}, \frac{800}{8000}$

PROBLEM CORNER

1. Is $\frac{3}{4}$ gallon the same as $\frac{4}{5}$ gallon? Why?

2. A student got 63 out of 78 questions correct on one test. On another test the student got 75 out of 100 questions correct. Did the student get the same fraction correct on each test? Why?

More Practice, page 162

COMPARING FRACTIONS

Notice that $\frac{3}{4}$ is greater than $\frac{2}{5}$. $\frac{3}{4} > \frac{2}{5}$

Also see that $\frac{2}{5}$ is less than $\frac{3}{4}$. $\frac{2}{5} < \frac{3}{4}$

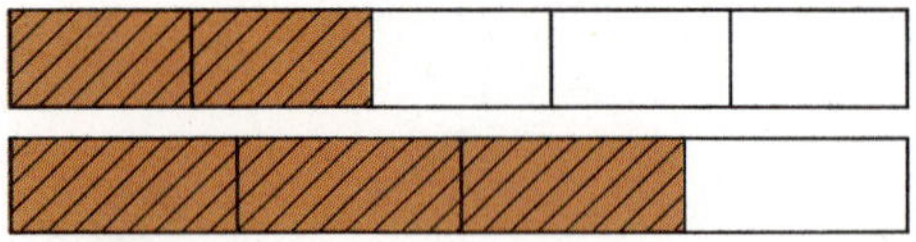

Compare. $\frac{2}{5}$ and $\frac{3}{4}$

$\frac{2}{5} \times \frac{3}{4}$	Write the fractions horizontally.
$2 \times 4 \quad 5 \times 3$	Find the cross products.
$8 < 15$	Compare.
$\frac{2}{5} < \frac{3}{4}$	Write the same symbol between the fractions.

EXAMPLE 1. Compare. $\frac{3}{5}$ and $\frac{1}{2}$

$\frac{3}{5} \times \frac{1}{2}$

$3 \times 2 \quad 5 \times 1$ *Find the cross products.*

$6 > 5$ *Compare.*

$\frac{3}{5} > \frac{1}{2}$ *Write the same symbol between the fractions.*

When the denominators are the same, we simply compare the numerators.

EXAMPLE 2. Compare. $\frac{4}{7}$ and $\frac{3}{7}$

$\frac{4}{7} \quad \frac{3}{7}$

$4 > 3$ *Compare the numerators.*

$\frac{4}{7} > \frac{3}{7}$ *Write the same symbol between the fractions.*

TRY THIS

Compare.

1. $\frac{2}{3}$ and $\frac{5}{8}$
2. $\frac{3}{5}$ and $\frac{7}{10}$
3. $\frac{2}{7}$ and $\frac{3}{8}$
4. $\frac{8}{12}$ and $\frac{3}{4}$
5. $\frac{1}{2}$ and $\frac{5}{12}$
6. $\frac{3}{9}$ and $\frac{4}{8}$
7. $\frac{4}{9}$ and $\frac{5}{9}$
8. $\frac{3}{14}$ and $\frac{1}{7}$
9. $\frac{2}{3}$ and $\frac{3}{6}$

PRACTICE

Compare.

1. $\frac{5}{8}$ and $\frac{6}{8}$
2. $\frac{7}{6}$ and $\frac{5}{6}$
3. $\frac{9}{10}$ and $\frac{7}{10}$
4. $\frac{4}{9}$ and $\frac{8}{9}$
5. $\frac{1}{3}$ and $\frac{1}{4}$
6. $\frac{1}{5}$ and $\frac{1}{6}$
7. $\frac{1}{6}$ and $\frac{1}{8}$
8. $\frac{1}{5}$ and $\frac{1}{4}$
9. $\frac{2}{3}$ and $\frac{5}{7}$
10. $\frac{5}{8}$ and $\frac{3}{4}$
11. $\frac{3}{5}$ and $\frac{4}{7}$
12. $\frac{3}{8}$ and $\frac{4}{9}$
13. $\frac{3}{4}$ and $\frac{7}{9}$
14. $\frac{5}{6}$ and $\frac{4}{5}$
15. $\frac{3}{4}$ and $\frac{5}{7}$
16. $\frac{3}{2}$ and $\frac{7}{5}$
17. $\frac{11}{16}$ and $\frac{2}{3}$
18. $\frac{11}{12}$ and $\frac{12}{13}$
19. $\frac{5}{6}$ and $\frac{13}{16}$
20. $\frac{11}{15}$ and $\frac{3}{4}$
21. $\frac{8}{10}$ and $\frac{79}{100}$
22. $\frac{7}{10}$ and $\frac{73}{100}$
23. $\frac{5}{10}$ and $\frac{49}{100}$
24. $\frac{3}{10}$ and $\frac{31}{100}$
25. $\frac{6}{100}$ and $\frac{8}{1000}$
26. $\frac{9}{1000}$ and $\frac{1}{100}$
27. $\frac{2109}{1000}$ and $\frac{21}{10}$
28. $\frac{21,006}{1000}$ and $\frac{2105}{100}$
29. $\frac{5}{9}$ and $\frac{7}{20}$
30. $\frac{8}{25}$ and $\frac{3}{9}$
31. $\frac{3}{25}$ and $\frac{2}{17}$
32. $\frac{4}{25}$ and $\frac{3}{19}$
33. $\frac{20}{21}$ and $\frac{18}{19}$
34. $\frac{23}{25}$ and $\frac{25}{26}$
35. $\frac{18}{19}$ and $\frac{13}{15}$
36. $\frac{17}{18}$ and $\frac{19}{20}$
37. $\frac{10}{9}$ and $\frac{10}{8}$
38. $\frac{20}{7}$ and $\frac{20}{8}$
39. $\frac{40}{3}$ and $\frac{40}{7}$
40. $\frac{30}{4}$ and $\frac{30}{5}$

PROBLEM CORNER

Andy got 16 out of 19 questions correct on a test. On the next test he got 18 out of 21 correct. On which test did Andy score higher?

More Practice, page 162

WRITING DECIMALS AS FRACTIONS

To convert 9.87 to a fraction, we can multiply by 100, then divide by 100.

$9.87 \times 100 = 987 \qquad 987 \div 100 = \frac{987}{100}$

Convert to a fraction. 4.867

Count the number of decimal places. → Move the decimal point that many places to the right. → Write the whole number over 10, 100, 1000, etc. depending on the number of 0's needed.

4.867 — 3 places   4.867 — 3 moves $\frac{4867}{1000}$ — three 0's

EXAMPLE 1. Convert to a fraction. 0.675

0.675 — *3 places*   0.675 — *3 moves* $0.675 = \frac{675}{1000}$ — *three 0's*

EXAMPLE 2. Convert to a fraction. 1.6

1.6 — *1 place*   1.6 — *1 move* $1.6 = \frac{16}{10}$ — *one 0*

EXAMPLE 3. Convert to a fraction. 21.43

21.43 — *2 places*   21.43 — *2 moves* $\frac{2143}{100}$ — *two 0's*

TRY THIS

Convert to a fraction.

1. 0.679 **2.** 3.7 **3.** 17.95 **4.** 0.3
5. 4.7732 **6.** 0.32 **7.** 1.05 **8.** 0.06

PRACTICE

Convert to a fraction.

1. 0.1	**2.** 0.3	**3.** 0.4	**4.** 0.8	**5.** 0.6
6. 1.3	**7.** 2.7	**8.** 1.4	**9.** 3.9	**10.** 4.8
11. 6.8	**12.** 7.2	**13.** 3.1	**14.** 9.7	**15.** 5.3
16. 0.17	**17.** 0.89	**18.** 0.27	**19.** 0.32	**20.** 0.64
21. 0.03	**22.** 0.01	**23.** 0.07	**24.** 0.09	**25.** 0.08
26. 17.98	**27.** 14.95	**28.** 21.18	**29.** 43.57	**30.** 52.91
31. 3.167	**32.** 1.732	**33.** 46.032	**34.** 45.916	**35.** 23.011
36. 1.007	**37.** 2.0118	**38.** 34.1212	**39.** 78.9011	**40.** 10.3456
41. 299.49	**42.** 191.3	**43.** 234.019	**44.** 0.0005	**45.** 0.0076
46. 10.4	**47.** 10.5	**48.** 2.05	**49.** 1.15	**50.** 14.3
51. 4.95	**52.** 7.96	**53.** 1.004	**54.** 2.015	**55.** 8.061

PROBLEM CORNER

1. Beth bought 8.5 gallons of gasoline. Convert the number of gallons to a fraction.

2. Kevin wants to paint a wall in his bedroom. The wall is 2.7 meters high and 3.9 meters long. Convert each measurement to a fraction.

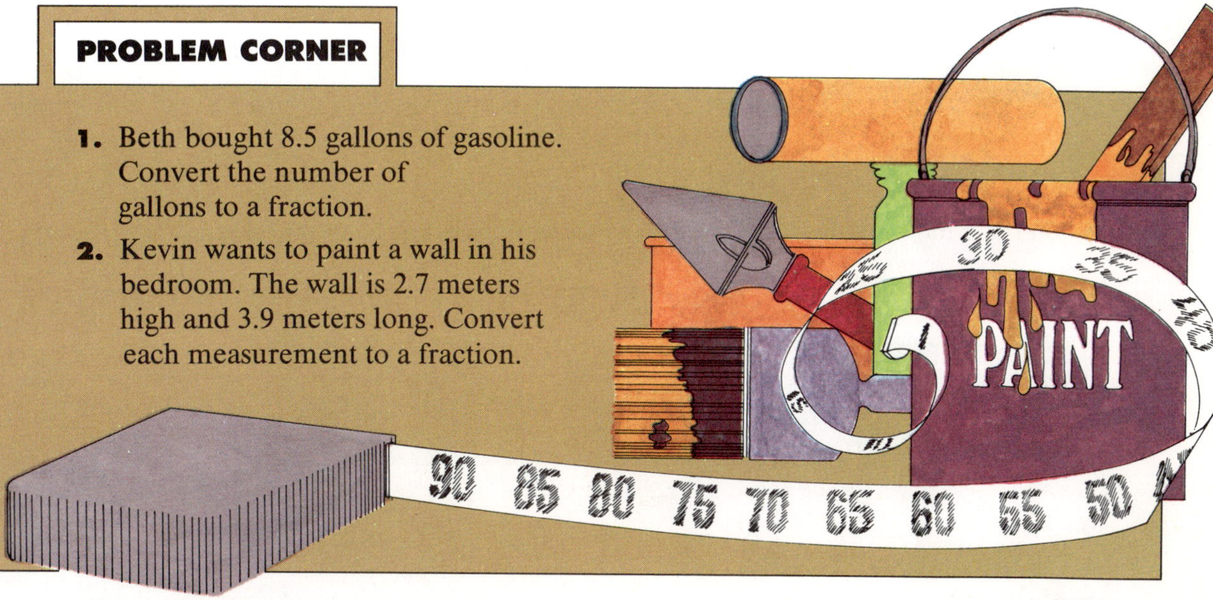

WRITING FRACTIONS AS DECIMALS

Convert to a decimal. $\frac{7}{20}$

Divide the numerator by the denominator. → Write the quotient.

$$20\overline{)7.00} = 0.35$$

$$\frac{7}{20} = 0.35$$

EXAMPLE 1. Convert to a decimal. $\frac{5}{8}$

$$8\overline{)5.000} = 0.625 \qquad \frac{5}{8} = 0.625$$

When a repeating decimal occurs, we divide to a certain decimal place and round. The stopping place is a matter of choice, but we will use ten-thousandths.

EXAMPLE 2. Convert to a decimal. $\frac{9}{11}$ Round to the nearest thousandth.

$$11\overline{)9.0000} = 0.8181 \approx 0.818$$

Divide to the ten-thousandths' place. Round to the nearest thousandth.

TRY THIS

Convert to a decimal. Round to the nearest thousandth.

1. $\frac{3}{8}$ **2.** $\frac{1}{16}$ **3.** $\frac{7}{12}$ **4.** $\frac{59}{12}$ **5.** $\frac{7}{11}$

When the denominator is 10, 100, 1000, etc., we use a shorter method.

Convert to a decimal. $\frac{693}{100}$

Count the number of 0's in the denominator. → Move the decimal point that many places to the left. → Write the decimal.

$\frac{693}{100}$ two 0's 6 9 3. 2 places $\frac{693}{100} = 6.93$

EXAMPLE 3. Convert to a decimal. $\frac{17}{1000}$

$$\frac{17}{1000} \longrightarrow 0\,0\,1\,7. \longrightarrow 0.0\,1\,7$$

Move the decimal point 3 places to the left. Write necessary 0's.

TRY THIS

Convert to a decimal.

6. $\frac{57}{100}$ **7.** $\frac{17}{10}$ **8.** $\frac{987}{10{,}000}$ **9.** $\frac{3}{10}$ 0.3 **10.** $\frac{406}{1000}$

PRACTICE

Convert to a decimal.

1. $\frac{1}{5}$	**2.** $\frac{7}{4}$	**3.** $\frac{11}{4}$	**4.** $\frac{9}{5}$	**5.** $\frac{1}{4}$	**6.** $\frac{7}{5}$
7. $\frac{11}{8}$	**8.** $\frac{13}{8}$	**9.** $\frac{1}{8}$	**10.** $\frac{53}{8}$	**11.** $\frac{7}{8}$	**12.** $\frac{43}{8}$
13. $\frac{17}{16}$	**14.** $\frac{15}{16}$	**15.** $\frac{3}{16}$	**16.** $\frac{35}{16}$	**17.** $\frac{13}{16}$	**18.** $\frac{53}{16}$
19. $\frac{21}{125}$	**20.** $\frac{17}{40}$	**21.** $\frac{123}{200}$	**22.** $\frac{5}{64}$	**23.** $\frac{553}{400}$	**24.** $\frac{7}{32}$

Convert to a decimal. Round to the nearest thousandth.

25. $\frac{1}{3}$	**26.** $\frac{2}{3}$	**27.** $\frac{5}{6}$	**28.** $\frac{1}{6}$	**29.** $\frac{2}{6}$	**30.** $\frac{4}{6}$
31. $\frac{2}{9}$	**32.** $\frac{4}{9}$	**33.** $\frac{4}{11}$	**34.** $\frac{3}{11}$	**35.** $\frac{5}{9}$	**36.** $\frac{7}{11}$
37. $\frac{25}{12}$	**38.** $\frac{17}{12}$	**39.** $\frac{19}{9}$	**40.** $\frac{8}{3}$	**41.** $\frac{21}{9}$	**42.** $\frac{5}{3}$

Convert to a decimal.

43. $\frac{8}{10}$	**44.** $\frac{3}{10}$	**45.** $\frac{29}{10}$	**46.** $\frac{51}{10}$	**47.** $\frac{44}{10}$	**48.** $\frac{2}{10}$
49. $\frac{99}{100}$	**50.** $\frac{9}{100}$	**51.** $\frac{246}{100}$	**52.** $\frac{1567}{100}$	**53.** $\frac{2321}{100}$	**54.** $\frac{467}{100}$
55. $\frac{2}{1000}$	**56.** $\frac{13}{1000}$	**57.** $\frac{6789}{1000}$	**58.** $\frac{14{,}012}{1000}$	**59.** $\frac{16{,}305}{1000}$	**60.** $\frac{5789}{1000}$

More Practice, pages 162 and 163

APPLICATION

MEASURING TIME

Time is given in many different measures. Sometimes we need to change from one measure to another.

1 minute (min) = 60 seconds (sec)	1 month (mo) = 30 days
1 hour (hr) = 60 minutes	1 year (yr) = 365 days
1 day = 24 hours	1 year = 52 weeks
1 week (wk) = 7 days	1 year = 12 months

To change from a larger to a smaller unit, we multiply.

EXAMPLE 1. How many minutes are there in 8 hours?

1 hour = 60 min
$8 \times 60 = 480$ *Think: larger to smaller—multiply.*
8 hr = 480 min

How many are there in each of the following?

1. hours in 4 days
2. weeks in 3 years
3. months in 4 years
4. Mel played soccer for 4 hours. How many minutes was that?
5. Jeremy bought a stereo and will make monthly payments for 2 years. How many payments will he make?

To change from a smaller unit to a larger, we divide.

EXAMPLE 2. How many hours are there in 180 minutes?

1 hour = 60 min
$180 \div 60 = 3$ *Think: smaller to larger—divide.*
180 min = 3 hr

How many are there in each of the following?

6. hours in 300 min
7. minutes in 420 sec
8. years in 1248 weeks
9. Vicki rode her bicycle to work for 28 days in a row. How many weeks did she ride her bike?
10. To train for the track team Audrey plans to run 5 miles each day for 90 days. How many months does she plan to run?

Sometimes we express time in two measures.

WHOLE WHEAT BREAD
2 pkg active dry yeast
$\frac{1}{4}$ c butter
$2\frac{1}{4}$ c warm water
8 c whole wheat flour
Bake 80 minutes

EXAMPLE 3. How many hours and minutes should the bread bake?

1 hr = 60 min
80 min ÷ 60 min = 1 hr, 20 min

11. Steven's younger sister is 30 months old. How old is his sister in years and months?
12. Stacey can run the mile in 310 seconds. What is her speed in minutes and seconds?

Often we need to add time.

EXAMPLE 4. Add. 3 hr, 25 min and 2 hr, 55 min

$$\begin{array}{rl} 3\text{ hr, }25\text{ min} & \\ +\ 2\text{ hr, }55\text{ min} & \\ \hline 5\text{ hr, }80\text{ min} & = 5\text{ hr} + 80\text{ min} \\ & = 5\text{ hr} + 1\text{ hr} + 20\text{ min} \\ & = 6\text{ hr, }20\text{ min} \end{array}$$

13. Jack practiced the piano 5 hr, 29 min one week. The next week he practiced 6 hr, 45 min. How long did he practice in all?
14. Sandy drove for 4 hr, 45 min. Martha drove for 3 hr, 35 min. How long did they drive altogether?

We can also subtract time.

EXAMPLE 5. Subtract. 3 min, 25 sec and 1 min, 40 sec

$$\begin{array}{rcr} 3\text{ min, }25\text{ sec} = 2\text{ min} + 60\text{ sec} + 25\text{ sec} = & & 2\text{ min, }85\text{ sec} \\ -\ 1\text{ min, }40\text{ sec} & & -\ 1\text{ min, }40\text{ sec} \\ \hline & & 1\text{ min, }45\text{ sec} \end{array}$$

15. Ralph has been playing the drums for 6 yr, 3 mo. Tony has been playing for 1 yr, 6 mo. How much longer has Ralph been playing?
16. In the 1920 Olympics Italy won the 4000 meter team cycling race with a time of 5 min, 20 sec. In 1960 they won it with a time of 4 min, 30 sec. By how much less time did they win in 1960?

WRITING MIXED NUMBERS AS FRACTIONS

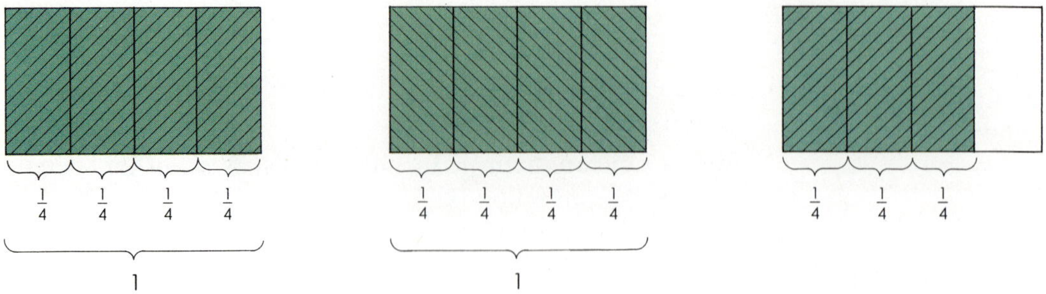

We can think of $2\frac{3}{4}$ as 2 whole units plus $\frac{3}{4}$ of a unit.

$2\frac{3}{4}$ is an example of a mixed number. It means $2 + \frac{3}{4}$.

EXAMPLES. Write as a mixed number.

1. $7 + \frac{2}{5} = 7\frac{2}{5}$ **2.** $\frac{3}{10} + 4 = 4\frac{3}{10}$

TRY THIS Write as a mixed number.

1. $8 + \frac{3}{4}$ **2.** $\frac{2}{3} + 12$ **3.** $\frac{7}{9} + 6$

Convert to a fraction. $4\frac{3}{10}$

Multiply the denominator by the whole number. → Add the numerator. → Write the sum over the denominator.

$4\frac{3}{10}$ $4\frac{3}{10}$ $\frac{43}{10}$

$10 \times 4 = 40$ $40 + 3 = 43$

To see why this method works, remember that $4 = \frac{40}{10}$.
Then $4\frac{3}{10} = 4 + \frac{3}{10} = \frac{40}{10} + \frac{3}{10} = \frac{43}{10}$.

EXAMPLES. Convert to a fraction.

Add. / *Multiply.*

3. $6\frac{2}{3} = \frac{20}{3}$

4. $10\frac{7}{8} = \frac{87}{8}$

TRY THIS

Convert to a fraction.

4. $4\frac{2}{5}$ **5.** $10\frac{1}{6}$ **6.** $9\frac{3}{10}$ **7.** $12\frac{3}{4}$

PRACTICE

Write as a mixed number.

1. $2 + \frac{1}{3}$ **2.** $3 + \frac{1}{4}$ **3.** $6 + \frac{3}{8}$ **4.** $5 + \frac{2}{3}$ **5.** $4 + \frac{2}{7}$

6. $\frac{2}{3} + 9$ **7.** $\frac{8}{9} + 7$ **8.** $\frac{4}{5} + 10$ **9.** $\frac{7}{8} + 12$ **10.** $\frac{1}{3} + 6$

Convert to a fraction.

11. $5\frac{2}{3}$ **12.** $4\frac{3}{5}$ **13.** $3\frac{5}{6}$ **14.** $7\frac{3}{4}$ **15.** $2\frac{4}{5}$

16. $8\frac{1}{2}$ **17.** $5\frac{1}{3}$ **18.** $6\frac{1}{4}$ **19.** $9\frac{1}{5}$ **20.** $4\frac{1}{6}$

21. $20\frac{2}{5}$ **22.** $30\frac{1}{4}$ **23.** $40\frac{1}{2}$ **24.** $50\frac{3}{4}$ **25.** $60\frac{3}{5}$

26. $7\frac{3}{10}$ **27.** $6\frac{9}{10}$ **28.** $5\frac{7}{10}$ **29.** $8\frac{3}{10}$ **30.** $4\frac{9}{10}$

31. $1\frac{5}{8}$ **32.** $1\frac{3}{5}$ **33.** $1\frac{5}{6}$ **34.** $1\frac{2}{3}$ **35.** $1\frac{3}{8}$

36. $12\frac{3}{4}$ **37.** $15\frac{2}{3}$ **38.** $16\frac{7}{8}$ **39.** $18\frac{8}{9}$ **40.** $13\frac{4}{5}$

41. $4\frac{8}{10}$ **42.** $5\frac{4}{10}$ **43.** $6\frac{2}{10}$ **44.** $8\frac{9}{10}$ **45.** $7\frac{3}{10}$

46. $2\frac{3}{100}$ **47.** $5\frac{1}{100}$ **48.** $14\frac{57}{100}$ **49.** $16\frac{95}{100}$ **50.** $6\frac{23}{100}$

More Practice, page 163

WRITING FRACTIONS AS MIXED NUMBERS

These circles are divided into thirds. Each part is $\frac{1}{3}$.

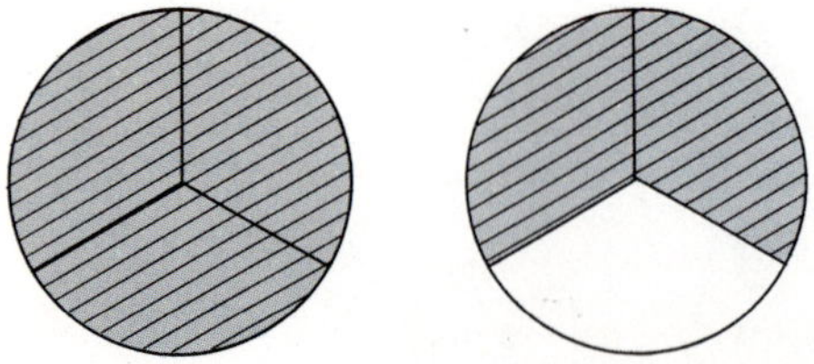

$$\frac{5}{3} = \underbrace{\frac{1}{3} + \frac{1}{3} + \frac{1}{3}}_{1} + \underbrace{\frac{1}{3} + \frac{1}{3}}_{\frac{2}{3}} = 1\frac{2}{3}$$

Convert to a mixed number. $\frac{74}{4}$

Divide. → The quotient is the whole number. The remainder over the divisor is the fraction.

$$\begin{array}{r} 18 \\ 4\overline{)74} \\ \underline{4} \\ 34 \\ \underline{32} \\ 2 \end{array}$$

$18\frac{2}{4} = 18\frac{1}{2}$ (18 is the quotient; Remainder: 2; Divisor: 4)

EXAMPLES. Convert to a mixed number.

1. $\frac{8}{5}$

$$\begin{array}{r} 1 \\ 5\overline{)8} \\ \underline{5} \\ 3 \end{array}$$

$\frac{8}{5} = 1\frac{3}{5}$

2. $\frac{122}{8}$

$$\begin{array}{r} 15 \\ 8\overline{)122} \\ \underline{120} \\ 2 \end{array}$$

$\frac{122}{8} = 15\frac{2}{8} = 15\frac{1}{4}$

TRY THIS

Convert to a mixed number.

1. $\frac{7}{3}$ **2.** $\frac{11}{10}$ **3.** $\frac{110}{6}$ **4.** $\frac{134}{8}$

PRACTICE

Convert to a mixed number.

1. $\frac{7}{5}$	**2.** $\frac{13}{4}$	**3.** $\frac{14}{3}$	**4.** $\frac{7}{6}$	**5.** $\frac{16}{5}$
6. $\frac{13}{4}$	**7.** $\frac{19}{8}$	**8.** $\frac{19}{5}$	**9.** $\frac{13}{3}$	**10.** $\frac{13}{6}$
11. $\frac{26}{4}$	**12.** $\frac{27}{6}$	**13.** $\frac{25}{2}$	**14.** $\frac{28}{8}$	**15.** $\frac{29}{9}$
16. $\frac{57}{10}$	**17.** $\frac{69}{10}$	**18.** $\frac{73}{10}$	**19.** $\frac{41}{10}$	**20.** $\frac{27}{10}$
21. $\frac{65}{10}$	**22.** $\frac{54}{10}$	**23.** $\frac{46}{10}$	**24.** $\frac{78}{10}$	**25.** $\frac{92}{10}$
26. $\frac{53}{7}$	**27.** $\frac{62}{5}$	**28.** $\frac{53}{8}$	**29.** $\frac{49}{4}$	**30.** $\frac{47}{6}$
31. $\frac{45}{6}$	**32.** $\frac{50}{8}$	**33.** $\frac{46}{4}$	**34.** $\frac{50}{6}$	**35.** $\frac{56}{5}$
36. $\frac{12}{8}$	**37.** $\frac{18}{4}$	**38.** $\frac{28}{6}$	**39.** $\frac{22}{4}$	**40.** $\frac{17}{2}$
41. $\frac{213}{100}$	**42.** $\frac{467}{100}$	**43.** $\frac{757}{100}$	**44.** $\frac{321}{100}$	**45.** $\frac{143}{100}$
46. $\frac{345}{8}$	**47.** $\frac{223}{4}$	**48.** $\frac{135}{6}$	**49.** $\frac{185}{3}$	**50.** $\frac{166}{5}$
51. $\frac{456}{16}$	**52.** $\frac{554}{12}$	**53.** $\frac{678}{15}$	**54.** $\frac{567}{31}$	**55.** $\frac{289}{22}$

PROBLEM CORNER

1. There are $\frac{365}{7}$ weeks in a year. Convert the number of weeks to a mixed number.

2. The Jefferson twins are making nut bread. Ada measures $\frac{7}{4}$ cups of sugar. Arthur measures $\frac{9}{2}$ cups of flour. Convert each measurement to a mixed number.

More Practice, page 163

MIXED NUMBERS AND DECIMALS

Convert to a mixed number. 8.95

Write as a sum.	→	Convert the decimal to a fraction.	→	Write the mixed number.
$8.95 = 8 + 0.95$	$=$	$8 + \frac{95}{100}$	$=$	$8\frac{95}{100}$

EXAMPLES. Convert to a mixed number.

1. $6.3 = 6 + 0.3 = 6 + \frac{3}{10} = 6\frac{3}{10}$

2. $12.007 = 12\frac{7}{1000}$

TRY THIS

Convert to a mixed number.

1. 4.17 **2.** 13.018 **3.** 1.9 **4.** 2.779

Convert to a decimal. $17\frac{3}{4}$

Write as a sum.	→	Convert the fraction to a decimal.	→	Write the decimal.
$17\frac{3}{4} = 17 + \frac{3}{4}$	$=$	$17 + 0.75$	$=$	17.75

EXAMPLES. Convert to a decimal.

3. $9\frac{1}{2} = 9 + \frac{1}{2} = 9 + 0.5 = 9.5$

4. $28\frac{97}{100} = 28.97$

TRY THIS

Convert to a decimal.

5. $5\frac{6}{10}$ **6.** $22\frac{4}{5}$ **7.** $2\frac{33}{1000}$ **8.** $14\frac{3}{5}$

PRACTICE

Convert to a mixed number.

1. 1.2 **2.** 3.7 **3.** 11.1 **4.** 99.7 **5.** 42.9

6. 8.98 **7.** 1.23 **8.** 24.76 **9.** 32.22 **10.** 12.46

11. 1.113 **12.** 7.097 **13.** 89.001 **14.** 12.999 **15.** 22.078

16. 4.7878 **17.** 5.6177 **18.** 11.4562 **19.** 13.1188 **20.** 16.3045

21. 124.9 **22.** 621.7 **23.** 134.56 **24.** 221.08 **25.** 242.69

Convert to a decimal.

26. $1\frac{9}{10}$ **27.** $3\frac{8}{10}$ **28.** $18\frac{1}{10}$ **29.** $23\frac{5}{10}$ **30.** $41\frac{7}{10}$

31. $1\frac{17}{100}$ **32.** $7\frac{59}{100}$ **33.** $13\frac{99}{100}$ **34.** $14\frac{23}{100}$ **35.** $29\frac{69}{100}$

36. $4\frac{3}{1000}$ **37.** $8\frac{1}{1000}$ **38.** $166\frac{213}{1000}$ **39.** $249\frac{188}{1000}$ **40.** $14\frac{28}{1000}$

41. $8\frac{1}{5}$ **42.** $7\frac{3}{5}$ **43.** $9\frac{2}{5}$ **44.** $6\frac{4}{5}$ **45.** $2\frac{3}{5}$

46. $13\frac{3}{8}$ **47.** $14\frac{5}{8}$ **48.** $19\frac{1}{8}$ **49.** $20\frac{7}{8}$ **50.** $18\frac{5}{8}$

51. $10\frac{1}{2}$ **52.** $17\frac{1}{4}$ **53.** $35\frac{3}{4}$ **54.** $29\frac{1}{4}$ **55.** $62\frac{3}{4}$

56. $23\frac{13}{16}$ **57.** $9\frac{11}{20}$ **58.** $103\frac{14}{25}$ **59.** $128\frac{23}{40}$ **60.** $204\frac{13}{20}$

PROBLEM CORNER

At service stations, pumps measure gasoline by one-tenth gallons. What will a pump read after filling an empty $16\frac{1}{5}$ gallon tank?

More Practice, page 163

APPLICATION

TIME ZONES

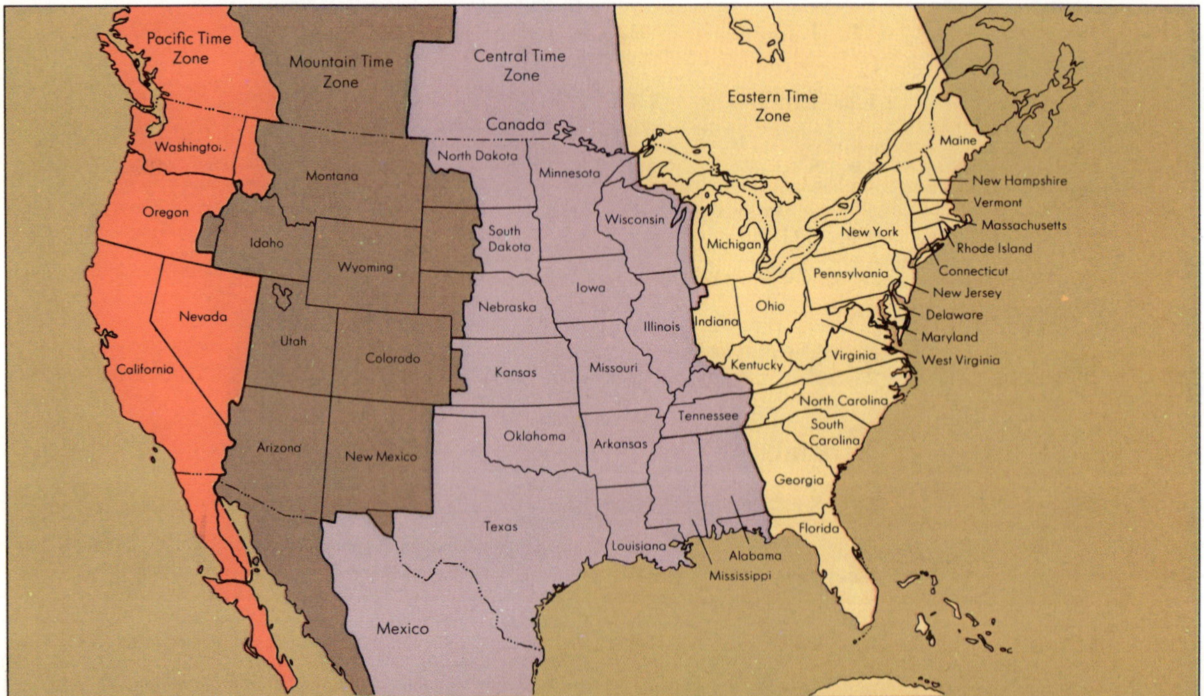

Most of the U.S. is divided into 4 time zones. It takes about 1 hour for the sun to pass over a zone.

A TV newscaster announces, "The tennis match will begin at 2 PM Eastern Standard Time (EST)." What time will this be in each zone?

1. Central (CST)
2. Mountain (MST)
3. Pacific (PST)

Almost half the states have a city called Springfield. Use the time zone map to find what time it is in each city when it is 12:00 PM (noon) in Springfield, Illinois.

4. Springfield, Oregon
5. Springfield, Maine
6. Springfield, Colorado

We can subtract to find how much time has passed.

EXAMPLE 1. School starts at 8:00 AM and ends at 2:30 PM. How long is the school day?

$$\begin{array}{r} 2{:}30 \text{ PM} \\ -\ 8{:}00 \text{ AM} \\ \hline \end{array} \longrightarrow \begin{array}{r} 14{:}30 \\ -\ 8{:}00 \\ \hline 6{:}30 \end{array} = 6 \text{ hr, } 30 \text{ min}$$

We passed from AM to PM, so add 12 to 2:30.

7. Soccer practice begins at 10:00 AM and ends at 1:15 PM. How long is practice?
8. The concert started at 11:30 AM and ended at 2:00 PM. How long was the concert?

Airlines give the time of departure and the time of arrival in local time.

EXAMPLE 2. A plane departs Boston, Massachusetts at 10:30 AM and arrives in Los Angeles, California at 1:20 PM. How long does the trip take?

$$\begin{array}{r} 1{:}20 \text{ PM (PST)} \\ -\ 10{:}30 \text{ AM (EST)} \\ \hline \end{array} \longrightarrow \begin{array}{r} 1{:}20 \text{ PM (PST)} \\ -\ 7{:}30 \text{ AM (PST)} \\ \hline \end{array}$$

Subtract departure from arrival. Change to the same time zone.

$$\begin{array}{r} 1{:}20 \text{ PM} \\ -\ 7{:}30 \text{ AM} \\ \hline \end{array} \longrightarrow \begin{array}{r} 13{:}20 \\ -\ 7{:}30 \\ \hline \end{array} \longrightarrow \begin{array}{r} 12{:}80 \\ -\ 7{:}30 \\ \hline 5{:}50 \end{array} = 5 \text{ hr, } 50 \text{ min}$$

Rename to subtract.

How long is each trip?

	Depart		Arrive	
9.	Chicago, Ill.	11:45 PM	New York City, N.Y.	2:37 AM
10.	Miami, Fla.	10:50 AM	Denver, Colo.	1:10 PM
11.	Philadelphia, Penn.	11:15 AM	Kansas City, Mo.	2:20 PM
12.	Seattle, Wash.	11:00 PM	Detroit, Mich.	6:25 AM

CAREER CAREER CAREER CAREER CAREER

OPERATIONS MANAGER

Miguel Romero is the operations manager for the Arden Plastic Company. He plans truck deliveries. He must think about loads, routes, time, and cost. The maximum legal weight of a loaded truck is 73,280 pounds.

1. Truck number 82 weighs 34,306 pounds. What is the heaviest load it can carry?
2. Truck number 103 weighs 32,495 pounds. What is the heaviest load it can carry?
3. Truck number 18 weighs 38,782 pounds. Can it carry a load which weighs 35,598 pounds?
4. How many more pounds can truck number 103 carry than truck number 18?

Miguel records the time it takes to load the trucks. How long does it take to load each truck?

Truck	Start Loading	Finish Loading
23	6:15 AM	8:32 AM
184	7:20 AM	10:15 AM
73	9:13 AM	11:27 AM
87	10:06 AM	12:30 PM
159	11:50 AM	2:17 PM
41	1:45 PM	4:22 PM

5. truck 23
6. truck 184
7. truck 73
8. truck 87
9. truck 159
10. truck 41

CAREER CAREER CAREER CAREER CAREER

Small trucks are used to make local deliveries. The drivers sign out and sign in to keep a record of hours driven. How much time did each driver take?

Name	Time Out	Time In
Sonny	6:35 AM	11:50 AM
Rhonda	7:10 AM	2:40 PM
Jacques	7:45 AM	1:45 PM
Mike	7:40 AM	1:30 PM
Red Eye	8:00 AM	2:45 PM
Buck	8:09 AM	3:12 PM

11. Sonny
12. Rhonda
13. Jacques
14. Mike
15. Red Eye
16. Buck

Arden Plastic Company ships products of many weights and sizes. Miguel must consider these measurements in deciding how to load the trucks, because there are weight limits for each truck.

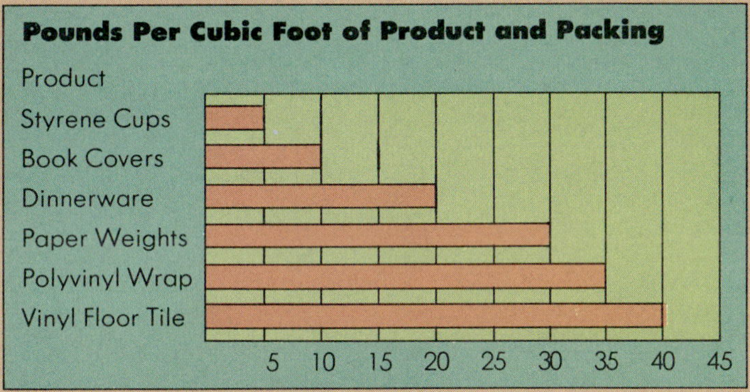

Pounds Per Cubic Foot of Product and Packing

How many pounds are there in one cubic foot of each of the following?

17. polyvinyl wrap
18. styrene cups

How many pounds are there in 250 cubic feet of each?

19. book covers
20. dinnerware
21. paper weights
22. floor tile

How many cubic feet are needed for 100 pounds of each?

23. styrene cups
24. dinnerware
25. polyvinyl wrap
26. floor tile

CHAPTER 5 REVIEW

1. What part is shaded? p. 136

2. What part is shaded?

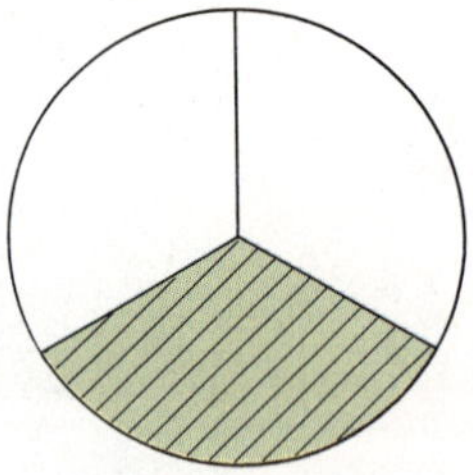

Divide, if possible. p. 138

3. $\frac{15}{1}$ **4.** $\frac{0}{7}$ **5.** $\frac{7}{7}$ **6.** $\frac{0}{0}$ **7.** $\frac{5}{5}$

Are the fractions equal? p. 140

8. $\frac{2}{4}, \frac{3}{6}$ **9.** $\frac{3}{5}, \frac{3}{7}$

Compare. p. 142

10. $\frac{6}{7}$ and $\frac{7}{8}$ **11.** $\frac{9}{11}$ and $\frac{4}{5}$ **12.** $\frac{3}{4}$ and $\frac{6}{8}$

Convert to a fraction. pp. 144, 150

13. 1.093 **14.** 0.07 **15.** $4\frac{1}{3}$ **16.** $10\frac{5}{8}$ **17.** $12\frac{3}{5}$

Convert to a decimal. p. 146

18. $\frac{9}{8}$ **19.** $\frac{37}{40}$ **20.** $\frac{13}{100}$ **21.** $\frac{9898}{1000}$ **22.** $\frac{3}{8}$

Convert to a decimal. Round to the nearest thousandth, if necessary.

23. $\frac{86}{7}$ **24.** $\frac{4}{3}$ **25.** $8\frac{1}{10}$ **26.** $118\frac{3}{5}$ **27.** $3\frac{7}{9}$

Convert to a mixed number. pp. 152, 154

28. $\frac{19}{3}$ **29.** $\frac{69}{7}$ **30.** 4.9 **31.** 8.013 **32.** 3.75

Application pp. 148, 156

33. Jerry was on the ski slopes for 4 hr, 15 min. Sally was there for 2 hr, 50 min. How much longer was Jerry on the ski slopes than Sally? **1 hr, 25 min**

34. When it is 7 PM in New York, what time is it in Los Angeles? **4 PM**

CHAPTER 5 TEST

1. What part is shaded?

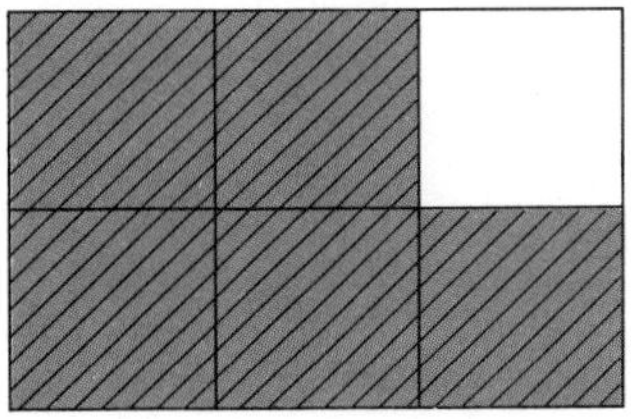

2. What part is shaded?

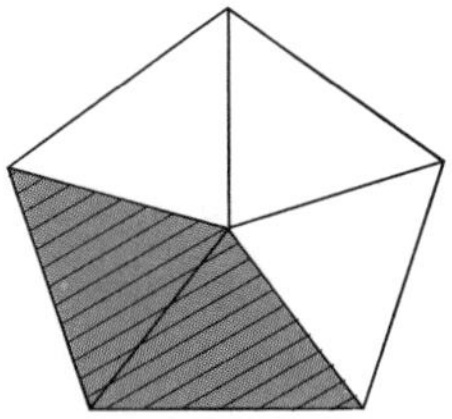

Divide, if possible.

3. $\frac{0}{23}$ **4.** $\frac{27}{1}$ **5.** $\frac{27}{0}$ **6.** $\frac{38}{38}$ **7.** $\frac{1}{1}$

Are the fractions equal?

8. $\frac{2}{6}, \frac{6}{18}$ **9.** $\frac{7}{8}, \frac{8}{9}$

Compare.

10. $\frac{7}{6}$ and $\frac{8}{7}$ **11.** $\frac{5}{12}$ and $\frac{3}{8}$ **12.** $\frac{5}{8}$ and $\frac{3}{4}$

Convert to a fraction.

13. 1.237 **14.** 0.3 **15.** $2\frac{1}{8}$ **16.** $11\frac{4}{5}$ **17.** $5\frac{7}{10}$

Convert to a decimal.

18. $\frac{11}{25}$ **19.** $\frac{2}{5}$ **20.** $\frac{7}{10}$ **21.** $\frac{7784}{1000}$ **22.** $\frac{1}{8}$

Convert to a decimal. Round to the nearest thousandth, if necessary.

23. $\frac{56}{13}$ **24.** $\frac{5}{11}$ **25.** $4\frac{9}{10}$ **26.** $29\frac{3}{4}$ **27.** $5\frac{2}{3}$

Convert to a mixed number.

28. $\frac{25}{4}$ **29.** $\frac{125}{9}$ **30.** 2.8 **31.** 120.67 **32.** 14.4

Application

33. Kevin spent 2 hr, 35 min mowing the lawn and 1 hr, 40 min trimming it. How much time did he take in all?

34. When it is 6 PM in Chicago, what time is it in New York?

MORE PRACTICE

What part is shaded? Use after page 137.

1. 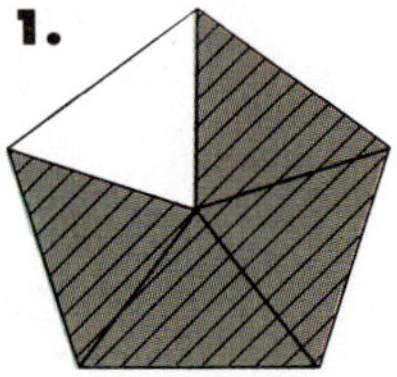

2. 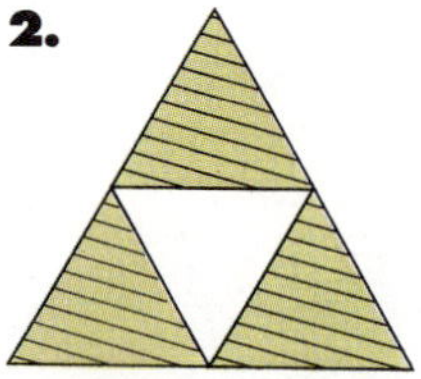

3. 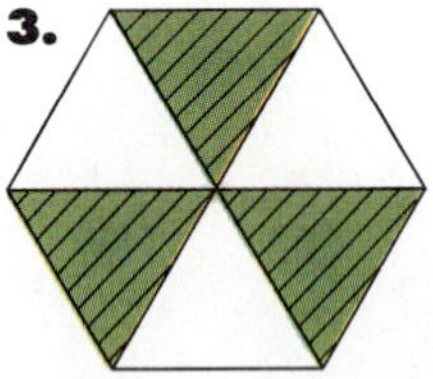

4. 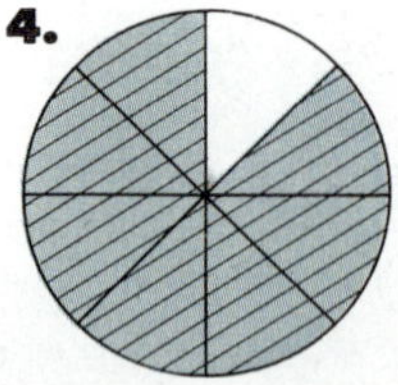

Divide, if possible. Use after page 139.

5. $\frac{39}{1}$ **6.** $\frac{39}{39}$ **7.** $\frac{43}{43}$ **8.** $\frac{43}{0}$

9. $\frac{17}{0}$ **10.** $\frac{17}{1}$ **11.** $\frac{0}{38}$ **12.** $\frac{0}{100}$

Tell whether the fractions are equal. Use after page 141.

13. $\frac{1}{4}, \frac{1}{3}$ **14.** $\frac{2}{3}, \frac{5}{8}$ **15.** $\frac{3}{2}, \frac{6}{4}$ **16.** $\frac{2}{3}, \frac{5}{7}$

17. $\frac{6}{16}, \frac{3}{8}$ **18.** $\frac{2}{3}, \frac{8}{12}$ **19.** $\frac{7}{25}, \frac{2}{7}$ **20.** $\frac{8}{40}, \frac{6}{30}$

Compare. Use after page 143.

21. $\frac{1}{4}$ and $\frac{1}{3}$ **22.** $\frac{2}{3}$ and $\frac{5}{8}$ **23.** $\frac{5}{7}$ and $\frac{2}{3}$ **24.** $\frac{9}{10}$ and $\frac{89}{100}$

25. $\frac{5}{16}$ and $\frac{1}{4}$ **26.** $\frac{2}{5}$ and $\frac{1}{3}$ **27.** $\frac{7}{12}$ and $\frac{5}{6}$ **28.** $\frac{3}{4}$ and $\frac{7}{10}$

Convert to a fraction. Use after page 145.

29. 9.3 **30.** 1.07 **31.** 0.013 **32.** 1.9132

33. 0.89 **34.** 0.9 **35.** 0.4133 **36.** 7.567

Convert to a decimal. Use after page 147.

37. $\frac{4}{5}$ **38.** $\frac{5}{4}$ **39.** $\frac{7}{8}$ **40.** $\frac{53}{20}$

41. $\frac{13}{16}$ **42.** $\frac{32}{25}$ **43.** $\frac{49}{40}$ **44.** $\frac{64}{125}$

Convert to a decimal. Round to the nearest thousandth. Use after page 147.

1. $\frac{1}{6}$ **2.** $\frac{8}{3}$ **3.** $\frac{4}{7}$ **4.** $\frac{10}{9}$ **5.** $\frac{7}{3}$

6. $\frac{12}{13}$ **7.** $\frac{19}{12}$ **8.** $\frac{14}{9}$ **9.** $\frac{127}{11}$ **10.** $\frac{21}{11}$

Convert to a decimal. Use after page 147.

11. $\frac{6}{10}$ **12.** $\frac{17}{10}$ **13.** $\frac{11}{1000}$ **14.** $\frac{5673}{1000}$ **15.** $\frac{306}{1000}$

16. $\frac{23}{100}$ **17.** $\frac{997}{100}$ **18.** $\frac{569}{10{,}000}$ **19.** $\frac{18{,}977}{10{,}000}$ **20.** $\frac{2659}{10{,}000}$

Convert to a fraction. Use after page 151.

21. $9\frac{1}{4}$ **22.** $7\frac{2}{3}$ **23.** $4\frac{5}{8}$ **24.** $10\frac{1}{2}$ **25.** $5\frac{2}{5}$

26. $14\frac{2}{7}$ **27.** $8\frac{3}{10}$ **28.** $6\frac{3}{100}$ **29.** $17\frac{8}{9}$ **30.** $2\frac{13}{100}$

31. $8\frac{7}{8}$ **32.** $13\frac{2}{5}$ **33.** $7\frac{5}{6}$ **34.** $19\frac{2}{7}$ **35.** $12\frac{4}{9}$

Convert to a mixed number. Use after page 153.

36. $\frac{11}{3}$ **37.** $\frac{9}{4}$ **38.** $\frac{29}{5}$ **39.** $\frac{35}{6}$ **40.** $\frac{7}{2}$

41. $\frac{79}{10}$ **42.** $\frac{457}{2}$ **43.** $\frac{577}{12}$ **44.** $\frac{492}{49}$ **45.** $\frac{161}{13}$

46. $\frac{15}{7}$ **47.** $\frac{98}{5}$ **48.** $\frac{29}{3}$ **49.** $\frac{137}{9}$ **50.** $\frac{73}{3}$

Convert to a mixed number. Use after page 155.

51. 5.3 **52.** 5.37 **53.** 1.012 **54.** 14.7 **55.** 12.25

56. 5.1137 **57.** 29.5142 **58.** 109.01 **59.** 123.789 **60.** 684.5026

Convert to a decimal. Use after page 155.

61. $5\frac{9}{10}$ **62.** $8\frac{41}{100}$ **63.** $78\frac{7883}{10{,}000}$ **64.** $14\frac{1}{1000}$ **65.** $61\frac{22}{1000}$

66. $9\frac{3}{5}$ **67.** $12\frac{3}{8}$ **68.** $10\frac{3}{4}$ **69.** $8\frac{17}{20}$ **70.** $7\frac{5}{8}$

CALCULATOR CORNER

MULTIPLICATION "IN BETWEEN"

Players two
Materials two calculators, paper and pencils

Sample Aaron gives Jodi a calculator with a number, such as 212, on the display. He tells Jodi that by multiplying 212 by a number, she must get a number on the display which is between 120 and 130.

Jodi gives Aaron a calculator with a number, such as 183, on the display. She tells Aaron that by multiplying 183 by a number, he must get a number on the display which is between 260 and 270.

JODI Jodi writes down 212 and the interval 120–130.

Play 1 Jodi multiplies 212 by 0.8. She records 0.8.
Result: 169.5–too large

Play 2 Jodi multiplies 212 by 0.7. She records 0.7.
Result: 148.4–too large

Play 3 Jodi multiplies 212 by 0.6. She records 0.6.
Result: 127.2–OK

It took Jodi 3 plays, so her score is 3.

AARON Aaron writes down 183 and the interval 260–270.

Play 1 Aaron multiplies 183 by 2. He records 2.
Result: 366–too large

Play 2 Aaron multiplies 183 by 1.5. He records 1.5.
Result: 274.5–too large

Play 3 Aaron multiplies 183 by 1.4. He records 1.4.
Result: 256.2–too small

Play 4 Aaron multiplies 183 by 1.45. He records 1.45.
Result: 265.35–OK

It took Aaron 4 plays, so his score is 4.

Play continues for 5 rounds. The player with the lowest score wins.

READY FOR CHAPTER 6

Multiply.

1. $\frac{1}{2} \cdot \frac{1}{5}$ **2.** $\frac{2}{3} \cdot \frac{4}{5}$ **3.** $6 \times \frac{2}{7}$

4. Find a fraction equivalent to $\frac{3}{4}$ with a denominator of 12.

All the fractions are equivalent. Which is simplest?

5. $\frac{6}{16}, \frac{3}{8}, \frac{12}{32}, \frac{15}{40}, \frac{18}{48}$ **6.** $\frac{8}{28}, \frac{6}{21}, \frac{2}{7}, \frac{4}{14}$

Simplify.

7. $\frac{12}{18}$ **8.** $\frac{8}{20}$ **9.** $\frac{30}{6}$

Multiply and simplify.

10. $\frac{3}{8} \times \frac{4}{9}$ **11.** $\frac{6}{7} \cdot \frac{5}{3}$ **12.** $\frac{4}{5} \cdot 20$

13. $9 \times 4\frac{1}{3}$ **14.** $6\frac{3}{4} \cdot \frac{2}{3}$ **15.** $3\frac{1}{3} \cdot 1\frac{3}{4}$

Find the reciprocal.

16. $\frac{3}{4}$ **17.** 8 **18.** $\frac{1}{10}$

Divide and simplify.

19. $\frac{5}{6} \div \frac{2}{3}$ **20.** $\frac{3}{5} \div \frac{1}{2}$ **21.** $\frac{2}{5} \div 6$

22. $33 \div 5\frac{1}{2}$ **23.** $2\frac{1}{3} \div 1\frac{1}{6}$ **24.** $2\frac{1}{12} \div 75$

Application

25. Rachel budgets $960 for transportation. By taking the bus to work, she can save $\frac{1}{3}$ of this expense. How much can she save?

26. For breakfast Alex eats the following. How many calories are in Alex's breakfast?

Food	Calories
1 cup of cereal	400 per cup
$\frac{1}{2}$ cup of milk	160 per cup
2 pieces of toast	100 per piece
$1\frac{1}{2}$ cups of orange juice	112 per cup

6
MULTIPLYING AND DIVIDING FRACTIONS

GETTING STARTED

FRACTION MULTO

WHAT'S NEEDED The 30 squares of paper with digits marked on them as in GOTCHA, Chapter 1.

THE RULES

1. On a sheet of paper each player writes this:

   $\frac{[\ ]}{[\ ]} \cdot \frac{[\ ]}{[\ ]}$

2. To play, a leader draws digits from the box, one at a time.
3. Each player writes the digit in any one of the four spaces. Once it has been written, it cannot be changed.
4. The leader continues to draw digits until four digits have been drawn.
5. Each player multiplies the resulting fractions. The winner is the player who has the largest product.

SAMPLE

Louis	Chris	Sally
$\frac{[2]}{[3]} \cdot \frac{[5]}{[4]} = \frac{10}{12}$	$\frac{[4]}{[5]} \cdot \frac{[3]}{[2]} = \frac{12}{10}$	$\frac{[4]}{[3]} \cdot \frac{[5]}{[2]} = \frac{20}{6}$

Sally wins because she has the largest product.

MULTIPLYING

Multiply. $\frac{2}{3} \times \frac{5}{7}$

Multiply the numerators. → Multiply the denominators.

$$\frac{2}{3} \times \frac{5}{7} = \frac{2 \times 5}{3 \times 7} = \frac{10}{21}$$

EXAMPLES. Multiply.

1. $\frac{3}{5} \times \frac{7}{8} = \frac{3 \times 7}{5 \times 8} = \frac{21}{40}$

2. $\frac{2}{5} \times \frac{4}{4} = \frac{8}{20}$ *Try to multiply mentally.*

3. $\frac{1}{3} \cdot \frac{1}{4} = \frac{1 \cdot 1}{3 \cdot 4} = \frac{1}{12}$ *We can also use a dot $\cdot$ for $\times$.*

TRY THIS

Multiply.

1. $\frac{2}{3} \times \frac{4}{5}$ **2.** $\frac{2}{3} \cdot \frac{2}{2}$ **3.** $\frac{2}{7} \times \frac{1}{6}$

4. $\frac{1}{2} \times \frac{1}{5}$ **5.** $\frac{7}{1} \cdot \frac{5}{6}$ **6.** $\frac{3}{4} \cdot \frac{2}{5}$

EXAMPLE 4. Multiply. $6 \times \frac{4}{5}$

$$6 \times \frac{4}{5} = \frac{6}{1} \times \frac{4}{5} = \frac{6 \times 4}{1 \times 5} = \frac{24}{5} = 4\frac{4}{5}$$

EXAMPLE 5. Multiply. $\frac{3}{8} \cdot 11$

$$\frac{3}{8} \cdot 11 = \frac{3}{8} \cdot \frac{11}{1} = \frac{33}{8} = 4\frac{1}{8}$$

TRY THIS

Multiply.

7. $10 \cdot \frac{2}{3}$ **8.** $\frac{2}{5} \times 16$ **9.** $5 \cdot \frac{1}{8}$ **10.** $3 \times \frac{2}{9}$

PRACTICE

Multiply.

1. $\frac{1}{4} \cdot \frac{1}{5}$	**2.** $\frac{1}{10} \cdot \frac{1}{4}$	**3.** $\frac{1}{5} \cdot \frac{1}{10}$	**4.** $\frac{1}{6} \cdot \frac{1}{8}$	**5.** $\frac{1}{7} \cdot \frac{1}{9}$
6. $\frac{2}{3} \cdot \frac{1}{5}$	**7.** $\frac{3}{4} \cdot \frac{1}{3}$	**8.** $\frac{2}{5} \cdot \frac{1}{7}$	**9.** $\frac{3}{5} \cdot \frac{1}{2}$	**10.** $\frac{5}{6} \cdot \frac{1}{3}$
11. $\frac{3}{4} \times \frac{2}{2}$	**12.** $\frac{4}{5} \times \frac{3}{3}$	**13.** $\frac{4}{4} \times \frac{2}{3}$	**14.** $\frac{8}{8} \times \frac{5}{6}$	**15.** $\frac{6}{6} \times \frac{2}{5}$
16. $\frac{2}{5} \times \frac{5}{6}$	**17.** $\frac{2}{3} \times \frac{7}{8}$	**18.** $\frac{3}{8} \times \frac{2}{3}$	**19.** $\frac{4}{5} \times \frac{2}{6}$	**20.** $\frac{3}{4} \times \frac{2}{7}$
21. $\frac{4}{9} \cdot \frac{5}{6}$	**22.** $\frac{2}{9} \cdot \frac{3}{4}$	**23.** $\frac{5}{6} \cdot \frac{3}{4}$	**24.** $\frac{2}{9} \cdot \frac{4}{9}$	**25.** $\frac{3}{8} \times \frac{5}{8}$
26. $\frac{1}{5} \cdot 3$	**27.** $\frac{1}{3} \cdot 2$	**28.** $\frac{1}{6} \cdot 5$	**29.** $\frac{1}{7} \cdot 4$	**30.** $\frac{1}{4} \times 9$
31. $8 \times \frac{3}{5}$	**32.** $1 \times \frac{2}{7}$	**33.** $8 \times \frac{2}{3}$	**34.** $9 \times \frac{3}{5}$	**35.** $5 \times \frac{4}{5}$
36. $\frac{7}{8} \cdot \frac{7}{8}$	**37.** $\frac{4}{5} \cdot \frac{4}{5}$	**38.** $\frac{5}{6} \cdot \frac{5}{6}$	**39.** $\frac{5}{8} \cdot \frac{5}{8}$	**40.** $\frac{3}{4} \cdot \frac{3}{4}$
41. $\frac{2}{3} \cdot 23$	**42.** $17 \cdot \frac{5}{6}$	**43.** $\frac{3}{4} \cdot 19$	**44.** $20 \cdot \frac{5}{3}$	**45.** $\frac{6}{5} \cdot 12$
46. $\frac{10}{11} \cdot 32$	**47.** $25 \cdot \frac{12}{13}$	**48.** $20 \cdot \frac{14}{15}$	**49.** $\frac{11}{12} \cdot 23$	**50.** $11 \cdot \frac{12}{15}$

PROBLEM CORNER

1. It takes $\frac{3}{4}$ of a yard of ribbon to make a bow. How much ribbon is needed to make 7 bows?

2. A granola recipe calls for $\frac{3}{4}$ of a cup of honey. How much honey is needed to make $\frac{1}{2}$ of a recipe.

More Practice, page 190

EQUIVALENT FRACTIONS

When we multiply a number by 1, we get the same number. When we multiply a fraction by 1, we get an equivalent fraction.

Find a fraction equivalent to $\frac{3}{4}$.

Multiply by 1. → Multiply the numerators. Multiply the denominators. → Write the equivalent fraction.

$$\frac{3}{4} \cdot \frac{2}{2} \quad = \quad \frac{3 \cdot 2}{4 \cdot 2} \quad = \quad \frac{6}{8}$$

Here we use $\frac{2}{2}$ for 1.

EXAMPLE 1. Find two other fractions equivalent to $\frac{3}{4}$.

$$\frac{3}{4} = \frac{3}{4} \cdot \frac{3}{3} = \frac{3 \cdot 3}{4 \cdot 3} = \frac{9}{12} \qquad \frac{3}{4} = \frac{3}{4} \cdot \frac{4}{4} = \frac{3 \cdot 4}{4 \cdot 4} = \frac{12}{16}$$

TRY THIS

Find three fractions equivalent to the given fraction.

1. $\frac{2}{3}$ **2.** $\frac{7}{8}$ **3.** $\frac{1}{2}$ **4.** $\frac{4}{7}$

Find a fraction equivalent to $\frac{5}{6}$ with a denominator of 24.

Divide to find what to use for 1. → Multiply by 1. → Multiply the numerators. Multiply the denominators. → Write the equivalent fraction.

$$24 \div 6 = 4 \qquad \frac{5}{6} \cdot \frac{4}{4} \quad = \quad \frac{5 \cdot 4}{6 \cdot 4} \quad = \quad \frac{20}{24}$$

EXAMPLE 2. Find a fraction equivalent to $\frac{2}{5}$ with a denominator of 35.

$$\frac{2}{5} = \frac{2}{5} \cdot \frac{7}{7} = \frac{2 \cdot 7}{5 \cdot 7} = \frac{14}{35}$$

Since $35 \div 5 = 7$, we multiply by $\frac{7}{7}$.

TRY THIS

Find a fraction equivalent to the given one, but with the indicated denominator.

5. $\frac{2}{3}$ [9] **6.** $\frac{9}{10}$ [20] **7.** $\frac{1}{4}$ [32] **8.** $\frac{4}{5}$ [20]

PRACTICE

Find a fraction equivalent to the given one, but with the indicated denominator.

1. $\frac{1}{2}$ [10]	**2.** $\frac{1}{3}$ [6]	**3.** $\frac{1}{2}$ [6]	**4.** $\frac{1}{4}$ [12]	**5.** $\frac{1}{6}$ [12]
6. $\frac{3}{4}$ [24]	**7.** $\frac{2}{9}$ [18]	**8.** $\frac{2}{3}$ [12]	**9.** $\frac{2}{5}$ [10]	**10.** $\frac{2}{7}$ [14]
11. $\frac{5}{6}$ [18]	**12.** $\frac{7}{8}$ [16]	**13.** $\frac{3}{5}$ [20]	**14.** $\frac{2}{9}$ [27]	**15.** $\frac{5}{8}$ [24]
16. $\frac{2}{3}$ [12]	**17.** $\frac{2}{3}$ [15]	**18.** $\frac{2}{3}$ [27]	**19.** $\frac{2}{3}$ [30]	**20.** $\frac{2}{3}$ [18]
21. $\frac{3}{4}$ [20]	**22.** $\frac{3}{4}$ [12]	**23.** $\frac{3}{4}$ [8]	**24.** $\frac{3}{4}$ [36]	**25.** $\frac{3}{4}$ [28]
26. $\frac{3}{8}$ [48]	**27.** $\frac{7}{8}$ [32]	**28.** $\frac{1}{6}$ [18]	**29.** $\frac{7}{18}$ [36]	**30.** $\frac{2}{5}$ [20]
31. $\frac{4}{5}$ [50]	**32.** $\frac{1}{6}$ [24]	**33.** $\frac{5}{8}$ [24]	**34.** $\frac{1}{9}$ [18]	**35.** $\frac{3}{7}$ [21]
36. $\frac{3}{8}$ [24]	**37.** $\frac{5}{6}$ [24]	**38.** $\frac{7}{22}$ [132]	**39.** $\frac{10}{21}$ [126]	**40.** $\frac{5}{12}$ [108]

PROBLEM CORNER

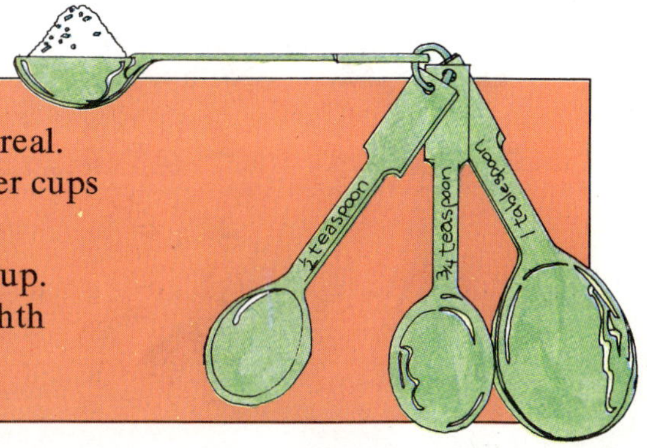

1. Tomas needs to measure $\frac{1}{2}$ c milk to make cereal. He only has a $\frac{1}{4}$ c measure. How many quarter cups should he use to get $\frac{1}{2}$ c?
2. Olivia needs to measure $\frac{3}{4}$ tsp salt to make soup. She only has a $\frac{1}{8}$ tsp measure. How many eighth teaspoons should she use to get $\frac{3}{4}$ tsp?

More Practice, page 190

SIMPLIFYING

All the following are equivalent to $\frac{3}{4}$. $\frac{3}{4} = \frac{6}{8} = \frac{9}{12} = \frac{12}{16} = \frac{15}{20} = \frac{18}{24}$

$\frac{3}{4}$ is simplest because:

1. All the other fractions have larger numerators and denominators.
2. 1 is the only number that divides evenly into both the numerator and denominator.

EXAMPLE 1. All the fractions are equivalent. Which is simplest?

$\frac{2}{4}, \frac{3}{6}, \frac{1}{2}, \frac{6}{12}, \frac{5}{10}$ $\quad$ $\frac{1}{2}$ is simplest.

TRY THIS

All the fractions are equivalent. Which is simplest?

1. $\frac{6}{10}, \frac{15}{25}, \frac{18}{30}, \frac{3}{5}, \frac{12}{20}$ **2.** $\frac{15}{40}, \frac{9}{24}, \frac{12}{32}, \frac{6}{16}, \frac{3}{8}$ **3.** $\frac{12}{18}, \frac{6}{9}, \frac{2}{3}, \frac{14}{21}, \frac{10}{15}$

Simplify. $\frac{18}{24}$

Divide the numerator and denominator by the same number. → Continue dividing until 1 is the only number that divides evenly into both the numerator and denominator. → The fraction is simplified.

$\frac{18 \div 3}{24 \div 3} = \frac{6}{8}$ $\quad$ $\frac{6 \div 2}{8 \div 2} = \frac{3}{4}$ $\quad$ $\frac{18}{24} = \frac{3}{4}$

To check we multiply by 1: $\frac{3}{4} = \frac{3}{4} \cdot \frac{2}{2} = \frac{6}{8} \cdot \frac{3}{3} = \frac{18}{24}$

EXAMPLES. Simplify.

2. $\frac{28}{16} = \frac{28 \div 2}{16 \div 2} = \frac{14 \div 2}{8 \div 2} = \frac{7}{4} = 1\frac{3}{4}$

3. $\frac{2}{6} = \frac{2 \div 2}{6 \div 2} = \frac{1}{3}$

4. $\frac{8}{20} = \frac{8 \div 4}{20 \div 4} = \frac{2}{5}$

By choosing the largest number, we only need to divide once.

5. $\frac{30}{6} = \frac{30 \div 6}{6 \div 6} = \frac{5}{1} = 5$

TRY THIS

Simplify.

4. $\frac{10}{12}$ **5.** $\frac{6}{2}$ **6.** $\frac{14}{21}$ **7.** $\frac{60}{12}$ **8.** $\frac{16}{96}$

9. $\frac{32}{20}$ **10.** $\frac{16}{40}$ **11.** $\frac{14}{8}$ **12.** $\frac{12}{48}$ **13.** $\frac{72}{9}$

14. $\frac{54}{6}$ **15.** $\frac{84}{9}$ **16.** $\frac{11}{22}$ **17.** $\frac{6}{8}$ **18.** $\frac{37}{5}$

PRACTICE

All the fractions are equivalent. Which is simplest?

1. $\frac{10}{15}, \frac{2}{3}, \frac{8}{12}, \frac{6}{9}, \frac{14}{21}, \frac{4}{6}$ **2.** $\frac{8}{10}, \frac{12}{15}, \frac{24}{30}, \frac{28}{35}, \frac{4}{5}, \frac{16}{20}$ **3.** $\frac{12}{40}, \frac{15}{50}, \frac{3}{10}, \frac{6}{20}, \frac{9}{30}, \frac{18}{60}$

4. $\frac{1}{3}, \frac{2}{6}, \frac{3}{9}, \frac{4}{12}, \frac{5}{15}, \frac{6}{18}$ **5.** $\frac{6}{48}, \frac{5}{40}, \frac{4}{32}, \frac{3}{24}, \frac{2}{16}, \frac{1}{8}$ **6.** $\frac{20}{24}, \frac{15}{18}, \frac{35}{42}, \frac{5}{6}, \frac{10}{12}, \frac{25}{30}$

Simplify.

7. $\frac{2}{4}$ **8.** $\frac{4}{12}$ **9.** $\frac{3}{6}$ **10.** $\frac{2}{10}$ **11.** $\frac{2}{8}$ **12.** $\frac{2}{12}$

13. $\frac{6}{8}$ **14.** $\frac{9}{12}$ **15.** $\frac{6}{12}$ **16.** $\frac{6}{10}$ **17.** $\frac{8}{12}$ **18.** $\frac{6}{15}$

19. $\frac{10}{6}$ **20.** $\frac{12}{8}$ **21.** $\frac{12}{10}$ **22.** $\frac{15}{6}$ **23.** $\frac{14}{6}$ **24.** $\frac{21}{12}$

25. $\frac{12}{3}$ **26.** $\frac{24}{8}$ **27.** $\frac{25}{5}$ **28.** $\frac{36}{4}$ **29.** $\frac{27}{3}$ **30.** $\frac{32}{4}$

31. $\frac{18}{24}$ **32.** $\frac{42}{48}$ **33.** $\frac{20}{50}$ **34.** $\frac{40}{50}$ **35.** $\frac{15}{18}$ **36.** $\frac{11}{22}$

37. $\frac{14}{24}$ **38.** $\frac{14}{16}$ **39.** $\frac{14}{32}$ **40.** $\frac{15}{25}$ **41.** $\frac{16}{24}$ **42.** $\frac{12}{32}$

43. $\frac{8}{6}$ **44.** $\frac{12}{10}$ **45.** $\frac{16}{14}$ **46.** $\frac{32}{14}$ **47.** $\frac{14}{12}$ **48.** $\frac{25}{20}$

49. $\frac{6}{16}$ **50.** $\frac{4}{24}$ **51.** $\frac{16}{48}$ **52.** $\frac{15}{24}$ **53.** $\frac{10}{12}$ **54.** $\frac{12}{40}$

55. $\frac{100}{20}$ **56.** $\frac{56}{7}$ **57.** $\frac{42}{6}$ **58.** $\frac{125}{25}$ **59.** $\frac{72}{9}$ **60.** $\frac{108}{18}$

61. $\frac{35}{40}$ **62.** $\frac{25}{60}$ **63.** $\frac{30}{45}$ **64.** $\frac{40}{55}$ **65.** $\frac{20}{24}$ **66.** $\frac{24}{30}$

67. $\frac{10}{8}$ **68.** $\frac{10}{4}$ **69.** $\frac{20}{6}$ **70.** $\frac{20}{8}$ **71.** $\frac{40}{16}$ **72.** $\frac{42}{10}$

More Practice, page 190

MULTIPLYING AND SIMPLIFYING

Multiply and simplify. $\frac{3}{8} \times \frac{4}{9}$

Multiply the numerators. Multiply the denominators.	→	Simplify.
$\frac{3}{8} \times \frac{4}{9} = \frac{12}{72}$		$\frac{12}{72} = \frac{12 \div 12}{72 \div 12} = \frac{1}{6}$

EXAMPLE 1. Multiply and simplify. $\frac{6}{7} \times \frac{5}{3}$

$$\frac{6}{7} \times \frac{5}{3} = \frac{30}{21}$$

Multiply numerators: 6 × 5 = 30
Multiply denominators: 7 × 3 = 21

$$= \frac{30 \div 3}{21 \div 3}$$

Simplify.

$$= \frac{10}{7}$$

$$= 1\frac{3}{7}$$

TRY THIS

Multiply and simplify.

1. $\frac{2}{3} \times \frac{7}{8}$ **2.** $\frac{4}{5} \cdot \frac{5}{12}$ **3.** $\frac{10}{3} \cdot \frac{4}{5}$ **4.** $\frac{8}{5} \times \frac{5}{4}$

EXAMPLES. Multiply and simplify.

2. $24 \cdot \frac{3}{8} = \frac{24}{1} \cdot \frac{3}{8} = \frac{72}{8} = \frac{72 \div 8}{8 \div 8} = \frac{9}{1} = 9$

3. $\frac{5}{6} \cdot 26 = \frac{5}{6} \cdot \frac{26}{1} = \frac{130}{6} = \frac{130 \div 2}{6 \div 2} = \frac{65}{3} = 21\frac{2}{3}$

TRY THIS

Multiply and simplify.

5. $12 \cdot \frac{2}{3}$ **6.** $\frac{3}{4} \times 14$ **7.** $10 \cdot \frac{3}{8}$

PRACTICE

Multiply and simplify.

1. $\frac{2}{3} \times \frac{1}{2}$	**2.** $\frac{4}{5} \times \frac{1}{4}$	**3.** $\frac{7}{8} \times \frac{1}{7}$	**4.** $\frac{5}{6} \times \frac{1}{5}$	**5.** $\frac{3}{5} \times \frac{2}{3}$
6. $\frac{1}{8} \times \frac{4}{5}$	**7.** $\frac{2}{5} \times \frac{1}{6}$	**8.** $\frac{1}{4} \times \frac{2}{3}$	**9.** $\frac{3}{6} \times \frac{1}{6}$	**10.** $\frac{1}{4} \times \frac{8}{9}$
11. $\frac{12}{5} \times \frac{9}{8}$	**12.** $\frac{16}{15} \times \frac{5}{4}$	**13.** $\frac{10}{9} \times \frac{7}{5}$	**14.** $\frac{25}{12} \times \frac{4}{3}$	**15.** $\frac{14}{5} \times \frac{16}{7}$
16. $9 \cdot \frac{1}{9}$	**17.** $4 \cdot \frac{1}{4}$	**18.** $\frac{1}{3} \cdot 3$	**19.** $\frac{1}{6} \cdot 6$	**20.** $5 \cdot \frac{1}{5}$
21. $\frac{7}{10} \cdot \frac{10}{7}$	**22.** $\frac{8}{9} \cdot \frac{9}{8}$	**23.** $\frac{7}{5} \cdot \frac{5}{7}$	**24.** $\frac{2}{11} \cdot \frac{11}{2}$	**25.** $\frac{4}{3} \cdot \frac{3}{4}$
26. $\frac{1}{4} \times 8$	**27.** $\frac{1}{6} \times 12$	**28.** $15 \times \frac{1}{3}$	**29.** $14 \times \frac{1}{2}$	**30.** $16 \times \frac{1}{8}$
31. $12 \times \frac{3}{4}$	**32.** $18 \times \frac{5}{6}$	**33.** $\frac{3}{8} \times 24$	**34.** $\frac{2}{9} \times 36$	**35.** $\frac{2}{3} \times 21$
36. $13 \times \frac{2}{5}$	**37.** $15 \times \frac{1}{6}$	**38.** $\frac{7}{10} \times 28$	**39.** $\frac{5}{8} \times 34$	**40.** $\frac{3}{4} \times 26$
41. $\frac{1}{6} \times 360$	**42.** $\frac{1}{3} \times 120$	**43.** $240 \times \frac{1}{8}$	**44.** $150 \times \frac{1}{5}$	**45.** $320 \times \frac{1}{4}$
46. $\frac{4}{10} \cdot \frac{5}{10}$	**47.** $\frac{7}{10} \cdot \frac{34}{150}$	**48.** $\frac{8}{10} \cdot \frac{45}{100}$	**49.** $\frac{3}{10} \cdot \frac{8}{10}$	**50.** $\frac{6}{10} \cdot \frac{5}{10}$
51. $\frac{11}{24} \cdot \frac{3}{5}$	**52.** $\frac{15}{22} \cdot \frac{4}{7}$	**53.** $\frac{10}{21} \cdot \frac{3}{4}$	**54.** $\frac{17}{18} \cdot \frac{3}{5}$	**55.** $\frac{15}{16} \cdot \frac{2}{7}$

PROBLEM CORNER

1. How much steak will be needed to serve 30 people if each person gets $\frac{2}{3}$ lb?

2. A gardener uses $\frac{2}{5}$ lb of peat moss for a rose bush. How much would be needed for 25 rose bushes?

More Practice, page 190

MULTIPLYING WITH MIXED NUMBERS

Multiply and simplify. $6 \times 2\frac{1}{2}$

Convert to fractions. → Multiply. Simplify.

$$6 \times 2\frac{1}{2} = \frac{6}{1} \times \frac{5}{2} = \frac{30}{2} = 15$$

EXAMPLE 1. Multiply and simplify. $4\frac{2}{3} \cdot 8$

$$4\frac{2}{3} \cdot 8 = \frac{14}{3} \cdot \frac{8}{1} = \frac{112}{3} = 37\frac{1}{3}$$

Convert to fractions.
Multiply. Simplify.

TRY THIS

Multiply and simplify.

1. $9 \times 4\frac{1}{3}$ **2.** $6\frac{2}{5} \cdot 2$ **3.** $3\frac{5}{7} \times 3$

EXAMPLE 2. Multiply and simplify. $3\frac{1}{2} \cdot \frac{3}{4}$

$$3\frac{1}{2} \cdot \frac{3}{4} = \frac{7}{2} \cdot \frac{3}{4} = \frac{21}{8} = 2\frac{5}{8}$$

Convert the mixed number to a fraction.

TRY THIS

Multiply and simplify.

4. $4\frac{1}{8} \cdot \frac{2}{3}$ **5.** $\frac{5}{6} \cdot 2\frac{1}{5}$ **6.** $3\frac{3}{4} \times \frac{2}{7}$

EXAMPLE 3. Multiply and simplify. $2\frac{3}{4} \cdot 3\frac{2}{5}$

$$2\frac{3}{4} \cdot 3\frac{2}{5} = \frac{11}{4} \cdot \frac{17}{5} = \frac{187}{20} = 9\frac{7}{20}$$

Convert the mixed numbers to fractions.

TRY THIS

Multiply and simplify.

7. $3\frac{1}{3} \cdot 2\frac{1}{2}$ **8.** $4\frac{1}{6} \cdot 2\frac{3}{5}$ **9.** $3\frac{4}{7} \times 3\frac{1}{2}$

PRACTICE

Multiply and simplify.

1. $9 \times 2\frac{1}{3}$
2. $8 \times 3\frac{1}{5}$
3. $4\frac{1}{2} \times 10$
4. $5\frac{1}{7} \times 4$
5. $3\frac{1}{5} \times \frac{2}{5}$
6. $3\frac{1}{8} \times \frac{2}{3}$
7. $1\frac{5}{8} \times \frac{2}{3}$
8. $6\frac{2}{3} \times \frac{1}{4}$
9. $6 \cdot 5\frac{2}{3}$
10. $4 \cdot 3\frac{5}{8}$
11. $2\frac{5}{6} \cdot 8$
12. $5 \cdot 3\frac{3}{4}$
13. $2\frac{1}{2} \cdot 3\frac{2}{3}$
14. $3\frac{1}{2} \cdot 2\frac{1}{3}$
15. $4\frac{1}{4} \cdot 3\frac{1}{2}$
16. $1\frac{1}{4} \cdot 3\frac{1}{2}$
17. $1\frac{3}{4} \cdot 1\frac{1}{2}$
18. $1\frac{3}{4} \cdot 2\frac{1}{3}$
19. $3\frac{2}{5} \cdot 5\frac{1}{3}$
20. $3\frac{3}{4} \cdot 2\frac{1}{8}$
21. $5\frac{3}{4} \times 1\frac{1}{3}$
22. $3\frac{2}{5} \times 1\frac{1}{4}$
23. $1\frac{3}{5} \times 3\frac{1}{3}$
24. $3\frac{1}{6} \times 2\frac{3}{4}$
25. $2\frac{2}{5} \cdot 2\frac{7}{8}$
26. $2\frac{1}{10} \cdot 6\frac{2}{3}$
27. $1\frac{3}{7} \cdot 2\frac{2}{5}$
28. $2\frac{3}{10} \cdot 4\frac{3}{5}$
29. $10\frac{1}{4} \cdot 50$
30. $20\frac{1}{3} \cdot 60$
31. $40 \cdot 30\frac{1}{6}$
32. $20 \cdot 80\frac{1}{10}$
33. $5\frac{1}{4} \times 28$
34. $40 \times 3\frac{5}{7}$
35. $16 \times 4\frac{7}{8}$
36. $20 \times 3\frac{2}{3}$
37. $10\frac{1}{2} \cdot 10\frac{1}{3}$
38. $10\frac{1}{4} \cdot 20\frac{1}{5}$
39. $20\frac{1}{2} \cdot 20\frac{1}{4}$
40. $20\frac{1}{2} \cdot 10\frac{1}{5}$

PROBLEM CORNER

1. A long-playing record makes $33\frac{1}{3}$ revolutions per minute. It plays 12 minutes. How many revolutions does it make?
2. A car travels on an interstate highway at 55 miles per hour for $3\frac{1}{2}$ hr. How far does it go?

More Practice, page 191

APPLICATION

BUDGETING

A budget is a plan which shows how money is to be spent and saved. Robert Carlson is planning his expenses. He has $ 2580 in his savings account and $ 868 in his checking account.

Income		**Expenses**	
Regular Job	$6500	Rent and Utilities	$3000
Part-time Job	$ 780	Food and Supplies	$2400
		Transportation	$1200
		Insurance	$ 500
		Personal	$1100

1. What is his total income?
2. What are his total expenses?
3. Will he have to use his savings account to pay his expenses?
4. What part of his expenses is for food?
5. If Robert has a roommate, he can save $\frac{1}{2}$ of his expenses for rent and utilities. How much can he save?
6. If Robert has a roommate, he can also save $\frac{3}{10}$ of his food and supplies expense. How much can he save?
7. How much in all can he save by having a roommate?

The federal government budgets how it spends its money. The graph shows how each government dollar was spent in a previous year. What part of the dollar went for each expense?

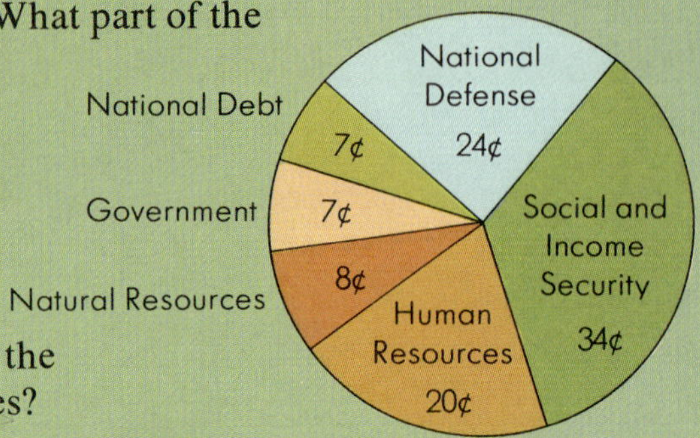

8. the national debt
9. national defense
10. human resources
11. social and income security
12. How much is spent annually by the government for natural resources?

Janet Thompson's budget for volleyball was $ 1000 for the season. The following expenses were estimated at the beginning of the year.

Equipment:	5 balls at $10 each
Uniforms:	24 of each: T-shirts, $3; shorts $5; pair of socks, $2
Towel Service:	10 weeks at $7.50 per week
Transportation:	$50 bus fee for each of 10 away games
Awards:	Remainder of budget money

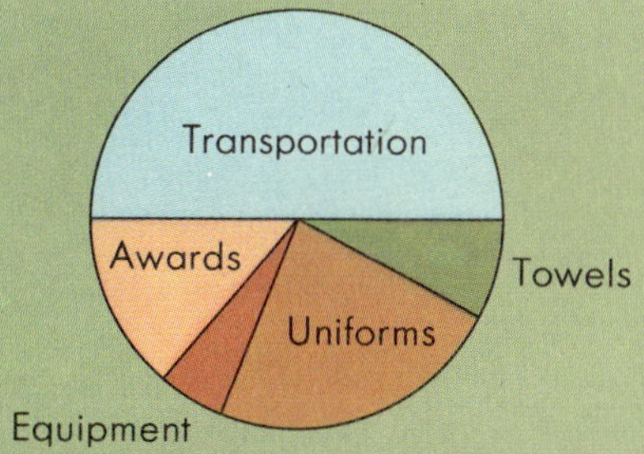

How much money is allowed for each budget item?

13. equipment 14. uniforms 15. towels 16. transportation

What part of the budget went for each of the following?

17. equipment 18. uniforms 19. towels 20. transportation
21. How much money was left for awards?
22. What part of the budget was left for awards?

The income from 10 games is estimated to be $ 184 per game. Of this, $\frac{2}{3}$ goes to the general athletic fund, $\frac{1}{4}$ goes for administration, and the rest goes into the volleyball award fund.

23. What is the total income from the games?
24. How much money goes to the general athletic fund?
25. How much money goes for administration?
26. How much money goes into the volleyball award fund?

RECIPROCALS AND DIVIDING

To find the reciprocal of a fraction, we interchange the numerator and denominator.

$\frac{2}{3} \rightarrow \frac{3}{2}$

EXAMPLES. Find the reciprocals.

1. $\frac{5}{4}$ The reciprocal of $\frac{5}{4}$ is $\frac{4}{5}$.

2. $\frac{1}{3}$ The reciprocal of $\frac{1}{3}$ is $\frac{3}{1}$, or 3.

3. 10. Think of 10 as $\frac{10}{1}$.
The reciprocal of 10 is $\frac{1}{10}$.

TRY THIS

Find the reciprocal.

1. $\frac{2}{5}$ **2.** $\frac{5}{2}$ **3.** $\frac{1}{5}$ **4.** 24 **5.** 16

Divide and simplify. $\frac{6}{5} \div \frac{3}{4}$

Find the reciprocal of the divisor. → Multiply by the reciprocal of the divisor. → Simplify.

The reciprocal of $\frac{3}{4}$ is $\frac{4}{3}$.

$$\frac{6}{5} \div \frac{3}{4} = \frac{6}{5} \cdot \frac{4}{3} = \frac{24}{15} = \frac{24 \div 3}{15 \div 3} = \frac{8}{5} = 1\frac{3}{5}$$

EXAMPLE 4. Divide and simplify. $\frac{5}{6} \div \frac{3}{2}$

$$\frac{5}{6} \div \frac{3}{2} = \frac{5}{6} \cdot \frac{2}{3} = \frac{10}{18} = \frac{10 \div 2}{18 \div 2} = \frac{5}{9}$$

The reciprocal of $\frac{3}{2}$ is $\frac{2}{3}$.

TRY THIS

Divide and simplify.

6. $\frac{6}{7} \div \frac{3}{4}$ **7.** $\frac{2}{3} \div \frac{1}{4}$ **8.** $\frac{4}{5} \div 8$ **9.** $\frac{2}{3} \div 6$

PRACTICE

Find the reciprocal.

1. $\frac{5}{6}$	**2.** $\frac{3}{5}$	**3.** $\frac{6}{9}$	**4.** $\frac{3}{8}$	**5.** $\frac{3}{4}$
6. 6	**7.** 2	**8.** 3	**9.** 5	**10.** 7
11. $\frac{1}{3}$	**12.** $\frac{1}{2}$	**13.** $\frac{1}{6}$	**14.** $\frac{1}{4}$	**15.** $\frac{1}{5}$

Divide and simplify.

16. $\frac{3}{5} \div \frac{3}{4}$	**17.** $\frac{2}{3} \div \frac{3}{4}$	**18.** $\frac{3}{5} \div \frac{9}{4}$	**19.** $\frac{6}{7} \div \frac{3}{5}$	**20.** $\frac{5}{6} \div \frac{1}{2}$
21. $\frac{4}{3} \div \frac{1}{3}$	**22.** $\frac{1}{3} \div \frac{1}{4}$	**23.** $\frac{9}{8} \div \frac{1}{2}$	**24.** $\frac{10}{9} \div \frac{1}{3}$	**25.** $\frac{8}{7} \div \frac{2}{7}$
26. $\frac{6}{5} \div 3$	**27.** $\frac{6}{5} \div 2$	**28.** $\frac{12}{7} \div 4$	**29.** $\frac{8}{7} \div 2$	**30.** $\frac{8}{9} \div 4$
31. $10 \div \frac{2}{3}$	**32.** $8 \div \frac{5}{4}$	**33.** $12 \div \frac{3}{2}$	**34.** $24 \div \frac{3}{8}$	**35.** $6 \div \frac{5}{6}$
36. $\frac{2}{3} \div \frac{2}{3}$	**37.** $\frac{3}{4} \div \frac{3}{4}$	**38.** $\frac{3}{8} \div \frac{3}{8}$	**39.** $\frac{2}{5} \div \frac{2}{5}$	**40.** $\frac{4}{7} \div \frac{4}{7}$
41. $\frac{3}{4} \div 1$	**42.** $\frac{2}{5} \div 1$	**43.** $\frac{3}{10} \div 1$	**44.** $\frac{5}{12} \div 1$	**45.** $\frac{8}{9} \div 1$
46. $\frac{8}{15} \div \frac{4}{5}$	**47.** $\frac{6}{13} \div \frac{3}{26}$	**48.** $\frac{9}{5} \div \frac{8}{10}$	**49.** $\frac{101}{5} \div 10$	**50.** $\frac{15}{6} \div 10$

PROBLEM CORNER

1. How many test tubes, each containing $\frac{3}{5}$ milliliter, can be filled from a container of 60 milliliters?

2. Each loop in a spring takes $\frac{3}{8}$ in. of wire. How many loops can be made from 120 in. of wire?

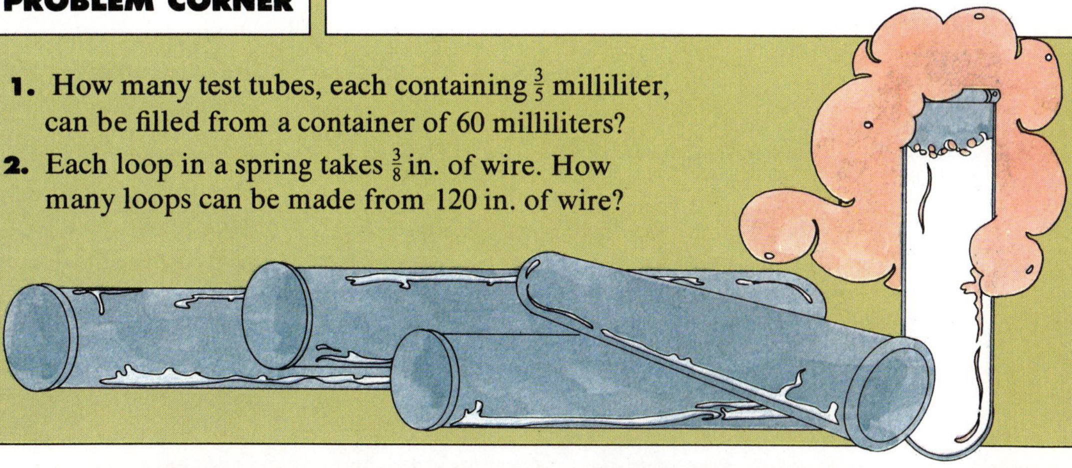

More Practice, page 191

DIVIDING WITH MIXED NUMBERS

Divide. $1\frac{1}{2} \div \frac{1}{6}$

Convert the mixed number to a fraction. → Multiply by the reciprocal of the divisor. → Simplify.

$$1\frac{1}{2} \div \frac{1}{6} = \frac{3}{2} \div \frac{1}{6} \quad = \quad \frac{3}{2} \cdot \frac{6}{1} \quad = \quad \frac{18}{2} = 9$$

EXAMPLE 1. Divide and simplify. $\frac{7}{8} \div 2\frac{1}{2}$

$$\frac{7}{8} \div 2\frac{1}{2} = \frac{7}{8} \div \frac{5}{2} = \frac{7}{8} \cdot \frac{2}{5} = \frac{14}{40} = \frac{7}{20}$$

Convert the mixed number to a fraction. Multiply by the reciprocal of the divisor. Simplify.

TRY THIS

Divide and simplify.

1. $2\frac{1}{3} \div \frac{3}{4}$ **2.** $\frac{9}{10} \div 3\frac{2}{5}$ **3.** $3\frac{1}{4} \div \frac{5}{6}$

EXAMPLE 2. Divide and simplify. $35 \div 4\frac{1}{3}$

$$35 \div 4\frac{1}{3} = \frac{35}{1} \div \frac{13}{3} = \frac{35}{1} \cdot \frac{3}{13} = \frac{105}{13} = 8\frac{1}{13}$$

Convert each number to a fraction. Multiply by the reciprocal of the divisor. Simplify.

EXAMPLE 3. Divide and simplify. $2\frac{1}{3} \div 1\frac{3}{4}$

$$2\frac{1}{3} \div 1\frac{3}{4} = \frac{7}{3} \div \frac{7}{4} = \frac{7}{3} \cdot \frac{4}{7} = \frac{28}{21} = 1\frac{1}{3}$$

Convert each mixed number to a fraction. Multiply by the reciprocal of the divisor. Simplify.

TRY THIS

Divide and simplify.

4. $3\frac{1}{5} \div 32$ **5.** $1\frac{3}{5} \div 3\frac{1}{3}$ **6.** $2\frac{2}{3} \div 1\frac{1}{5}$

PRACTICE

Divide and simplify.

1. $6 \div 4\frac{1}{2}$	**2.** $20 \div 3\frac{1}{5}$	**3.** $9 \div 2\frac{1}{4}$	**4.** $18 \div 2\frac{1}{4}$
5. $42 \div 3\frac{1}{5}$	**6.** $34 \div 5\frac{1}{2}$	**7.** $63 \div 2\frac{1}{3}$	**8.** $52 \div 3\frac{1}{2}$
9. $4\frac{3}{4} \div 1\frac{1}{3}$	**10.** $2\frac{1}{2} \div 1\frac{1}{4}$	**11.** $5\frac{4}{5} \div 2\frac{1}{2}$	**12.** $3\frac{1}{2} \div 2\frac{2}{3}$
13. $1\frac{7}{8} \div 1\frac{2}{3}$	**14.** $7\frac{1}{3} \div 1\frac{1}{9}$	**15.** $6\frac{7}{8} \div 1\frac{2}{3}$	**16.** $4\frac{3}{8} \div 2\frac{5}{6}$
17. $2\frac{2}{3} \div 3\frac{1}{5}$	**18.** $3\frac{1}{3} \div 5\frac{1}{2}$	**19.** $2\frac{1}{2} \div 2\frac{2}{3}$	**20.** $3\frac{1}{3} \div 5\frac{1}{2}$
21. $\frac{3}{5} \div 1\frac{1}{3}$	**22.** $\frac{2}{3} \div 3\frac{1}{4}$	**23.** $\frac{1}{2} \div 2\frac{1}{3}$	**24.** $\frac{5}{8} \div 3\frac{1}{3}$
25. $2\frac{1}{7} \div 5$	**26.** $5\frac{1}{4} \div 6$	**27.** $8\frac{2}{5} \div 7$	**28.** $3\frac{3}{8} \div 3$
29. $3\frac{1}{4} \div \frac{1}{2}$	**30.** $4\frac{1}{5} \div 10$	**31.** $5\frac{1}{2} \div 22$	**32.** $4\frac{1}{3} \div 26$
33. $125 \div 12\frac{1}{2}$	**34.** $86 \div 14\frac{1}{4}$	**35.** $84 \div 8\frac{2}{5}$	**36.** $130 \div 9\frac{3}{8}$
37. $10\frac{1}{3} \div 3\frac{2}{5}$	**38.** $3\frac{1}{5} \div 1\frac{1}{3}$	**39.** $12\frac{1}{3} \div 6\frac{7}{8}$	**40.** $11\frac{1}{4} \div 2\frac{1}{2}$

PROBLEM CORNER

1. A truck carries $18\frac{1}{3}$ tons of sand. How many trips must the truck make to carry 550 tons?

2. The weight of water is $62\frac{1}{2}$ lb per cubic foot. How many cubic feet would be occupied by 250 lb of water?

More Practice, page 191

APPLICATION

FOOD AND NUTRITION

Our bodies need food for energy, growth, and maintenance of health. The following is a table showing the Recommended Daily Allowances (RDA) of calories and other nutrients.

RDA for a 15-Year-Old to 18-Year-Old Male

Calories	Protein	Niacin	Riboflavin	Thiamin	Magnesium	Iodine
3000	54 g	20 mg	1.8 mg	1.5 mg	400 mg	150 micrograms

The RDA for a 15-year-old to 18-year-old female is found by multiplying by the fraction shown after each of the items.

Tammy wants to find how many calories she should have in her diet.

calories $\left(\frac{7}{10}\right) \longrightarrow 3000 \times \frac{7}{10} = 2100$ calories

Find each RDA.

1. Protein $\left(\frac{8}{9}\right)$
2. Niacin $\left(\frac{7}{10}\right)$
3. Magnesium $\left(\frac{3}{4}\right)$
4. Iodine $\left(\frac{23}{30}\right)$
5. Riboflavin $\left(\frac{7}{9}\right)$
6. Thiamin $\left(\frac{11}{15}\right)$

Jesse compares some of the nutrients in his daily diet to the RDA.

	Calories	Protein	Vitamin C	Calcium	Iron
RDA	3000	54 g	45 mg	1200 mg	18 mg
Jesse's Diet	4200	46 g	20 mg	1600 mg	12 mg

He divides to find the part of the RDA in his diet.

$4200 \div 3000 = 1\frac{2}{5}$ of the RDA of calories

What part of the RDA of each nutrient does Jesse consume daily?

7. Protein
8. Vitamin C
9. Calcium
10. Iron

Carbohydrates and calories are important factors in our diets.

Food	**Unit of Measure**	**Carbohydrates**	**Calories**
Broccoli	1 cup	8.2	44
Spaghetti	1 cup	60	308
Bread, whole wheat	1 slice	11.3	55
Chicken	5 oz	0	159
Brown rice	$\frac{1}{2}$ cup	26.9	100
Milk	1 cup	12	160
Apple juice	1 cup	34.4	124
Corn	1 cup	33.3	140
Ice cream	$\frac{1}{2}$ cup	14.6	145
Peaches	1 cup	12	46

11. Wednesday Verne ate 2 cups of spaghetti and Elise ate $1\frac{1}{2}$ cups. How many more calories did Verne eat than Elise? carbohydrates?
12. Yoko had $1\frac{1}{2}$ cups of peaches for dessert, while Earl had $1\frac{1}{2}$ cups of ice cream. How many fewer calories did Yoko eat than Earl? carbohydrates?

For dinner Irene had 5 ounces of chicken, 1 cup of brown rice, 1 cup of corn, and $1\frac{1}{2}$ cups of milk. Yuki had $1\frac{1}{2}$ cups of spaghetti, 2 slices of bread, 1 cup of broccoli, and 2 cups of apple juice.

13. How many calories were in Irene's dinner? Yuki's dinner?
14. How many carbohydrates were in Irene's dinner? Yuki's dinner?

Carbohydrates provide energy for the muscles in the body. The day before the track meet the track team members want to eat foods high in carbohydrates for energy.

15. Does 1 cup of brown rice or 2 slices of bread contain more carbohydrates?
16. To receive more carbohydrates should the members eat corn or broccoli?
17. Should the team members eat chicken or spaghetti?
18. Does 1 cup of milk or 1 cup of apple juice contain more carbohydrates? How much more?

HOME ECONOMIST

Lisa Hutchins is a home economist. She teaches day and evening classes at an adult education center.

Travis and Emma Johnson need to cut expenses in order to save more money. Lisa shows the Johnsons how to cut out wasteful spending. She helps them plan a new budget for food, personal items, housing, and transportation.

How much will Travis and Emma save given the following monthly expenses?

1. food: $234
2. personal items: $116
3. housing: $364
4. transportation: $80

How much will Travis and Emma spend for the following expenses in their new budget?

5. food
6. personal items
7. housing
8. transportation

Food planning is an important part of home economy. Lisa helps Roger Sherman plan how much food he needs in order to serve a group of people. Roger plans to use $\frac{3}{8}$ pound of ground beef per serving. How many pounds of ground beef are needed to serve the following?

9. 4 people
10. 2 people
11. 10 people
12. 5 people

Sometimes it is necessary to change a recipe in order to serve more people or fewer people. Lisa's cooking class changes the Banana Nut Bread recipe. Then each person bakes a loaf of bread according to the new recipe.

Banana Nut Bread	Serves 6
$1\frac{3}{4}$ C sifted flour	$\frac{2}{3}$ C sugar
$2\frac{3}{4}$ tsp baking powder	2 eggs
$\frac{1}{2}$ tsp salt	4 ripe bananas
$\frac{1}{3}$ C shortening	$\frac{1}{2}$ C nuts

13. How much flour is used to make Banana Nut Bread for 3 people? 8 people?
14. To make bread for 12 people, how much sugar is needed? How many bananas?
15. What amount of nuts is used to make bread for 4 people?

In a sewing class, Melba Cole wants to make pillows for her living room. Lisa helps Melba find how much one pillow will cost. Then Melba will decide how many pillows she can afford.

16. Find the total cost to make 1 pillow.
17. How much does it cost to make 4 pillows?
18. How many pillows can be made for $ 20?

	Amount per pillow	Unit Cost
Fabric	$\frac{1}{2}$ yd	$6.98 per yd
Trim	1 yd	$1.99 per yd
Filling	$\frac{1}{3}$ lb	$1.20 per lb

CHAPTER 6 REVIEW

Multiply. p. 168

1. $\frac{1}{3} \cdot \frac{1}{4}$ **2.** $\frac{3}{5} \cdot \frac{2}{5}$ **3.** $5 \times \frac{4}{7}$ **4.** $\frac{3}{4} \times 9$

5. Find a fraction equivalent to $\frac{7}{8}$ with a denominator of 40. p. 170 $\frac{35}{40}$

All the fractions are equivalent. Which is simplest? p. 172

6. $\frac{4}{10}, \frac{6}{15}, \frac{2}{5}, \frac{8}{20}, \frac{12}{30}$ **7.** $\frac{3}{18}, \frac{5}{30}, \frac{2}{12}, \frac{4}{24}, \frac{1}{6}$ **8.** $\frac{10}{15}, \frac{18}{27}, \frac{6}{9}, \frac{2}{3}, \frac{12}{18}$

Simplify. p. 172

9. $\frac{12}{24}$ **10.** $\frac{42}{7}$ **11.** $\frac{2}{18}$ **12.** $\frac{3}{12}$

Multiply and simplify. pp. 174–176

13. $\frac{4}{3} \cdot 24$ **14.** $\frac{3}{5} \cdot \frac{1}{6}$ **15.** $\frac{2}{3} \cdot \frac{15}{4}$ **16.** $14 \cdot \frac{3}{4}$

17. $10 \times 8\frac{4}{5}$ **18.** $7\frac{1}{2} \cdot \frac{4}{5}$ **19.** $8\frac{1}{3} \cdot 5\frac{1}{4}$ **20.** $1\frac{1}{4} \cdot 3\frac{1}{2}$

Find the reciprocal. p. 180

21. $\frac{1}{4}$ **22.** 18 **23.** $\frac{5}{8}$ **24.** $\frac{5}{4}$

Divide and simplify. pp. 180–182

25. $\frac{3}{8} \div \frac{5}{4}$ **26.** $\frac{1}{5} \div \frac{1}{8}$ **27.** $12 \div \frac{2}{3}$ **28.** $\frac{9}{8} \div \frac{1}{2}$

29. $20 \div 5\frac{1}{3}$ **30.** $7\frac{1}{2} \div 5\frac{1}{4}$ **31.** $7\frac{1}{2} \div 30$ **32.** $2\frac{3}{8} \div 2$

Application pp. 178, 184

33. The government spends 7¢ of each dollar for the national debt. What part of the dollar is used to pay for the national debt?

34. Susan needs $\frac{8}{9}$ as much protein in her diet as Jack. Jack should have 54 grams of protein each day. How much protein should Susan have?

CHAPTER 6 TEST

Multiply.

1. $\frac{1}{2} \cdot \frac{1}{6}$ **2.** $\frac{3}{5} \cdot \frac{4}{7}$ **3.** $8 \times \frac{2}{3}$

4. Find a fraction equivalent to $\frac{5}{6}$ with a denominator of 12.

All the fractions are equivalent. Which is simplest?

5. $\frac{4}{6}, \frac{6}{9}, \frac{8}{12}, \frac{2}{3}, \frac{10}{15}$ **6.** $\frac{9}{12}, \frac{3}{4}, \frac{6}{8}, \frac{15}{20}, \frac{12}{16}$

Simplify.

7. $\frac{12}{18}$ **8.** $\frac{3}{15}$ **9.** $\frac{54}{6}$

Multiply and simplify.

10. $\frac{2}{5} \cdot \frac{5}{8}$ **11.** $\frac{3}{4} \times 36$ **12.** $\frac{2}{3} \cdot \frac{9}{16}$

13. $20 \times 5\frac{1}{3}$ **14.** $5\frac{1}{2} \cdot \frac{4}{5}$ **15.** $9\frac{3}{5} \cdot 2\frac{1}{4}$

Find the reciprocal.

16. 16 **17.** $\frac{7}{9}$ **18.** $\frac{1}{20}$

Divide and simplify.

19. $36 \div \frac{1}{2}$ **20.** $\frac{1}{4} \div \frac{2}{3}$ **21.** $\frac{1}{8} \div \frac{1}{6}$

22. $10 \div 8\frac{4}{5}$ **23.** $4\frac{1}{3} \div 6\frac{1}{2}$ **24.** $7\frac{1}{3} \div 44$

Application

25. The Arrow Soccer Club spends $1000 for uniforms each year. Their total budget is $3500. What part of the total budget does the club spend for uniforms?

26. There are 35 calories in $\frac{1}{2}$ cup of red raspberries. After school Janet ate $\frac{1}{2}$ cup of raspberries, and Jason ate $1\frac{1}{2}$ cups. How many more calories did Jason eat than Janet?

MORE PRACTICE

Multiply. Use after page 169.

1. $\frac{1}{6} \cdot \frac{1}{3}$	**2.** $\frac{2}{3} \cdot \frac{1}{5}$	**3.** $\frac{3}{4} \cdot \frac{5}{7}$	**4.** $9 \times \frac{5}{8}$
5. $\frac{2}{5} \times \frac{4}{9}$	**6.** $\frac{7}{10} \times \frac{7}{8}$	**7.** $\frac{6}{11} \cdot 6$	**8.** $3 \times \frac{9}{16}$

Find a fraction equivalent to the given one, but with the indicated denominator. Use after page 171.

9. $\frac{1}{2}$ [8]	**10.** $\frac{2}{3}$ [12]	**11.** $\frac{2}{3}$ [15]	**12.** $\frac{9}{10}$ [100]
13. $\frac{7}{8}$ [48]	**14.** $\frac{7}{12}$ [60]	**15.** $\frac{13}{16}$ [80]	**16.** $\frac{11}{32}$ [96]
17. $\frac{3}{4}$ [28]	**18.** $\frac{9}{16}$ [48]	**19.** $\frac{5}{14}$ [42]	**20.** $\frac{7}{8}$ [56]

Simplify. Use after page 173.

21. $\frac{2}{4}$	**22.** $\frac{6}{18}$	**23.** $\frac{63}{72}$	**24.** $\frac{9}{12}$
25. $\frac{23}{46}$	**26.** $\frac{6}{54}$	**27.** $\frac{54}{81}$	**28.** $\frac{62}{93}$
29. $\frac{425}{575}$	**30.** $\frac{96}{84}$	**31.** $\frac{32}{20}$	**32.** $\frac{48}{36}$

Multiply and simplify. Use after page 175.

33. $\frac{3}{4} \cdot \frac{1}{3}$	**34.** $\frac{2}{5} \cdot \frac{5}{2}$	**35.** $\frac{3}{5} \times \frac{1}{6}$	**36.** $\frac{4}{5} \times \frac{15}{16}$
37. $6 \times \frac{7}{8}$	**38.** $20 \times \frac{1}{5}$	**39.** $40 \times \frac{3}{4}$	**40.** $17 \times \frac{2}{5}$
41. $9 \times \frac{1}{9}$	**42.** $56 \cdot \frac{1}{8}$	**43.** $\frac{7}{10} \cdot \frac{8}{10}$	**44.** $\frac{9}{10} \cdot \frac{2}{3}$
45. $\frac{16}{25} \times \frac{5}{8}$	**46.** $\frac{2}{5} \times 125$	**47.** $\frac{7}{8} \times 64$	**48.** $\frac{9}{7} \times \frac{2}{3}$
49. $\frac{15}{8} \cdot \frac{4}{3}$	**50.** $\frac{5}{6} \times \frac{13}{5}$	**51.** $\frac{8}{9} \cdot \frac{3}{5}$	**52.** $\frac{7}{2} \times \frac{5}{14}$

Multiply and simplify. Use after page 177.

1. $8 \times 3\frac{1}{4}$ **2.** $5 \times 2\frac{1}{2}$ **3.** $3\frac{1}{4} \cdot \frac{2}{5}$ **4.** $5\frac{2}{3} \cdot \frac{5}{6}$

5. $4\frac{1}{3} \cdot 5\frac{1}{2}$ **6.** $4\frac{1}{4} \cdot 6\frac{3}{8}$ **7.** $9\frac{3}{10} \cdot 12$ **8.** $1\frac{3}{7} \times 2\frac{5}{8}$

9. $3\frac{1}{2} \cdot 3\frac{1}{2}$ **10.** $2\frac{1}{3} \cdot 2\frac{1}{3}$ **11.** $5\frac{3}{4} \times 5\frac{3}{4}$ **12.** $3\frac{2}{5} \times \frac{2}{17}$

13. $5\frac{2}{3} \times 8$ **14.** $3\frac{1}{2} \cdot 4\frac{2}{3}$ **15.** $\frac{1}{2} \cdot 4\frac{5}{8}$ **16.** $6\frac{1}{10} \cdot 1\frac{3}{8}$

17. $\frac{3}{4} \cdot 2\frac{1}{2}$ **18.** $1\frac{5}{7} \times 1\frac{5}{9}$ **19.** $8\frac{3}{4} \times 10$ **20.** $2\frac{1}{3} \times 3\frac{2}{5}$

Find the reciprocal. Use after page 181.

21. $\frac{7}{8}$ **22.** $\frac{9}{10}$ **23.** $\frac{1}{12}$ **24.** $\frac{11}{3}$

25. 23 **26.** $\frac{1}{37}$ **27.** $\frac{23}{16}$ **28.** 135

Divide and simplify. Use after page 181.

29. $\frac{1}{2} \div \frac{1}{4}$ **30.** $\frac{1}{3} \div 6$ **31.** $\frac{1}{3} \div \frac{1}{5}$ **32.** $\frac{7}{8} \div 7$

33. $12 \div \frac{2}{3}$ **34.** $\frac{7}{8} \div \frac{7}{8}$ **35.** $\frac{3}{4} \div \frac{5}{6}$ **36.** $\frac{2}{3} \div \frac{4}{9}$

37. $\frac{4}{5} \div \frac{5}{4}$ **38.** $8 \div \frac{2}{9}$ **39.** $\frac{6}{5} \div \frac{7}{6}$ **40.** $\frac{5}{8} \div \frac{3}{2}$

Divide and simplify. Use after page 183.

41. $1\frac{1}{2} \div 3$ **42.** $9 \div 4\frac{1}{2}$ **43.** $1\frac{3}{5} \div \frac{2}{3}$ **44.** $3\frac{1}{4} \div \frac{5}{6}$

45. $16\frac{1}{2} \div 4\frac{1}{4}$ **46.** $5\frac{1}{4} \div 10$ **47.** $2\frac{1}{10} \div 1\frac{1}{5}$ **48.** $100 \div 33\frac{1}{3}$

49. $10\frac{1}{2} \div 5\frac{1}{2}$ **50.** $14\frac{1}{2} \div 3\frac{5}{8}$ **51.** $30\frac{1}{4} \div 121$ **52.** $15\frac{3}{4} \div 9\frac{1}{8}$

53. $6\frac{3}{5} \div 4$ **54.** $6\frac{5}{6} \div \frac{3}{4}$ **55.** $3\frac{2}{5} \div 2\frac{4}{7}$ **56.** $\frac{1}{2} \div 2\frac{2}{3}$

CALCULATOR CORNER

GET 100

Players one or more
Materials one calculator

* Suppose a calculator has 127 on the display.
How many ways can we get 100 on the display?

Try these on your calculator.

1. Press [−] [2] [7] [=]
2. Press [−] [1] [0] [−] [1] [0] [−] [7] [=]
3. Press [−] [1] [2] [7] [+] [1] [0] [0] [=]

Start with 127. Find 5 more ways to get 100 on the display. Write down your steps.

**Suppose 127 is still on the display. This time we will get 100 on the display but must press [×] at least once.

Try these on your calculator.

4. Press [×] [2] [−] [1] [2] [7] [−] [2] [7] [=]
5. Press [×] [1] [−] [2] [7] [=]

Start with 127. Find 5 more ways to get 100 on the display. You must press [×] at least once. Write down your steps.

EXERCISES Write down the steps for each of the following.

1. Start with 154. Find 10 ways to get 100 on the display.
2. Start with 136. Find 10 ways to get 100 on the display. You must press [×] at least once.
3. Start with 182. Find 10 ways to get 100 on the display. You must press [÷] at least once.

READY FOR CHAPTER 7

Add and simplify.

1. $\frac{1}{2} + \frac{5}{2}$ **2.** $\frac{3}{8} + \frac{2}{8}$ **3.** $\frac{5}{12} + \frac{11}{12} + \frac{2}{12}$

Find the LCD.

4. $\frac{5}{12}, \frac{7}{16}$ **5.** $\frac{1}{3}, \frac{2}{5}$ **6.** $\frac{1}{2}, \frac{1}{4}, \frac{1}{3}$

Add and simplify.

7. $\frac{7}{8} + \frac{2}{3}$ **8.** $\frac{7}{10} + \frac{9}{100}$ **9.** $\frac{1}{2} + \frac{3}{4} + \frac{5}{6}$

Add and simplify.

10. $6\frac{2}{5} + 7\frac{4}{5}$ **11.** $12\frac{5}{8} + 7$ **12.** $9\frac{1}{4} + 5\frac{1}{6}$

13. $10\frac{1}{8} + 5\frac{1}{2} + 7\frac{3}{4}$ **14.** $\frac{3}{8} + 2\frac{5}{6}$ **15.** $7\frac{1}{12} + 4\frac{3}{4}$

Subtract and simplify.

16. $\frac{5}{6} - \frac{3}{6}$ **17.** $\frac{5}{6} - \frac{3}{4}$ **18.** $\frac{17}{24} - \frac{5}{8}$

Subtract and simplify.

19. $7\frac{5}{8} - 5\frac{1}{8}$ **20.** $15\frac{3}{4} - 9$ **21.** $7\frac{5}{6} - 2\frac{1}{3}$

22. $10\frac{1}{6} - 5\frac{7}{8}$ **23.** $14 - 7\frac{5}{6}$ **24.** $16\frac{1}{3} - 7\frac{2}{3}$

Application

25. George drank the following amounts of water one day. How much did he drink in all?

Home	School	Restaurant
$4\frac{1}{2}$ cups	$\frac{3}{4}$ cup	$1\frac{3}{4}$ cups

26. A carpenter cuts $2\frac{3}{4}$ ft off a board which is $7\frac{1}{2}$ ft long. How much is left?

7
ADDING AND SUBTRACTING FRACTIONS

GETTING STARTED

FRACTION ADDO

WHAT'S NEEDED The 30 squares of paper with digits marked on them as in GOTCHA, Chapter 1.

THE RULES

1. On a sheet of paper each player writes this:

   $$\frac{[\ ]}{[\ ]} + \frac{[\ ]}{[\ ]}$$

2. To play, a leader draws digits from the box, one at a time.
3. Each player writes the digit in any one of the four spaces. Once it has been written, it cannot be changed.
4. The leader continues to draw digits until four digits have been drawn.
5. Each player adds the resulting fractions. The winner is the player who has the largest sum.

Sample

Louis	**Chris**	**Sally**
$\frac{[2]}{[3]} + \frac{[5]}{[4]} = \frac{23}{12}$	$\frac{[4]}{[5]} + \frac{[3]}{[2]} = \frac{23}{10}$	$\frac{[4]}{[3]} + \frac{[5]}{[2]} = \frac{23}{6}$

Sally wins because she has the largest sum.

ADDING WITH THE SAME DENOMINATORS

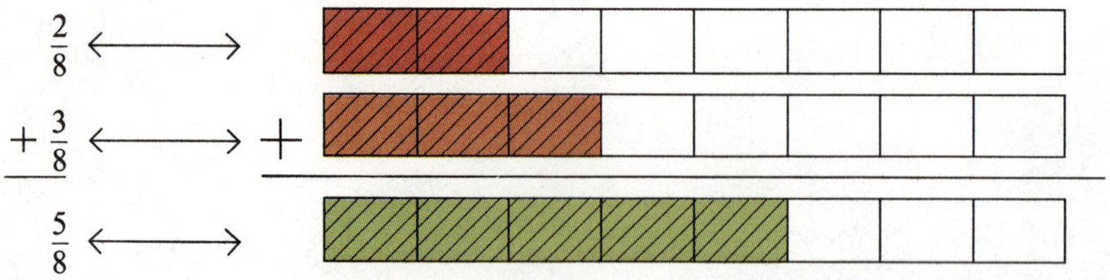

Add and simplify. $\frac{2}{3} + \frac{4}{3}$

Are the denominators the same? → If so, add the numerators. → Keep the same denominator. → Simplify.

Are the denominators the same?	If so, add the numerators.	Keep the same denominator.	Simplify.
$\begin{array}{r} \frac{2}{3} \\ + \frac{4}{3} \\ \hline \end{array}$ Yes.	$\begin{array}{r} \frac{2}{3} \\ + \frac{4}{3} \\ \hline 6 \end{array}$	$\begin{array}{r} \frac{2}{3} \\ + \frac{4}{3} \\ \hline \frac{6}{3} \end{array}$	$\frac{6}{3} = 2$

EXAMPLE 1. Add and simplify. $\frac{2}{5} + \frac{1}{5}$

$\frac{2}{5}$ *Are the denominators the same? Yes.*

$+\frac{1}{5}$ *Add the numerators.*

$\frac{3}{5}$ *Keep the same denominator.*

EXAMPLE 2. Add and simplify. $\frac{3}{12} + \frac{5}{12} + \frac{1}{12}$

$\frac{3}{12}$ *Are the denominators the same? Yes.*

$\frac{5}{12}$ *Add the numerators.*

$+\frac{1}{12}$ *Keep the same denominator.*

$\frac{9}{12} = \frac{3}{4}$ *Simplify.*

TRY THIS

Add and simplify.

1. $\frac{1}{3} + \frac{2}{3}$
2. $\frac{5}{12} + \frac{2}{12}$
3. $\frac{9}{16} + \frac{3}{16} + \frac{2}{16}$
4. $\frac{5}{10} + \frac{6}{10} + \frac{9}{10}$

PRACTICE

Add and simplify.

1. $\frac{4}{5} + \frac{1}{5}$
2. $\frac{1}{4} + \frac{3}{4}$
3. $\frac{7}{8} + \frac{1}{8}$
4. $\frac{1}{6} + \frac{5}{6}$
5. $\frac{3}{8} + \frac{4}{8}$
6. $\frac{1}{5} + \frac{2}{5}$
7. $\frac{3}{10} + \frac{4}{10}$
8. $\frac{3}{12} + \frac{2}{12}$
9. $\frac{2}{9} + \frac{1}{9} + \frac{5}{9}$
10. $\frac{1}{10} + \frac{3}{10} + \frac{4}{10}$
11. $\frac{5}{16} + \frac{7}{16} + \frac{3}{16}$
12. $\frac{3}{8} + \frac{1}{8} + \frac{2}{8}$
13. $\frac{3}{2} + \frac{5}{2}$
14. $\frac{5}{3} + \frac{4}{3}$
15. $\frac{11}{10} + \frac{9}{10}$
16. $\frac{11}{12} + \frac{13}{12}$
17. $\frac{5}{12} + \frac{9}{12}$
18. $\frac{11}{10} + \frac{1}{10}$
19. $\frac{13}{16} + \frac{5}{16}$
20. $\frac{11}{8} + \frac{3}{8}$
21. $\frac{3}{10} + \frac{1}{10} + \frac{9}{10}$
22. $\frac{5}{6} + \frac{11}{6} + \frac{1}{6}$
23. $\frac{3}{5} + \frac{2}{5} + \frac{4}{5}$
24. $\frac{7}{8} + \frac{5}{8} + \frac{3}{8}$
25. $\frac{13}{15} + \frac{12}{15}$
26. $\frac{13}{14} + \frac{15}{14}$
27. $\frac{11}{12} + \frac{11}{12}$
28. $\frac{21}{16} + \frac{21}{16}$
29. $\frac{15}{100} + \frac{31}{100}$
30. $\frac{24}{100} + \frac{22}{100}$
31. $\frac{51}{100} + \frac{14}{100}$
32. $\frac{71}{100} + \frac{14}{100}$
33. $\frac{21}{50} + \frac{9}{50}$
34. $\frac{7}{60} + \frac{5}{60}$
35. $\frac{48}{80} + \frac{12}{80}$
36. $\frac{45}{90} + \frac{15}{90}$

PROBLEM CORNER

1. Marti ran $\frac{5}{10}$ of a mile and then $\frac{3}{10}$ of a mile. How far did she run in all?
2. Lou bought $\frac{1}{3}$ lb of peanuts and $\frac{2}{3}$ lb of cashews. How many lbs of nuts did he buy in all?

More Practice, page 218

LEAST COMMON DENOMINATOR

The LCD is the smallest number that is divisible by the denominators.

What is the LCD of $\frac{3}{4}$ and $\frac{1}{8}$?

Look at the denominators. →	Is the larger denominator divisible by the smaller one? →	If so, then the larger denominator is the LCD.
4 and 8	Is 8 divisible by 4? Yes.	The LCD is 8.

EXAMPLE 1. What is the LCD of $\frac{2}{3}$, $\frac{5}{6}$, and $\frac{1}{2}$?

The denominators are 3, 6, and 2.
Is 6 divisible by 3 and divisible by 2? Yes. The LCD is 6.

TRY THIS

Find the LCD.

1. $\frac{5}{6}, \frac{7}{18}$ **2.** $\frac{2}{3}, \frac{3}{4}, \frac{7}{12}$ **3.** $\frac{1}{10}, \frac{7}{20}, \frac{3}{5}$ **4.** $\frac{1}{2}, \frac{2}{7}, \frac{5}{14}$

What is the LCD of $\frac{1}{12}$ and $\frac{2}{15}$?

Is the larger denominator divisible by the smaller denominator? →	If not, multiply the larger denominator by 2, then 3, then 4, . . . →	Decide each time if the answer is divisible by the smaller denominator. →	If it is, then the answer is the LCD.
Is 15 divisible by 12? No.	$15 \times 2 = 30$ $15 \times 3 = 45$ $15 \times 4 = 60$	Is 30 divisible by 12? No. Is 45 divisible by 12? No. Is 60 divisible by 12? Yes.	The LCD is 60.

EXAMPLE 2. What is the LCD of $\frac{1}{6}$, $\frac{1}{3}$, and $\frac{3}{4}$?

The denominators are 6, 3, and 4.
Is 6 divisible by 3 and divisible by 4? No.
$6 \times 2 = 12$ Is 12 divisible by 3 and divisible by 4? Yes.
The LCD is 12.

TRY THIS

Find the LCD.

5. $\frac{3}{8}, \frac{5}{6}$ **6.** $\frac{2}{5}, \frac{1}{3}$ **7.** $\frac{2}{3}, \frac{3}{4}$ **8.** $\frac{1}{7}, \frac{4}{5}$

9. $\frac{1}{2}, \frac{1}{5}, \frac{1}{4}$ **10.** $\frac{4}{9}, \frac{2}{3}, \frac{1}{6}$ **11.** $\frac{3}{4}, \frac{1}{6}, \frac{3}{8}$ **12.** $\frac{3}{8}, \frac{6}{7}, \frac{1}{2}$

PRACTICE

Find the LCD.

1. $\frac{2}{3}, \frac{5}{6}$ **2.** $\frac{2}{3}, \frac{1}{9}$ **3.** $\frac{1}{2}, \frac{5}{6}$ **4.** $\frac{1}{4}, \frac{3}{8}$ **5.** $\frac{3}{10}, \frac{1}{5}$

6. $\frac{1}{8}, \frac{1}{6}$ **7.** $\frac{1}{4}, \frac{1}{6}$ **8.** $\frac{1}{9}, \frac{1}{6}$ **9.** $\frac{1}{8}, \frac{1}{12}$ **10.** $\frac{1}{3}, \frac{1}{4}$

11. $\frac{2}{5}, \frac{1}{10}$ **12.** $\frac{2}{3}, \frac{1}{12}$ **13.** $\frac{4}{5}, \frac{7}{10}$ **14.** $\frac{3}{4}, \frac{1}{12}$ **15.** $\frac{2}{7}, \frac{3}{14}$

16. $\frac{5}{12}, \frac{3}{8}, \frac{1}{4}$ **17.** $\frac{4}{3}, \frac{5}{6}, \frac{1}{8}$ **18.** $\frac{7}{8}, \frac{1}{16}, \frac{1}{12}$ **19.** $\frac{2}{9}, \frac{7}{12}, \frac{1}{6}$ **20.** $\frac{1}{2}, \frac{3}{7}, \frac{5}{6}$

21. $\frac{3}{20}, \frac{3}{4}$ **22.** $\frac{2}{15}, \frac{2}{5}$ **23.** $\frac{1}{18}, \frac{2}{3}$ **24.** $\frac{3}{4}, \frac{3}{16}$ **25.** $\frac{5}{7}, \frac{9}{14}$

26. $\frac{1}{8}, \frac{7}{12}$ **27.** $\frac{3}{8}, \frac{5}{6}$ **28.** $\frac{3}{4}, \frac{5}{6}$ **29.** $\frac{5}{6}, \frac{7}{9}$ **30.** $\frac{3}{8}, \frac{5}{12}$

31. $\frac{5}{12}, \frac{2}{15}$ **32.** $\frac{3}{10}, \frac{5}{12}$ **33.** $\frac{3}{16}, \frac{1}{12}$ **34.** $\frac{9}{10}, \frac{11}{16}$ **35.** $\frac{3}{10}, \frac{1}{4}$

36. $\frac{3}{4}, \frac{2}{5}, \frac{11}{20}$ **37.** $\frac{1}{8}, \frac{2}{3}, \frac{15}{24}$ **38.** $\frac{2}{3}, \frac{3}{4}, \frac{5}{12}$ **39.** $\frac{1}{5}, \frac{1}{3}, \frac{4}{15}$ **40.** $\frac{1}{7}, \frac{3}{11}, \frac{5}{77}$

41. $\frac{4}{5}, \frac{2}{3}, \frac{1}{6}$ **42.** $\frac{4}{5}, \frac{3}{10}, \frac{1}{4}$ **43.** $\frac{3}{8}, \frac{5}{6}, \frac{1}{2}$ **44.** $\frac{3}{4}, \frac{1}{6}, \frac{5}{8}$ **45.** $\frac{1}{2}, \frac{3}{8}, \frac{3}{4}$

46. $\frac{3}{8}, \frac{7}{10}$ **47.** $\frac{3}{10}, \frac{1}{15}$ **48.** $\frac{5}{12}, \frac{11}{18}$ **49.** $\frac{11}{12}, \frac{3}{16}$ **50.** $\frac{1}{25}, \frac{3}{10}$

51. $\frac{5}{16}, \frac{3}{24}$ **52.** $\frac{13}{18}, \frac{1}{24}$ **53.** $\frac{11}{24}, \frac{15}{36}$ **54.** $\frac{12}{15}, \frac{9}{20}$ **55.** $\frac{2}{9}, \frac{1}{24}$

More Practice, page 218

ADDING WITH DIFFERENT DENOMINATORS

Add and simplify. $\frac{1}{6} + \frac{3}{4}$

Are the denominators the same? → If not, find the LCD. → Find equivalent fractions with the LCD. → Add. Simplify, if possible.

$$\begin{array}{r} \frac{1}{6} \\ +\frac{3}{4} \\ \hline \end{array} \quad \text{No.}$$

The LCD is 12.

$$\begin{array}{rl} \frac{1}{6} &= \frac{2}{12} \\ +\frac{3}{4} &= \frac{9}{12} \\ \hline \end{array}$$

$$\begin{array}{rl} \frac{1}{6} &= \frac{2}{12} \\ +\frac{3}{4} &= \frac{9}{12} \\ \hline & \frac{11}{12} \end{array}$$

EXAMPLES

1. Add and simplify. $\frac{4}{5} + \frac{7}{10}$

$$\begin{array}{rl} \frac{4}{5} &= \frac{8}{10} \\ +\frac{7}{10} &= \frac{7}{10} \\ \hline & \frac{15}{10} = \frac{3}{2} = 1\frac{1}{2} \end{array}$$

Are the denominators the same? No.
If not, find the LCD. The LCD is 10.
Find equivalent fractions with the LCD.
Add. Simplify.

2. Add and simplify. $\frac{5}{12} + \frac{7}{8}$

$$\begin{array}{rl} \frac{5}{12} &= \frac{10}{24} \\ +\frac{7}{8} &= \frac{21}{24} \\ \hline & \frac{31}{24} = 1\frac{7}{24} \end{array}$$

3. Add. $\frac{5}{9} + \frac{7}{18} + \frac{5}{6}$

$$\begin{array}{rl} \frac{5}{9} &= \frac{10}{18} \\ \frac{7}{18} &= \frac{7}{18} \\ +\frac{5}{6} &= \frac{15}{18} \\ \hline & \frac{32}{18} = \frac{16}{9} \\ & = 1\frac{7}{9} \end{array}$$

TRY THIS

Add and simplify.

1. $\frac{3}{8} + \frac{5}{6}$ **2.** $\frac{1}{6} + \frac{7}{18}$ **3.** $\frac{1}{2} + \frac{2}{5} + \frac{6}{10}$ **4.** $\frac{1}{12} + \frac{3}{4} + \frac{1}{6}$

PRACTICE

Add and simplify.

1. $\frac{2}{3} + \frac{5}{6}$ **2.** $\frac{2}{3} + \frac{1}{9}$ **3.** $\frac{1}{2} + \frac{5}{6}$ **4.** $\frac{1}{4} + \frac{3}{8}$

5. $\frac{2}{5} + \frac{1}{10}$ **6.** $\frac{2}{3} + \frac{1}{12}$ **7.** $\frac{4}{5} + \frac{7}{10}$ **8.** $\frac{3}{4} + \frac{1}{12}$

9. $\frac{5}{12} + \frac{3}{8}$ **10.** $\frac{4}{3} + \frac{5}{6}$ **11.** $\frac{7}{8} + \frac{1}{16}$ **12.** $\frac{2}{9} + \frac{7}{12}$

13. $\frac{1}{8} + \frac{7}{12}$ **14.** $\frac{3}{8} + \frac{5}{6}$ **15.** $\frac{3}{4} + \frac{5}{6}$ **16.** $\frac{5}{6} + \frac{7}{9}$

17. $\frac{3}{10} + \frac{1}{100}$ **18.** $\frac{1}{10} + \frac{7}{100}$ **19.** $\frac{2}{10} + \frac{6}{100}$ **20.** $\frac{9}{10} + \frac{3}{100}$

21. $\frac{3}{10} + \frac{7}{40}$ **22.** $\frac{4}{10} + \frac{3}{20}$ **23.** $\frac{9}{10} + \frac{11}{30}$ **24.** $\frac{1}{10} + \frac{3}{50}$

25. $\frac{5}{12} + \frac{2}{15}$ **26.** $\frac{3}{10} + \frac{5}{12}$ **27.** $\frac{3}{16} + \frac{1}{12}$ **28.** $\frac{9}{10} + \frac{11}{16}$

29. $\frac{4}{5} + \frac{3}{25}$ **30.** $\frac{7}{20} + \frac{4}{5}$ **31.** $\frac{3}{8} + \frac{15}{16}$ **32.** $\frac{3}{4} + \frac{1}{20}$

33. $\frac{1}{8} + \frac{1}{6} + \frac{1}{2}$ **34.** $\frac{1}{4} + \frac{1}{6} + \frac{1}{3}$ **35.** $\frac{1}{9} + \frac{1}{6} + \frac{1}{3}$

36. $\frac{3}{20} + \frac{3}{4} + \frac{1}{5}$ **37.** $\frac{2}{15} + \frac{2}{5} + \frac{2}{3}$ **38.** $\frac{1}{18} + \frac{2}{3} + \frac{1}{6}$

39. $\frac{3}{4} + \frac{2}{3} + \frac{1}{12}$ **40.** $\frac{1}{8} + \frac{2}{3} + \frac{1}{24}$ **41.** $\frac{2}{3} + \frac{3}{4} + \frac{5}{12}$

42. $\frac{5}{10} + \frac{31}{100} + \frac{1}{1000}$ **43.** $\frac{3}{10} + \frac{27}{100} + \frac{5}{1000}$ **44.** $\frac{1}{10} + \frac{11}{100} + \frac{3}{1000}$

PROBLEM CORNER

1. Sara walked $\frac{3}{4}$ of a mile to a friend's house, and then $\frac{7}{6}$ of a mile to class. How far did Sara walk?

2. Allen bought $\frac{1}{4}$ lb of bean sprouts, $\frac{1}{2}$ lb of mushrooms, and $\frac{2}{3}$ lb of carrots. How many pounds of vegetables did Allen buy?

More Practice, page 218

ADDING WITH MIXED NUMBERS

Add and simplify. $12\frac{1}{2} + 7\frac{3}{10}$

Are the denominators the same?	→	If not, find the LCD.	→	Find equivalent fractions with the LCD.	→	Add the fractions. Add the whole numbers. Simplify.

$$\begin{array}{r} 12\frac{1}{2} \\ +\ 7\frac{3}{10} \\ \hline \end{array} \text{ No.}$$

The LCD is 10.

$$\begin{array}{rcl} 12\frac{1}{2} &=& 12\frac{5}{10} \\ +\ 7\frac{3}{10} &=& 7\frac{3}{10} \\ \hline \end{array}$$

$$\begin{array}{r} 12\frac{5}{10} \\ +\ 7\frac{3}{10} \\ \hline 19\frac{8}{10} = 19\frac{4}{5} \end{array}$$

EXAMPLES. Add and simplify.

1. $$\begin{array}{rcr} 4\frac{1}{2} &=& 4\frac{3}{6} \\ +\ 7\frac{1}{3} &=& 7\frac{2}{6} \\ \hline && 11\frac{5}{6} \end{array}$$

2. $$\begin{array}{r} 7\frac{5}{8} \\ +\ 3\frac{1}{8} \\ \hline 10\frac{6}{8} = 10\frac{3}{4} \end{array}$$

3. $$\begin{array}{r} 19\frac{3}{5} \\ +\ 6 \\ \hline 25\frac{3}{5} \end{array}$$

TRY THIS Add and simplify.

1. $8\frac{2}{5} + 3\frac{4}{10}$ **2.** $16\frac{2}{9} + 7\frac{4}{9}$ **3.** $12 + 9\frac{7}{8}$

EXAMPLE 4. Add and simplify. $5\frac{2}{3} + 3\frac{5}{6} + 2\frac{1}{4}$

$$\begin{array}{rcl} 5\frac{2}{3} &=& 5\frac{8}{12} \\ 3\frac{5}{6} &=& 3\frac{10}{12} \\ +\ 2\frac{1}{4} &=& 2\frac{3}{12} \\ \hline \end{array}$$

$$10\frac{21}{12} = 10 + \frac{21}{12} = 10 + \frac{7}{4}$$

$$= 10 + 1\frac{3}{4} = 11\frac{3}{4}$$

Simplify. Convert the fraction to a mixed number.

TRY THIS

Add and simplify.

4. $9\frac{3}{4} + 3\frac{5}{6}$ **5.** $3\frac{1}{2} + 5\frac{1}{4} + 7\frac{3}{10}$ **6.** $2\frac{2}{3} + 1\frac{1}{6} + 1\frac{4}{9}$

PRACTICE

Add and simplify.

1. $1\frac{2}{3} + 5\frac{1}{3}$ **2.** $4\frac{5}{6} + 3\frac{5}{6}$ **3.** $4\frac{1}{6} + 3\frac{5}{6}$ **4.** $8\frac{2}{3} + 5\frac{2}{3}$

5. $1\frac{1}{4} + 1\frac{2}{3}$ **6.** $1\frac{2}{5} + 2\frac{1}{4}$ **7.** $2\frac{3}{4} + 2\frac{2}{5}$ **8.** $2\frac{1}{6} + 1\frac{1}{5}$

9. $4\frac{3}{8} + 6\frac{5}{12}$ **10.** $3\frac{2}{9} + 4\frac{1}{6}$ **11.** $8\frac{3}{4} + 5\frac{1}{6}$ **12.** $5\frac{3}{8} + 10\frac{5}{6}$

13. $\frac{5}{8} + 1\frac{5}{6}$ **14.** $14\frac{5}{8} + \frac{1}{4}$ **15.** $5\frac{1}{4} + \frac{1}{10}$ **16.** $2\frac{3}{16} + \frac{1}{2}$

17. $5\frac{3}{8} + 12$ **18.** $6\frac{2}{3} + 14$ **19.** $15 + 4\frac{5}{6}$ **20.** $18 + 2\frac{1}{8}$

21. $16\frac{1}{4} + 25\frac{3}{10}$ **22.** $14\frac{1}{6} + 12\frac{2}{3}$ **23.** $24\frac{1}{3} + 12\frac{2}{9}$ **24.** $14\frac{3}{5} + 16\frac{1}{10}$

25. $\frac{3}{4} + 2\frac{1}{4} + 3\frac{1}{4}$ **26.** $\frac{2}{3} + 5\frac{1}{3} + 6\frac{2}{3}$ **27.** $9\frac{4}{5} + \frac{1}{5} + 1\frac{2}{5}$

28. $3\frac{5}{16} + 6\frac{1}{4} + 2\frac{1}{2}$ **29.** $3\frac{5}{6} + 2\frac{5}{12} + 5\frac{1}{3}$ **30.** $4\frac{1}{3} + 2\frac{5}{8} + 3\frac{1}{4}$

31. $4\frac{7}{8} + \frac{1}{2} + 5\frac{3}{4}$ **32.** $4\frac{1}{6} + 2\frac{2}{3} + 7\frac{1}{2}$ **33.** $2\frac{5}{6} + 10\frac{4}{9} + 9\frac{1}{3}$

PROBLEM CORNER

1. One day a stock opened at $63\frac{7}{8}$ and gained $1\frac{3}{4}$. What was the closing price?

2. A fabric store sold three pieces of burlap. One was $6\frac{1}{4}$ yd, one was $4\frac{1}{2}$ yd, and one was $10\frac{5}{6}$ yd long. What was the total length of the burlap?

APPLICATION

FOOD VALUES

Nutrients in food promote growth and repair tissues of the body.

Food	Serving Size	Calories	Protein (gm)	Calcium (mg)	Iron (mg)	Vitamin C (mg)
Whole Milk	1 cup	160	9	288	0.2	3
Skimmed Milk	1 cup	90	9	303	0.2	3
Whole Wheat Bread	1 slice	55	2	22	0.5	0
Peanut Butter	1 tbs	90	4	12	0.3	0
Swiss Cheese	1 oz	105	8	262	0.3	0
Spinach (cooked)	1 cup	45	6	223	3.6	54
Vanilla Ice Cream	$\frac{1}{7}$ qt	165	3	100	0.1	1

1. Compare the Vitamin C in whole and skimmed milk. Which has more Vitamin C? How much more?
2. How many tablespoons of peanut butter does it take to get the same quantity of protein as in 1 oz of Swiss cheese?
3. How many tablespoons of peanut butter does it take to get the same quantity of calcium as in 1 oz of Swiss cheese?
4. How many slices of whole wheat bread would it take to get the same amount of calcium as in 1 cup of whole milk?
5. How many slices of whole wheat bread would it take to get the same amount of calcium as in $\frac{1}{7}$ qt of vanilla ice cream?

Yesterday Barry and Joan drank these quantities of skimmed milk.

	Home	**School**	**Restaurant**
Barry	$3\frac{3}{8}$ cups	$3\frac{1}{2}$ cups	$1\frac{1}{8}$ cups
Joan	$3\frac{1}{2}$ cups	$2\frac{3}{4}$ cups	$1\frac{2}{3}$ cups

6. How much skimmed milk did each person drink?
7. How many calories were in the milk that each person drank?
8. How many grams of protein did Barry and Joan each receive?
9. How many milligrams of calcium did each receive?
10. How many milligrams of Vitamin C did each receive?

On Tuesday Juanita had one serving of each food in the table. How much of each of the following did she receive?

11. calories
12. protein
13. calcium
14. iron
15. vitamin C

The following show that some foods are high in certain nutrients.

16. For each item in the table find the amount of iron divided by the number of calories. Which food is highest in iron?
17. For each item in the table find the amount of protein divided by the number of calories. Which food is highest in protein?
18. For each item in the table find the amount of calcium divided by the number of calories. Which food is highest in calcium?
19. For each item in the table find the amount of Vitamin C divided by the number of calories. Which food is highest in Vitamin C?

SUBTRACTING

Subtract and simplify. $\frac{7}{10} - \frac{3}{10}$

Are the denominators the same? → Subtract the numerators. → Keep the same denominator. → Simplify.

$$\begin{array}{r} \frac{7}{10} \\ -\frac{3}{10} \\ \hline \end{array} \quad \text{Yes.} \qquad \begin{array}{r} \frac{7}{10} \\ -\frac{3}{10} \\ \hline 4 \end{array} \qquad \begin{array}{r} \frac{7}{10} \\ -\frac{3}{10} \\ \hline \frac{4}{10} \end{array} \qquad \frac{4}{10} = \frac{2}{5}$$

EXAMPLE 1. Subtract and simplify. $\frac{11}{12} - \frac{1}{12}$

$$\begin{array}{r} \frac{11}{12} \\ -\frac{1}{12} \\ \hline \frac{10}{12} = \frac{5}{6} \end{array}$$

Are the denominators the same? Yes.
Subtract the numerators.
Keep the same denominator.

Simplify.

TRY THIS

Subtract and simplify.

1. $\frac{7}{8} - \frac{5}{8}$ **2.** $\frac{12}{16} - \frac{4}{16}$ **3.** $\frac{8}{12} - \frac{4}{12}$ **4.** $\frac{7}{9} - \frac{4}{9}$

When the denominators are different, we find equivalent fractions with the LCD.

Subtract and simplify. $\frac{3}{4} - \frac{1}{6}$

Are the denominators the same? → If not, find the LCD. → Find equivalent fractions with the LCD. → Subtract. Simplify.

$$\begin{array}{r} \frac{3}{4} \\ -\frac{1}{6} \\ \hline \end{array} \quad \text{No.}$$

The LCD is 12.

$$\begin{array}{r} \frac{3}{4} = \frac{9}{12} \\ -\frac{1}{6} = \frac{2}{12} \\ \hline \end{array} \qquad \begin{array}{r} \frac{3}{4} = \frac{9}{12} \\ -\frac{1}{6} = \frac{2}{12} \\ \hline \frac{7}{12} \end{array}$$

EXAMPLE 2. Subtract and simplify. $\frac{5}{6} - \frac{7}{12}$

$$\begin{array}{rcl} \frac{5}{6} & = & \frac{10}{12} \\ -\frac{7}{12} & = & \frac{7}{12} \\ \hline & & \frac{3}{12} = \frac{1}{4} \end{array}$$

Are the denominators the same? No.
The LCD is 12.
Find equivalent fractions with the LCD.

Subtract. Simplify.

TRY THIS

Subtract and simplify.

5. $\frac{4}{5} - \frac{3}{10}$ **6.** $\frac{5}{6} - \frac{1}{9}$ **7.** $\frac{5}{12} - \frac{1}{4}$ **8.** $\frac{5}{6} - \frac{1}{3}$

PRACTICE

Subtract and simplify.

1. $\frac{5}{6} - \frac{1}{6}$	**2.** $\frac{3}{4} - \frac{1}{4}$	**3.** $\frac{7}{5} - \frac{2}{5}$	**4.** $\frac{4}{3} - \frac{2}{3}$	**5.** $\frac{3}{7} - \frac{2}{7}$
6. $\frac{11}{12} - \frac{2}{12}$	**7.** $\frac{9}{10} - \frac{3}{10}$	**8.** $\frac{15}{16} - \frac{1}{16}$	**9.** $\frac{17}{18} - \frac{5}{18}$	**10.** $\frac{13}{14} - \frac{6}{14}$
11. $\frac{3}{4} - \frac{1}{8}$	**12.** $\frac{2}{3} - \frac{1}{8}$	**13.** $\frac{3}{4} - \frac{1}{2}$	**14.** $\frac{3}{5} - \frac{1}{4}$	**15.** $\frac{5}{6} - \frac{1}{18}$
16. $\frac{2}{3} - \frac{1}{9}$	**17.** $\frac{5}{6} - \frac{2}{3}$	**18.** $\frac{3}{8} - \frac{1}{4}$	**19.** $\frac{5}{6} - \frac{1}{2}$	**20.** $\frac{1}{2} - \frac{1}{8}$
21. $\frac{4}{3} - \frac{5}{6}$	**22.** $\frac{5}{12} - \frac{3}{8}$	**23.** $\frac{7}{12} - \frac{2}{9}$	**24.** $\frac{7}{8} - \frac{1}{16}$	**25.** $\frac{5}{8} - \frac{1}{16}$
26. $\frac{2}{5} - \frac{2}{15}$	**27.** $\frac{3}{4} - \frac{3}{20}$	**28.** $\frac{3}{4} - \frac{4}{16}$	**29.** $\frac{2}{3} - \frac{1}{18}$	**30.** $\frac{3}{4} - \frac{1}{20}$
31. $\frac{5}{6} - \frac{3}{8}$	**32.** $\frac{7}{12} - \frac{1}{8}$	**33.** $\frac{5}{6} - \frac{7}{9}$	**34.** $\frac{5}{6} - \frac{3}{4}$	**35.** $\frac{7}{12} - \frac{3}{8}$
36. $\frac{5}{12} - \frac{3}{10}$	**37.** $\frac{5}{12} - \frac{2}{15}$	**38.** $\frac{9}{10} - \frac{11}{16}$	**39.** $\frac{3}{16} - \frac{1}{12}$	**40.** $\frac{9}{10} - \frac{5}{12}$
41. $\frac{2}{3} - \frac{1}{8}$	**42.** $\frac{3}{4} - \frac{2}{5}$	**43.** $\frac{1}{3} - \frac{1}{5}$	**44.** $\frac{3}{4} - \frac{2}{3}$	**45.** $\frac{2}{3} - \frac{3}{8}$
46. $\frac{4}{5} - \frac{7}{20}$	**47.** $\frac{4}{5} - \frac{3}{25}$	**48.** $\frac{3}{4} - \frac{1}{20}$	**49.** $\frac{15}{16} - \frac{3}{8}$	**50.** $\frac{4}{5} - \frac{1}{20}$

More Practice, page 219

SUBTRACTING WITH MIXED NUMBERS

Subtract and simplify. $9\frac{4}{5} - 3\frac{1}{2}$

Are the denominators the same?	→ If not, find the LCD.	→ Find equivalent fractions with the LCD.	→ Subtract the fractions.	→ Subtract the whole numbers. Simplify, if possible.
$\begin{array}{r} 9\frac{4}{5} \\ -\,3\frac{1}{2} \\ \hline \end{array}$ No.	The LCD is 10.	$\begin{array}{r} 9\frac{4}{5} = 9\frac{8}{10} \\ -\,3\frac{1}{2} = 3\frac{5}{10} \\ \hline \end{array}$	$\begin{array}{r} 9\frac{8}{10} \\ -\,3\frac{5}{10} \\ \hline \frac{3}{10} \end{array}$	$\begin{array}{r} 9\frac{8}{10} \\ -\,3\frac{5}{10} \\ \hline 6\frac{3}{10} \end{array}$

EXAMPLE 1. Subtract and simplify. $7\frac{3}{4} - 2\frac{1}{4}$

$$\begin{array}{r} 7\frac{3}{4} = 7\frac{3}{4} \\ -\,2\frac{1}{4} = 2\frac{1}{4} \\ \hline 5\frac{2}{4} = 5\frac{1}{2} \end{array}$$

Are the denominators the same? Yes.
Subtract the fractions.
Subtract the whole numbers.
Simplify.

EXAMPLE 2. Subtract and simplify. $16\frac{3}{7} - 9$

$$\begin{array}{r} 16\frac{3}{7} \\ -\,9 \\ \hline 7\frac{3}{7} \end{array}$$

Are the denominators the same. Yes.
Think of 9 as $9\frac{0}{7}$.

EXAMPLE 3. Subtract and simplify. $15\frac{5}{6} - 8\frac{1}{2}$

$$\begin{array}{r} 15\frac{5}{6} = 15\frac{5}{6} \\ -\,8\frac{1}{2} = 8\frac{3}{6} \\ \hline 7\frac{2}{6} = 7\frac{1}{3} \end{array}$$

Are the denominators the same? No.
The LCD is 6.
Find equivalent fractions with the LCD.
Subtract. Simplify.

TRY THIS

Subtract and simplify.

1. $11\frac{5}{8} - 6\frac{3}{8}$ **2.** $21\frac{7}{8} - 9$ **3.** $8\frac{2}{3} - 5\frac{1}{2}$ **4.** $\frac{2}{3} - \frac{1}{6}$

PRACTICE

Subtract and simplify.

1. $4\frac{3}{5} - 2\frac{1}{5}$ **2.** $8\frac{7}{9} - 2\frac{1}{9}$ **3.** $6\frac{5}{8} - 3\frac{1}{8}$ **4.** $9\frac{7}{10} - 4\frac{1}{10}$

5. $14\frac{5}{6} - 9\frac{5}{6}$ **6.** $23\frac{7}{8} - 8\frac{7}{8}$ **7.** $14\frac{3}{7} - 5\frac{3}{7}$ **8.** $31\frac{2}{5} - 7\frac{2}{5}$

9. $6\frac{3}{5} - 2\frac{1}{2}$ **10.** $8\frac{2}{5} - 4\frac{1}{3}$ **11.** $7\frac{2}{3} - 6\frac{1}{2}$ **12.** $8\frac{4}{5} - 1\frac{1}{4}$

13. $7\frac{1}{3} - 3\frac{1}{6}$ **14.** $9\frac{1}{4} - 5\frac{1}{8}$ **15.** $10\frac{1}{3} - 6\frac{1}{9}$ **16.** $9\frac{1}{2} - 2\frac{1}{4}$

17. $5\frac{7}{8} - 3$ **18.** $9\frac{1}{4} - 2$ **19.** $5\frac{5}{6} - 4$ **20.** $9\frac{4}{5} - 7$

21. $27\frac{4}{5} - 19$ **22.** $26\frac{5}{8} - 13$ **23.** $32\frac{3}{4} - 16$ **24.** $29\frac{1}{6} - 11$

25. $24\frac{5}{6} - 2\frac{5}{12}$ **26.** $23\frac{5}{16} - 6\frac{1}{4}$ **27.** $34\frac{5}{8} - 2\frac{1}{3}$ **28.** $28\frac{2}{3} - 2\frac{1}{6}$

29. $9\frac{5}{6} - \frac{1}{6}$ **30.** $10\frac{3}{4} - \frac{1}{4}$ **31.** $4\frac{9}{10} - \frac{7}{10}$ **32.** $8\frac{9}{12} - \frac{5}{12}$

33. $14\frac{3}{4} - \frac{1}{8}$ **34.** $12\frac{1}{3} - \frac{1}{4}$ **35.** $20\frac{5}{8} - \frac{1}{4}$ **36.** $16\frac{5}{6} - \frac{1}{3}$

PROBLEM CORNER

1. Joel bicycled $11\frac{7}{10}$ miles on Saturday and $4\frac{1}{10}$ miles on Monday. How much farther did he bicycle on Saturday than on Monday?

2. A pattern calls for $3\frac{3}{8}$ yards of fabric and $2\frac{1}{4}$ yards of elastic. How much more fabric is needed than elastic?

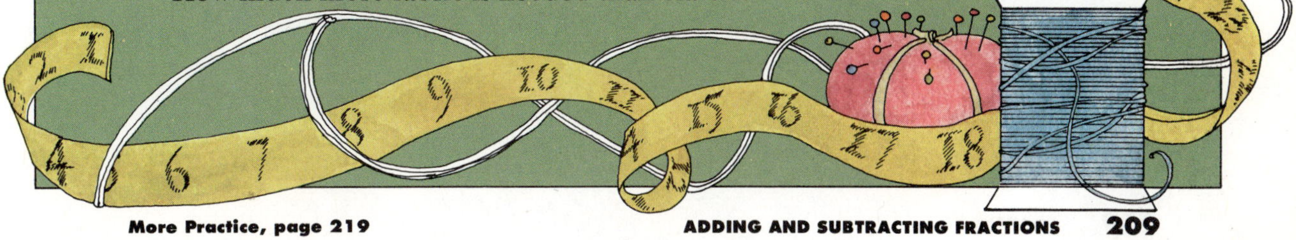

SUBTRACTING WITH RENAMING

Subtract and simplify. $12 - 9\frac{3}{8}$

Can the fraction be subtracted? → Rename. → Subtract the fractions. → Subtract the whole numbers.

$$\begin{array}{r} 12 \\ -\ 9\frac{3}{8} \\ \hline \end{array} \quad \begin{array}{r} \frac{0}{8} \\ -\frac{3}{8} \\ \hline \end{array} \text{ No.}$$

$$\begin{aligned} 12 &= 11 + 1 \\ &= 11 + \frac{8}{8} \\ &= 11\frac{8}{8} \end{aligned}$$

$$\begin{array}{r} 11\frac{8}{8} \\ -\ 9\frac{3}{8} \\ \hline \frac{5}{8} \end{array} \qquad \begin{array}{r} 11\frac{8}{8} \\ -\ 9\frac{3}{8} \\ \hline 2\frac{5}{8} \end{array}$$

EXAMPLE 1. Subtract and simplify. $24 - 17\frac{2}{5}$

$$\begin{array}{rcr} 24 & = & 23\frac{5}{5} \\ -\ 17\frac{2}{5} & = & 17\frac{2}{5} \\ \hline & & 6\frac{3}{5} \end{array}$$

$24 = 23 + 1$

$24 = 23 + \frac{5}{5}$

$24 = 23\frac{5}{5}$

TRY THIS

Subtract and simplify.

1. $10 - 3\frac{1}{6}$ **2.** $21 - 14\frac{5}{8}$ **3.** $8 - 2\frac{3}{4}$

Subtract and simplify. $7\frac{1}{6} - 2\frac{1}{4}$

Find equivalent fractions with the LCD. → Can the fractions be subtracted? → Rename. → Subtract the fractions and whole numbers. Simplify.

$$\begin{array}{rcl} 7\frac{1}{6} & = & 7\frac{2}{12} \\ -\ 2\frac{1}{4} & = & 2\frac{3}{12} \\ \hline \end{array}$$

The LCD is 12.

No, $\frac{2}{12}$ is smaller than $\frac{3}{12}$.

$$\begin{aligned} 7\frac{2}{12} &= 6 + \frac{12}{12} + \frac{2}{12} \\ &= 6\frac{14}{12} \end{aligned}$$

$$\begin{array}{r} 6\frac{14}{12} \\ -\ 2\frac{3}{12} \\ \hline 4\frac{11}{12} \end{array}$$

EXAMPLE 2. Subtract and simplify. $10\frac{1}{3} - 7\frac{5}{6}$

$$\begin{array}{rcrcr} 10\frac{1}{3} & = & 10\frac{2}{6} & = & 9\frac{8}{6} \\ -\ 7\frac{5}{6} & = & 7\frac{5}{6} & = & 7\frac{5}{6} \\ \hline & & & & 2\frac{3}{6} = 2\frac{1}{2} \end{array}$$

Rename. $10\frac{2}{6} = 9 + \frac{6}{6} + \frac{2}{6} = 9\frac{8}{6}$

TRY THIS

Subtract.

4. $5\frac{1}{8} - 2\frac{1}{6}$ **5.** $3\frac{1}{6} - 1\frac{1}{2}$

6. $13\frac{1}{12} - 3\frac{3}{4}$ **7.** $10\frac{1}{6} - 5\frac{2}{5}$

PRACTICE

Subtract and simplify.

1. $5 - 1\frac{1}{3}$ **2.** $4 - 2\frac{1}{8}$ **3.** $9 - 1\frac{1}{6}$ **4.** $6 - 2\frac{1}{5}$

5. $14 - 8\frac{5}{6}$ **6.** $17 - 9\frac{4}{9}$ **7.** $23 - 6\frac{3}{4}$ **8.** $31 - 4\frac{7}{8}$

9. $40 - 15\frac{3}{8}$ **10.** $20 - 16\frac{3}{5}$ **11.** $30 - 9\frac{7}{10}$ **12.** $40 - 12\frac{3}{7}$

13. $8\frac{2}{9} - 2\frac{7}{9}$ **14.** $4\frac{3}{5} - 2\frac{4}{5}$ **15.** $4\frac{1}{6} - 1\frac{5}{6}$ **16.** $5\frac{1}{8} - 2\frac{3}{8}$

17. $9\frac{1}{4} - 2\frac{1}{2}$ **18.** $10\frac{2}{9} - 6\frac{1}{3}$ **19.** $9\frac{1}{8} - 5\frac{1}{4}$ **20.** $7\frac{1}{6} - 3\frac{5}{6}$

21. $8\frac{1}{4} - 1\frac{4}{5}$ **22.** $7\frac{1}{2} - 6\frac{2}{3}$ **23.** $18\frac{1}{6} - 5\frac{3}{4}$ **24.** $10\frac{5}{6} - 5\frac{3}{8}$

25. $31 - 19\frac{11}{12}$ **26.** $18 - 16\frac{2}{9}$ **27.** $22 - 14\frac{7}{8}$ **28.** $24 - 13\frac{5}{6}$

29. $14\frac{1}{8} - \frac{3}{4}$ **30.** $12\frac{1}{4} - \frac{1}{3}$ **31.** $20\frac{1}{4} - \frac{5}{8}$ **32.** $16\frac{1}{3} - \frac{5}{6}$

33. $28\frac{2}{3} - 2\frac{1}{6}$ **34.** $34\frac{1}{3} - 2\frac{5}{8}$ **35.** $23\frac{1}{4} - 7\frac{5}{16}$ **36.** $20\frac{5}{12} - 2\frac{5}{6}$

More Practice, page 219

APPLICATION

HOUSEHOLD PROBLEMS

Linda Torres plans to build a bookcase. She makes a sketch to use during the project.

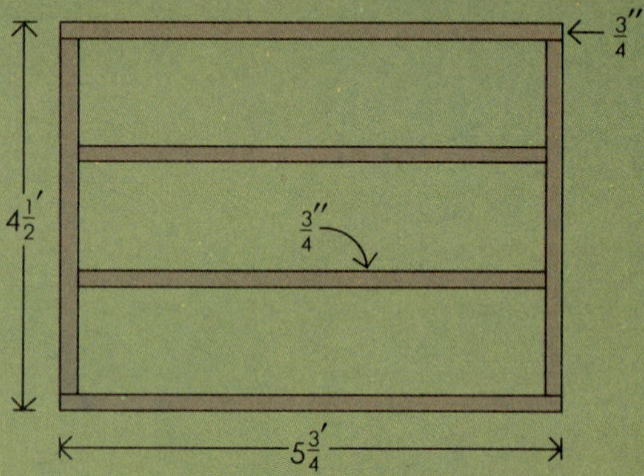

$4\frac{1}{2}'$ means $4\frac{1}{2}$ ft

$\frac{3}{4}''$ means $\frac{3}{4}$ in.

1. How many inches long is each side piece?
2. How long is each middle shelf?
3. How many feet of wood will be needed?
4. How many inches are there between each shelf?
5. What important measure is left off the drawing?
6. To make the bookcase she buys 4 boards, each 10 ft long. They cost \$3.25 each. What is the total cost?
7. Estimate how much lumber will be left over.

Pete O'Brien wants to put up a clothesline to save electricity by not using the electric dryer.

How many feet of clothesline are used for each of the following?

8. the $1\frac{5}{6}'$ square
9. the $2\frac{2}{3}'$ square
10. the $3\frac{1}{2}'$ square
11. the $4\frac{1}{3}'$ square
12. How many feet of clothesline are used in all?
13. Clothesline comes in a 50 ft roll. How much of the roll will not be used?

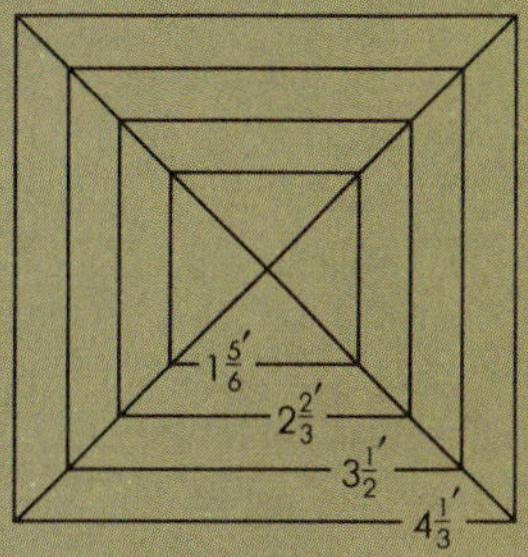

The Andersons want to refurnish a room.

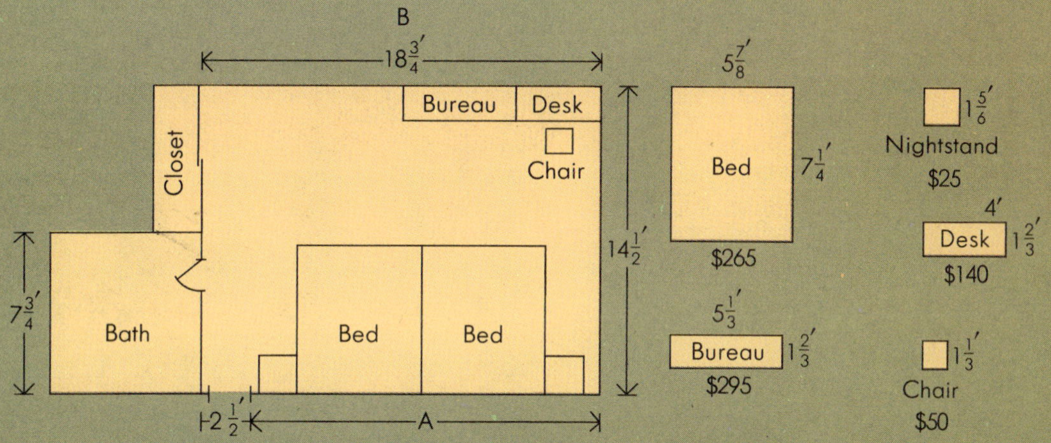

Room Dimensions (feet)

14. What is the length of wall A?
15. How long is the closet?
16. How much room do the two beds and nightstands take up along wall A?
17. How much room is left on wall B after putting in the bureau and desk?
18. Find the cost of furnishing the room.
19. Curtains and lamps for the room cost \$ 525. What is the cost of furnishing the room if these items are included?
20. What is the cost of furnishing the room if only one bed and one nightstand are used?

VAN CUSTOMIZING

Mark Ryan owns Custom Van Works. The employees specialize in upholstery, carpentry, auto painting, and body work.

The Wilsons want benches and a table built into their van. Mark's estimate must consider the following.

How Much Will It Cost?		
Materials:	Fabric and Foam	\$70
	Wood	\$20
	Laminated Plastic	\$10
	Trim and Hardware	\$16
Labor: 10 hours at \$15 per hour		

1. How much will the materials cost?
2. How much will the labor cost?
3. What is the total cost of the job to the Wilsons?
4. Mark makes a profit of $\frac{1}{4}$ on the cost of the materials. How much is this?
5. Mark pays his workers \$ 8 per hour. What profit does he make on the labor charges?
6. What is Mark's total profit?

The various types of furniture Mark puts in a van require different amounts of fabric.

Captain's Chair	$\frac{1}{5}$ bolt
Armless Chair	$\frac{1}{6}$ bolt
Double Sofa Seat	$\frac{5}{6}$ bolt
Sofa Bed	$\frac{7}{10}$ bolt

How many bolts of fabric would be needed for each job?

7. 1 captain's chair
   1 armless chair
   1 double sofa seat
   1 sofa bed

8. 2 captain's chairs
   2 sofa beds

9. 1 captain's chair
   2 armless chairs
   2 double sofa seats

10. 1 captain's chair
    3 armless chairs
    1 sofa bed

Mark charges $ 0.72 per square foot to put plain sun screen film on windows. How much would each of these jobs cost?

11. 1 rear window, 8.3 square feet
12. 2 rear windows, 3.4 square feet each
13. 2 side windows: 1 is 6.8 square feet, and the other is 9.4 square feet

Mark must know how long a job will take. Employees who refinish paneling work at different rates according to their skill.

Jan took 3 hours to refinish 216 square feet of paneling. She was paid $ 5 per hour. Chiang took 4 hours to refinish 384 square feet of paneling. He was paid $ 6.75 per hour.

14. How much did Mark pay each person?
15. At that rate, how many square feet can each person do in 1 hour?
16. To the nearest cent, what is Mark's cost per square foot to have each employee refinish paneling?

CHAPTER 7 REVIEW

Add and simplify. p. 196

1. $\frac{1}{4} + \frac{7}{4}$ **2.** $\frac{5}{8} + \frac{1}{8}$ **3.** $\frac{7}{10} + \frac{9}{10} + \frac{6}{10}$ **4.** $\frac{1}{12} + \frac{5}{12} + \frac{9}{12}$

Find the LCD. p. 198

5. $\frac{4}{5}, \frac{7}{10}$ **6.** $\frac{5}{6}, \frac{7}{8}$ **7.** $\frac{4}{9}, \frac{2}{3}, \frac{1}{6}$ **8.** $\frac{2}{5}, \frac{1}{4}, \frac{7}{10}$

Add and simplify. p. 200

9. $\frac{5}{6} + \frac{7}{8}$ **10.** $\frac{4}{5} + \frac{1}{10}$ **11.** $\frac{2}{3} + \frac{3}{4} + \frac{1}{6}$ **12.** $\frac{2}{5} + \frac{2}{3} + \frac{5}{6}$

Add and simplify. p. 202

13. $5\frac{1}{8} + 4\frac{1}{2}$ **14.** $19\frac{3}{4} + 8$ **15.** $16\frac{1}{4} + 7\frac{2}{3}$ **16.** $12\frac{1}{3} + 4\frac{1}{6}$

17. $5\frac{1}{3} + 18\frac{1}{2} + 4\frac{5}{6}$ **18.** $\frac{5}{8} + 4\frac{1}{3}$ **19.** $6\frac{3}{4} + 8\frac{3}{16}$ **20.** $2\frac{7}{8} + 4\frac{3}{4}$

Subtract and simplify. p. 206

21. $\frac{8}{5} - \frac{3}{5}$ **22.** $\frac{7}{8} - \frac{1}{12}$ **23.** $\frac{87}{100} - \frac{3}{10}$ **24.** $\frac{5}{6} - \frac{1}{8}$

Subtract and simplify. pp. 208–210

25. $9\frac{7}{10} - 2\frac{1}{10}$ **26.** $12\frac{7}{8} - 7$ **27.** $8\frac{3}{4} - 7\frac{1}{2}$ **28.** $5\frac{7}{8} - 2\frac{1}{2}$

29. $15\frac{1}{4} - 3\frac{1}{2}$ **30.** $32 - 8\frac{2}{5}$ **31.** $17\frac{3}{8} - 9\frac{5}{8}$ **32.** $6\frac{1}{10} - 1\frac{1}{5}$

Application pp. 204, 212

33. Les drank the following amounts of milk one day. How much did he drink in all?

Home	School	Restaurant
$2\frac{3}{4}$ cups	$1\frac{2}{3}$ cups	$2\frac{1}{2}$ cups

34. How much clothesline would it take to make a rectangle $8\frac{1}{2}$ yds by $7\frac{3}{4}$ yds?

CHAPTER 7 TEST

Add and simplify.

1. $\frac{2}{3} + \frac{7}{3}$ **2.** $\frac{6}{12} + \frac{2}{12}$ **3.** $\frac{7}{8} + \frac{5}{8} + \frac{2}{8}$

Find the LCD.

4. $\frac{5}{6}, \frac{2}{3}$ **5.** $\frac{7}{12}, \frac{1}{8}$ **6.** $\frac{3}{4}, \frac{1}{2}, \frac{5}{6}$

Add and simplify.

7. $\frac{5}{6} + \frac{2}{3}$ **8.** $\frac{7}{12} + \frac{1}{8}$ **9.** $\frac{1}{3} + \frac{3}{4} + \frac{1}{8}$

Add and simplify.

10. $10\frac{1}{6} + 2\frac{1}{3}$ **11.** $14\frac{2}{3} + 9$ **12.** $4\frac{1}{6} + 6\frac{5}{8}$

13. $7\frac{1}{5} + 4\frac{1}{2} + 8\frac{3}{10}$ **14.** $\frac{3}{4} + 7\frac{1}{5}$ **15.** $9\frac{2}{3} + 6\frac{5}{12}$

Subtract and simplify.

16. $\frac{7}{8} - \frac{3}{8}$ **17.** $\frac{7}{8} - \frac{1}{4}$ **18.** $\frac{8}{9} - \frac{5}{6}$

Subtract and simplify.

19. $11\frac{5}{6} - 8\frac{1}{6}$ **20.** $12\frac{3}{10} - 9$ **21.** $7\frac{4}{9} - 2\frac{1}{3}$

22. $10\frac{2}{3} - 8\frac{4}{5}$ **23.** $29 - 5\frac{3}{8}$ **24.** $14\frac{3}{10} - 6\frac{9}{10}$

Application

25. Selma drank the following amounts of milk one day. How much did she drink in all?

Home	School	Restaurant
$3\frac{1}{8}$ cups	$2\frac{1}{4}$ cups	$1\frac{3}{4}$ cups

26. How much clothesline would it take to make a square $5\frac{2}{3}$ ft on a side?

MORE PRACTICE

Add and simplify. Use after page 197.

1. $\frac{1}{8}+\frac{7}{8}$ **2.** $\frac{5}{10}+\frac{2}{10}$ **3.** $\frac{2}{9}+\frac{4}{9}$ **4.** $\frac{9}{12}+\frac{15}{12}$

5. $\frac{2}{7}+\frac{9}{7}$ **6.** $\frac{3}{5}+\frac{9}{5}$ **7.** $\frac{8}{3}+\frac{7}{3}$ **8.** $\frac{5}{4}+\frac{1}{4}$

Find the LCD. Use after page 199.

9. $\frac{1}{3},\frac{1}{5}$ **10.** $\frac{3}{4},\frac{2}{5}$ **11.** $\frac{5}{8},\frac{2}{3}$ **12.** $\frac{7}{10},\frac{1}{3}$

13. $\frac{1}{2},\frac{3}{4}$ **14.** $\frac{5}{6},\frac{1}{3}$ **15.** $\frac{3}{10},\frac{2}{5}$ **16.** $\frac{5}{6},\frac{7}{12}$

17. $\frac{3}{4},\frac{1}{6},\frac{2}{3}$ **18.** $\frac{7}{8},\frac{5}{6},\frac{1}{2}$ **19.** $\frac{3}{8},\frac{5}{12},\frac{2}{9}$ **20.** $\frac{7}{10},\frac{3}{100},\frac{41}{50}$

Add and simplify. Use after page 201.

21. $\frac{1}{3}+\frac{1}{5}$ **22.** $\frac{3}{4}+\frac{2}{5}$ **23.** $\frac{5}{8}+\frac{2}{3}$ **24.** $\frac{7}{10}+\frac{1}{3}$

25. $\frac{1}{2}+\frac{3}{4}$ **26.** $\frac{1}{6}+\frac{1}{3}$ **27.** $\frac{7}{10}+\frac{2}{5}$ **28.** $\frac{5}{6}+\frac{7}{12}$

29. $\frac{3}{4}+\frac{1}{6}$ **30.** $\frac{7}{8}+\frac{5}{6}$ **31.** $\frac{3}{8}+\frac{5}{12}$ **32.** $\frac{7}{10}+\frac{5}{100}$

Add and simplify. Use after page 203.

33. $13\frac{2}{3}+9$ **34.** $15\frac{2}{3}+7\frac{3}{4}$ **35.** $8\frac{1}{2}+9\frac{1}{4}$ **36.** $5\frac{3}{4}+7\frac{3}{10}$

37. $12\frac{2}{5}+8\frac{1}{10}$ **38.** $8\frac{1}{6}+9\frac{1}{8}$ **39.** $5\frac{7}{8}+6\frac{3}{4}$ **40.** $17\frac{7}{10}+6\frac{3}{10}$

41. $18+5\frac{5}{9}$ **42.** $6\frac{2}{3}+5\frac{2}{3}$ **43.** $7\frac{1}{2}+9$ **44.** $14\frac{4}{5}+8\frac{1}{3}$

45. $1\frac{2}{3}+9\frac{5}{6}+7\frac{1}{2}$ **46.** $4\frac{1}{5}+10\frac{3}{4}+9\frac{1}{2}$ **47.** $7\frac{1}{10}+6\frac{2}{5}+14\frac{3}{4}$

48. $21\frac{1}{3}+9\frac{5}{6}+12\frac{1}{2}$ **49.** $14\frac{1}{6}+9\frac{2}{3}+6\frac{3}{4}$ **50.** $16\frac{4}{5}+2\frac{1}{6}+8\frac{3}{10}$

Subtract and simplify. Use after page 207.

1. $\frac{7}{8} - \frac{1}{8}$
2. $\frac{11}{12} - \frac{1}{12}$
3. $\frac{3}{5} - \frac{1}{2}$
4. $\frac{2}{3} - \frac{1}{4}$
5. $\frac{3}{4} - \frac{1}{6}$
6. $\frac{7}{10} - \frac{3}{5}$
7. $\frac{5}{12} - \frac{3}{8}$
8. $\frac{8}{9} - \frac{5}{6}$
9. $\frac{7}{12} - \frac{1}{10}$
10. $\frac{5}{6} - \frac{3}{8}$
11. $\frac{7}{10} - \frac{3}{100}$
12. $\frac{5}{16} - \frac{1}{12}$
13. $\frac{4}{5} - \frac{3}{4}$
14. $\frac{7}{9} - \frac{2}{3}$
15. $\frac{4}{7} - \frac{1}{2}$
16. $\frac{5}{8} - \frac{3}{5}$

Subtract and simplify. Use after page 209.

17. $8\frac{7}{8} - 5\frac{5}{8}$
18. $11\frac{9}{10} - 7\frac{3}{10}$
19. $15\frac{4}{5} - 8$
20. $12\frac{5}{6} - 11$
21. $13\frac{5}{6} - 7\frac{1}{3}$
22. $12\frac{9}{10} - 4\frac{2}{5}$
23. $11\frac{3}{8} - 4\frac{1}{6}$
24. $15\frac{8}{9} - 11\frac{5}{6}$
25. $14\frac{3}{4} - 6$
26. $11\frac{2}{3} - 7$
27. $12\frac{5}{6} - 5\frac{1}{6}$
28. $16\frac{11}{12} - 9\frac{5}{12}$
29. $21\frac{1}{2} - 18\frac{1}{3}$
30. $17\frac{5}{6} - 9\frac{3}{4}$
31. $8\frac{5}{9} - 3\frac{1}{4}$
32. $23\frac{6}{7} - 15\frac{5}{14}$
33. $32\frac{4}{5} - 6\frac{2}{15}$
34. $15\frac{7}{8} - 4\frac{1}{2}$
35. $29\frac{7}{10} - 17\frac{9}{20}$
36. $10\frac{7}{9} - 4\frac{1}{6}$

Subtract and simplify. Use after page 211.

37. $7 - 2\frac{1}{4}$
38. $10 - 5\frac{1}{8}$
39. $20 - 4\frac{7}{10}$
40. $52 - 39\frac{5}{6}$
41. $10\frac{1}{6} - 3\frac{5}{6}$
42. $12\frac{3}{8} - 4\frac{5}{8}$
43. $24\frac{1}{10} - 13\frac{9}{10}$
44. $18\frac{5}{12} - 3\frac{7}{12}$
45. $50\frac{1}{2} - 37\frac{3}{4}$
46. $11\frac{1}{2} - 8\frac{3}{5}$
47. $9\frac{1}{8} - 7\frac{1}{6}$
48. $12\frac{1}{10} - 8\frac{3}{4}$
49. $12 - 5\frac{2}{3}$
50. $21 - 8\frac{1}{2}$
51. $30 - 7\frac{3}{8}$
52. $16 - 6\frac{7}{11}$
53. $10\frac{2}{7} - 4\frac{3}{7}$
54. $15\frac{1}{3} - 9\frac{2}{3}$
55. $8\frac{1}{5} - 3\frac{4}{5}$
56. $20\frac{2}{9} - 15\frac{8}{9}$
57. $14\frac{2}{3} - 7\frac{7}{8}$
58. $10\frac{1}{6} - 3\frac{7}{12}$
59. $28\frac{1}{2} - 8\frac{2}{9}$
60. $17\frac{3}{5} - 6\frac{2}{3}$

CALCULATOR CORNER

A STRANGE CALCULATOR

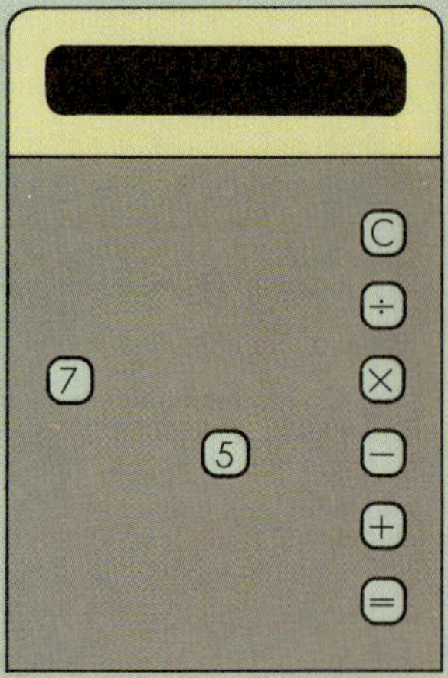

This strange calculator has only the digits [5] and [7]. It also has [C], [=], [+], [−], [×], and [÷].

Try these on your calculator.

1. To get a 6 on the display press [7] [÷] [7] [+] [5] [=]
2. To get a 9 on the display press [5] [÷] [5] [=] [+] [=] [+] [7] [=]
3. To get an 80 on the display press [7] [×] [5] [=] [+] [=] [+] [5] [=] [=]

Use the strange calculator described above. Tell how to get these numbers on the display. Write down your steps.

1. 30 2. 16 3. 50 4. 90

Suppose you had a calculator with only [4], [6], [C], [+], [−], [×], [÷] and [=].

Try these on your calculator.

1. To get a 5 on the display press [6] [÷] [6] [+] [4] [=]
2. To get 0.5 on the display press [6] [÷] [6] [+] [=] [÷] [4] [=]
3. To get 100 on the display press [6] [+] [4] [=] [×] [=]

Use this calculator for the following. Tell how to get these numbers on the display. Write down your steps.

1. 22 2. 34 3. 94 4. 0.25

READY FOR CHAPTER 8

Write the ratio. Simplify, if possible.

1. 6 problems in 4 minutes

2. 4 green apples out of a total of 7 apples

3. 2 blue marbles, 3 red marbles

4. 9 hits in 15 times at bat

Find the unit rate.

5. A man walks 8 miles in 2 hours. How many miles per hour is this?

6. If a car travels 96 kilometers on 8 liters of gas, how many kilometers per liter is this?

7. If 5 items cost 74¢, what is the cost per item? Round to tenths.

8. If a car travels 419 kilometers in 8 hours, how many kilometers per hour is this? Round to tenths.

Tell whether each proportion is true.

9. $\frac{12}{15} = \frac{6}{7}$

10. $\frac{5}{6} = \frac{15}{18}$

Solve.

11. $\frac{2}{3} = \frac{12}{x}$

12. $\frac{9}{8} = \frac{x}{24}$

13. $\frac{3}{7} = \frac{4}{x}$

14. $\frac{7}{5} = \frac{x}{8}$

15. If 5 items cost 85¢, how much would 9 items cost?

16. If a bicycle travels 45 kilometers in 9 hours, how many kilometers would be covered in 4 hours?

The figures are similar. Find x.

17.

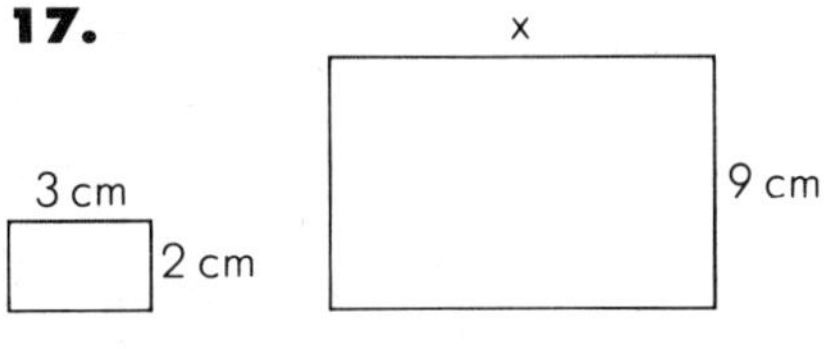

18.

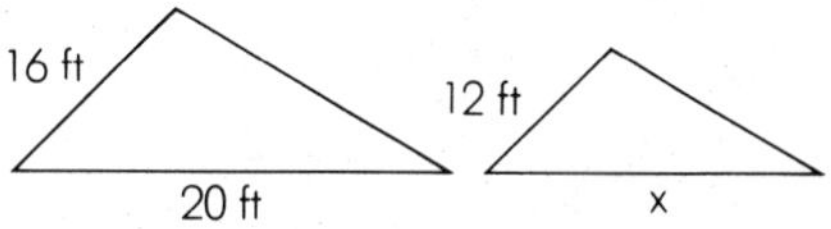

Application

19. Two liters of milk cost $ 0.95. Four liters cost $ 1.83. Which is the more economical buy?

20. A pack of 3 cassette tapes costs $ 8.99. The shipping cost is $ 1.08. What is the total cost?

8
RATIO AND PROPORTION

GETTING STARTED

THE GREATEST RATIO

WHAT'S NEEDED The 30 squares of paper with digits marked on them as in GOTCHA, Chapter 1. Take out the 0's. Leave the other 27 digits in the box.

THE RULES

1. On a sheet of paper each player writes the following. __ __ __ __ to __
2. A leader draws a digit from the box. Each player writes the digit on any one of the blanks. Once it is written it cannot be changed or moved.
3. The leader draws four more digits, one at a time. Write each digit on a blank as it is drawn.
4. Each player now has a ratio.
5. Who has the greatest ratio? Find out by dividing the last digit into the first four.

SAMPLE 7 6 4 5 to 3

```
    2 5 4 8.3
 3)7 6 4 5.0
   6
   1 6
   1 5
     1 4
     1 2
       2 5
       2 4
         1 0
           9
           1
```

The player with the largest quotient has the greatest ratio and is the winner.

RATIOS

The ratio of wins to losses for San Francisco is 57 to 38. We write the ratio as $\frac{57}{38}$.

The ratio of losses to wins for Cincinnati is 39 to 54. We write the ratio as $\frac{39}{54}$.

The ratio of wins to games played (wins plus losses) for Los Angeles is 54 to 94. We write $\frac{54}{94}$.

National League West

	Won	Lost
San Francisco	57	38
Cincinnati	54	39
Los Angeles	54	40

EXAMPLES. Write the ratio.

1. pennies to dimes $\frac{4}{3}$ (4 to 3)

2. dimes to pennies $\frac{3}{4}$ (3 to 4)

3. pennies to coins $\frac{4}{7}$ (4 to 7)

TRY THIS

Write the ratio.

1. Circles to squares

2. Squares to circles

3. Squares to figures

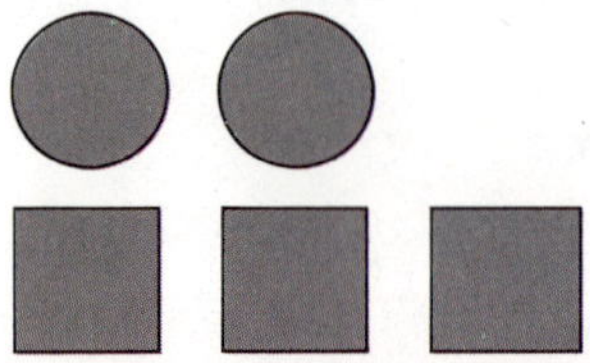

EXAMPLES. Write the ratio.

4. 7 problems in 5 minutes $\frac{7}{5}$

5. 17 breaths per 60 seconds $\frac{17}{60}$

6. 3 apples for 78¢ $\frac{3}{78} = \frac{1}{26}$

7. 210 people on 7 buses $\frac{210}{7} = \frac{30}{1}$ *Leave 1 as the denominator.*

TRY THIS

Write the ratio. Simplify, if possible.

4. 12 miles in 5 hours
5. 7 points to 10 points
6. 72 heartbeats per 60 seconds
7. 400 pennies for 4 dollars
8. 25 pennies for 5 nickels
9. 16 baskets in 20 tries
10. 8 hits in 32 times at bat
11. 5 miles per 30 minutes

PRACTICE

Write the ratio.

Boston: Won 62, Lost 28

1. Wins to losses
2. Losses to wins
3. Wins to games played

New York: Won 48, Lost 38

4. Wins to losses
5. Losses to wins
6. Losses to games played

4 dogs, 3 cats

7. Dogs to cats
8. Cats to dogs
9. Dogs to animals

3 dimes, 5 nickels

10. Dimes to nickels
11. Nickels to dimes
12. Nickels to coins

Write the ratio. Simplify, if possible.

13. 210 kilometers in 3 hours
14. 2 parts hydrogen for 1 part oxygen
15. 200 books on 5 shelves
16. 2 eggs per 3 servings
17. 8 hours sleep in 24 hours
18. 2 teachers for 50 students
19. 27 hits in 100 times at bat
20. 2000 grams in 2 kilograms
21. 100 miles on 5 gallons of gas
22. 10 nickels for 2 quarters
23. 17 right out of 20 problems
24. 3 wrong out of 20 problems

More Practice, page 242

UNIT RATES

A **rate** is a ratio used to compare two measures. A **unit rate** is the rate for (per) one.

Maria drives 162 miles in 3 hours. How many miles per hour does she travel?

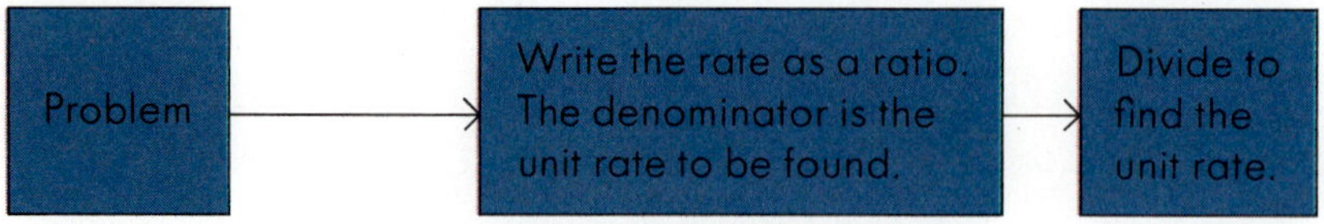

162 miles in 3 hours $\qquad \dfrac{162 \text{ miles}}{3 \text{ hours}} \qquad 3\overline{)162}$ = 54 $\qquad$ Maria travels 54 mph.

EXAMPLE 1. A car travels 272 kilometers on 34 liters of gas. How many kilometers per liter is this?

$\dfrac{272 \text{ km}}{34 \text{ L}} \qquad 34\overline{)272}$ = 8 $\qquad$ *Write the rate. Divide.*
The unit rate is 8 kilometers per liter.

EXAMPLE 2. A car travels 272 kilometers on 34 liters of gas. How many liters per kilometer is this?

$\dfrac{34 \text{ L}}{272 \text{ km}}$

```
        0.1 2 ≈ 0.1
2 7 2)3 4.0 0
      2 7 2
        6 8 0
        5 4 4
        1 3 6
```

Write the rate.
Divide. Round to tenths.
The car uses 0.1 liter per kilometer.

The unit rate for money is often called the unit cost or unit price.

EXAMPLE 3. 30 tablets cost 80 cents. Find the cost per tablet.

$\dfrac{80¢}{30 \text{ tablets}}$

```
      2.6 6 ≈ 2.7
3 0)8 0.0 0
    6 0
    2 0 0
    1 8 0
      2 0 0
      1 8 0
        2 0
```

Write the rate.
Divide. Round to tenths.
The cost is 2.7¢ per tablet.

TRY THIS

Find the unit rate.

1. A car travels 98 miles on 5 gallons of gas. How many miles per gallon is this? Round to the nearest tenth.
2. 7 oranges cost 84¢. What is the unit price?
3. Pierre receives $ 26.75 for 5 hours of work. What is his hourly wage?

PRACTICE

Find the unit rate.

1. Kay drives 1000 kilometers in 20 hours. How many kilometers per hour does she drive?
2. A car travels 96 miles on 6 gallons of gas. How many miles per gallon is this?
3. 10 pounds of grapes cost $ 3.75. What is the unit price? Round to the nearest tenth.
4. Charlie sprints 100 meters in 16 seconds. How many meters per second is this? Round to the nearest tenth.
5. John sprints 100 yards in 10 seconds. How many yards per second is this?
6. John sprints 100 yards in 10 seconds. How many seconds per yard is this? Round to the nearest tenth.
7. Miko serves 5 kilograms of hamburger to 15 people. How many kilograms per person is this? Round to the nearest tenth.
8. Miko serves 5 kilograms of hamburger to 15 people. How many people per kilogram is this?
9. If 5 apples cost 94¢, how much does 1 apple cost? Round to the nearest tenth.
10. If 3 ties cost $ 28.74, how much does 1 tie cost?
11. Lori receives $ 29.95 for working 9 hours. What is her hourly wage?
12. Jeff reads 102 pages in 4 hours. How many pages per hour is this? Round to the nearest tenth.

More Practice, page 242

APPLICATION

COMPARISON SHOPPING

We use unit prices to find the most economical buy.

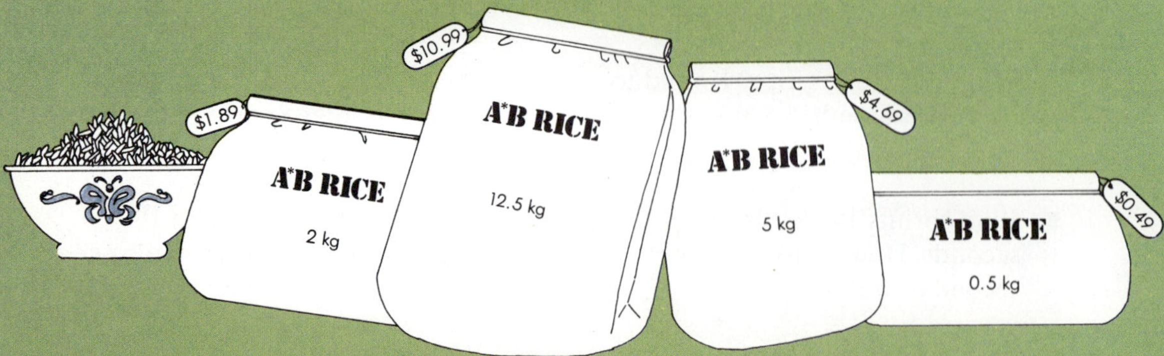

Which package of rice is the more economical: the 2 kg package or the 0.5 kg package? To find the unit price, divide the price by the amount.

2 kg $\longrightarrow$ \$ 1.89 $\div$ 2 = \$ 0.95 per kg   0.5 kg $\longrightarrow$ \$ 0.49 $\div$ 0.5 = \$ 0.98 per kg

The 2 kg package is the more economical.

Which is the more economical buy on rice?

1. the 12.5 kg package or the 5 kg package
2. the 5 kg package or the 2 kg package
3. the 12.5 kg package or the 0.5 kg package

What is the unit price of each liquid detergent? Round to the nearest tenth.

4. 12 oz for 79¢
5. 16 oz for \$ 1.09
6. 22 oz for \$ 1.39
7. 48 oz for \$ 2.49
8. Which size is the most economical?
9. Which is the least economical?

Sometimes the unit price is the cost per unit of prepared product.

10. Find the cost per prepared quart of each lemon drink.
11. Assume the quality is equal. Which is the best buy?

12. Find the cost per liter for each batter.
13. If the quality is equal, which is the better buy?

Add for the total cost.

14. Find the cost per ounce of each soup.
15. How many ounces of soup can be made from the condensed soup?
16. From which can would you get the greatest number of ounces of prepared soup?
17. Find the cost per prepared ounce of each soup.
18. If the quality is equal, which is the better buy?

PROPORTIONS

A **proportion** is a sentence that says two ratios are equal. If the cross products are the same, the proportion is true.

Tell whether the proportion is true. $\frac{2}{3} = \frac{6}{9}$

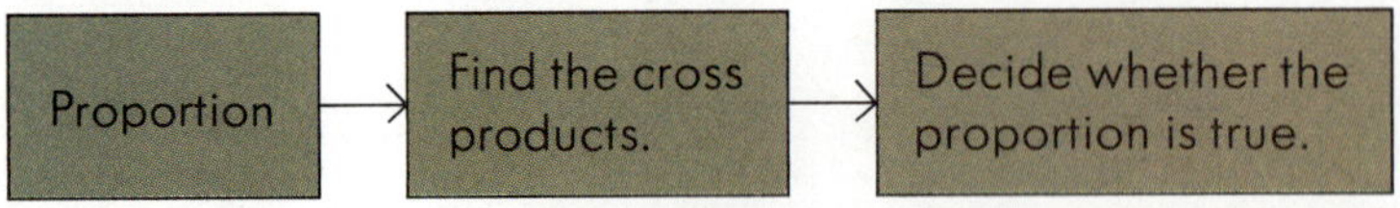

$\frac{2}{3} = \frac{6}{9}$ $\qquad$ $\frac{2}{3} = \frac{6}{9}$

$2 \cdot 9 = 18$
$3 \cdot 6 = 18$
$18 = 18$
The proportion is true.

The ratios 2 to 3 and 6 to 9 are equal. We say 2 is to 3 as 6 is to 9.

EXAMPLES. Tell whether each proportion is true.

1. $\frac{6}{8} = \frac{3}{4}$ *Find the cross products.*
$6 \cdot 4 = 24$
$8 \cdot 3 = 24$
$24 = 24$ *The proportion is true.*

2. $\frac{4}{6} = \frac{8}{10}$
$4 \cdot 10 = 40$
$6 \cdot 8 = 48$
$40 \neq 48$ *The proportion is not true.*

TRY THIS Tell whether each proportion is true.

1. $\frac{1}{2} = \frac{2}{3}$ **2.** $\frac{5}{10} = \frac{2}{4}$ **3.** $\frac{6}{10} = \frac{3}{5}$ **4.** $\frac{5}{6} = \frac{4}{5}$

EXAMPLE 3. Solve. $\frac{5}{8} = \frac{10}{x}$

$5 \cdot x = 8 \cdot 10$ *Find the cross products.*
$5x = 80$
$x = \frac{80}{5}$ *Since 80 is 5 times x, we can find x by dividing by 5.*
$x = 16$ *Solve.*

EXAMPLES. Solve.

4. $\frac{x}{5} = \frac{2}{3}$

$x \cdot 3 = 5 \cdot 2$

$3x = 10$

$x = \frac{10}{3} = 3\frac{1}{3}$

5. $\frac{x}{7} = \frac{3}{4}$

$x \cdot 4 = 7 \cdot 3$

$4x = 21$

$x = \frac{21}{4} = 5\frac{1}{4}$ or 5.25

TRY THIS

Solve.

5. $\frac{3}{5} = \frac{12}{x}$ **6.** $\frac{9}{x} = \frac{3}{8}$ **7.** $\frac{4}{5} = \frac{x}{25}$

8. $\frac{1}{2} = \frac{x}{5}$ **9.** $\frac{7}{6} = \frac{35}{x}$ **10.** $\frac{9}{1} = \frac{36}{x}$

11. $\frac{x}{4} = \frac{7}{6}$ **12.** $\frac{5}{1} = \frac{x}{6}$ **13.** $\frac{x}{7} = \frac{4}{9}$

PRACTICE

Tell whether each proportion is true.

1. $\frac{1}{2} = \frac{5}{10}$ **2.** $\frac{2}{3} = \frac{4}{6}$ **3.** $\frac{5}{2} = \frac{10}{4}$ **4.** $\frac{8}{6} = \frac{4}{3}$ **5.** $\frac{3}{4} = \frac{4}{7}$

6. $\frac{4}{5} = \frac{20}{25}$ **7.** $\frac{2}{7} = \frac{6}{21}$ **8.** $\frac{75}{25} = \frac{1}{3}$ **9.** $\frac{50}{1} = \frac{100}{2}$ **10.** $\frac{42}{3} = \frac{51}{2}$

Solve.

11. $\frac{1}{2} = \frac{x}{6}$ **12.** $\frac{1}{2} = \frac{6}{x}$ **13.** $\frac{1}{4} = \frac{16}{x}$ **14.** $\frac{1}{4} = \frac{x}{16}$ **15.** $\frac{1}{5} = \frac{x}{10}$

16. $\frac{x}{5} = \frac{2}{10}$ **17.** $\frac{5}{x} = \frac{10}{20}$ **18.** $\frac{8}{12} = \frac{4}{x}$ **19.** $\frac{25}{75} = \frac{1}{x}$ **20.** $\frac{x}{20} = \frac{4}{2}$

21. $\frac{1}{2} = \frac{x}{7}$ **22.** $\frac{1}{3} = \frac{x}{7}$ **23.** $\frac{1}{6} = \frac{7}{x}$ **24.** $\frac{3}{8} = \frac{7}{x}$ **25.** $\frac{3}{4} = \frac{x}{100}$

26. $\frac{5}{12} = \frac{x}{6}$ **27.** $\frac{9}{10} = \frac{x}{5}$ **28.** $\frac{5}{x} = \frac{9}{11}$ **29.** $\frac{9}{x} = \frac{2}{3}$ **30.** $\frac{2}{5} = \frac{8}{x}$

31. $\frac{1}{2} = \frac{x}{9}$ **32.** $\frac{1}{4} = \frac{x}{9}$ **33.** $\frac{8}{4} = \frac{5}{x}$ **34.** $\frac{2}{7} = \frac{5}{x}$ **35.** $\frac{3}{4} = \frac{x}{9}$

36. $\frac{x}{4} = \frac{5}{8}$ **37.** $\frac{x}{5} = \frac{8}{6}$ **38.** $\frac{7}{x} = \frac{4}{5}$ **39.** $\frac{2}{x} = \frac{5}{9}$ **40.** $\frac{3}{8} = \frac{x}{4}$

More Practice, page 242

SOLVING PROBLEMS USING PROPORTIONS

How far is it from Griffith to West Lafayette?

The legend on the map tells us the ratio: 0.5 centimeters to 25 kilometers.

We measure and find that on the map it is 3 centimeters between the cities.

Think: 0.5 cm is to 25 km as 3 cm is to x km.

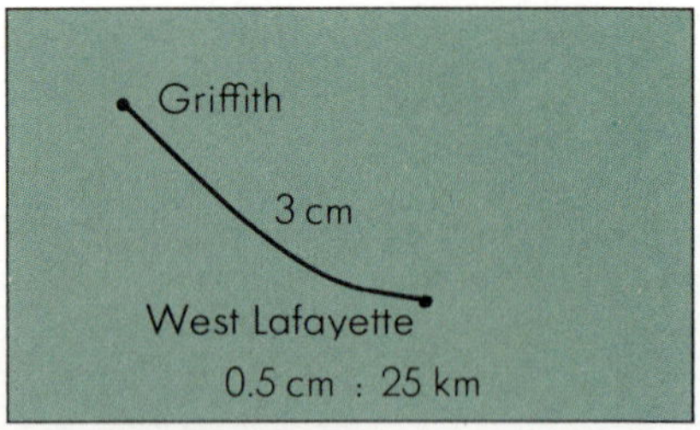

Write the proportion. → Find the cross products. → Solve.

$$\frac{0.5}{25} = \frac{3}{x}$$

$$0.5 \cdot x = 25 \cdot 3$$
$$0.5x = 75$$

$$x = \frac{75}{0.5}$$
$$x = 150$$

It is 150 km from Griffith to West Lafayette.

EXAMPLE 1. A car can travel 100 kilometers on 8 liters of gas. How far can it travel on 10 liters?

Think: 100 km is to 8 L as x km is to 10 L.

$\frac{100}{8} = \frac{x}{10}$ *Write the proportion.*

$100 \cdot 10 = 8 \cdot x$ *Find the cross products.*

$1000 = 8x$

$\frac{1000}{8} = x$ *Solve.*

$125 = x$ The car can travel 125 km on 10 liters of gas.

EXAMPLE 2. Potatoes cost \$ 1.80 for 3 kilograms. How much did 7 kilograms cost?

Think: 3 kg is to \$ 1.80 as 7 kg is to x.

$\frac{3}{180} = \frac{7}{x}$ *Write the proportion.*

$3 \cdot x = 7 \cdot 180$ *Find the cross products.*

$3x = 1260$

$x = \frac{1260}{3} = 420$ 7 kg of potatoes cost \$ 4.20.

TRY THIS

1. Oranges were 8 for 96 cents. How many could you buy for 60 cents?
2. A car traveled 159 miles in 3 hours. At this rate, how far did it travel in 8 hours?

PRACTICE

Solve.

1. A football team has won 3 out of its first 4 games. At this rate how many will it win in 16 games?
2. On a map $\frac{1}{2}$ inch represents 100 miles. What distance does $2\frac{1}{2}$ inches represent?
3. If you can do 3 problems in 15 seconds, how long will it take to do 10 problems?
4. If 8 tickets cost \$ 6.00, how many can you buy for \$ 9.00?
5. A pitcher struck out 3.4 batters per 9 innings. At this rate how many would be struck out in 297 innings?
6. The property tax on a home is \$ 5 per \$ 1000 assessed evaluation. What is the tax on a \$ 45,000 home?
7. If 9 pounds of hamburger cost \$ 14.58, how much do 6 pounds cost?
8. A clock loses 3 minutes every 12 hours. How much will it lose in 72 hours?
9. A company car was driven 4200 miles in the first 4 months. At this rate, how far will it be driven in 12 months?
10. Mary had 7 hits in her first 20 times at bat. At this rate how many hits will she get in 340 times at bat?
11. Jerry earned \$ 32.80 for 8 hours of work. If he earned \$ 82.00, how many hours did he work?
12. A car travels 150 kilometers on 10 liters of gas. How many liters of gas are needed to travel 500 kilometers?
13. If 3 records cost \$ 16.50, how much do 5 records cost?
14. If 6 boxes of raisins cost \$ 2, how many boxes can you buy for \$ 4.80?
15. If 2 cans of tennis balls cost \$ 6.50, how much do 5 cans cost?
16. If you make 3 baskets in 5 tries, how many baskets can you make in 20 tries?

More Practice, page 243

SIMILAR FIGURES AND PROPORTIONS

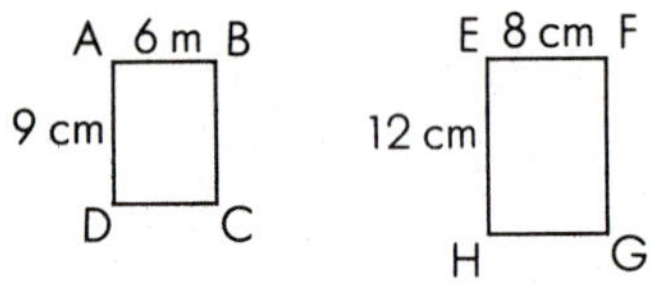

Figure $ABCD$ is similar to figure $EFGH$. Sides $\overline{AB}$ and $\overline{EF}$ are corresponding. Sides $\overline{AD}$ and $\overline{EH}$ are corresponding. In similar figures corresponding lengths are proportional. $\frac{6}{9} = \frac{8}{12}$

The figures are similar. Find x.

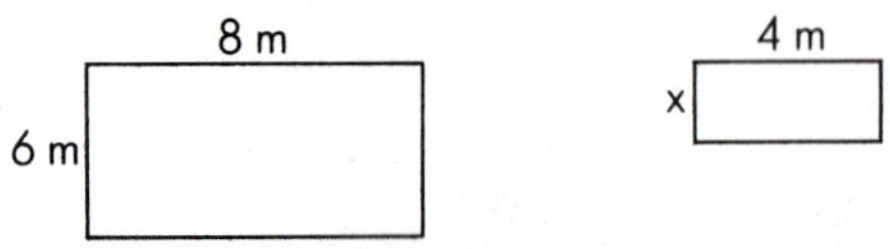

Write the proportion. → Find the cross products. → Solve.

$$\frac{8}{6} = \frac{4}{x}$$

$$8 \cdot x = 6 \cdot 4$$
$$8x = 24$$

$$x = \frac{24}{8}$$
$$x = 3$$

The side is 3 m long.

EXAMPLE 1. The figures are similar. Find x.

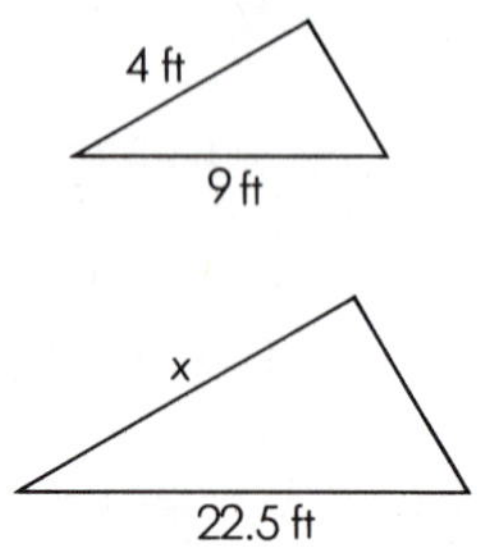

$\frac{4}{9} = \frac{x}{22.5}$ *Write the proportion.*

$4 \cdot 22.5 = 9 \cdot x$ *Find the cross products.*

$90 = 9x$

$\frac{90}{9} = x$ *Solve.*

$10 = x$ *The side is 10 feet long.*

TRY THIS

The figures are similar. Find x.

1.

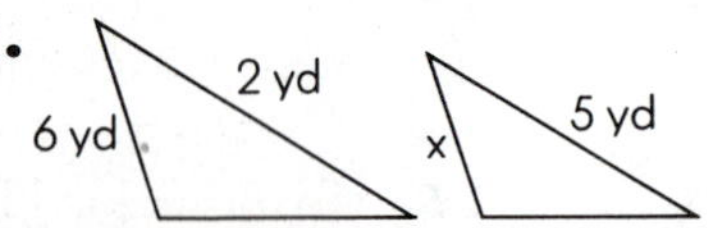

2.

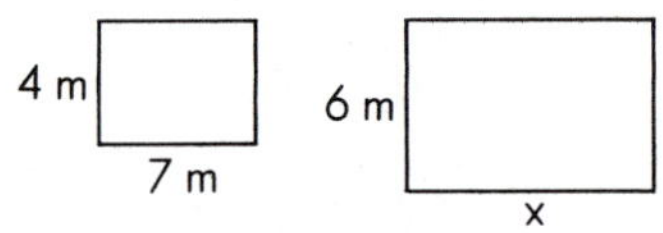

PRACTICE

The figures are similar. Find *x*.

1.

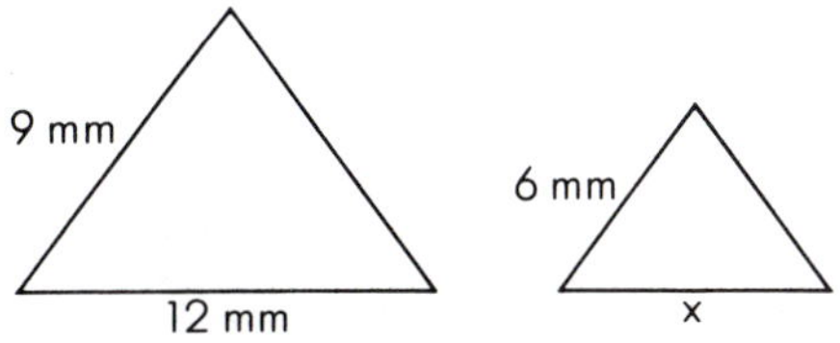

2.

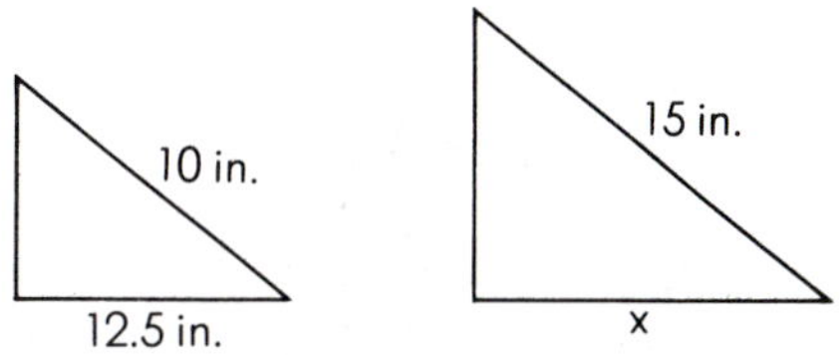

3.

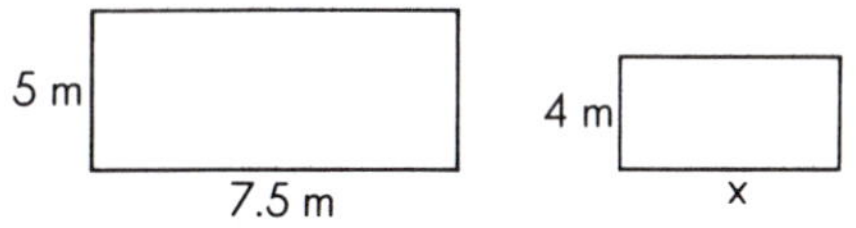

4.

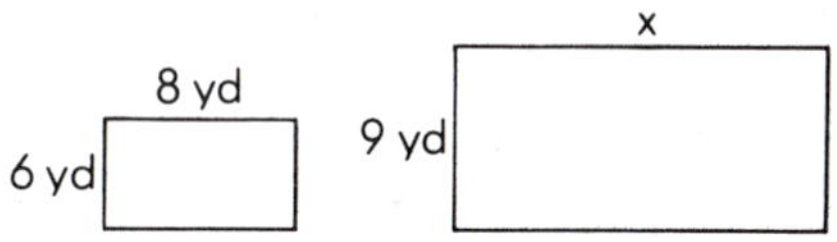

5.

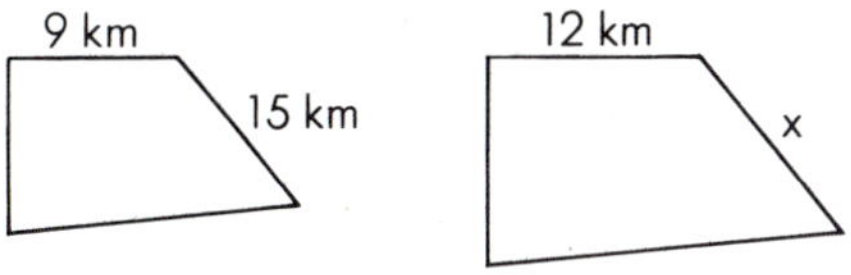

6.

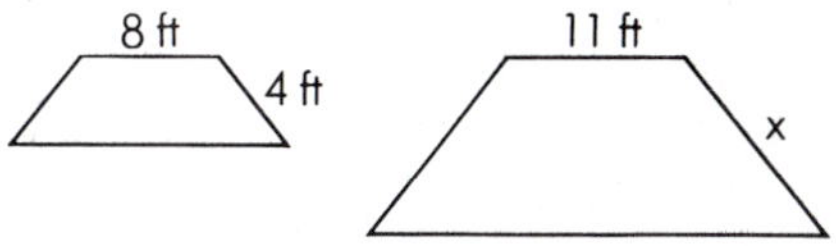

7.

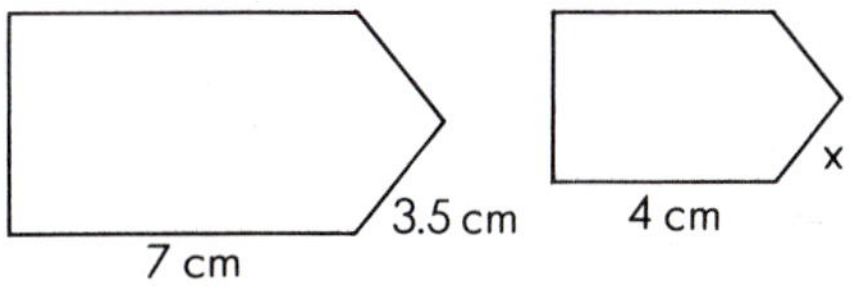

8.

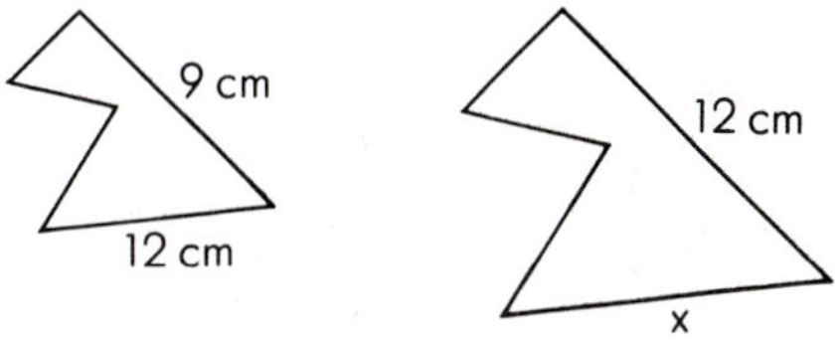

PROBLEM CORNER

How tall is the flagpole?

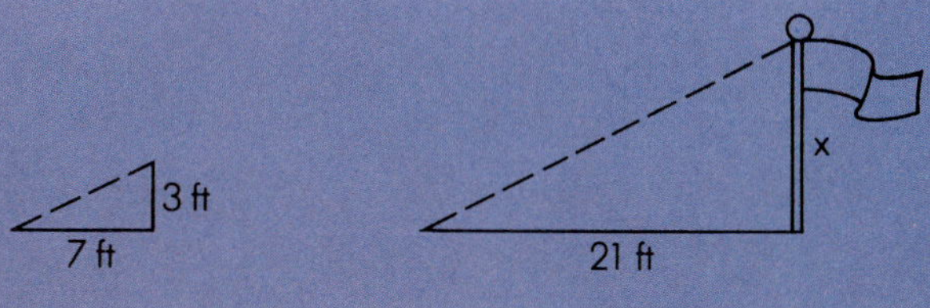

More Practice, page 243

APPLICATION

CATALOG SHOPPING

Many items can be purchased from a catalog. Product and shipping information is written in code.

KNIT SHIRTS IN 3 COLORS

Polyester and Cotton

11 Red 12 Blue 13 Green
Machine wash and dry

KNIT SHIRTS **Polyester and Cotton**

Child sizes 8-10-12-14-16-18

A. 12D-3211-T **B.** 12D-3212-T
C. 12D-3213-T

(style ↗ and color ↖ label parts of the code)

Shipping weight 7 oz $5.99 pack of 3 $16.95

Adult sizes S-M-L-XL-XXL

A. 12D-3611-T **B.** 12D-3612-T **C.** 12D-3613-T

Shipping weight 12 oz $6.99 pack of 3 $19.65

Find the cost and shipping weight of the following.

1. 1 child's shirt, size 12
2. 1 adult's shirt, size XL
3. 1 pack of 3 children's shirts, size 16
4. 1 pack of 3 adults' shirts, size M

How much do you save per shirt by buying a pack of 3 shirts?

5. Child sizes
6. Adult sizes

Consumer Hint

The unit cost of shipping ($ per ounce) decreases as the weight increases. When possible plan ahead and purchase several items in one order.

Shipping charges are added to the cost of the items in the order. Find the shipping cost for each order.

APPROXIMATE DISTANCE FROM CATALOG DISTRIBUTION CENTER								
Shipping Weight	Local Zone	Zones 1 and 2 Not over 150 Miles	Zone 3 151 to 300 Miles	Zone 4 301 to 600 Miles	Zone 5 601 to 1000 Miles	Zone 6 1001 to 1400 Miles	Zone 7 1401 to 1800 Miles	Zone 8 Over 1800 Miles
1 oz to 8 oz	$0.28	$0.36	$0.36	$0.38	$0.43	$ 0.46	$ 0.51	$ 0.56
9 oz to 15 oz	0.68	0.74	0.74	0.74	0.81	0.90	0.99	1.08
1 lb to 2 lbs	0.75	0.85	0.90	1.10	1.20	1.30	1.40	1.50
2 lbs 1 oz to 3 lbs	0.80	1.00	1.05	1.20	1.30	1.40	1.60	1.75
3 lbs 1 oz to 5 lbs	0.90	1.15	1.25	1.35	1.50	1.75	1.95	2.05
5 lbs 1 oz to 10 lbs	1.10	1.40	1.50	1.75	2.10	2.45	2.85	3.00
10 lbs 1 oz to 15 lbs	1.30	1.80	2.00	2.25	2.85	3.30	3.75	4.20

7. 3 lb 10 oz, 200 miles
8. 10 lb 8 oz, 1700 miles
9. 14 oz, 10 miles
10. 1 lb 12 oz, New York to Los Angeles
11. 130 oz, 1000 miles (16 oz = 1 lb)

Find the cost and weight of each item.

COST = QUANTITY × UNIT PRICE WEIGHT = QUANTITY × UNIT WEIGHT

	Quantity	Number	Item-Size	Unit Price	Unit Weight	Cost	Weight
12.	3	623BF	blouse-M	$4.95	9 oz		
13.	2	3794BR	pants-28	$9.95	1 lb 2 oz		
14.	1	5321SM	art brushes set of 6	$7.62	12 oz		
15.	5	2634PX	tube of oil paint	$1.99	12 oz		
					Subtotal		
					Shipping Cost		
					Total Cost		

16. Find the subtotal cost and total weight of the four items.
17. Use the shipping chart to find the shipping cost to zone 3.
18. Find the total cost.

What is the actual unit cost of the following items? (Catalog cost + shipping cost)

19. 1 pair of pants sent to zone 1
20. 1 tube of oil paint sent to zone 6

MERCHANDISING

Betsy Rathman owns the Melody Music Store. She sells records, musical instruments, and provides music lessons.

Betsy finds that about $\frac{1}{6}$ of all Americans play a musical instrument. Find the number of instrument players in these areas.

1. New England: population, 12,600,000
2. West Coast: population, 28,620,000
3. Southwest: population, 18,610,000
4. Great Lakes: population, 42,250,000

A cash discount is given for each item purchased at some sales at the Melody Music store.

For each of the following find the sale price. Round to the nearest cent.

5. $4.95 6. $12.95
7. $7.95 8. $21.95

Nov. 5, 1983 City Gazette

MELODY MUSIC SPECIAL
1/4 Off All Records, Tapes, Cassettes
1 WEEK ONLY NOV. 6–12

Melody Music employs teachers to give music lessons.

ORGAN AND PIANO Private Lessons	GUITAR Private Lessons
\$9.50 -- $\frac{1}{2}$ hour \$17.50 -- 1 hour	\$8.50 -- $\frac{1}{2}$ hour \$14 -- 1 hour
50 $\frac{1}{2}$ - hour lessons -- \$425	50 $\frac{1}{2}$ - hour lessons -- \$325

9. For each type of lesson which rate is the best value?
10. For which set of 50 lessons is a larger discount given? How much would each of the 50 lessons cost?
11. Find the cost of the best value for $\frac{1}{2}$-hour piano lessons, 1 per week for 1 year.
12. Teachers suggest practicing 15 minutes daily for each $\frac{1}{2}$-hour lesson per week. For a 1-hour lesson each week, how much practice time per week is suggested?

Summary Sheet			
Item	Quantity	Wholesale	Retail
LP Singles	1820	\$7134.40	\$15,690.50
LP Albums	365	\$2151.30	\$ 4,732.80
45 Singles	2360	\$1132.80	\$ 2,832.00
Classical Album Sets	20	\$ 532.60	\$ 1,270.75

For each department Betsy keeps inventory records she calls summary sheets. These tell hei when to reorder. They also give the wholesale and retail value of the stock. The **markup** is the difference between the wholesale and retail costs. The **rate of markup** is the retail cost divided by the wholesale cost.

13. For each group of items find the markup.
14. For each item what is the rate of markup?
15. Which of the four items has the highest rate of markup?

CHAPTER 8 REVIEW

Write the ratio. Simplify, if possible. p. 224

1. 4 peaches to 3 plums

2. 9 questions in 12 minutes

3. 4 cups per 2 servings

4. 2 dollars for 8 quarters

Find the unit rate. p. 226

5. If 3 items cost 93¢, what is the cost per item?

6. If 5 hours of lessons cost $ 16.25, what is the cost per hour?

7. If a person drives 187 kilometers in 4 hours, how many kilometers per hour is this? Round to tenths.

8. If a car travels 386 miles on 20 gallons of gas, how many miles per gallon is this? Round to tenths.

Tell whether each proportion is true. p. 230

9. $\frac{7}{8} = \frac{6}{7}$

10. $\frac{6}{8} = \frac{3}{4}$

11. $\frac{15}{18} = \frac{5}{6}$

12. $\frac{15}{10} = \frac{3}{2}$

Solve. p. 232

13. $\frac{5}{6} = \frac{15}{x}$

14. $\frac{2}{x} = \frac{6}{21}$

15. $\frac{4}{3} = \frac{x}{15}$

16. $\frac{13}{x} = \frac{6}{7}$

17. $\frac{3}{4} = \frac{7}{x}$

18. $\frac{x}{7} = \frac{4}{8}$

19. $\frac{7}{5} = \frac{x}{9}$

20. $\frac{7}{x} = \frac{4}{5}$

21. If there are 14 books for 21 students, how many are there for 15 students?

22. If 3 items cost $ 7.50, how much would 5 items cost?

The figures are similar. Find x. p. 234

23.

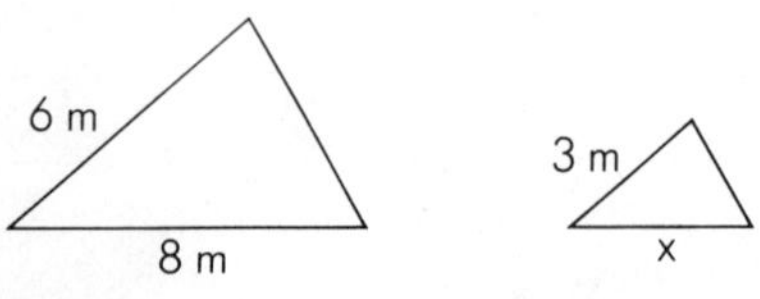

24.

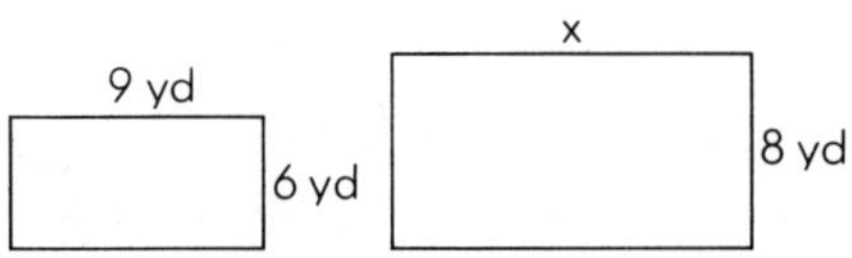

Application pp. 228, 236

25. A 15-ounce can of tomato sauce costs $ 0.47. A 29-ounce can costs $ 0.78. Which is the more economical buy?

26. A box of golf balls costs $ 15.39. The shipping cost is $ 0.90. What is the total cost?

CHAPTER 8 TEST

Write the ratio. Simplify, if possible.

1. 7 pages per 2 hours

2. 9 chairs to 6 desks

3. 12 pens to 9 pencils

4. 8 hits in 23 times at bat

Find the unit rate.

5. If a woman's wages are $ 18.75 for 3 hours of work, what is the amount per hour?

6. If a car travels 184 miles on 8 gallons of gas, how many miles per gallon is this?

7. If a man drives 798 kilometers in 4 hours, how many kilometers per hour is this?

8. If 3 items cost 77¢, what is the cost per item?

Tell whether each proportion is true.

9. $\frac{9}{15} = \frac{3}{5}$

10. $\frac{4}{3} = \frac{12}{18}$

Solve.

11. $\frac{6}{3} = \frac{12}{x}$

12. $\frac{7}{8} = \frac{x}{24}$

13. $\frac{5}{6} = \frac{9}{x}$

14. $\frac{3}{8} = \frac{x}{20}$

15. If 8 items cost $ 20.00, what is the cost for 6 items?

16. If Joe jogs 28 miles in 4 hours, how far will he jog in 9 hours?

The figures are similar. Find x.

17.

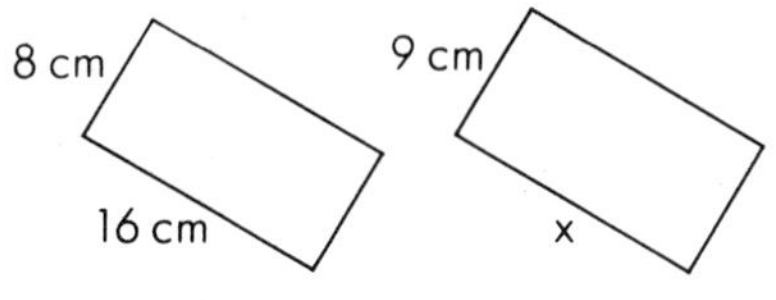

18.

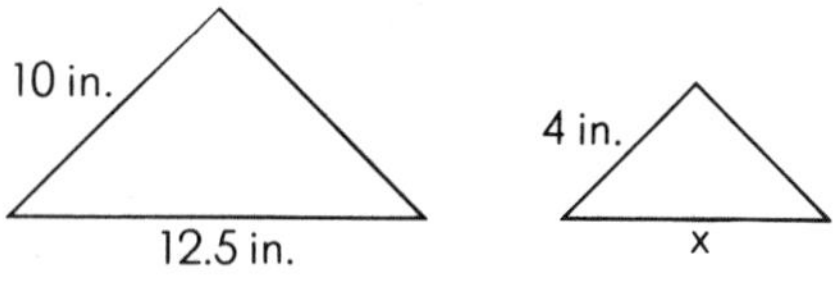

Application

19. A 12-ounce can of juice concentrate costs $ 1.39. The concentrate must be mixed with 3 cans of water. Find the cost per prepared ounce of juice.

20. A sleeping bag costs $ 32.98. The shipping cost is $ 1.95. What is the total cost?

MORE PRACTICE

Write the ratio. Simplify, if possible. Use after page 225.

1. 10 tires to 2 cars

2. 8 wheel covers to 2 cars

3. 9 stitches to 3 centimeters

4. 3 dresses in 6 hours

5. 8 carries for 40 yards

6. 12 hits in 36 times at bat

7. 56 goals in 8 games

8. 5 kicks for 170 yards

Find the unit rate. Use after page 227.

9. If a dozen eggs cost 96¢, how much does 1 egg cost?

10. 10 pounds of potatoes cost $ 1.50. What is the price per pound?

11. Ann receives $ 23.75 for working 5 hours. How much is this per hour?

12. Lester drives 458.4 miles in 8 hours. How many miles per hour is this?

13. Nora drives 179 kilometers in 2 hours. How many kilometers per hour is this? Round to the nearest tenth.

14. A car travels 77 miles on 3 gallons of gas. How many miles per gallon is this? Round to the nearest tenth.

15. Bonnie reads 185 pages in 3 hours. How many pages per hour does she read? Round to the nearest tenth.

16. A van travels 148 kilometers on 9 liters of gas. How many kilometers per liter is this? Round to the nearest tenth.

Tell whether each proportion is true. Use after page 231.

17. $\frac{2}{3} = \frac{3}{4}$ **18.** $\frac{7}{8} = \frac{8}{7}$ **19.** $\frac{5}{6} = \frac{10}{12}$ **20.** $\frac{1}{3} = \frac{3}{1}$

21. $\frac{4}{6} = \frac{6}{9}$ **22.** $\frac{9}{12} = \frac{6}{9}$ **23.** $\frac{6}{7} = \frac{7}{8}$ **24.** $\frac{4}{8} = \frac{7}{14}$

Solve. Use after page 231

25. $\frac{7}{8} = \frac{x}{16}$ **26.** $\frac{5}{4} = \frac{25}{x}$ **27.** $\frac{3}{9} = \frac{x}{3}$ **28.** $\frac{12}{16} = \frac{6}{x}$

29. $\frac{1}{2} = \frac{x}{5}$ **30.** $\frac{3}{4} = \frac{10}{x}$ **31.** $\frac{6}{x} = \frac{5}{7}$ **32.** $\frac{x}{9} = \frac{8}{7}$

Solve. Use after page 233.

1. If 8 pencils cost 48¢, how many pencils can be purchased for 72¢?

2. If 8 pencils cost 48¢, how much will 10 pencils cost?

3. Ken receives $ 11.25 for 3 hours of work. At that rate how much will he receive for working 2 hours?

4. Alice is paid $ 11.25 for working 3 hours. How many hours does she work to receive $ 18.75?

5. There are 70 ounces of juice in 5 bottles. How many ounces are there in 7 bottles?

6. If 5 bottles contain 60 ounces, how many bottles contain 96 ounces?

7. There are 165 minutes in 3 class periods. How many minutes are there in 5 class periods?

8. If 3 class periods are 165 minutes long, how many class periods are there in 110 minutes?

The figures are similar. Find x. Use after page 235.

9.

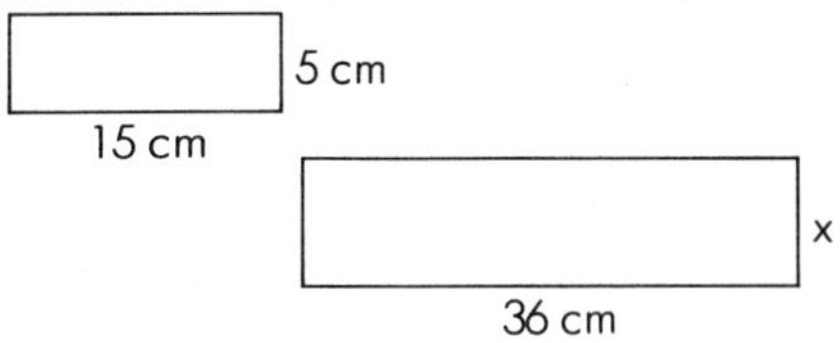

10.

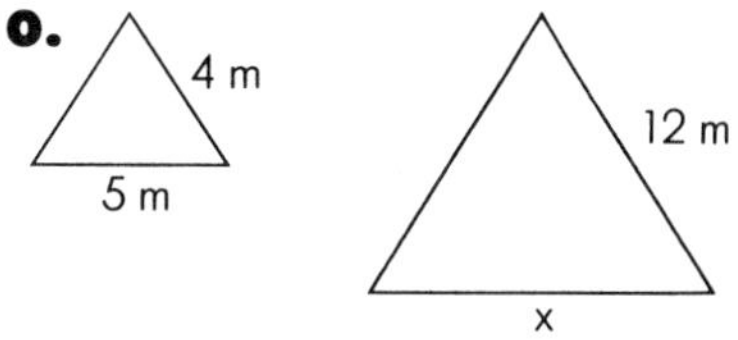

11.

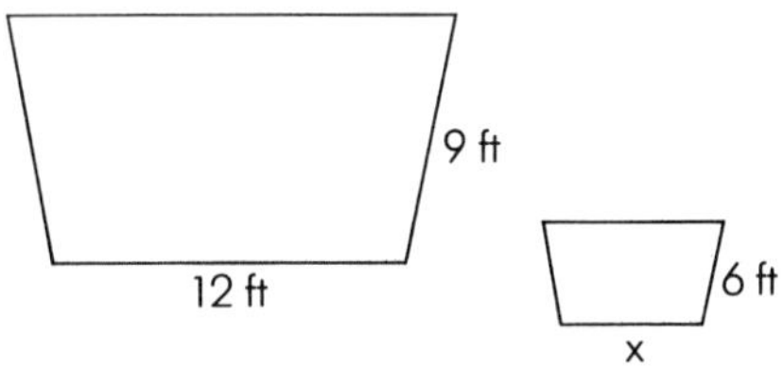

12.

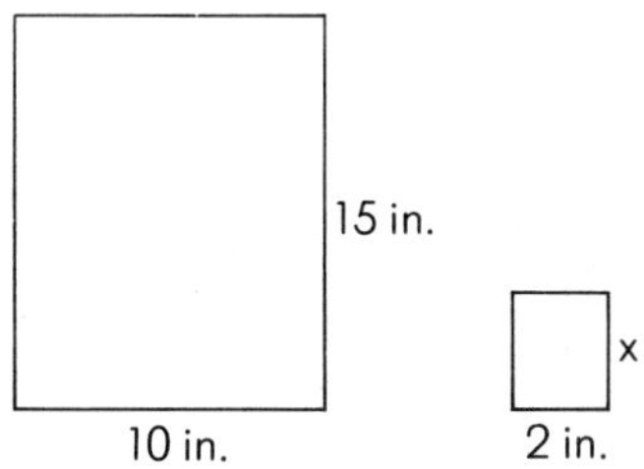

13.

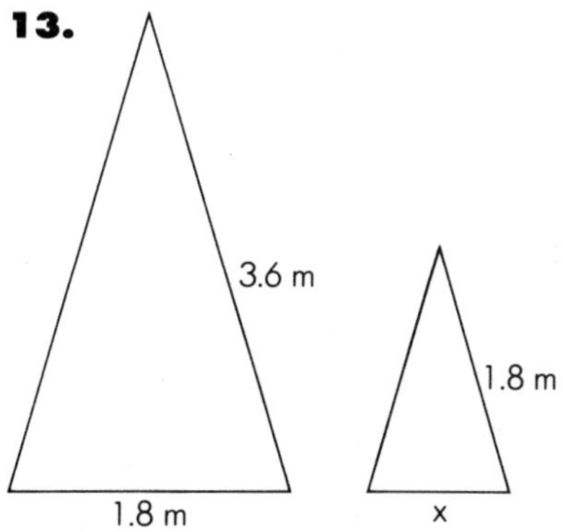

14.

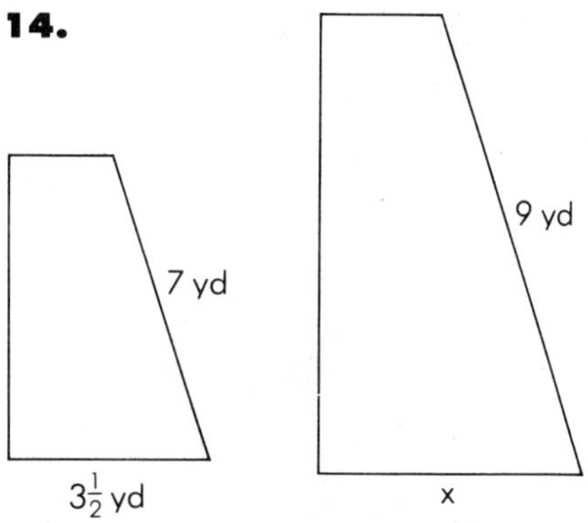

CALCULATOR CORNER

WIPE OUT

Players two
Materials two calculators, paper and pencils

SAMPLE Chin enters a five-digit whole number, such as 59,762, on the display. Kristen enters a five-digit whole number, such as 48,829, on the display.

They exchange calculators. Each player must now attempt to "wipe out" the number and get 0 on the display.

To do so they may add, subtract, multiply, or divide by any one or two-digit whole number or decimal except 0.

Kristen		Result
Play 1	59,762 − 62	59,700
Play 2	×0.01	597
Play 3	−97	500
Play 4	×0.1	50
Play 5	−50	0

Chin		Result
Play 1	48,829 − 99	48,730
Play 2	−30	48,700
Play 3	×0.01	487
Play 4	−87	400
Play 5	×0.01	4
Play 6	−4	0

Kristen's score is 5. Chin's score is 6.

Kristen wins the game, because she reaches 0 in fewer plays.

READY FOR CHAPTER 9

Convert to a decimal.

1. 37% **2.** 136% **3.** 9% **4.** 4.2%

Convert to a percent.

5. 0.17 **6.** 2.47 **7.** 0.03 **8.** 0.004

Convert to a fraction, mixed number, or whole number.

9. 37% **10.** 45% **11.** 120% **12.** $12\frac{1}{2}\%$

Convert to a percent.

13. $\frac{4}{5}$ **14.** $\frac{19}{20}$ **15.** $3\frac{1}{2}$ **16.** 6

17. $\frac{2}{3}$ **18.** $\frac{1}{8}$ **19.** $2\frac{1}{3}$ **20.** $1\frac{3}{7}$

Translate. Do not solve.

21. 18% of 38 is what number?

22. 42 is 75% of what number?

23. 5 is what percent of 90?

24. What percent of 10 is 3?

Solve.

25. What is 25% of 40?

26. 20% of 32 is what number?

27. 8 is what percent of 12?

28. What percent of 9 is 3?

29. 4 is 20% of what number?

30. 36 is $16\frac{2}{3}\%$ of what number?

Find the percent of increase or decrease.

31. From 10 to 12

32. From 12 to 10

Application

33. Matt earns \$ 300 per month plus 9% of sales over \$ 1000. How much does he earn if his total sales are \$ 3300 in a month?

34. A tennis racquet regularly sells for \$ 35. It is on sale for 20% off. What is the sale price?

9
PERCENT

GETTING STARTED

MAXIMIZE

WHAT'S NEEDED The 30 squares of paper with digits marked on them as in GOTCHA, Chapter 1.

THE RULES

1. On a sheet of paper each player writes the following.
   ___ ___% of ___ ___ ___ is what number?
2. A leader draws a digit from the box. Each player writes the digit on any one of the blanks. Once it is written, it cannot be changed or moved.
3. The leader draws four more digits, one at a time. Write each digit on a blank as it is drawn.
4. Each player is now ready to solve the problem.

SAMPLE 3 8% of 9 4 0 is what number?

5. Who has the largest number? Find out by multiplying.

SAMPLE

$$\begin{array}{r} 940 \\ \times\, 0.38 \\ \hline 7520 \\ 28200 \\ \hline 357.20 \end{array}$$

The player with the largest answer is the winner.

CONVERTING PERCENTS AND DECIMALS

Percent means hundredths: $n\%$ means $n \times 0.01$

Convert 87% to a decimal.

Percent → Drop the percent sign. Move the decimal point 2 places to the left. → Write the decimal.

87 %   87.   0.87

EXAMPLES. Convert to a decimal.

1. 247% ⟶ 2.47. ⟶ 2.47   **2.** 8% ⟶ 0.08. ⟶ 0.08

3. 7.6% ⟶ 0.07.6 ⟶ 0.076   **4.** $3\frac{1}{3}\%$ ⟶ $0.03.\frac{1}{3}$ ⟶ $0.03\frac{1}{3}$

TRY THIS

1. 25%   **2.** 100%   **3.** 4%   **4.** $4\frac{1}{2}\%$ **5.** 6.9%

Decimals can be written as percents: $n \times 0.01$ means $n\%$

Convert 0.79 to a percent.

Decimal → Move the decimal point 2 places to the right. → Write the % sign.

0.79   0.79   79 %

EXAMPLES. Convert to a percent.

5. 0.006 ⟶ 0.00.6 ⟶ 0.6 %   **6.** $0.02\frac{1}{2}$ ⟶ $0.02.\frac{1}{2}$ ⟶ $2\frac{1}{2}\%$

7. 0.3 ⟶ 0.30. ⟶ 30 %   **8.** 6 ⟶ 6.00. ⟶ 600 %

TRY THIS

Convert to a percent.

6. 0.72 **7.** 7 **8.** 7.2 **9.** $0.66\frac{2}{3}$ **10.** 0.04

PRACTICE

Convert to a decimal.

1. 15%	**2.** 241%	**3.** 42%	**4.** 312%	**5.** 154%
6. 6%	**7.** 1%	**8.** 7%	**9.** 3%	**10.** 4%
11. 3.9%	**12.** 8.7%	**13.** 1.2%	**14.** 5.5%	**15.** 6.8%
16. 0.8%	**17.** 0.7%	**18.** 0.1%	**19.** 0.9%	**20.** 0.4%
21. 0.04%	**22.** 0.18%	**23.** 0.09%	**24.** 0.42%	**25.** 0.06%
26. $33\frac{1}{3}\%$	**27.** $5\frac{1}{2}\%$	**28.** $9\frac{2}{3}\%$	**29.** $18\frac{3}{4}\%$	**30.** $11\frac{2}{3}\%$

Convert to a percent.

31. 0.99	**32.** 1.82	**33.** 0.71	**34.** 2.04	**35.** 0.75
36. 0.01	**37.** 0.09	**38.** 0.05	**39.** 0.04	**40.** 0.08
41. 0.001	**42.** 0.039	**43.** 0.008	**44.** 0.082	**45.** 0.043
46. $0.05\frac{1}{2}$	**47.** $0.07\frac{2}{3}$	**48.** $0.43\frac{1}{3}$	**49.** $1.36\frac{3}{4}$	**50.** $2.17\frac{4}{5}$
51. 0.5	**52.** 1.7	**53.** 0.1	**54.** 3.9	**55.** 1.8

PROBLEM CORNER

1. By the age of 14 (age 16 for boys), the average girl has attained 98.3% of her total height. Write the decimal for this percent.

2. Water makes up about 0.7 of a person's weight. Write the percent for this.

More Practice, page 272

CONVERTING PERCENTS TO FRACTIONS

Percent means hundredths: $n\%$ means $n \times \frac{1}{100}$

Convert 25% to a fraction.

Percent → Replace % with $\times \frac{1}{100}$. → Multiply. → Simplify.

$$25\% = 25 \times \frac{1}{100} = \frac{25}{100} = \frac{1}{4}$$

EXAMPLES. Convert to a fraction, mixed number, or whole number.

1. $19\% = 19 \times \frac{1}{100} = \frac{19}{100}$

2. $125\% = 125 \times \frac{1}{100} = \frac{125}{100} = 1\frac{25}{100} = 1\frac{1}{4}$

TRY THIS

Convert to a fraction, mixed number, or whole number.

1. 31% **2.** 75% **3.** 400% **4.** 150%

Percents may contain fractions, mixed numbers, or decimals.

EXAMPLES. Convert to a fraction.

3. $\frac{2}{3}\% = \frac{2}{3} \times \frac{1}{100} = \frac{2}{300} = \frac{1}{150}$

4. $33\frac{1}{3}\% = \frac{100}{3}\% = \frac{100}{3} \times \frac{1}{100} = \frac{100}{300} = \frac{1}{3}$

Rename the mixed numbers as fractions. Replace % with $\times \frac{1}{100}$. *Multiply. Simplify.*

TRY THIS

Convert to a fraction.

5. $\frac{1}{2}\%$ **6.** $16\frac{2}{3}\%$ **7.** $12\frac{1}{2}\%$ **8.** $37\frac{1}{2}\%$

EXAMPLES. Convert to a fraction.

5. $13.9\% = 0.139 = \frac{139}{1000}$

6. $0.8\% = 0.008 = \frac{8}{1000} = \frac{1}{125}$

Convert the percent to a decimal. Convert the decimal to a fraction. Simplify.

TRY THIS

Convert to a fraction.

9. 10.3% **10.** 0.4% **11.** 0.04% **12.** 62.5%

PRACTICE

Convert to a fraction, mixed number, or whole number.

1. 49%	**2.** 3%	**3.** 99%	**4.** 9%	**5.** 77%
6. 10%	**7.** 20%	**8.** 30%	**9.** 90%	**10.** 80%
11. 5%	**12.** 15%	**13.** 85%	**14.** 95%	**15.** 76%
16. 2%	**17.** 14%	**18.** 48%	**19.** 98%	**20.** 6%
21. 100%	**22.** 300%	**23.** 500%	**24.** 1000%	**25.** 1200%
26. 130%	**27.** 175%	**28.** 180%	**29.** 225%	**30.** 350%

Convert to a fraction.

31. $\frac{1}{3}\%$	**32.** $\frac{1}{4}\%$	**33.** $\frac{3}{4}\%$	**34.** $\frac{4}{5}\%$	**35.** $\frac{2}{7}\%$
36. $\frac{3}{8}\%$	**37.** $\frac{7}{8}\%$	**38.** $\frac{2}{3}\%$	**39.** $\frac{7}{10}\%$	**40.** $\frac{5}{12}\%$
41. $66\frac{2}{3}\%$	**42.** $83\frac{1}{3}\%$	**43.** $14\frac{2}{7}\%$	**44.** $28\frac{4}{7}\%$	**45.** $33\frac{1}{3}\%$
46. $37\frac{1}{2}\%$	**47.** $62\frac{1}{2}\%$	**48.** $44\frac{4}{9}\%$	**49.** $77\frac{7}{9}\%$	**50.** $87\frac{1}{2}\%$
51. 15.3%	**52.** 99.9%	**53.** 0.5%	**54.** 5.7%	**55.** 6.2%
56. 7.75%	**57.** 8.25%	**58.** 0.05%	**59.** 9.50%	**60.** 4.05%

PROBLEM CORNER

1. China has about 30% of the world's population. Write a fraction for this.

2. About $66\frac{2}{3}\%$ of the Earth is covered with water. Write a fraction for this.

More Practice, page 272

CONVERTING FRACTIONS TO PERCENTS

Convert $\frac{1}{2}$ to a percent.

Fraction	→ Convert to a decimal. Divide to hundredths.	→ Convert to a percent.
$\frac{1}{2}$	$2\overline{)1.00}$ = 0.50	50%

EXAMPLE. Convert to a percent.

1. $\frac{3}{4} \longrightarrow 4\overline{)3.00} = 0.75 \longrightarrow 75\%$

2. $1\frac{1}{2} \longrightarrow \frac{3}{2} \longrightarrow 2\overline{)3.00} = 1.50 \longrightarrow 150\%$

Convert to a decimal. Divide to hundredths. Convert to a percent.

For mixed numbers, change to a fraction.

TRY THIS

Convert to a percent.

1. $\frac{1}{4}$ 2. $\frac{3}{10}$ 3. $1\frac{2}{5}$ 4. $2\frac{3}{10}$ 5. $3\frac{3}{5}$

When we do not have a 0 remainder, we write a decimal or a fraction.

EXAMPLES. Convert to a percent.

3. $\frac{1}{3} \longrightarrow 3\overline{)1.00}$ = $0.33\frac{1}{3} = 33\frac{1}{3}\%$

$$\begin{array}{r} 9 \\ \hline 10 \\ 9 \\ \hline 1 \end{array}$$

4. $\frac{3}{8} \longrightarrow 8\overline{)3.000}$ = $0.375 = 37.5\%$

$$\begin{array}{r} 2\,4 \\ \hline 60 \\ 56 \\ \hline 40 \\ 40 \\ \hline \end{array}$$

TRY THIS

Convert to a percent.

6. $\frac{2}{3}$ 7. $\frac{5}{8}$ 8. $\frac{1}{8}$

9. $1\frac{5}{6}$ 10. $2\frac{7}{8}$ 11. $4\frac{1}{6}$

PRACTICE

Convert to a percent.

1. $\frac{2}{5}$	**2.** $\frac{3}{5}$	**3.** $\frac{3}{10}$	**4.** $\frac{7}{10}$	**5.** $\frac{1}{5}$	**6.** $\frac{1}{2}$
7. $\frac{17}{20}$	**8.** $\frac{6}{25}$	**9.** $\frac{3}{50}$	**10.** $\frac{38}{100}$	**11.** $\frac{8}{25}$	**12.** $\frac{23}{100}$
13. $1\frac{1}{4}$	**14.** $1\frac{3}{4}$	**15.** $2\frac{1}{2}$	**16.** $1\frac{1}{5}$	**17.** $1\frac{3}{10}$	**18.** $3\frac{4}{5}$
19. 3	**20.** 2	**21.** 1	**22.** 4	**23.** 8	**24.** 9
25. $\frac{1}{6}$	**26.** $\frac{2}{6}$	**27.** $\frac{4}{6}$	**28.** $\frac{5}{6}$	**29.** $\frac{3}{6}$	**30.** $\frac{6}{6}$
31. $\frac{1}{8}$	**32.** $\frac{3}{7}$	**33.** $\frac{5}{7}$	**34.** $\frac{7}{8}$	**35.** $\frac{3}{8}$	**36.** $\frac{2}{7}$
37. $1\frac{1}{6}$	**38.** $1\frac{2}{3}$	**39.** $2\frac{1}{9}$	**40.** $3\frac{1}{3}$	**41.** $3\frac{2}{9}$	**42.** $1\frac{5}{6}$
43. $1\frac{3}{7}$	**44.** $2\frac{2}{3}$	**45.** $1\frac{5}{9}$	**46.** $1\frac{5}{6}$	**47.** $2\frac{4}{7}$	**48.** $3\frac{7}{9}$
49. $\frac{1}{3}$	**50.** $\frac{5}{8}$	**51.** $\frac{2}{7}$	**52.** $\frac{4}{7}$	**53.** $\frac{6}{7}$	**54.** $\frac{1}{8}$
55. $\frac{1}{9}$	**56.** $\frac{8}{9}$	**57.** $\frac{1}{11}$	**58.** $\frac{9}{11}$	**59.** $\frac{1}{13}$	**60.** $\frac{9}{13}$
61. $1\frac{1}{3}$	**62.** $3\frac{2}{3}$	**63.** $2\frac{5}{6}$	**64.** $1\frac{4}{6}$	**65.** $1\frac{1}{5}$	**66.** $1\frac{3}{5}$
67. $2\frac{1}{6}$	**68.** $1\frac{1}{8}$	**69.** $1\frac{7}{8}$	**70.** $2\frac{3}{8}$	**71.** $1\frac{5}{8}$	**72.** $1\frac{1}{4}$

PROBLEM CORNER

1. The human body is about $\frac{2}{3}$ water. Write a percent for this.
2. About $\frac{1}{3}$ of the human body is not water. Write a percent for this.
3. About $\frac{4}{5}$ of all households own one or more cars. Write a percent for this.

More Practice, page 272

TRANSLATING TO NUMBER SENTENCES

To solve problems with percents, it is helpful to translate to number sentences.

Translate. 36% of 97 is what number?

$$36\% \times 97 = a$$

Of translates to ×. *Is* translates to =.
What or *what percent* or *what number* translates to any letter.

EXAMPLES. Translate.

1. What is 19% of 27?

$$n = 19\% \times 27$$

2. What percent of 76 is 8?

$$n \times 76 = 8$$

TRY THIS

Translate.

1. What is 60% of 70?

2. 76 is 19% of what number?

3. 18 is what percent of 36?

We often use the following percent, decimal, and fraction equivalents. They should be memorized. Some decimals have been rounded to the nearest thousandth.

Percent	Decimal	Fraction	Percent	Decimal	Fraction
10%	0.1	$\frac{1}{10}$	50%	0.5	$\frac{1}{2}$
$12\frac{1}{2}\%$	0.125	$\frac{1}{8}$	$66\frac{2}{3}\%$	0.667	$\frac{2}{3}$
$16\frac{2}{3}\%$	0.167	$\frac{1}{6}$	75%	0.75	$\frac{3}{4}$
20%	0.2	$\frac{1}{5}$	$37\frac{1}{2}\%$	0.375	$\frac{3}{8}$
25%	0.25	$\frac{1}{4}$	$62\frac{1}{2}\%$	0.625	$\frac{5}{8}$
$33\frac{1}{3}\%$	0.333	$\frac{1}{3}$	$87\frac{1}{2}\%$	0.875	$\frac{7}{8}$

EXAMPLES. Translate.

3. Convert the percent to a decimal.

10% of 48 is what number?

$$0.1 \times 48 = n$$

4. Convert the percent to a fraction.

56 is $12\frac{1}{2}$% of what number?

$$56 = \frac{1}{8} \times n$$

TRY THIS

4. Translate. Convert the percent to a decimal.
What is 50% of 48?

5. Translate. Convert the percent to a fraction.
$62\frac{1}{2}$% of what number is 32?

PRACTICE

Translate.

1. 17% of 36 is what number?

2. What is 42% of 98?

3. 36 is $12\frac{1}{2}$% of what number?

4. $66\frac{2}{3}$% of what number is 9?

5. 8.7 is what percent of 19.6?

6. What percent of $7\frac{1}{2}$ is $3\frac{1}{4}$?

Translate. Convert each percent to a decimal.

7. $87\frac{1}{2}$% of what number is 80?

8. 48 is 75% of what number?

9. What is 70% of 98?

10. 50% of 94.2 is what number?

11. 10% of what number is 84?

12. 3.7 is 25% of what number?

Translate. Convert each percent to a fraction.

13. What is 25% of 8?

14. $16\frac{2}{3}$% of 18 is what number?

15. $12\frac{1}{2}$% of what number is 16?

16. 40 is 10% of what number?

17. What is $87\frac{1}{2}$% of 80?

18. $62\frac{1}{2}$% of 48 is what number?

More Practice, page 272

FINDING A PERCENT OF A NUMBER

When we solve percent problems, we can write the percent either as a decimal or as a fraction.

USING A DECIMAL		USING A FRACTION
25% of 75 is what number?	Problem	25% of 75 is what number?
$0.25 \times 75 = n$	Translate.	$\frac{1}{4} \times 75 = n$
$0.25 \times 75 = 18.75$	Multiply.	$\frac{1}{4} \times \frac{75}{1} = \frac{75}{4} = 18\frac{3}{4}$
$18.75 = n$		$18\frac{3}{4} = n$
25% of 75 is 18.75.		25% of 75 is $18\frac{3}{4}$.

EXAMPLES. Solve. Convert each percent to a decimal.

1. What is 38% of 95?
$n = 0.38 \times 95$
$n = 36.10$
36.10 is 38% of 95.

2. 3.8% of 9.5 is what number?
$0.038 \times 9.5 = n$
$0.3610 = n$
3.8% of 9.5 is 0.3610.

TRY THIS Solve. Convert each percent to a decimal.

1. 8% of 159 is what number?
2. What is 32% of 75?
3. What is 0.5% of 37?
4. 0.7% of 33 is what number?

EXAMPLES. Solve. Convert each percent to a fraction.

3. $12\frac{1}{2}$% of 96 is what?

$\frac{1}{8} \times 96 = n$

$\frac{1}{8} \times \frac{96}{1} = \frac{96}{8} = 12$

$12 = n$

$12\frac{1}{2}$% of 96 is 12.

4. What is 75% of 40?

$n = \frac{3}{4} \times 40$

$\frac{3}{4} \times \frac{40}{1} = \frac{120}{4} = 30$

$n = 30$

30 is 75% of 40.

TRY THIS

Solve. Convert each percent to a fraction.

5. $16\frac{2}{3}\%$ of 50 is what number?

6. What is 50% of 36?

PRACTICE

Solve. Convert each percent to a decimal.

1. 97% of 500 is what number?

2. 1% of 73 is what number?

3. What is 3% of 10?

4. What is 128% of 100?

5. 49% of 7 is what number?

6. 146% of 5 is what number?

7. 3.1% of 10 is what number?

8. 9.7% of 20 is what number?

9. What is 12.7% of 88?

10. What is 0.6% of 52?

11. What is 4.9% of 87.3?

12. What is 59.7% of 34.8?

Solve. Convert each percent to a fraction.

13. 10% of 98 is what number?

14. $87\frac{1}{2}\%$ of 16 is what number?

15. $12\frac{1}{2}\%$ of 40 is what number?

16. 25% of 86 is what number?

17. What is $62\frac{1}{2}\%$ of 10?

18. What is $16\frac{2}{3}\%$ of 100?

19. What is 50% of 84?

20. What is $37\frac{1}{2}\%$ of 120?

21. 20% of 8 is what number?

22. 75% of 218 is what number?

PROBLEM CORNER

1. About 55% of Americans wear glasses. In a group of 50 people, about how many would wear glasses?

2. About $66\frac{2}{3}\%$ of a person's weight is water. How much water is there in a person who weighs 120 pounds?

More Practice, page 273

WHAT PERCENT A NUMBER IS OF ANOTHER

Solve.

6 is what percent of 8?

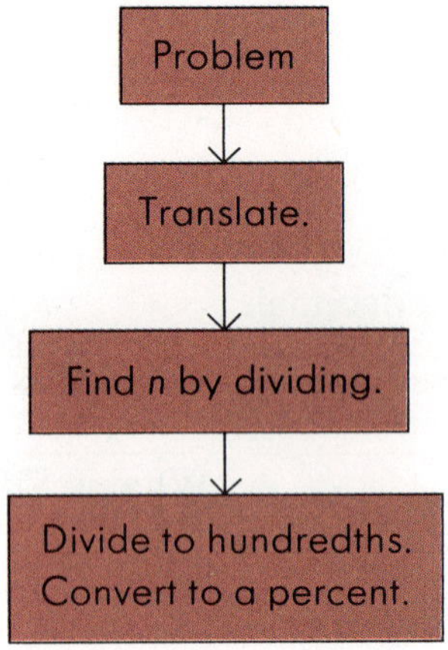

$6 = n \times 8$

$\frac{6}{8} = n$

$$\begin{array}{r} 0.75 = 75\% \\ 8\overline{)6.00} \end{array}$$

$75\% = n$ 6 is 75% of 8.

EXAMPLES. Solve.

1. 6 is what percent of 9?

$6 = n \times 9$ *Translate.*

$\frac{6}{9} = n$ *Find n by dividing.*

$$\begin{array}{r} 0.66\frac{6}{9} = 0.66\frac{2}{3} = 66\frac{2}{3}\% \\ 9\overline{)6.00} \\ \underline{5\,4} \\ 60 \\ \underline{54} \\ 6 \end{array}$$

Divide to hundredths.
Convert to a percent.

$66\frac{2}{3}\% = n$ 6 is $66\frac{2}{3}\%$ of 9.

2. 9 is what percent of 6?

$9 = n \times 6$ *Translate.*

$\frac{9}{6} = n$ *Find n by dividing.*

$$\begin{array}{r} 1.50 = 150\% \\ 6\overline{)9.00} \end{array}$$

Divide to hundredths.
Convert to a percent.

$150\% = n$ 9 is 150% of 6.

3. What percent of 8 is 3?

$n \times 8 = 3$ *Translate.*

$n = \frac{3}{8}$ *Find n by dividing.*

$n = 37\frac{1}{2}\%$ $37\frac{1}{2}\%$ of 8 is 3.

TRY THIS

Solve.

1. 8 is what percent of 16?

2. 10 is what percent of 6?

3. What percent of 9 is 7?

4. What percent of 11 is 5?

PRACTICE

Solve.

1. 1 is what percent of 2?

2. 3 is what percent of 4?

3. 1 is what percent of 3?

4. 2 is what percent of 3?

5. 4 is what percent of 6?

6. 7 is what percent of 8?

7. 6 is what percent of 4?

8. 8 is what percent of 7?

9. 10 is what percent of 2?

10. 8 is what percent of 1?

11. 5 is what percent of 5?

12. 12 is what percent of 8?

13. What percent of 6 is 2?

14. What percent of 6 is 6?

15. What percent of 8 is 3?

16. What percent of 8 is 4?

17. What percent of 20 is 16?

18. What percent of 30 is 25?

19. What percent of 4 is 10?

20. What percent of 4 is 12?

21. What percent of 9 is 12?

22. What percent of 6 is 24?

23. What percent of 100 is 63?

24. What percent of 100 is 75?

More Practice, page 273

APPLICATION

WORKING FOR COMMISSIONS AND TIPS

Some workers are paid a commission rather than a fixed salary. Commission may be based on the number of items sold.

1. Rico Torrez receives 8¢ for each pack of raisins he sells. He ordered 500 packs and sold 80% of them. How much is his commission?

Commission is usually based on a percent of sale.

Nancy WhiteEagle earns 7% commission on her monthly sales. How much did she earn these months?

2. January—total sales = $ 8820.00
3. June—total sales = $ 10,150.00

Some salespeople are paid a salary plus a commission.

Martin Herbert earns \$ 500 per month plus 2% commission on sales over \$ 1000. How much will he earn these months?

4. January—total sales = \$ 2450.00
5. September—total sales = \$ 4165.00

Some workers are paid a graduated commission. The commission rate increases as sales increase. Each month Marion Reesey earns $12\frac{1}{2}\%$ of sales on the first \$ 500, 15% on sales up to \$ 1000 and 10% on sales over \$ 1000. Find each percentage of commission and add to find each total monthly salary.

6.
7.
8.
9.
10.
11.

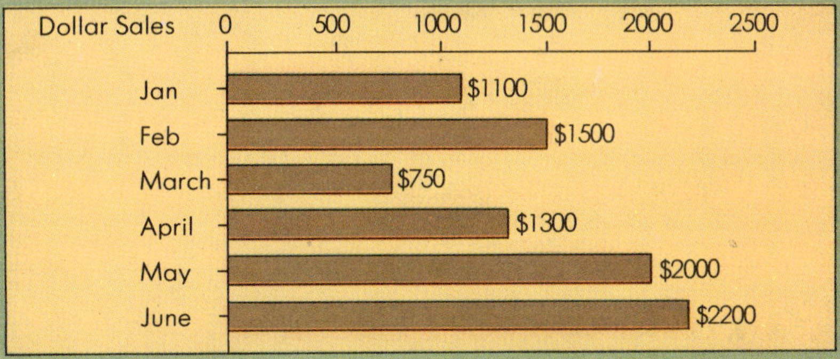

Marilyn Fisher earns \$ 3.50 an hour in salary as a waitress. She averages \$ 4.50 per hour in tips in an 8-hour day. Round each percent to the nearest tenth.

12. What are Marilyn's total daily wages?
13. What percent of her daily wages is in tips?
14. What percent of her daily wages is in salary?

Last year Marilyn earned \$ 16,000 for the year. At the above rate find how much she earned in each of the following.

15. salary   16. tips

Restaurant tips are often based on the amount of the customer's check. Many diners tip 15% of the total check. Find how much that would be for the following checks.

17. \$ 20   18. \$ 46.50
19. \$ 9.45   20. \$ 58.20

FINDING A NUMBER GIVEN A PERCENT

USING A DECIMAL

6 is 75% of what number?

$6 = 0.75 \times n$

$\frac{6}{0.75} = n$

$$\begin{array}{r} 8. \\ 0.75\overline{)\,6.00} \\ \underline{600} \\ 0 \end{array}$$

$8 = n$

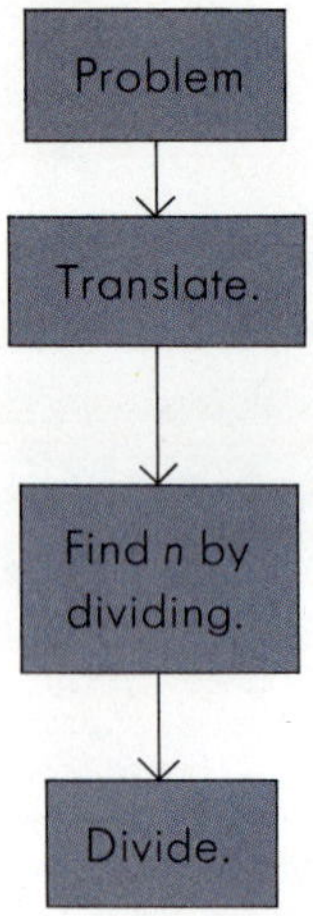

USING A FRACTION

6 is 75% of what number?

$6 = \frac{3}{4} \times n$

$\frac{6}{\frac{3}{4}} = n$

$6 \div \frac{3}{4} = \frac{6}{1} \times \frac{4}{3} = \frac{24}{3} = 8$

$8 = n$

6 is 75% of 8.

EXAMPLE 1. Solve. Convert the percent to a decimal.

9 is 36% of what number?

$9 = 0.36 \times n$	*Translate. 36% = 0.36*
$\frac{9}{0.36} = n$	*Divide.*
$\begin{array}{r} 25. \\ 0.36\overline{)\,9.00} \end{array}$	
$25 = n$	9 is 36% of 25.

Sometimes it is easier to use fractions.

EXAMPLE 2. Solve. Convert the percent to a fraction.

42 is $33\frac{1}{3}$% of what number?

$42 = \frac{1}{3} \times n$	*Translate.*
$\frac{42}{\frac{1}{3}} = n$	*Divide.*
$42 \div \frac{1}{3} = \frac{42}{1} \times \frac{3}{1} = 126$	
$126 = n$	42 is $33\frac{1}{3}$% of 126.

TRY THIS

Solve.

1. Convert the percent to a decimal.
   30 is 15% of what number?
2. Convert the percent to a fraction.
   12 is $66\frac{2}{3}$% of what number?

PRACTICE

Solve. Convert each percent to a decimal.

1. 7 is 70% of what number?
2. 9 is 30% of what number?
3. 6 is 15% of what number?
4. 8 is 16% of what number?
5. 2 is 6% of what number?
6. 4 is 8% of what number?
7. 2 is 60% of what number?
8. 4 is 80% of what number?
9. 8 is 4% of what number?
10. 8 is 40% of what number?

Solve. Convert each percent to a fraction.

11. 4 is $12\frac{1}{2}$% of what number?
12. 21 is $16\frac{2}{3}$% of what number?
13. 8 is 20% of what number?
14. 9 is $12\frac{1}{2}$% of what number?
15. 10 is $16\frac{2}{3}$% of what number?
16. 4 is 50% of what number?
17. 9 is 25% of what number?
18. 72 is 20% of what number?
19. 1 is $33\frac{1}{3}$% of what number?
20. 48 is 75% of what number?

PROBLEM CORNER

1. Three students in the room have red hair. This is 10% of the total number of students. How many students are in the room?
2. 40% of a person's weight is muscle. A person has 80 pounds of muscle. What is the person's total weight?

More Practice, page 273

PERCENT OF INCREASE OR DECREASE

Find the percent of increase: from 80 to 100

Find the increase. → Write a fraction. The numerator is the increase. The denominator is the starting number. Simplify. → Divide to hundredths. → Convert to a percent.

$$\begin{array}{r} 100 \\ -\ \ 80 \\ \hline 20 \end{array} \qquad \text{increase} \longrightarrow \frac{20}{80} = \frac{1}{4} \quad (\text{starting number} \longrightarrow 80) \qquad 4\overline{)1.00} = 0.25 \qquad 25\%$$

EXAMPLE 1. Find the percent of increase: from 30 to 40

$$\begin{array}{r} 40 \\ -\ 30 \\ \hline 10 \end{array}$$

Find the increase.

$$\text{increase} \longrightarrow \frac{10}{30} = \frac{1}{3} \quad (\text{starting number} \longrightarrow 30)$$

Write a fraction.
Simplify.

$$3\overline{)1.00} = 0.33\tfrac{1}{3} = 33\tfrac{1}{3}\%$$

Divide to hundredths.
Convert to a percent.

EXAMPLE 2. Find the percent of decrease: from 30 to 25

$$\begin{array}{r} 30 \\ -\ 25 \\ \hline 5 \end{array}$$

Find the decrease.

$$\text{decrease} \longrightarrow \frac{5}{30} = \frac{1}{6} \quad (\text{starting number} \longrightarrow 30)$$

Write a fraction.
Simplify.

$$\begin{array}{r} 0.16\tfrac{4}{6} \\ 6\overline{)1.00} \\ 6 \\ \hline 40 \\ 36 \\ \hline 4 \end{array} \quad = 0.16\tfrac{2}{3} = 16\tfrac{2}{3}\%$$

Divide to hundredths.
Convert to a percent.

TRY THIS

Find the percent of increase or decrease.

1. From 12 to 20
2. From 100 to 105
3. From 16 to 20
4. From 15 to 12
5. From 100 to 60
6. From 30 to 13

PRACTICE

Find the percent of increase or decrease.

1. From 8 to 10
2. From 12 to 15
3. From 20 to 25
4. From 10 to 8
5. From 24 to 18
6. From 40 to 30
7. From 10 to 15
8. From 6 to 9
9. From 5 to 7.5
10. From 20 to 10
11. From 34 to 17
12. From 16 to 8
13. From 5 to 10
14. From 10 to 20
15. From 50 to 100
16. From 9 to 3
17. From 42 to 35
18. From 100 to 87
19. From 6 to 8
20. From 9 to 12
21. From 12 to 16
22. From 80 to 60
23. From 45 to 35
24. From 50 to 30
25. From 10 to 11
26. From 16 to 22
27. From 45 to 50
28. From 3 to 2
29. From 4 to 3
30. From 5 to 4

PROBLEM CORNER

1. Jose weighed 100 pounds. A year later he weighed 120 pounds. What was his percent of increase in weight?
2. Maria lowered the thermostat setting in her house. The heating bill was $ 80 per month. Now it is $ 75. What was the percent of decrease?

More Practice, page 273

APPLICATION

BUYING ON SALE

Sporting Equipment on Sale—15% to 50% OFF the regular price.

Find the sale price. Find the amount of savings.

1. **BASEBALL GLOVES**
   Regularly $15.95
   NOW 20% Off

2. **TENNIS BALLS**
   50% Off Our Regular Price of $5.25

3. 25% Off
   PRO SKATEBOARDS
   Regularly $10.95

Find the percent of decrease.

4. TENNIS RACQUETS
   Were $36
   NOW $32

5. **FOOTBALLS**
   Formerly $12.00
   NOW $10

6. SOCCER BALLS
   Were $8.95
   NOW $5.37

Find the percent off. Find the sale price.

7. 1/3 OFF Were $10.98
   SIDEWALK SKATES
   NOW ________

8. 1/2 OFF
   PING-PONG BALLS
   Were $2.98 per box
   NOW ________ box

9. 1/4 Off
   SKI PANTS
   Were $24.88
   NOW ________

Find the original price.

10. **FISHING RODS**
    10% Off
    NOW $21.60

11. **WARM-UP SUITS**
    1/4 OFF
    NOW $29.25

12. **SOCCER SHORTS**
    30% Off
    NOW $7.00

A popular type of sale offer:

BUY 2 GET 1 FREE

You really pay for 2 items and get 3 items.

Each item costs $\frac{2}{3}$ of the regular price. The savings is $\frac{1}{3}$ or $33\frac{1}{3}\%$.

Find the sale price for each item.

BUY 2 GET 1 Free

13. Soap 29¢ each bar
14. Soup 55¢ per can
15. Dog Food 43¢ per can
16. Toothpaste 89¢ each tube

Read the ad carefully. Find the unit sale price. Find the percent of savings.

2 FOR 1 SALE
GET 2 FOR THE PRICE OF 1

17. Paper towels 66¢ each roll
18. Note pads 48¢ each
19. Auto oil \$ 1.08 per qt
20. Soap powder \$ 4.26 per box

Assume the quality is equal. Find the sale price for each size. Which is the most economical?

21.

SOAP

20 oz size, regularly 60¢, now 25% off
40 oz size, regularly \$1.08, now 20% off
104 oz size, regularly \$2.55, now 15% off

22.

APPLE JUICE

8 oz can, regularly 8/\$1.00, now 15/\$1.25
24 oz can, regularly 3/\$1.00, now 4/\$1.00
64 oz can, regularly 2/\$1.59, now 2/\$1.35

How much change would you receive from a \$ 20 bill for each purchase?

23. Buy 6 pairs of socks at 25% off. The regular price is \$ 1.78 per pair.
24. Buy 1 dozen drinking glasses. The regular cost is 69¢ each. On sale you can buy 3 for the price of 2.
25. Buy 15 rolls of paper towels at $\frac{1}{3}$ off. The regular price is 3 for \$ 1.00.
26. Buy 2 pounds of meat at $\frac{2}{3}$ of the regular price of \$ 1.47 per pound.
27. Buy 5 pounds of meat at $\frac{1}{3}$ off. The regular price is \$ 1.98 per pound.

BEAUTICIAN

Bill Oberle owns The Beauty Culture Studio. Among the specialists working for him are hair stylists, manicurists, and facial treatment operators.

Formula for Mrs. Darlene Spies	
95 mL of #1 red	25 mL of #4 drab
Mix with 200 mL peroxide	

Hair coloring requires knowledge of chemistry and proportions. Each customer has a special color formula.

1. What percent of the solution is #1 red?
2. What percent of the solution is #4 drab?
3. What is the ratio of red to drab?

Bill used only 10 mL of #4 drab to make a small amount of solution.

4. How much of #1 red should be used?
5. How much peroxide should be used?

Bill pays $ 8.00 for a 5 L jug of shampoo concentrate. 60 mL of the concentrate is used for each customer's shampoo.

6. How many shampoos can be done from a jug of concentrate?
7. How much does the concentrate for one shampoo cost?

Bill's employees earn a weekly salary. They also get tips and a percentage of their weekly gross sales.

Income for One Week—5 Day Week				
Operator	Salary	Percentage	Gross Sales	Tips
Sally	$ 50	20%	$ 425	$ 65
Joel	$ 75	25%	$ 675	$ 95
Eiko	$100	$37\frac{1}{2}$%	$ 842	$120
Jorge	$125	15%	$1250	$175

Find the income for one week for each operator. (Income = salary + % of gross + tips)

8. Sally 9. Joel 10. Eiko 11. Jorge

Tips are related to gross sales. Many patrons tip according to their bill for services.
Express the ratio of tips to sales to the nearest 1%.

12. Sally 13. Joel 14. Eiko 15. Jorge

Joel's goal is to increase his gross sales to $ 783 per week.

16. What percent of increase in gross sales does he want?
17. How much does he estimate his tips to be? (See problem 13.)
18. How much will Joel earn weekly?

CHAPTER 9 REVIEW

Convert to a decimal. p. 248

1. 43%   **2.** 120%   **3.** 8%   **4.** 6.3%

Convert to a percent. p. 248

5. 0.19   **6.** 1.78   **7.** 0.05   **8.** 0.004

Convert to a fraction, mixed number, or whole number. p. 250

9. 49%   **10.** 24%   **11.** 145%   **12.** $33\frac{1}{3}\%$

Convert to a percent. p. 252

13. $\frac{3}{5}$ **14.** $\frac{4}{25}$ **15.** $2\frac{1}{4}$ **16.** 5

17. $\frac{1}{3}$ **18.** $\frac{3}{8}$ **19.** $2\frac{2}{3}$ **20.** $1\frac{2}{7}$

Translate. Do not solve. p. 254

21. 17% of 35 is what number?

22. What percent of 20 is 4?

23. 54 is what percent of 96?

24. 7 is $33\frac{1}{3}\%$ of what number?

Solve. pp. 256–262

25. 36% of 50 is what number?

26. What is 90% of 120?

27. What percent of 12 is 6?

28. 7 is what percent of 21?

29. 40 is $33\frac{1}{3}\%$ of what number?

30. 9 is 75% of what number?

Find the percent of increase or decrease. p. 264

31. From 15 to10

32. From 10 to 15

Application pp. 260, 266

33. Sandra earns 8% commission on her monthly sales. How much commission does she earn on total monthly sales of $ 8450?

34. Skateboards usually cost $ 24. On sale they cost $ 18. Find the percent of decrease in price.

CHAPTER 9 TEST

Convert to a decimal.

1. 97% **2.** 326% **3.** 7% **4.** 2.7%

Convert to a percent.

5. 0.37 **6.** 1.62 **7.** 0.03 **8.** 0.007

Convert to a fraction, mixed number, or whole number.

9. 67% **10.** 15% **11.** 125% **12.** $66\frac{2}{3}\%$

Convert to a percent.

13. $\frac{4}{5}$ **14.** $\frac{7}{20}$ **15.** $1\frac{3}{4}$ **16.** 4

17. $\frac{3}{8}$ **18.** $\frac{3}{7}$ **19.** $1\frac{1}{6}$ **20.** $2\frac{5}{9}$

Translate. Do not solve.

21. 15 is what percent of 45?

22. 76% of 19 is what number?

23. 9 is 50% of what number?

24. What percent of 9 is 5?

Solve.

25. What is 25% of 32?

26. $16\frac{2}{3}\%$ of 42 is what number?

27. 9 is what percent of 45?

28. What percent of 18 is 12?

29. 50 is 25% of what number?

30. 7 is 75% of what number?

Find the percent of increase or decrease.

31. From 20 to 25

32. From 25 to 20

Application

33. Russ earns \$ 3.25 an hour in salary. He averages \$ 4.75 an hour in tips. How much does he earn in an 8-hour day?

34. Ski jackets are on sale for $\frac{1}{4}$ off. The regular price is \$ 30. Find the sale price.

MORE PRACTICE

Convert to a decimal. Use after page 249.

1. 44% **2.** 38% **3.** 78% **4.** 96%

5. 130% **6.** 240% **7.** 350% **8.** 680%

9. 9% **10.** 7% **11.** 1% **12.** 5%

13. 7.8% **14.** 9.2% **15.** 0.3% **16.** 0.7%

Convert to a percent. Use after page 249.

17. 0.26 **18.** 0.98 **19.** 0.39 **20.** 0.46

21. 1.12 **22.** 1.37 **23.** 2.85 **24.** 4.32

25. 0.01 **26.** 0.07 **27.** 0.09 **28.** 0.06

29. 0.003 **30.** 0.008 **31.** 0.001 **32.** 0.007

Convert to a fraction, mixed number, or whole number. Use after page 251.

33. 77% **34.** 13% **35.** 45% **36.** 75%

37. 175% **38.** 250% **39.** 400% **40.** 700%

Convert to a percent. Use after page 253.

41. $\frac{4}{5}$ **42.** $\frac{7}{10}$ **43.** $\frac{7}{25}$ **44.** $\frac{4}{50}$

45. $3\frac{3}{4}$ **46.** $1\frac{1}{2}$ **47.** 2 **48.** 6

49. $\frac{2}{3}$ **50.** $\frac{1}{8}$ **51.** $\frac{5}{8}$ **52.** $\frac{5}{6}$

53. $1\frac{1}{3}$ **54.** $3\frac{3}{8}$ **55.** $2\frac{1}{6}$ **56.** $4\frac{7}{8}$

Translate. Do not solve. Use after page 255.

57. 18% of 36 is what number? **58.** What is 19% of 76?

59. What percent of 40 is 8? **60.** What percent of 70 is 6?

Solve. Use after page 257.

1. 37% of 10 is what number?

2. 75% of 12 is what number?

3. 90% of 20 is what number?

4. $12\frac{1}{2}$% of 16 is what number?

5. What is 80% of 70?

6. What is 25% of 20?

7. What is $87\frac{1}{2}$% of 24?

8. What is $66\frac{2}{3}$% of 30?

9. $33\frac{1}{3}$% of 60 is what number?

10. What is 50% of 70?

Solve. Use after page 259.

11. What percent of 20 is 10?

12. What percent of 15 is 3?

13. What percent of 100 is 39?

14. What percent of 90 is 9?

15. 6 is what percent of 30?

16. 9 is what percent of 36?

17. 12 is what percent of 36?

18. 18 is what percent of 9?

19. What percent of 50 is 25?

20. 25 is what percent of 75?

Solve. Use after page 263.

21. 6 is 10% of what number?

22. 7 is 20% of what number?

23. 8 is $33\frac{1}{3}$% of what number?

24. 32 is 50% of what number?

25. 30 is 75% of what number?

26. 16 is $66\frac{2}{3}$% of what number?

27. 9 is 100% of what number?

28. 50 is 100% of what number?

29. 6 is $12\frac{1}{2}$% of what number?

30. 50 is 50% of what number?

Find the percent of increase or decrease. Use after page 265.

31. From 18 to 25

32. From 15 to 18

33. From 35 to 30

34. From 30 to 35

35. From 100 to 120

36. From 120 to 100

SKILLS REVIEW

Add. Refer to pages 14–17.

1. $\begin{array}{r} 298 \\ +\ 472 \\ \hline \end{array}$ **2.** $\begin{array}{r} 6497 \\ +\ 8584 \\ \hline \end{array}$ **3.** $\begin{array}{r} 398{,}108 \\ +\ 169{,}942 \\ \hline \end{array}$ **4.** $\begin{array}{r} 473{,}493 \\ +\ 87{,}875 \\ \hline \end{array}$

5. $\begin{array}{r} 4298 \\ 7394 \\ 6285 \\ +\ 4073 \\ \hline \end{array}$ **6.** $\begin{array}{r} 17{,}478 \\ 29{,}407 \\ 33{,}987 \\ +\ 49{,}785 \\ \hline \end{array}$ **7.** $\begin{array}{r} 496{,}784 \\ 748{,}147 \\ 286{,}595 \\ +\ 985{,}589 \\ \hline \end{array}$ **8.** $\begin{array}{r} 429{,}784 \\ 68{,}498 \\ 8{,}747 \\ +\ 9{,}606 \\ \hline \end{array}$

Add. Refer to page 22.

9. $\begin{array}{r} 9.9 \\ +\ 3.6 \\ \hline \end{array}$ **10.** $\begin{array}{r} 6.72 \\ +\ 3.98 \\ \hline \end{array}$ **11.** $\begin{array}{r} 14.784 \\ +\ 9.608 \\ \hline \end{array}$ **12.** $\begin{array}{l} \ \ 5.78 \\ +\ 6.9 \\ \hline \end{array}$

13. $\begin{array}{l} \ \ 3.97 \\ \ \ 0.4 \\ +\ 0.827 \\ \hline \end{array}$ **14.** $\begin{array}{l} \ \ 0.984 \\ \ \ 7.38 \\ +\ 1.8 \\ \hline \end{array}$ **15.** $\begin{array}{l} \ \ 7.673 \\ \ \ 0.8 \\ \ \ 0.87 \\ +\ 8.107 \\ \hline \end{array}$ **16.** $\begin{array}{l} \ \ 6.984 \\ \ \ 8.746 \\ \ \ 9.98 \\ +\ 0.7 \\ \hline \end{array}$

17. 4.39 + 15.8 **18.** 0.6 + 9.7 **19.** 2.4 + 0.3 + 5.71 **20.** 9.52 + 11.6

Subtract. Refer to pages 40–45.

21. $\begin{array}{r} 749 \\ -\ 298 \\ \hline \end{array}$ **22.** $\begin{array}{r} 37{,}385 \\ -\ 19{,}496 \\ \hline \end{array}$ **23.** $\begin{array}{r} 742{,}781 \\ -\ 647{,}398 \\ \hline \end{array}$ **24.** $\begin{array}{r} 725{,}684 \\ -\ 39{,}708 \\ \hline \end{array}$

25. $\begin{array}{r} 800 \\ -\ 147 \\ \hline \end{array}$ **26.** $\begin{array}{r} 3007 \\ -\ 2988 \\ \hline \end{array}$ **27.** $\begin{array}{r} 74{,}000 \\ -\ 28{,}146 \\ \hline \end{array}$ **28.** $\begin{array}{r} 90{,}085 \\ -\ 18{,}198 \\ \hline \end{array}$

Subtract. Refer to page 50.

29. $\begin{array}{r} 8.7 \\ -\ 1.9 \\ \hline \end{array}$ **30.** $\begin{array}{r} 9.37 \\ -\ 8.19 \\ \hline \end{array}$ **31.** $\begin{array}{r} 6.784 \\ -\ 0.928 \\ \hline \end{array}$ **32.** $\begin{array}{r} 6.085 \\ -\ 4.276 \\ \hline \end{array}$

33. $\begin{array}{l} \ \ 9.742 \\ -\ 1.8 \\ \hline \end{array}$ **34.** $\begin{array}{l} \ \ 8.072 \\ -\ 3.86 \\ \hline \end{array}$ **35.** $\begin{array}{l} \ \ 7.4 \\ -\ 1.874 \\ \hline \end{array}$ **36.** $\begin{array}{l} \ \ 6 \\ -\ 1.72 \\ \hline \end{array}$

37. 13.68 − 6 **38.** 22 − 8.4 **39.** 5.6 − 0.89 **40.** 24.1 − 7.93

41. 18 − 9.3 **42.** 16.2 − 6.43 **43.** 19.71 − 4 **44.** 20.3 − 7.28

Multiply. Refer to pages 68–73.

1. $\begin{array}{r} 88 \\ \times\ 7 \\ \hline \end{array}$	**2.** $\begin{array}{r} 427 \\ \times\ 8 \\ \hline \end{array}$	**3.** $\begin{array}{r} 7264 \\ \times\ 6 \\ \hline \end{array}$	**4.** $\begin{array}{r} 87 \\ \times\ 25 \\ \hline \end{array}$
5. $\begin{array}{r} 426 \\ \times\ 37 \\ \hline \end{array}$	**6.** $\begin{array}{r} 758 \\ \times\ 643 \\ \hline \end{array}$	**7.** $\begin{array}{r} 4986 \\ \times\ 278 \\ \hline \end{array}$	**8.** $\begin{array}{r} 6720 \\ \times\ 807 \\ \hline \end{array}$

Multiply. Refer to page 82.

9. $\begin{array}{r} 9.8 \\ \times\ 6 \\ \hline \end{array}$	**10.** $\begin{array}{r} 7.32 \\ \times\ 5 \\ \hline \end{array}$	**11.** $\begin{array}{r} 9.376 \\ \times\ 4 \\ \hline \end{array}$	**12.** $\begin{array}{r} 7.8 \\ \times\ 3.2 \\ \hline \end{array}$
13. $\begin{array}{r} 8.32 \\ \times\ 6.9 \\ \hline \end{array}$	**14.** $\begin{array}{r} 9.78 \\ \times\ 3.28 \\ \hline \end{array}$	**15.** $\begin{array}{r} 0.3 \\ \times\ 0.2 \\ \hline \end{array}$	**16.** $\begin{array}{r} 0.04 \\ \times\ 0.1 \\ \hline \end{array}$
17. $\begin{array}{r} 4.7 \\ \times\ 1.3 \\ \hline \end{array}$	**18.** $\begin{array}{r} 2.9 \\ \times\ 0.9 \\ \hline \end{array}$	**19.** $\begin{array}{r} 5.82 \\ \times\ 1.3 \\ \hline \end{array}$	**20.** $\begin{array}{r} 12.7 \\ \times\ 2.86 \\ \hline \end{array}$

Divide. Refer to pages 98–105.

21. $59 \div 8$	**22.** $75 \div 4$	**23.** $359 \div 6$	**24.** $962 \div 7$
25. $4\overline{)5732}$	**26.** $9\overline{)6354}$	**27.** $21\overline{)75}$	**28.** $84\overline{)19{,}864}$
29. $73\overline{)19{,}710}$	**30.** $846\overline{)6768}$	**31.** $379\overline{)8768}$	**32.** $842\overline{)42{,}100}$

Divide. Refer to pages 108–117.

33. $12.32 \div 7$	**34.** $2.96 \div 8$	**35.** $243.2 \div 64$	**36.** $381.80 \div 92$
37. $7\overline{)0.14}$	**38.** $9\overline{)2.7}$	**39.** $8\overline{)2}$	**40.** $3.1\overline{)105.4}$
41. $7.3\overline{)70.08}$	**42.** $0.3\overline{)6}$	**43.** $0.08\overline{)64}$	**44.** $0.12\overline{)6}$

Find each quotient to the nearest tenth. Refer to page 118.

45. $4 \div 18$	**46.** $59 \div 9$	**47.** $2.5 \div 6$	**48.** $11.8 \div 7$
49. $5.6 \div 3.4$	**50.** $4 \div 0.7$	**51.** $6 \div 1.4$	**52.** $1.5 \div 0.91$
53. $8\overline{)14}$	**54.** $9\overline{)6}$	**55.** $0.6\overline{)0.5}$	**56.** $0.09\overline{)1.7}$
57. $2.6\overline{)4.5}$	**58.** $0.23\overline{)10}$	**59.** $0.3\overline{)2}$	**60.** $4.8\overline{)3}$
61. $8\overline{)4.6}$	**62.** $0.03\overline{)0.47}$	**63.** $4.2\overline{)9.36}$	**64.** $15\overline{)13}$

What part is shaded? Refer to page 136.

1. 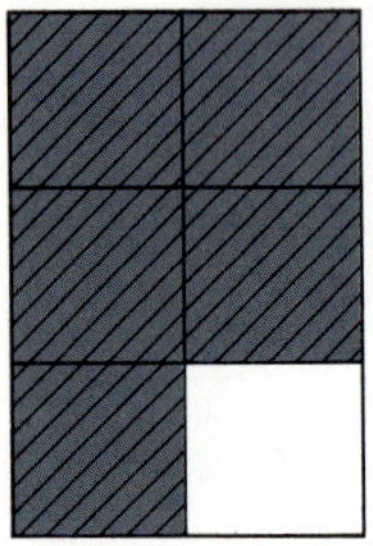

2. 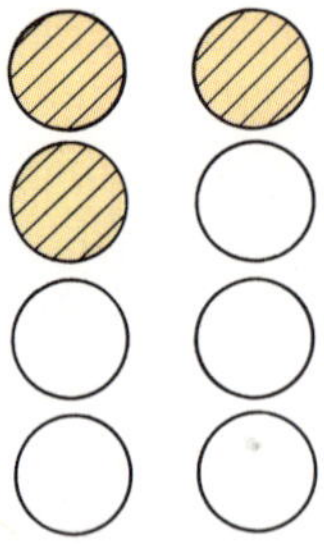

3. 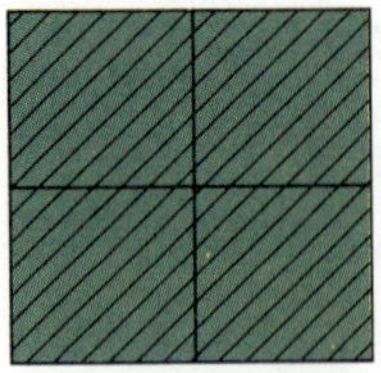

4. 

Compare. Use $>$ or $<$. Refer to page 142.

5. $\frac{2}{3}$ and $\frac{3}{4}$ **6.** $\frac{5}{6}$ and $\frac{4}{5}$ **7.** $\frac{7}{8}$ and $\frac{9}{10}$ **8.** $\frac{3}{2}$ and $\frac{4}{3}$

9. $\frac{5}{8}$ and $\frac{6}{8}$ **10.** $\frac{9}{10}$ and $\frac{8}{10}$ **11.** $\frac{7}{4}$ and $\frac{5}{4}$ **12.** $\frac{9}{8}$ and $\frac{12}{8}$

Convert to a fraction. Refer to page 150.

13. $2\frac{2}{3}$ **14.** $3\frac{3}{4}$ **15.** $1\frac{7}{8}$ **16.** $2\frac{1}{2}$

17. $7\frac{5}{8}$ **18.** $6\frac{1}{3}$ **19.** $3\frac{2}{5}$ **20.** $3\frac{7}{10}$

Convert to a mixed number. Refer to page 152.

21. $\frac{7}{4}$ **22.** $\frac{9}{8}$ **23.** $\frac{13}{2}$ **24.** $\frac{17}{4}$

25. $\frac{10}{3}$ **26.** $\frac{16}{5}$ **27.** $\frac{23}{10}$ **28.** $\frac{17}{12}$

Convert to a decimal. Refer to page 154.

29. $6\frac{3}{10}$ **30.** $7\frac{25}{100}$ **31.** $6\frac{327}{1000}$ **32.** $5\frac{3}{100}$

33. $9\frac{7}{8}$ **34.** $4\frac{1}{4}$ **35.** $87\frac{2}{5}$ **36.** $36\frac{9}{20}$

Multiply and simplify. Refer to page 174.

37. $\frac{3}{8} \cdot \frac{2}{3}$ **38.** $\frac{1}{2} \cdot \frac{8}{9}$ **39.** $\frac{7}{8} \cdot \frac{2}{3}$ **40.** $\frac{4}{12} \cdot \frac{3}{4}$

41. $\frac{1}{2} \cdot 8$ **42.** $9 \cdot \frac{2}{3}$ **43.** $3 \cdot \frac{1}{9}$ **44.** $\frac{3}{4} \cdot 12$

45. $\frac{11}{9} \cdot \frac{3}{5}$ **46.** $\frac{5}{8} \cdot \frac{4}{3}$ **47.** $\frac{12}{5} \times \frac{7}{3}$ **48.** $\frac{8}{3} \times \frac{7}{4}$

Multiply and simplify. Refer to page 176.

1. $6 \times 2\frac{1}{2}$ **2.** $9 \times 3\frac{1}{3}$ **3.** $3\frac{3}{4} \times 8$ **4.** $1\frac{2}{5} \times 10$

5. $3\frac{1}{6} \times 1\frac{1}{2}$ **6.** $4\frac{3}{8} \times 3\frac{2}{5}$ **7.** $6\frac{3}{10} \times 1\frac{2}{3}$ **8.** $2\frac{1}{2} \times 2\frac{1}{2}$

Divide and simplify. Refer to page 180.

9. $12 \div \frac{1}{2}$ **10.** $10 \div \frac{3}{5}$ **11.** $\frac{3}{4} \div 2$ **12.** $\frac{5}{8} \div 5$

13. $\frac{2}{3} \div \frac{2}{3}$ **14.** $\frac{3}{4} \div \frac{1}{2}$ **15.** $\frac{1}{8} \div \frac{2}{3}$ **16.** $\frac{7}{8} \div \frac{3}{10}$

Divide and simplify. Refer to page 182.

17. $10 \div 2\frac{1}{2}$ **18.** $9 \div 3\frac{2}{3}$ **19.** $7\frac{1}{2} \div 2$ **20.** $4\frac{3}{4} \div 19$

21. $6\frac{2}{3} \div 3\frac{1}{3}$ **22.** $7\frac{5}{6} \div 2\frac{1}{2}$ **23.** $9\frac{3}{4} \div \frac{1}{4}$ **24.** $8\frac{3}{10} \div \frac{3}{10}$

Add and simplify. Refer to pages 196–202.

25. $\frac{3}{4} + \frac{3}{4}$ **26.** $\frac{7}{8} + \frac{1}{8}$ **27.** $\frac{5}{6} + \frac{5}{6}$ **28.** $\frac{7}{10} + \frac{1}{10}$

29. $\frac{3}{12} + \frac{3}{12}$ **30.** $\frac{1}{2} + \frac{1}{2}$ **31.** $\frac{7}{10} + \frac{5}{10}$ **32.** $\frac{7}{3} + \frac{6}{3}$

33. $\frac{1}{4} + \frac{1}{2}$ **34.** $\frac{3}{4} + \frac{1}{2}$ **35.** $\frac{7}{8} + \frac{3}{4}$ **36.** $\frac{7}{10} + \frac{3}{5}$

37. $\frac{1}{3} + \frac{1}{2} + \frac{1}{4}$ **38.** $\frac{3}{8} + \frac{7}{8} + \frac{1}{2}$ **39.** $\frac{9}{10} + \frac{1}{5} + \frac{5}{6}$ **40.** $\frac{3}{4} + \frac{3}{5} + \frac{1}{2}$

41. $5\frac{1}{2} + 5\frac{1}{4}$ **42.** $7\frac{2}{3} + 1\frac{2}{3}$ **43.** $3\frac{1}{3} + 1\frac{5}{6}$ **44.** $8\frac{1}{2} + 6\frac{6}{12}$

45. $9\frac{4}{5} + 8$ **46.** $8 + 12\frac{3}{8}$ **47.** $2\frac{1}{2} + 2\frac{1}{4} + 2\frac{1}{3}$ **48.** $7\frac{3}{4} + 2\frac{1}{6} + 3\frac{5}{12}$

Subtract and simplify. Refer to page 206.

49. $\frac{9}{12} - \frac{3}{12}$ **50.** $\frac{7}{8} - \frac{1}{8}$ **51.** $\frac{6}{10} - \frac{1}{10}$ **52.** $\frac{5}{6} - \frac{1}{6}$

53. $\frac{7}{12} - \frac{1}{3}$ **54.** $\frac{8}{9} - \frac{2}{3}$ **55.** $\frac{1}{2} - \frac{3}{8}$ **56.** $\frac{4}{5} - \frac{7}{10}$

57. $\frac{7}{8} - \frac{5}{6}$ **58.** $\frac{3}{5} - \frac{4}{9}$ **59.** $\frac{4}{15} - \frac{1}{6}$ **60.** $\frac{2}{3} - \frac{9}{16}$

61. $\frac{3}{10} - \frac{1}{7}$ **62.** $\frac{11}{12} - \frac{3}{8}$ **63.** $\frac{5}{6} - \frac{2}{9}$ **64.** $\frac{1}{2} - \frac{1}{5}$

Subtract and simplify. Refer to page 208.

1. $9\frac{7}{12} - 6\frac{1}{12}$ **2.** $8\frac{7}{8} - 1\frac{6}{8}$ **3.** $7\frac{2}{3} - 3$ **4.** $12\frac{7}{10} - 7$

5. $4\frac{4}{5} - 3\frac{1}{10}$ **6.** $8\frac{5}{6} - 1\frac{2}{3}$ **7.** $13\frac{7}{8} - 4\frac{1}{2}$ **8.** $17\frac{1}{3} - 8\frac{1}{4}$

Subtract and simplify. Refer to page 210.

9. $9 - 2\frac{1}{2}$ **10.** $6 - 3\frac{3}{4}$ **11.** $7 - 5\frac{3}{10}$ **12.** $6 - 5\frac{2}{3}$

13. $10\frac{1}{6} - 3\frac{5}{6}$ **14.** $4\frac{3}{8} - 1\frac{7}{8}$ **15.** $5\frac{1}{2} - 3\frac{3}{4}$ **16.** $6\frac{1}{10} - 3\frac{1}{5}$

Write the ratio. Simplify, if possible. Refer to page 224.

17. 8 pencils for 4 people

18. 6 problems in 18 minutes

19. 5 servings per 5 people

20. 8 dimes for 16 nickels

Find the unit rate. Refer to page 226.

21. 5 items cost 45¢. What is the cost per item?

22. \$ 16.60 is received for 4 hours of work. What is the amount per hour?

23. A bus driver travels 3200 miles in 6 hours. How many miles per hour is this? Round to tenths.

24. A car travels 186 kilometers on 8 liters of gas. How many kilometers per liter is this? Round to tenths.

Solve. Refer to page 230.

25. $\frac{2}{3} = \frac{x}{15}$ **26.** $\frac{3}{4} = \frac{15}{x}$ **27.** $\frac{3}{4} = \frac{x}{9}$ **28.** $\frac{5}{6} = \frac{12}{x}$

29. $\frac{7}{x} = \frac{5}{8}$ **30.** $\frac{x}{3} = \frac{9}{4}$ **31.** $\frac{9}{x} = \frac{2}{7}$ **32.** $\frac{x}{10} = \frac{5}{6}$

Solve. Refer to page 232.

33. 4 pens are needed for every 2 students. How many are needed for 10 students?

34. \$ 9.75 is paid for 3 hours of work. How much is paid for 2 hours of work?

35. If 12 pounds of fruit cost 96¢, how much do 10 pounds cost?

36. If it takes 15 minutes to answer 5 questions, how long does it take to answer 7 questions?

Convert to a decimal. Refer to page 248.

1. 73% **2.** 140% **3.** 6% **4.** 7.3%

5. 17% **6.** 25% **7.** 150% **8.** 13.7%

Convert to a percent. Refer to page 248.

9. 0.37 **10.** 0.19 **11.** 0.06 **12.** 0.60

13. 1.35 **14.** 0.003 **15.** 0.99 **16.** 1.78

Convert to a fraction, mixed number, or whole number. Refer to page 250.

17. 27% **18.** 79% **19.** 35% **20.** 80%

21. $66\frac{2}{3}\%$ **22.** $12\frac{1}{2}\%$ **23.** 175% **24.** 300%

Convert to a percent. Refer to page 252.

25. $\frac{4}{5}$ **26.** $\frac{7}{25}$ **27.** $3\frac{3}{4}$ **28.** 4

29. $\frac{7}{8}$ **30.** $1\frac{1}{3}$ **31.** $3\frac{3}{7}$ **32.** $\frac{5}{6}$

Solve. Refer to pages 256–263.

33. 75% of 10 is what number?

34. 90% of 20 is what number?

35. What is 83% of 100?

36. What is 25% of 40?

37. What percent of 14 is 7?

38. What percent of 40 is 4?

39. 9 is what percent of 27?

40. 30 is what percent of 40?

41. 50 is 50% of what number?

42. 12 is 25% of what number?

43. 6 is $33\frac{1}{3}\%$ of what number?

44. 7 is $16\frac{2}{3}\%$ of what number?

Find the percent of increase or decrease. Refer to page 264.

45. From 2 to 3

46. From 3 to 2

47. From 15 to 20

48. From 20 to 15

SOLVING PROBLEMS

When solving a problem, we sometimes need to decide whether to do a mental estimate, a paper and pencil calculation, or use a calculator. For each of the following problem situations, tell whether you would use a calculator, do a paper and pencil calculation, or make a mental estimate.

SITUATION 1 You are at the ballgame. Do you have enough money to buy two hot dogs and a bag of peanuts?

SITUATION 2 You are completing your income tax return.

SITUATION 3 You are deciding about how much it will cost to buy furniture for a bedroom.

SITUATION 4 You are planning a vacation. From a map you want to estimate the approximate number of kilometers you will drive.

SITUATION 5 You are balancing your checkbook.

SITUATION 6 You are in a pastry shop with $ 6. You are to purchase a dozen pastries for a party. Pastries sell for $ 1.20/dozen, but there is a sign telling that there is a 10% discount for the purchase of 5 dozen or more. Should you take advantage of the discount and buy 5 dozen?

READY FOR CHAPTER 10

Which is the likely measure?

1. The length of a swimming pool: 20 mm, 20 cm, 20 m, 20 km

2. The height of a chair: 40 mm, 40 cm, 40 m, 40 km

3. A glass of water: 300 mL or 300 L

4. A storage tank: 500 mL or 500 L

5. An apple: 100 mg, 100 g, 100 kg, 100 t

6. A box of apples: 5 mg, 5 g, 5 kg, 5 t

7. A chilly day: 40°C or 40°F

8. A hot bath: 40°C or 40°F

Make these changes.

9. 19 m = _____ cm

10. 3284 m = _____ km

11. 72 cm = _____ m

12. 8.7 cm = _____ mm

13. 6.2 L = _____ mL

14. 9.87 g = _____ mg

15. 8240 g = _____ kg

16. 3.2 t = _____ kg

17. 9 yd = _____ ft

18. 4 qt = _____ pt

19. 80 oz = _____ lb

20. 60 in. = _____ ft

Find the measure of each angle.

21.

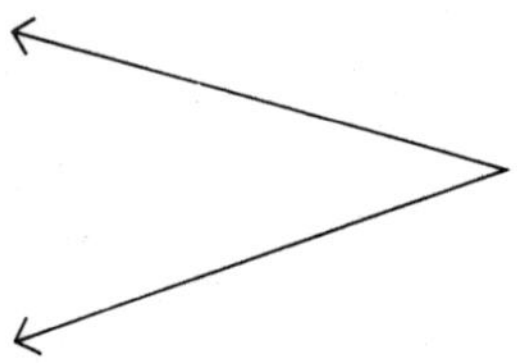

22.

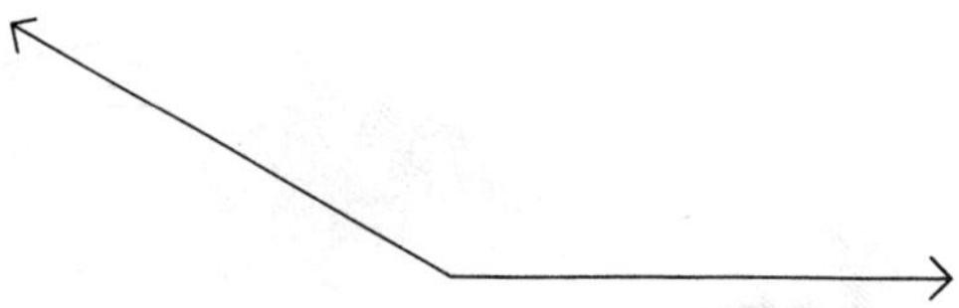

Application

23. A bus leaves New River, Arizona at 4:35 AM and arrives in Flagstaff, Arizona at 7:07 AM. How long does the trip take?

24. Sales tax is $6\frac{1}{2}$%. How much is the sales tax for a calculator which costs $ 29?

10 MEASUREMENT

GETTING STARTED

METRIC GUESTIMATION

A meter (m) is about the distance from the tip of your nose to the end of your outstretched hand, or 1 long step.

1. Estimate the length of your classroom's longest wall in meters.
2. Measure the wall. Was your estimate close?

A centimeter (cm) is about the width of your little finger.

3. Estimate the length of the long side of this book in centimeters.
4. Measure the side. Was your estimate close?

A millimeter (mm) is about the thickness of a dime.

5. Estimate the thickness of this book from cover to cover in millimeters.
6. Measure the thickness. Was your estimate close?

A kilometer (km) is about 1000 long steps.

7. Think of something one kilometer away. Estimate how long it would take to walk there.
8. It takes about 1 minute to walk 100 steps. Approximately how long would it take to walk 1 kilometer? Was your estimate close?

METRIC LENGTH

The meter (m) is the basic unit of length.

Unit	Size	Example
millimeter (mm)	0.001 m	Thickness of a dime
centimeter (cm)	0.01 m	Width of a pen
meter (m)	1 m	1 long step
kilometer (km)	1000 m	1000 long steps

EXAMPLES. To measure would you likely use mm, cm, m, or km?

1. The thickness of your book mm
2. The length of your pencil cm
3. The distance to the principal's office m
4. The distance to the North Pole km

TRY THIS

To measure would you likely use mm, cm, m, or km?

1. The distance from New York to Chicago
2. The length of a pin
3. The length of a sheet of paper
4. The length of an athletic field

EXAMPLES. Which is the likely measure?

5. The height of a 3-year-old child: 1 mm, 1 cm, 1 m, 1 km 1 m
6. The length of a river: 350 mm, 350 cm, 350 m, 350 km 350 km

TRY THIS

Which is the likely measure?

5. The length of a desk top: 30 mm, 30 cm, 30 m, 30 km
6. The distance around a room: 40 mm, 40 cm, 40 m, 40 km

We can use a metric ruler to measure length.

EXAMPLE 7. Give the length in mm and cm.

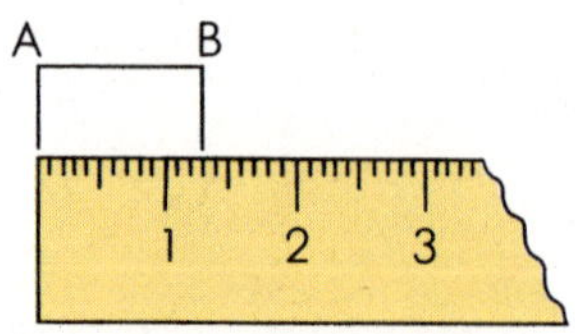

13 mm Count the mm from A to B.
1.3 cm 1 cm + 3 mm = 1.3 cm

TRY THIS

7. Give the length in mm and cm.

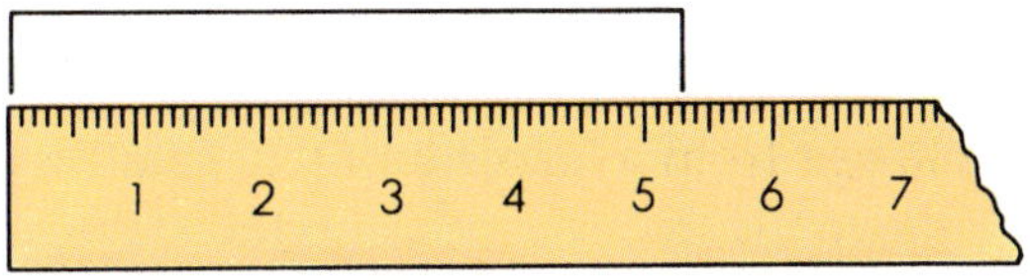

PRACTICE

To measure would you likely use mm, cm, m, or km?

1. The height of a building

2. The length of a picture

3. A running dash

4. A marathon run

5. The thickness of a picture frame

6. The height of a chair

Which is the likely measure?

7. The height of a person:
180 mm, 180 cm, 180 m, 180 km

8. The length of a baseball bat:
85 mm, 85 cm, 85 m, 85 km

9. The length of a tennis court:
20 mm, 20 cm, 20 m, 20 km

10. The length of a key:
20 mm, 20 cm, 20 m, 20 km

11. The distance per liter of gas:
10 mm, 10 cm, 10 m, 10 km

12. The length of a swimming pool:
50 mm, 50 cm, 50 m, 50 km

Give the length in mm and cm.

13.

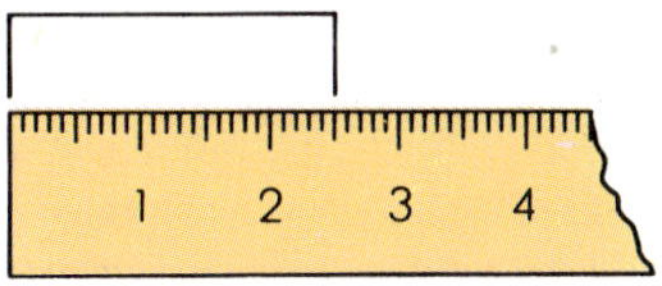

14.

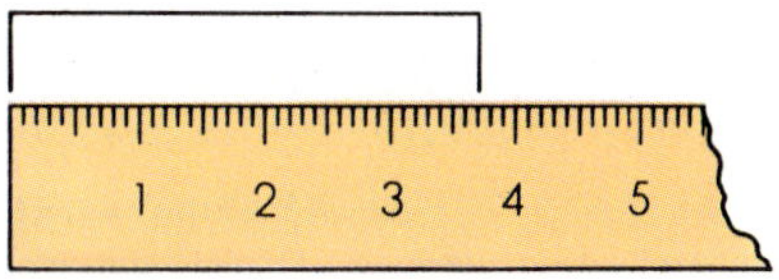

15.

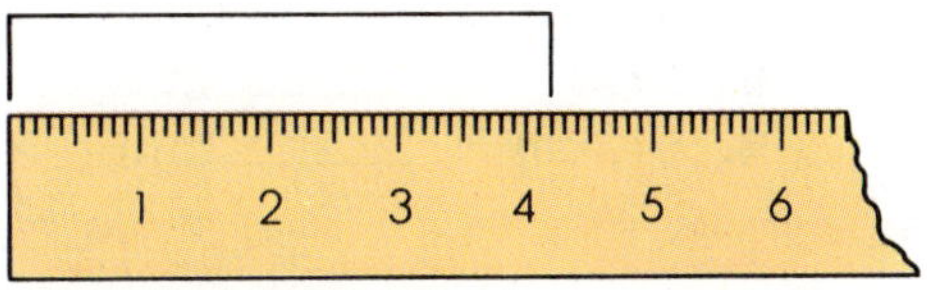

16.

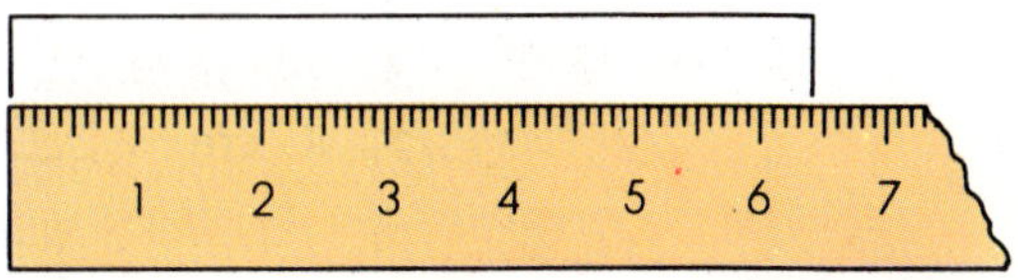

More Practice, page 306

CHANGING METRIC UNITS OF LENGTH

1 km = 1000 m	To change from a larger unit to a smaller unit, we multiply.
1 m = 100 cm	We move the decimal point to the right.
1 cm = 10 mm	To change from a smaller unit to a larger unit, we divide.
	We move the decimal point to the left.

Make this change. 5 km = ______ m

Decide whether to multiply or divide. →	To multiply or divide by 10, 100, or 1000, move the decimal point. →	Write the length.
5 km = ______ m Larger to smaller: Multiply.	1 km = 1000 m Multiply by 1000. 5.0 0 0	5 km = 5000 m

EXAMPLES. Make these changes.

1. 7.5 m = ______ cm

7.5 0

7.5 m = 7 5 0 cm

Think: Larger to smaller. Multiply. 1 m = 100 cm Multiply by 100. Move the decimal point 2 places to the right.

2. 4 mm = ______ cm

4.

4 mm = 0.4 cm

Think: Smaller to larger. Divide. 1 cm = 10 mm Divide by 10. Move the decimal point 1 place to the left.

3. 5 2 7 8 m = ______ km

5 2 7 8.

5 2 7 8 m = 5.2 7 8 km

Think: Smaller to larger. Divide. 1 m = 1000 km Divide by 1000. Move the decimal point 3 places to the left.

TRY THIS

Make these changes.

1. 18.7 cm = ______ m **2.** 32 cm = ______ m

3. 9.62 m = ______ cm **4.** 0.46 m = ______ cm

5. 296 m = ______ km **6.** 5932 m = ______ km

7. 87 cm = ______ mm **8.** 613 cm = ______ mm

PRACTICE

Make these changes.

1. 2 km = ______ m
2. 12 km = ______ m
3. 24 km = ______ m
4. 6.4 km = ______ m
5. 32.7 km = ______ m
6. 91.6 km = ______ m
7. 8.76 km = ______ m
8. 42.58 km = ______ m
9. 3.82 km = ______ m
10. 9 m = ______ cm
11. 15 m = ______ cm
12. 24 m = ______ cm
13. 8.6 m = ______ cm
14. 24.7 m = ______ cm
15. 99.9 m = ______ cm
16. 9 mm = ______ cm
17. 3 mm = ______ cm
18. 2 mm = ______ cm
19. 38 mm = ______ cm
20. 97 mm = ______ cm
21. 21 mm = ______ cm
22. 498 mm = ______ cm
23. 1284 mm = ______ cm
24. 2345 mm = ______ cm
25. 876 cm = ______ m
26. 378 cm = ______ m
27. 426 cm = ______ m
28. 78 cm = ______ m
29. 42 cm = ______ m
30. 65 cm = ______ m
31. 9.86 m = ______ cm
32. 32.04 m = ______ cm
33. 52.09 m = ______ cm
34. 8 cm = ______ m
35. 5 cm = ______ mm
36. 6 cm = ______ mm
37. 4 cm = ______ m
38. 5892 m = ______ km
39. 496 m = ______ km
40. 12,476 m = ______ km
41. 24 cm = ______ mm
42. 9856 m = ______ km
43. 947 m = ______ km
44. 8.7 cm = ______ mm
45. 4.5 cm = ______ mm

PROBLEM CORNER

1. A track race is $1\frac{1}{2}$ kilometers. How many meters is this?

2. A swimming race is 300 meters long. What part of a kilometer is this?

More Practice, page 306

APPLICATION

TIME SCHEDULES

AIRLINES SCHEDULE

SAN FRANCISCO/(S)
OAKLAND/(K) SAN JOSE(J)

	Leave	Arrive	Flight	Service	Via
To Birmingham, Ala. (CST)					
	7:00a (S)	2:35p	344	🍸✕	1
	1:00p (S)	9:05p	782 / 896	🍸✕	Lax
From Birmingham, Ala. (CST)					
67	8:45a	12:38p (S)	887 / 1431	🍸✕	Lax
	8:45a	1:08p (S)	887	🍸✕	1
	7:40p	12:06a (S)	637 / 546	🍸✕	Lax
To Honolulu, Hi. (HST)					
	8:45a (K)	12:10p (H)	105	🎥🍸✕	0
	9:00a (S)	**12:20p (H)**	**181**	🎥🍸✕★	0
	10:00a (J)	1:25p (H)	69	🎥🍸✕	0
	1:30p (S)	**4:45p (H)**	**35**	🎥🍸✕★	0
	4:35p (S)	**7:40p (I)**	**121**	🎥🍸✕●	0
	4:35p (S)	**9:25p (H)**	**121**	🎥🍸✕●	1
	7:10p (S)	**10:25p (H)**	**189**	🎥🍸✕★	0
From Honolulu, Hi. (HST)					
	9:30a (H)	**4:20p (S)**	**180**	🎥🍸✕★	0
	12:45p (H)	**7:35p (S)**	**96**	🎥🍸✕★	0
	1:15p (H)	8:10p (J)	68	🎥🍸✕	0
	1:25p (H)	**9:55p (S)**	**186**	🎥🍸✕●	1
	2:30p (H)	9:25p (K)	106	🎥🍸✕	0
	3:20p (I)	**9:55p (S)**	**186**	🎥🍸✕●	0
	11:50p (H)	**6:40a (S)**	**22**	♫ 🍸▬★	0
To Providence, R.I. (EST)					
	7:05a (J)	**8:29p**	630 / **240** 562	§ 🍸✕●	**Den** /Chi
	10:00a (S)	8:29p	**126** / 562	§ 🍸✕●	**Chi**
From Providence, R.I. (EST)					
	7:45a	12:20p (J)	445 / 943	🍸✕	Chi
	7:45a	**12:40p (S)**	445 / **123**	§ 🍸✕●	**Chi**
	2:20p	**9:00p (J)**	937 / **157** 921	🍸✕●	**Cle** /Chi
X6	3:55p	9:00p (J)	809 / 921	🍸✕	Chi
X6	3:55p	**9:05p (S)**	809 / **135**	§ 🍸✕●	**Chi**
X6	3:55p	9:15p (K)	809 / 841	🍸✕	Chi
To Vancouver, B.C. (PST)					
	11:30a (S)	3:14p	696 / 308	🍸✕	Sea
		Above Eff. Oct. 29			
	6:35p (K)	10:29p	468 / 342	🍸✕	Sea
	7:30p (S)	10:29p	246 / 342	🍸✕	Sea
From Vancouver, B.C. (PST)					
	7:10a	10:45a (S)	479 / 603	🍸✕	Sea
	4:00p	7:45p (S)	641 / 339	🍸✕	Sea
	4:00p	9:11p (J)	641 / 807	🍸✕	Sea
	4:00p	9:59p (K)	641 / 807	🍸✕	Sea
		Above Eff. Feb. 8			

✕ Meal ▬ Snacks 🍸 Cocktails
🎥 Movies ♫ Audio § Short-Subject

(I) – Hilo/Hawaii (S) – San Francisco
(K) – Oakland (H) – Honolulu
(J) – San Jose

1–Monday 2–Tuesday 3–Wednesday
4–Thursday 5–Friday 6–Saturday
7–Sunday X–Except

★ Indicates 747 ▲ indicates DC-10 to 747
● indicates DC-10 jet ⊕ indicates Night Coach may apply
♦ indicates 747 to DC-10

Joins Flight #896 Los Angeles to Birmingham

Indicates the airport when there is more than one major airport in an area.

Which flight or flights can you take for each?

1. to arrive in Birmingham after 6 PM
2. to return to San Francisco from Vancouver in the morning
3. to go to Honolulu on a 747

On flights with a stop, flight time includes time of layover. How long is each?

4. Flight #937/157 from Providence
5. Flight #22 from Honolulu?
6. Which flight to Vancouver takes the least time?
7. Which flight from Birmingham takes the most time?
8. Which flight from Providence does not operate on Saturday?

Montana Highway Bus Service						
Billings to Butte Read down			Billings—Butte Service	Butte to Billings Read up		
Daily	Mon-Fri	Mon-Fri		Daily	Mon-Fri	Mon-Fri
am + pm 47/67	am + pm 22/32	91	← Bus Number →	am + pm 147/167	am + pm 122/132	191
6:00	4:35	9:50a	dep BILLINGS arr	12:17	10:25	3:15p
6:35	↓	↓	↓ LAUREL ↑	11:43	↑	↑
7:25	↓	↓	↓ COLUMBUS ↑	10:53	↑	↑
7:59	↓	↓	↓ BIG TIMBER ↑	10:18	↑	↑
8:45	7:00	12:15a	↓ LIVINGSTON ↑	9:32	8:02	12:48p
9:15	7:30	↓	↓ BOZEMAN ↑	9:00	7:30	↑
10:06	↓	↓	↓ THREE FORKS ↑	8:05	↑	↑
10:35	↓	↓	↓ WHITE HALL ↑	7:24	↑	↑
11:30	9:30	2:40p	arr BUTTE dep	6:45	5:30	10:30a
See local schedules for town to town times. Bus remains 5 minutes at each scheduled stop before departure.						

Which bus can you take for each trip?

9. from Columbus to Butte?
10. from Boseman to Livingston

How long does it take for each trip?

11. from Billings to Big Timber on #47
12. from Columbus to Billings
13. from White Hall to Laurel
14. from Columbus to Three Forks

How much longer does it take for each trip?

15. from Billings to Livingston on #47 than on #91
16. from Butte to Billings on #147 than on #122
17. Suzie Plumhoff lives in Livingston and works in Bozeman. She rides the #47 bus to work. She rides a local bus home. The local takes 35% longer than the non-stop bus. How much longer does the local bus take?
18. Suzie buys a commuter ticket book of 20 rides for $ 36. The regular fare is $ 2.30 per ride. What percent does she save on each ride?

METRIC CAPACITY

The amount a container will hold is called capacity.
The **liter** is the basic unit of capacity.

Unit	Size	Example
milliliter (mL)	0.001 L	An eyedropper will hold 1 mL
liter (L)	1000 mL	About 4 glasses of water

EXAMPLES. Which is the likely measure?

1. A glass of milk: 200 mL or 200 L 200 mL
2. Gas for a car: 20 mL or 20 L 20 L

TRY THIS

Which is the likely measure?

1. A doctor's prescription: 10 mL or 10 L
2. A barrel of oil: 400 mL or 400 L

EXAMPLES. Make these changes:

3. 7 L = _______ mL

7.0 0 0

7 L = 7 0 0 0 mL

1 L = 1000 mL
Multiply by 1000.
Move the decimal point 3 places to the right.

4. 3 7 5 mL = _______ L

3 7 5.

3 7 5 mL = 0.3 7 5 L

1 L = 1000 mL
Divide by 1000.
Move the decimal point 3 places to the left.

5. 0.7 8 L = _______ mL

0.7 8 0

0.7 8 L = 7 8 0 mL

Move the decimal point 3 places to the right.

TRY THIS

Make these changes.

3. 3.2 L = _______ mL **4.** 16 L = _______ mL
5. 0.5 L = _______ mL **6.** 461 mL = _______ L
7. 250 mL = _______ L **8.** 7849 mL = _______ L

PRACTICE

Which is the likely measure?

1. A can of soda:
   360 mL or 360 L
2. A carton of cream:
   200 mL or 200 L
3. Water in a swimming pool:
   10,000 mL or 10,000 L
4. A cup of soup:
   150 mL or 150 L
5. A bathtub:
   300 mL or 300 L
6. A bottle of medicine:
   150 mL or 150 L
7. A bottle of perfume:
   100 mL or 100 L
8. A bucket:
   10 mL or 10 L

Make these changes.

9. 8 L = ______ mL
10. 12 L = ______ mL
11. 13 L = ______ mL
12. 3.7 L = ______ mL
13. 4.9 L = ______ mL
14. 2.8 L = ______ mL
15. 2598 mL = ______ L
16. 4782 mL = ______ L
17. 5893 mL = ______ L
18. 8000 mL = ______ L
19. 14,000 mL = ______ L
20. 7000 mL = ______ L
21. 0.49 L = ______ mL
22. 478 mL = ______ L
23. 0.57 L = ______ mL
24. 750 mL = ______ L
25. 0.98 L = ______ mL
26. 598 mL = ______ L
27. 0.3 L = ______ mL
28. 87 mL = ______ L
29. 0.2 L = ______ mL

PROBLEM CORNER

1. Some containers are labeled $\frac{1}{2}$ liter. How many milliliters is this?
2. Jonnie bought $\frac{1}{4}$ of a liter of juice. How many milliliters is this?
3. A container held 750 milliliters. What part of a liter is this?

More Practice, page 306

METRIC MASS

The basic unit of mass is the **kilogram.**

Unit	Size	Example
milligram (mg)	0.001 g	An aspirin is about 500 mg
gram (g)	0.001 kg	An average paper clip
kilogram (kg)	1 kg	The mass of this book
metric ton (t)	1000 kg	A truckload of coal

EXAMPLES. Which is the likely measure?

1. A potato chip: 400 mg, 400 g, 400 kg, 400 t   400 mg
2. A potato: 50 mg, 50 g, 50 kg, 50 t   50 g
3. A sack of potatoes: 5 mg, 5 g, 5 kg, 5 t   5 kg
4. A truckload of potatoes: 2 mg, 2 g, 2 kg, 2 t   2 t

TRY THIS

Which is the likely measure?

1. A mouse: 50 mg, 50 g, 50 kg, 50 t
2. A man: 70 mg, 70 g, 70 kg, 70 t
3. An elephant: 2 mg, 2 g, 2 kg, 2 t
4. A fly: 250 mg, 250 g, 250 kg, 250 t

EXAMPLES. Make these changes.

5. 4 2 8 6 mg = ______ g

   4 2 8 6.

   4 2 8 6 mg = 4.2 8 6 g

   1 g = 1000 mg
   Move the decimal point 3 places to the left.

6. 7.8 5 kg = ______ g

   7.8 5 0

   7.8 5 kg = 7 8 5 0 g

   1 kg = 1000 g
   Move the decimal point 3 places to the right.

TRY THIS

Make these changes.

5. 3786 g = ______ kg
6. 7.86 g = ______ mg
7. 1286 kg = ______ t
8. 730 mg = ______ g

PRACTICE

Which is the likely measure?

1. A pin:
250 mg, 250 g, 250 kg, 250 t

2. A pencil:
6 mg, 6 g, 6 kg, 6 t

3. A ping pong ball:
750 mg, 750 g, 750 kg, 750 t

4. A bowling ball:
7 mg, 7 g, 7 kg, 7t

5. A truckload of bricks:
2 mg, 2 g, 2 kg, 2 t

6. A wheelbarrow of bricks:
60 mg, 60 g, 60 kg, 60 t

Make these changes.

7. 8479 mg = ______ g

8. 250 mg = ______ g

9. 421 mg = ______ g

10. 5482 g = ______ kg

11. 100 g = ______ kg

12. 4895 g = ______ kg

13. 9 t = ______ kg

14. 7.47 t = ______ kg

15. 8.2 t = ______ kg

16. 6 kg = ______ g

17. 3.12 kg = ______ g

18. 5.47 kg = ______ g

19. 8249 kg = ______ t

20. 912 kg = ______ t

21. 1932 kg = ______ t

22. 3 g = ______ mg

23. 4.75 g = ______ mg

24. 9.81 g = ______ mg

25. 768 mg = ______ g

26. 7 t = ______ kg

27. 9 t = ______ kg

28. 7.4 kg = ______ g

29. 7.9 g = ______ mg

30. 4.7 g = ______ mg

PROBLEM CORNER

1. A large animal looked as if it weighed half a metric ton. How many kilograms is this?

2. A tablet was labeled 250 milligrams. What part of a gram is this?

3. A prescription called for $\frac{3}{4}$ grams. How many milligrams is this?

More Practice, page 306

APPLICATION

RATE SCHEDULES

STATE DEPARTMENT OF TAXATION											
3 1/2% BRACKET CARD											
Transaction			Tax	Transaction			Tax	Transaction			Tax
$0.01	to	$0.14	$0.00	$16.72	to	$16.99	$0.59	$33.58	to	$33.85	$1.18
0.15	to	0.38	0.01	17.00	to	17.28	0.60	33.86	to	34.14	1.19
0.39	to	0.64	0.02	17.29	to	17.57	0.61	34.15	to	34.42	1.20
0.65	to	0.88	0.03	17.58	to	17.85	0.62	34.43	to	34.71	1.21
0.89	to	1.18	0.04	17.86	to	18.14	0.63	34.72	to	34.99	1.22
1.19	to	1.57	0.05	18.15	to	18.42	0.64	35.00	to	35.28	1.23
1.58	to	1.85	0.06	18.43	to	18.71	0.65	35.29	to	35.57	1.24
1.86	to	2.14	0.07	18.72	to	18.99	0.66	35.58	to	35.85	1.25
2.15	to	2.42	0.08	19.00	to	19.28	0.67	35.86	to	36.14	1.26
2.43	to	2.71	0.09	19.29	to	19.57	0.68	36.15	to	36.42	1.27
2.72	to	2.99	0.10	19.58	to	19.85	0.69	36.43	to	36.71	1.28
3.00	to	3.28	0.11	19.86	to	20.14	0.70	36.72	to	36.99	1.29

In most states, customers pay sales tax on their purchases. Sales clerks use a chart to find sales tax.

Use the table. How much tax is there on these sales?

1. $ 17.56 2. $ 2.18 3. 45¢ 4. $ 36.28

Nevada charges $ 1.18 for all sales between $ 33.58 and $ 33.85.

5. For what exact sales total is the $3\frac{1}{2}\%$ tax closest to $ 1.18?

Compute the sales tax for these sales.

6. Connecticut–7%; sale–$ 29.85
7. Missouri–$3\frac{1}{8}\%$; sale–$ 158.90
8. Maryland–5%; sale–$ 253.87
9. California–$4\frac{3}{4}\%$; sale–$ 18.78

Allan bought a fish tank for $ 12.95. He paid a state tax of $3\frac{1}{2}\%$.

10. What was the total charge?
11. He gave the clerk a twenty-dollar bill. How much change did he get back?

The Act of 1789 created the U.S. Post Office. The first postage rate law was enacted in 1792. It established these rates for single-sheet letters.

Rates in 1792

Distance in miles	0–30	30–60	60–100	100–150	150–200	200–250	250–350	350–450	over 450
Rate	6¢	8¢	10¢	$12\frac{1}{2}$¢	15¢	17¢	20¢	22¢	25¢

Double sheets cost double, and so forth.

How much would it have cost to mail these items in 1792?

12. a 1-page letter mailed 220 miles
13. a 1-page letter mailed 650 miles
14. 2 pages mailed 40 miles
15. a 5-page letter
16. Why might this rate schedule have caused some disputes?

In 1879 the rates were simplified. The 4-class mail system was invented and is still used today. Postage rates depend on the weight and type of item being mailed.

Rates in 1879

1st class	3¢ per $\frac{1}{2}$ ounce (letters); 1¢ (post cards)
2nd class	special rate for publishers
3rd class	1¢ per 2 ounces (circulars, books, newspapers, etc.)
4th class	1¢ per ounce (merchandise)

How much would it have cost to mail these items in 1879?

17. a 2-ounce letter
18. books weighing $4\frac{1}{2}$ pounds
19. a $\frac{3}{4}$-pound hat
20. an 8-ounce newspaper
21. What is the current postage rate for each of the four classes?
22. How much would it cost to mail each of the items today?

TEMPERATURE

Temperatures can be measured on the Celsius (C) scale or the Fahrenheit (F) scale. Both use the degree symbol (°) to represent their unit of measure.

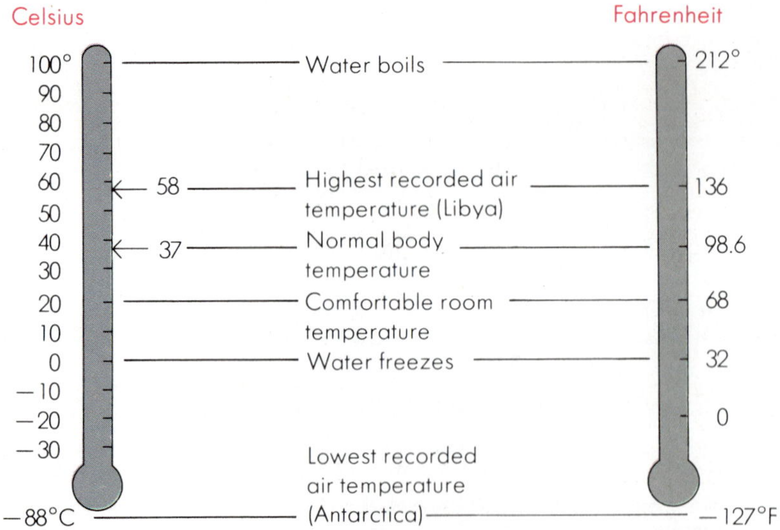

EXAMPLES. Is the temperature likely to be C or F?

1. A pleasant summer day: 23° C
2. A cold winter day: 15° F

TRY THIS Is the temperature likely to be C or F?

1. Baking bread: 350°
2. Freezing ice cream: −10°

EXAMPLES. Which is the likely temperature?

3. Warm soup: 40° C or 40° F 40° C
4. A chilled glass of milk: 38° C or 38° F 38° F

TRY THIS Which is the likely temperature?

3. A high fever: 100° C or 100° F
4. A hot day: 40° C or 40° F

PRACTICE

Is the temperature likely to be C or F?

1. A hot summer day: 42°

2. Bathtub water: 85°

3. Cold enough to have the heater on: 40°

4. Warm enough to have the air conditioner on: 40°

5. A snowball fight: 28°

6. Sunbathing: 35°

7. Water freezes: 32°

8. Water boils: 100°

9. Swimming pool water: 75°

10. A teapot steaming: 90°

Which is the likely temperature?

11. A warm stove: 50° C or 50° F

12. A hot stove: 75° C or 75° F

13. Cold enough for a jacket: 35° C or 35° F

14. Warm enough to go without a jacket: 30° C or 30° F

15. A summer day in the Sahara Desert: 55° C or 55° F

16. A winter day at the North Pole: −20° C or −20° F

17. A rainstorm: 20° C or 20° F

18. A snowstorm: 20° C or 20° F

19. Snow skiing: 40° C or 40° F

20. Water skiing: 30° C or 30° F

PROBLEM CORNER

1. During a storm the temperature dropped from 87° F to 73° F. How many degrees did it drop?

2. The temperature was 15° C in the morning. It was 32° C at noon. How many degrees did it rise?

CUSTOMARY MEASURES

Length
12 inches (in.) = 1 foot (ft)
3 feet = 1 yard (yd)
36 inches = 1 yard
5280 feet = 1 mile (mi)
1760 yards = 1 mile

Capacity
8 fluid ounces (fl oz) = 1 cup (c)
2 cups = 1 pint (pt)
2 pints = 1 quart (qt)
4 quarts = 1 gallon (gal)

Weight
16 ounces (oz) = 1 pound (lb)
2000 pounds = 1 ton

EXAMPLES. Make these changes.

1. 8 ft = ______ in.
8 ft = 96 in.
Think: Larger to smaller. Multiply.
1 ft = 12 in. $8 \times 12 = 96$

2. $4\frac{1}{2}$ gal = ______ qt
$4\frac{1}{2}$ gal = 18 qt
Think: Larger to smaller. Multiply.
1 gal = 4 qt $4\frac{1}{2} \times 4 = 18$

3. 20 oz = ______ lbs
20 oz = $1\frac{1}{4}$ lbs
Think: Smaller to larger. Divide.
16 oz = 1 lb $20 \div 16 = 1\frac{1}{4}$

4. 13,200 ft = ______ mi
13,200 ft = $2\frac{1}{2}$ mi
Think: Smaller to larger. Divide.
5280 ft = 1 mi
$13{,}200 \div 5280 = 2\frac{1}{2}$

TRY THIS

Make these changes.

1. 5 yd = ______ ft
2. $2\frac{1}{4}$ tons = ______ lbs
3. 20 pt = ______ qts
4. 40 in. = ______ yds

PRACTICE

Make these changes.

1. 5 ft = ____ in.
2. 3 yd = ____ in.
3. 14 cups = ____ pt

4. 2 mi = ____ yd
5. 4 pt = ____ cups
6. 8 oz = ____ lb

7. 5 qt = ____ pt
8. 4 ton = ____ lb
9. 2 mi = ____ ft

10. $5\frac{2}{3}$ yd = ____ ft
11. $1\frac{1}{4}$ c = ____ fl oz
12. 20 pt = ____ qt

13. $7\frac{3}{4}$ gal = ____ qt
14. $2\frac{3}{4}$ lb = ____ oz
15. 4000 lb = ____ ton

16. 12 ft = ____ yd
17. 64 fl oz = ____ c
18. 4 gal = ____ qt

19. 56 qt = ____ gal
20. 18 in. = ____ ft
21. 3 pt = ____ cups

22. 54 in. = ____ yd
23. 5 cups = ____ pt
24. 4 ft = ____ yd

25. 10 qt = ____ gal
26. 40 oz = ____ lb
27. 10 ft = ____ in

28. 7 yd = ____ ft
29. 6 gal = ____ qt
30. 6 qt = ____ pt

31. 18 pt = ____ qt
32. $2\frac{1}{2}$ yd = ____ in.
33. 32 oz = ____ lb

34. 3000 lb = ____ ton
35. 24 fl oz = ____ c
36. 3 mi = ____ yd

37. 3 mi = ____ ft
38. 108 in. = ____ yd
39. 2 c = ____ fl oz

40. 3 lb = ____ oz
41. 24 in. = ____ ft
42. 3 ton = ____ lb

PROBLEM CORNER

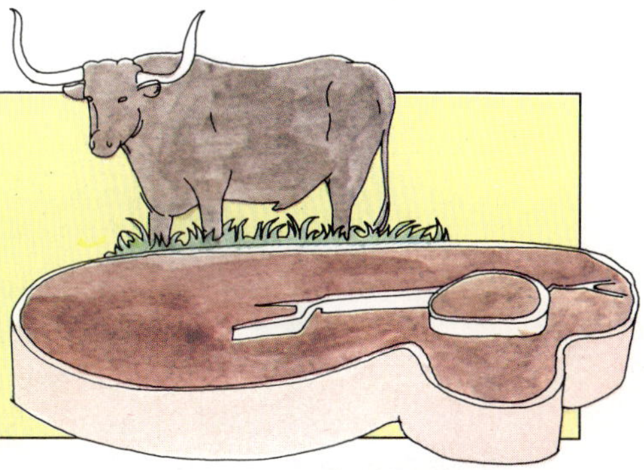

1. Jerry jumped 16 feet. George jumped $5\frac{2}{3}$ yards. Who jumped farther?

2. Mrs. Harold bought a 20-ounce steak for $ 4.35. Mr. Harold bought a $1\frac{1}{4}$ pound steak for $ 4.35. Which was the more economical buy?

More Practice, page 307

MEASURING ANGLES

A protractor measures angles in degrees (°).

Find the measure of this angle.

1. Place the center of the protractor where the sides of the angle meet.
2. Lay the line of the protractor along the side of the angle.
3. Read the degrees: 50°

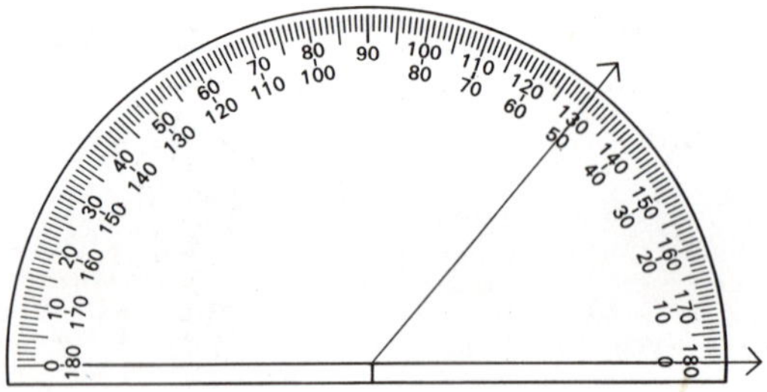

Notice the angle that makes a square corner.
This angle measures 90°. It is called a right angle.

EXAMPLES. Tell whether each angle is a right angle, larger than a right angle, or smaller than a right angle.

1. 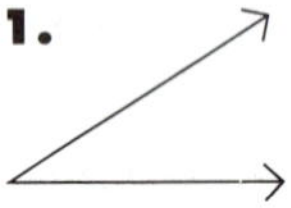 smaller

2. 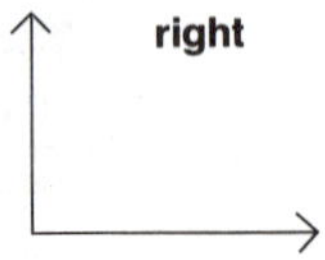 right

3. 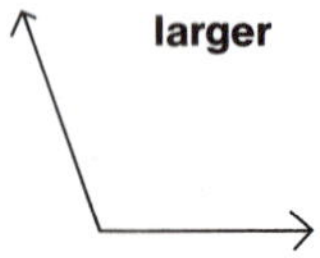 larger

TRY THIS

Tell whether each angle is a right angle, larger than a right angle, or smaller than a right angle.

1. 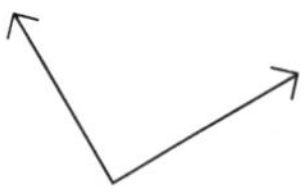

2. 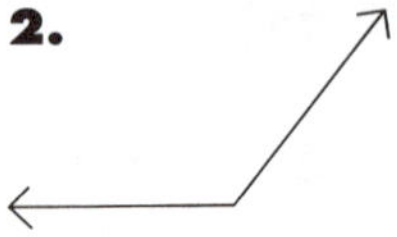

3. 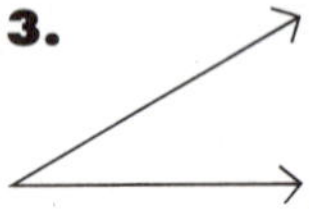

EXAMPLES. Find the measure of each angle.

4. The angle is smaller than a right angle, so the measure is less than 90°. Read the numbers less than 90°. The measure is 45°.

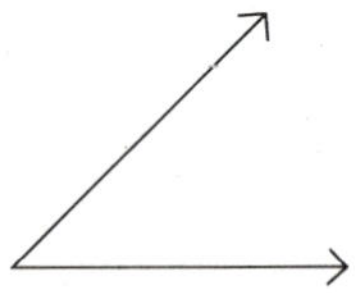

5. The angle is larger than a right angle, so the measure is greater than 90°. Read the numbers greater than 90°. The measure is 130°.

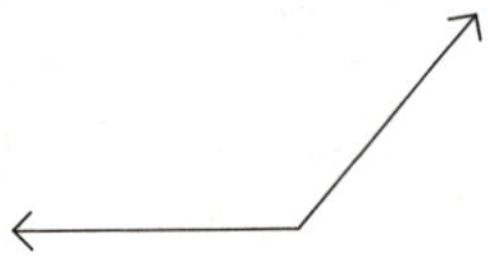

TRY THIS

Find the measure of each angle.

4. 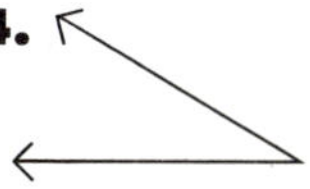

5. 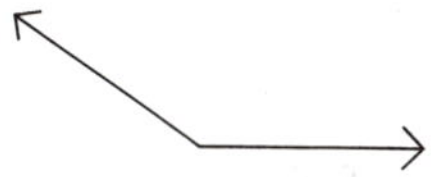

PRACTICE

Tell whether each is a right angle, larger than a right angle, or smaller than a right angle.

1. 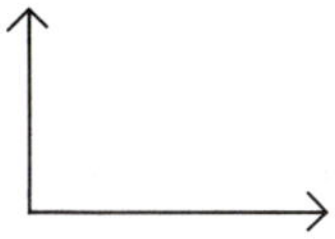

2. 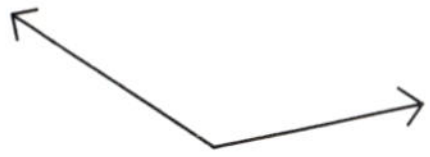

3. 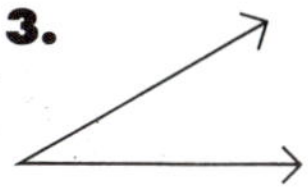

4. 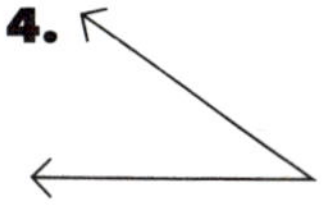

5. 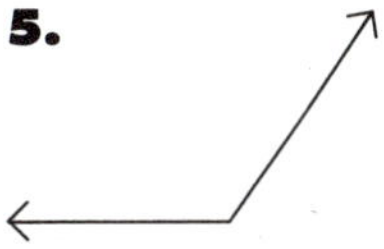

6. 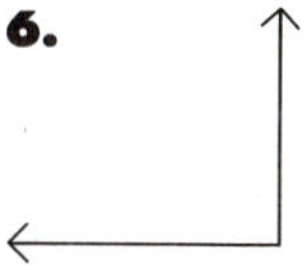

7. 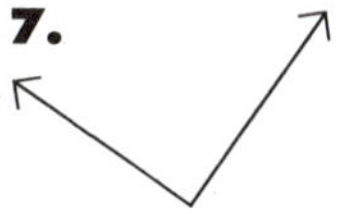

8. 

Find the measure of each angle.

9. 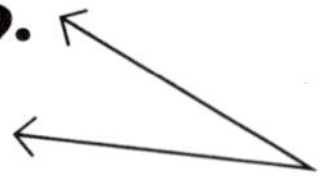

10. 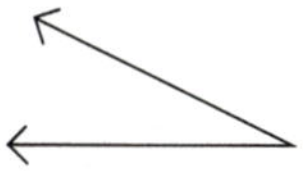

11. 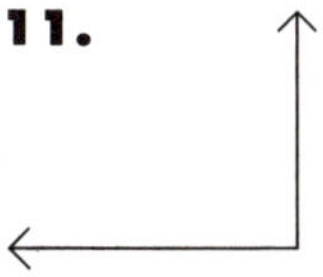

12. 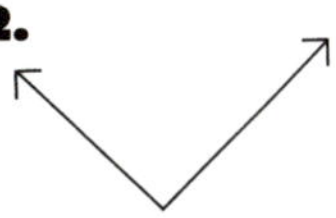

13. 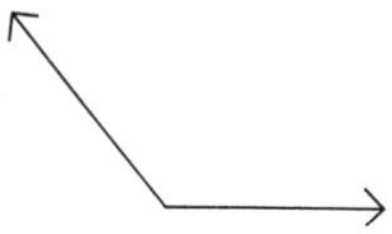

14. 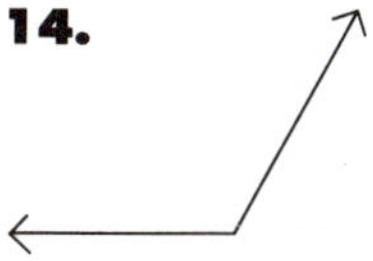

15. 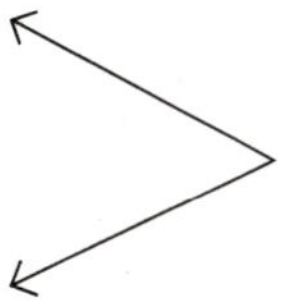

16. 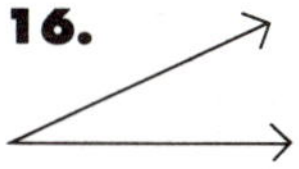

17. 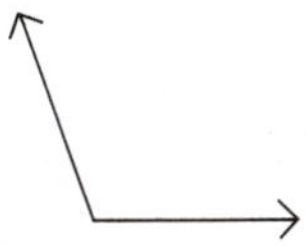

More Practice, page 307

CAREER CAREER CAREER CAREER CAREER

CIVIL SERVICE

The U.S. Government employs about one million people in Civil Service jobs. Applicants must take a Civil Service Exam which contains problems similar to the following. Which letter indicates the correct answer?

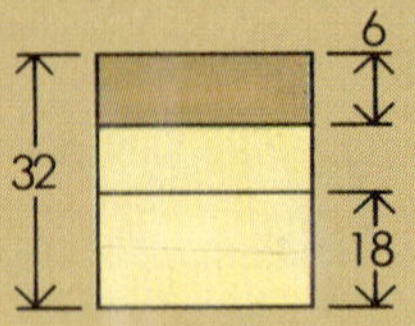

1. What part of the whole piece of wood is shaded?

   A. $\frac{1}{3}$ B. $\frac{3}{16}$ C. $\frac{1}{4}$ D. $\frac{3}{4}$ E. None of these

2. What is $\frac{1}{4}$ of $\frac{1}{8}$?

   A. $\frac{1}{2}$ B. $\frac{1}{16}$ C. $\frac{1}{32}$ D. $\frac{1}{4}$ E. 2

3. Which is the smallest amount?

   A. $\frac{1}{4}$ B. 0.4×0.4 C. 0.4 D. $\frac{0.4}{4}$ E. $\frac{0.4}{0.4}$

4. $5413.68 \div 36$ is:

   A. 15.038 B. 150.38 C. 150.382 D. 1503.8 E. None of these

5. A reamer cuts 3 holes per minute. There are 1125 holes to be cut. How many hours will it take?

   A. $6\frac{5}{12}$ B. $6\frac{1}{4}$ C. 37.5 D. $3\frac{3}{4}$ E. $18\frac{3}{4}$

6. Two bottles, X and Y, are the same size. Bottle X is $\frac{1}{4}$ full. Bottle Y is $\frac{1}{3}$ full. Pour the contents of X into Y. How full is Y?

A. $\frac{1}{12}$ B. $\frac{2}{7}$ C. $\frac{3}{4}$ D. $\frac{7}{12}$ E. None of these

7. Nails cost $0.76 per kilogram. How much do 4.75 kg of nails cost?

A. $6.17 B. $3.61 C. $36.10 D. $3.55 E. None of these

8. A sheet of metal is 1 m^2. About how many strips 2.5 cm by 1 m can be cut from the sheet of metal?

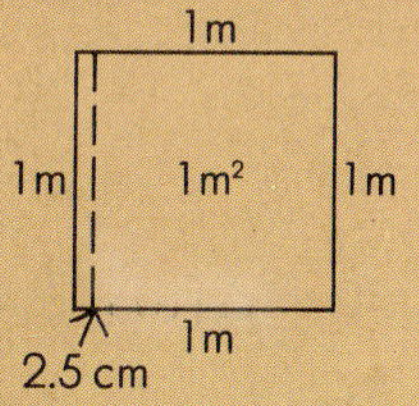

A. 4 B. 40 C. 400
D. 4000 E. None of these

9. A sheet of plywood is 4′ × 8′. How many 2′ × 2′ pieces can be cut from the sheet?

A. 8 B. 16 C. 4
D. 10 E. None of these

10. A rod is 15.3 cm long. A piece 58 mm long is cut off. How long is the piece that is left?

A. 9.5 mm B. 9.5 cm C. 95 cm
D. 42.7 mm E. None of these

11. Between which two speeds does the rate of gas consumption decrease the fastest?

A. 30 to 40 B. 40 to 50
C. 50 to 60 D. 60 to 70
E. None of these

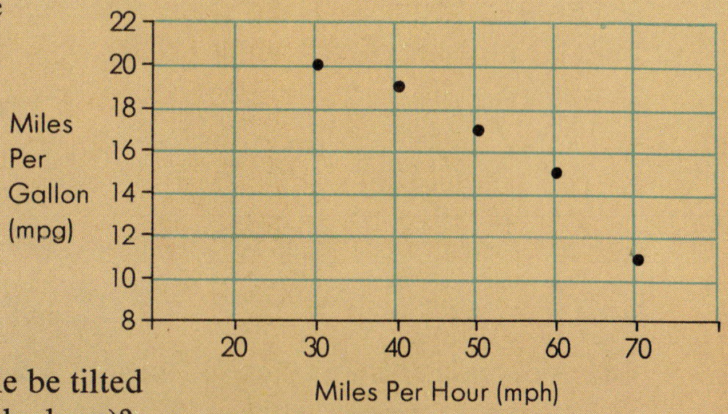

12. How many degrees must this pole be tilted to stand erect (perpendicular to the base)?

A. 44° B. 0.4° C. 40°
D. 8° E. None of these

CHAPTER 10 REVIEW

Which is the likely measure? pp. 284–296

1. The height of a roof: 8 mm, 8 cm, 8 m, 8 km
2. The thickness of a pancake: 250 mm, 250 cm, 250 m, 250 km
3. A bottle of pickles: 365 mL or 365 L
4. Water to take a shower: 60 mL or 60 L
5. A lion: 300 mg, 300 g, 300 kg, 300 t
6. A canary: 300 mg, 300 g, 300 kg, 300 t
7. Baseball season: 75° C or 75° F
8. Football season: 40° C or 40° F

Make these changes. pp. 286-298

9. 4349 m = ______ km
10. 9.6 cm = ______ mm
11. 85 cm = ______ m
12. 36 m = ______ cm
13. 8.62 L = ______ mL
14. 7.24 g = ______ mg
15. 5147 kg = ______ t
16. 9215 g = ______ kg
17. 8 ft = ______ in.
18. 6 gal = ______ qt
19. 4000 lb = ______ ton
20. 27 ft = ______ yd

Find the measure of each angle. p. 300

21. 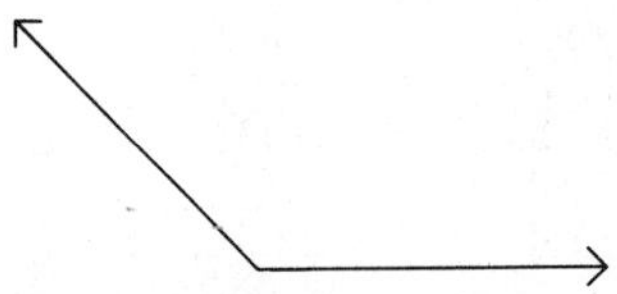

22. 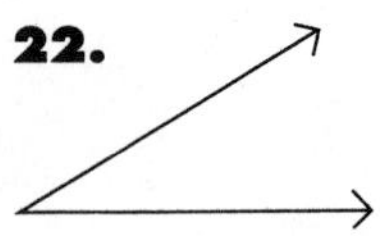

APPLICATION pp. 288, 294

23. A commuter train takes 20% less time than the regular train. If a trip takes 50 minutes on the regular train, how long does it take on the commuter train?

24. Sales tax is 5%. What is the total price for a stereo which costs $ 235?

CHAPTER 10 TEST

Which is the likely measure?

1. The length of a hall:
25 mm, 25 cm, 25 m, 25 km

2. The length of a river:
30 mm, 30 cm, 30 m, 30 km

3. A truck's gas tank:
200 mL or 200 L

4. A glass of juice:
400 mL or 400 L

5. A box of cereal:
425 mg, 425 g, 425 kg, 425 t

6. A truck:
10 mg, 10 g, 10 kg, 10 t

7. Ice skating:
25° C or 25° F

8. Water skiing:
28° C or 28° F

Make these changes.

9. 95 cm = ______ m

10. 6280 m = ______ km

11. 9.6 cm = ______ mm

12. 87 m = ______ cm

13. 398 mL = ______ L

14. 2387 kg = ______ t

15. 9.31 g = ______ mg

16. 3984 g = ______ kg

17. 36 yd = ______ ft

18. 7 qt = ______ pt

19. 108 in. = ______ ft

20. 160 oz = ______ lb

Find the measure of each angle.

21.

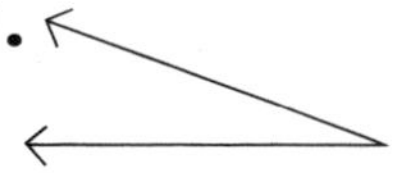

22.

APPLICATION

23. A plane leaves Honolulu, Hawaii at 10:18 PM and arrives in Hilo, Hawaii at 11:03 PM. How long does the trip take?

24. Sales tax is $5\frac{1}{2}$%. How much is the sales tax for a bicycle which costs $ 98?

MORE PRACTICE

Give the length in mm and cm. Use after page 285.

1.

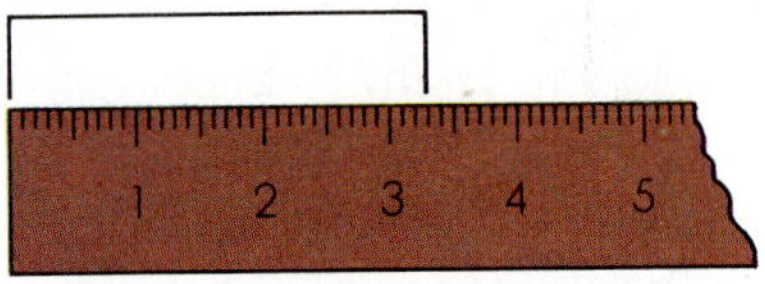

2.

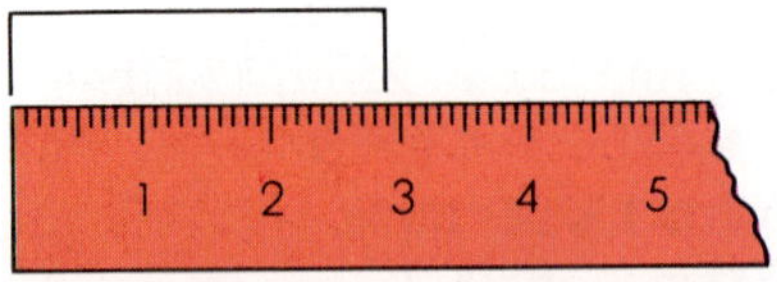

3.

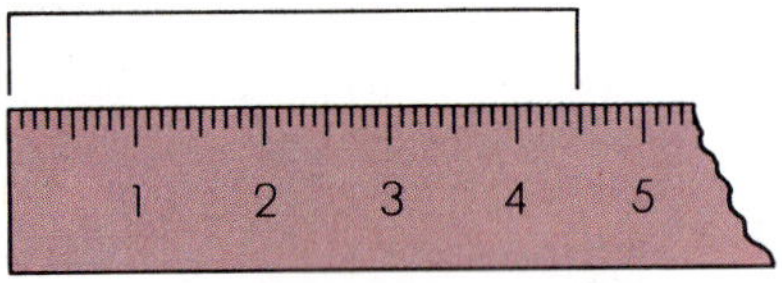

4.

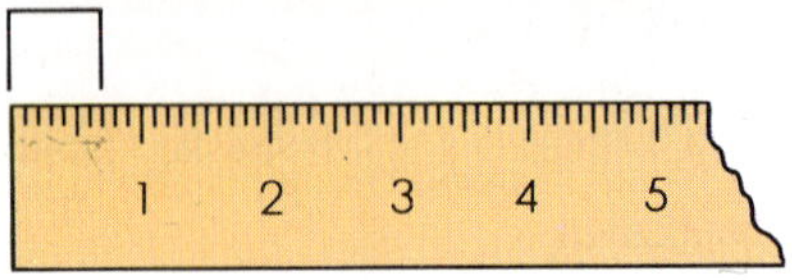

Make these changes. Use after page 287.

5. 3 km = ______ m

6. 96 cm = ______ m

7. 48 mm = ______ cm

8. 7 m = ______ cm

9. 9.6 km = ______ m

10. 6284 m = ______ km

11. 764 mm = ______ cm

12. 6.7 cm = ______ mm

Make these changes. Use after page 291.

13. 9 L = ______ mL

14. 747 mL = ______ L

15. 3462 mL = ______ L

16. 0.7 L = ______ mL

17. 4.9 L = ______ mL

18. 64 mL = ______ L

19. 7000 mL = ______ L

20. 450 mL = ______ L

Make these changes. Use after page 293.

21. 6 t = ______ kg

22. 478 g = ______ kg

23. 492 mg = ______ g

24. 6.3 kg = ______ g

25. 4.7 t = ______ kg

26. 6123 kg = ______ t

27. 7305 g = ______ kg

28. 6 g = ______ mg

Make these changes. Use after page 299.

1. 6 ft = ______ in. **2.** $4\frac{1}{2}$ ft = ______ in. **3.** 10 yd = ______ ft

4. $6\frac{1}{3}$ yd = ______ ft **5.** 4 yd = ______ in. **6.** $3\frac{1}{2}$ yd = ______ in.

7. 5 mi = ______ ft **8.** 5 mi = ______ yd **9.** 4 c = ______ fl oz

10. $7\frac{1}{2}$ c = ______ fl oz **11.** 7 pt = ______ cups **12.** $2\frac{1}{2}$ qt = ______ pt

13. $3\frac{1}{2}$ gal = ______ qt **14.** 6 lb = ______ oz **15.** 2 ton = ______ lb

16. $1\frac{1}{2}$ ton = ______ lb **17.** 60 in. = ______ ft **18.** 72 in. = ______ yd

19. 20 cups = ______ pt **20.** 40 qt = ______ gal **21.** 30 in. = ______ ft

22. 14 qt = ______ gal **23.** 44 in. = ______ yd **24.** 7000 lb = ______ ton

Find the measure of each angle. Use after page 301.

25.

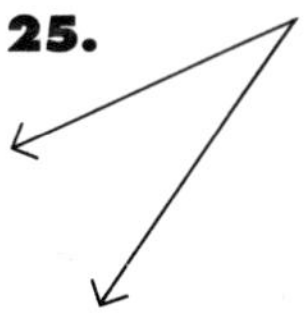

26.

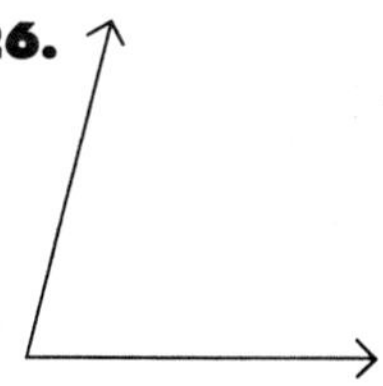

27.

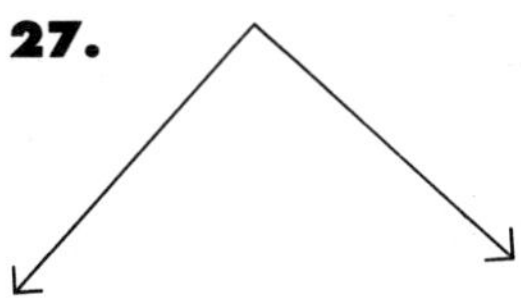

28.

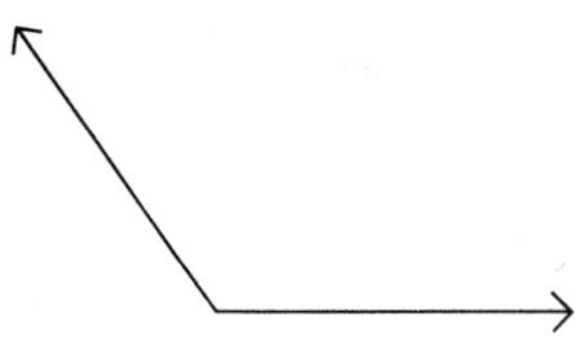

29.

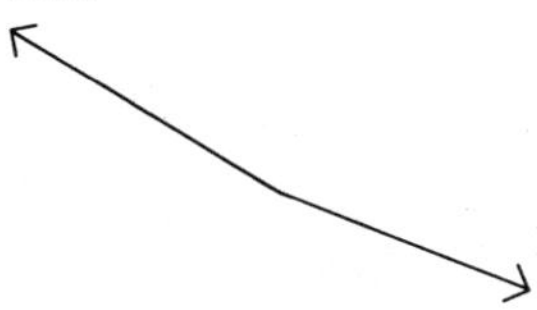

30.

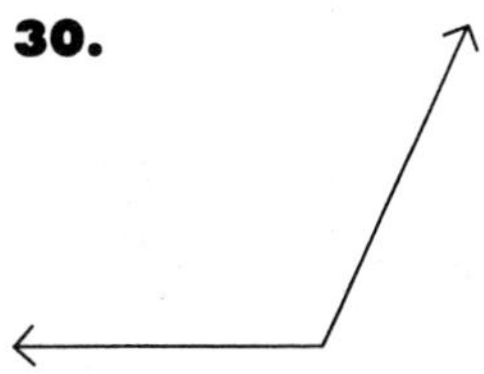

CALCULATOR CORNER

PUZZLING PATTERNS

Pattern A Make a pattern according to these rules.

Rule 1 If the number is even, divide by 2.
Rule 2 If the number is odd, multiply by 3 and add 1.

Try this with your calculator.
Start with 9.
9 is odd, so multiply by 3 and add 1. Write the result: 28
28 is even, so divide by 2. Write the result: 14
14 is even, so divide by 2. Write the result: 7
Continue until you make a pattern.

Result: 9, 28, 14, 7, 22, 11, 34, 17, 52, 26,
13, 40, 20, 10, 5, 16, 8, 4, 2, 1, . .

Use the following as starting numbers. Complete the pattern.

1. 10   2. 18   3. 15   4. 13   5. 84

Pattern B In each of the following do the first four calculations using your calculator. Write down your results. Look for a pattern. Then do the last calculation without your calculator.

6. 4×6
44×6
444×6
4444×6
44444×6

7. 9×9
99×89
999×889
9999×8889
99999×88889

8. 77×78
777×78
7777×78
77777×78
777777×78

9. 3×37
6×37
9×37
12×37
15×37
18×37
21×37

READY FOR CHAPTER 11

Find the perimeter.

1.

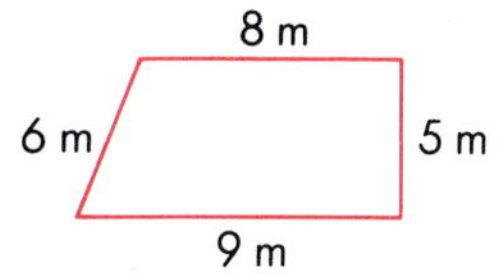

2.

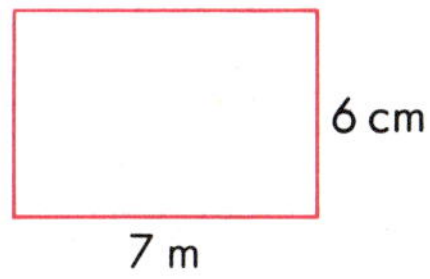

3.

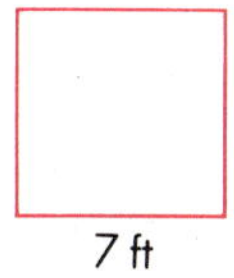

Find the circumference.

4.

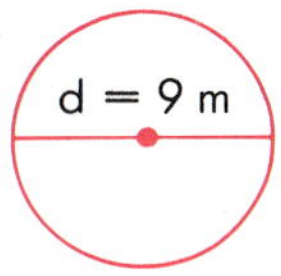

5.

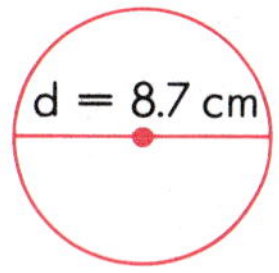

6.

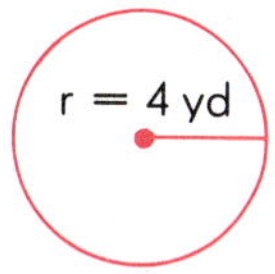

Find the area.

7.

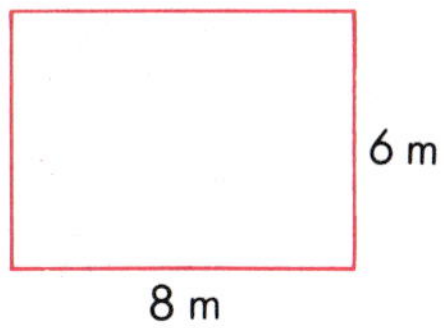

8.

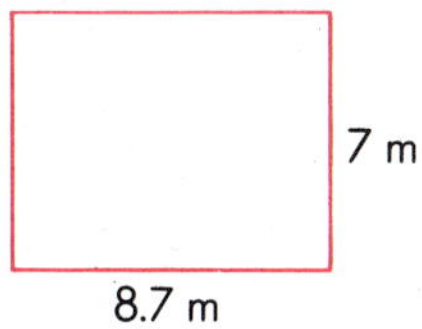

9.

10. Parallelogram
$b = 12$ cm
$h = 8$ cm

11. Triangle
$b = 9$ m
$h = 7$ m

12. Triangle
$b = 12$ ft
$h = 8$ ft

13. Trapezoid
$h = 4$ m
$b_1 = 13$ m
$b_2 = 9$ m

14. Trapezoid
$h = 3$ yd
$b_1 = 4$ yd
$b_2 = 8$ yd

15. Trapezoid
$h = 5$ cm
$b_1 = 20$ cm
$b_2 = 10$ cm

16. Circle
$d = 12$ cm

17. Circle
$r = 6$ m

18. Circle
$r = 20$ in.

Application

19. Darlene budgets 22% of her net income for housing. Her net income is \$ 550 per month. How much does Darlene budget for housing?

20. A house is 50 feet long and 25 feet wide. Construction costs \$ 30 per square foot. What is the cost of constructing the house?

11
PERIMETER AND AREA

GETTING STARTED

MATH PI

The **circumference** of a circle is the distance around it. The **diameter** is the distance across the circle through the center.

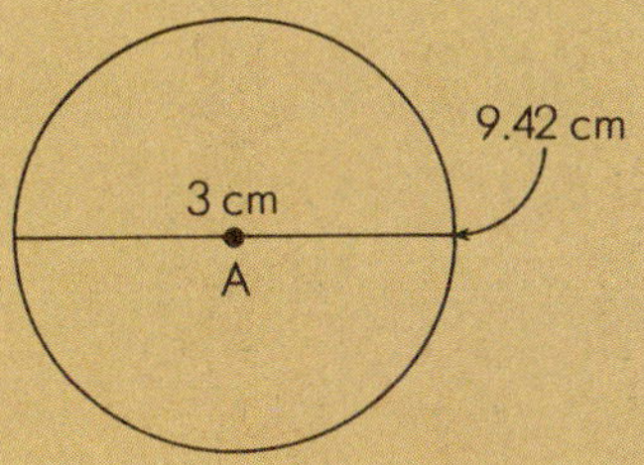

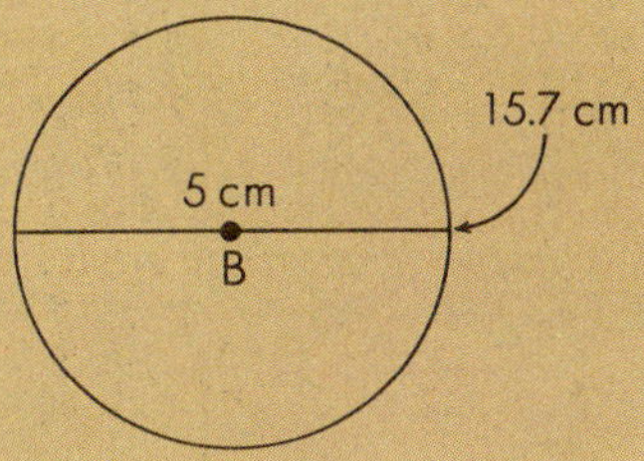

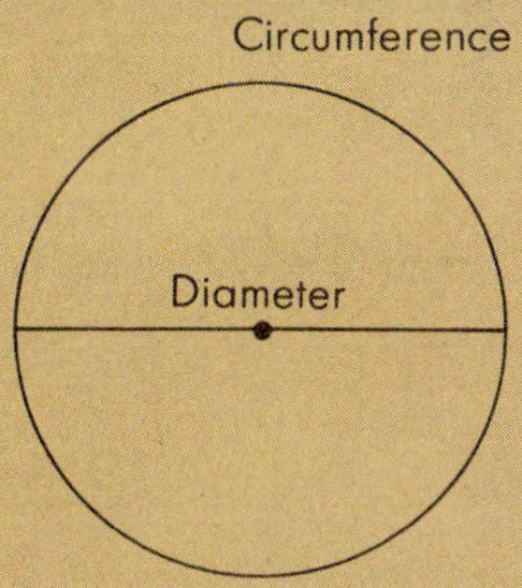

1. What is the circumference of circle A?
2. What is the diameter?
3. Divide the circumference by the diameter. $9.42 \div 3$ What is the quotient?
4. Divide the circumference of circle B by the diameter. Divide to hundredths.
5. Compare the quotient to problem 3. What do you find?

For all circles, the circumference divided by the diameter gives the same number. We use the Greek letter π (pronounced pi) for this number. It is approximately 3.14.

6. Find a circular object. Estimate its circumference. Wrap a string around it to find the circumference. Measure the string on a meter stick to the nearest millimeter.
7. Measure the diameter with a meter stick.
8. Divide the circumference by the diameter. Is the quotient close to π?

PERIMETER

The **perimeter** (P) of a polygon is the distance around the figure. To find the perimeter we add the lengths of the sides.

EXAMPLE 1. Find the perimeter.

$P = 3 + 4 + 4 + 3 + 5$
$P = 19$

The perimeter is 19 cm.

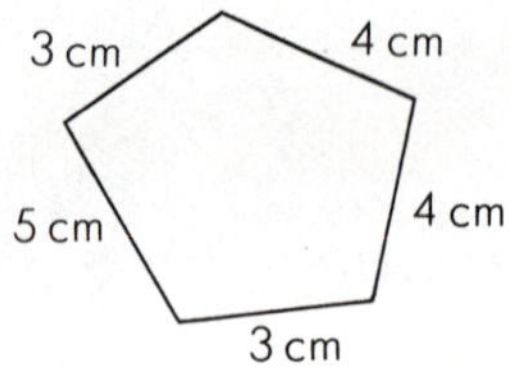

Formulas can be used to find perimeters of rectangles and squares.

EXAMPLE 2. Find the perimeter of the rectangle.

Perimeter = 2 × length + 2 × width
$P = 2l + 2w$
$P = (2 \times 8) + (2 \times 4)$
$P = 16 + 8$
$P = 24$

The perimeter is 24 cm.

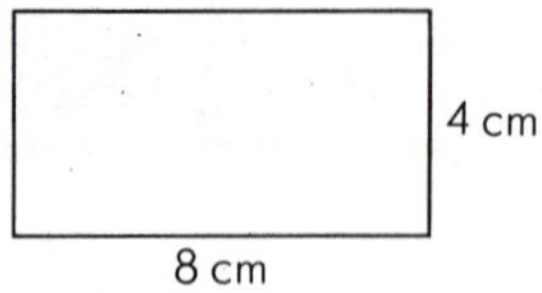

EXAMPLE 3. Find the perimeter of the square.

Perimeter = 4 × length of a side
$P = 4s$
$P = 4 \times 5$
$P = 20$

The perimeter is 20 m.

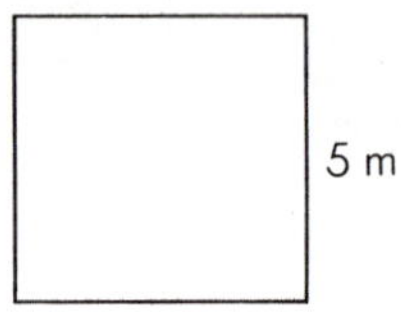

TRY THIS

Find the perimeter.

1.

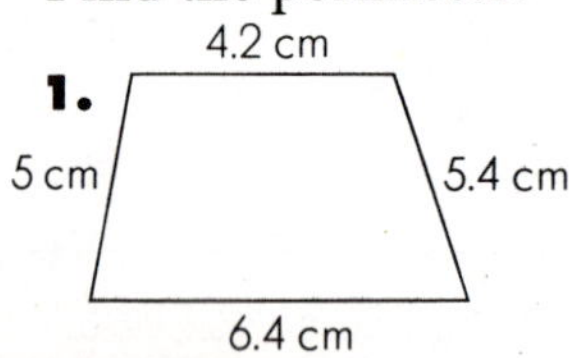

2.

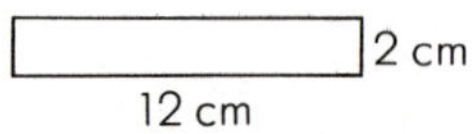

3.

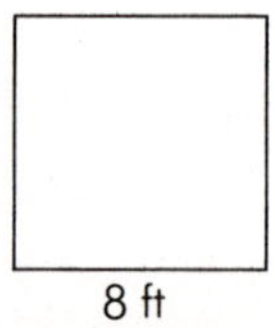

PRACTICE

Find the perimeter.

1. 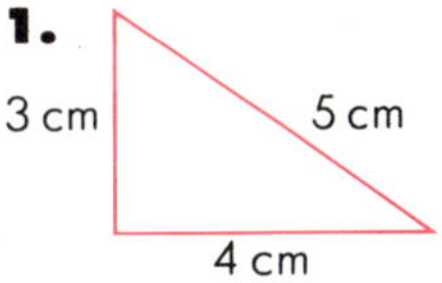

2. 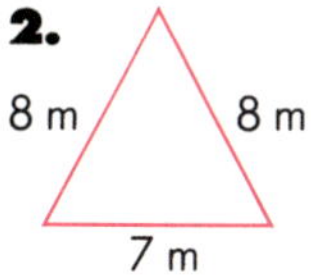

3. 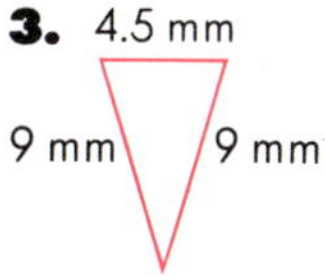

4. 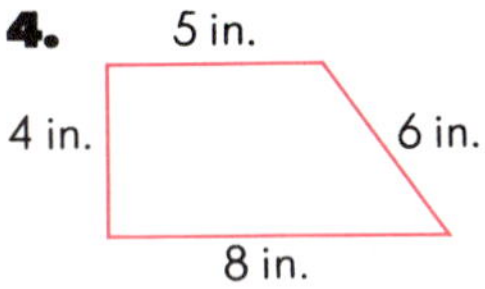

5. 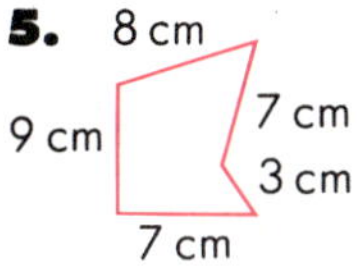

6. 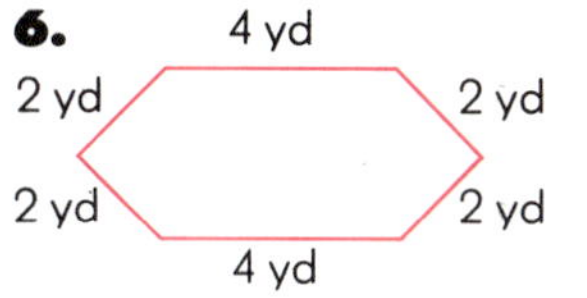

7. 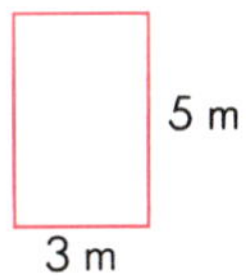

8. 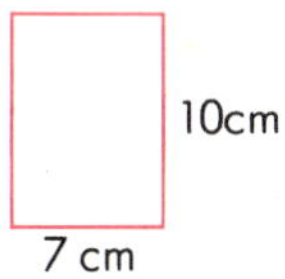

9. 

10. Rectangle
$l = 9.1$ cm
$w = 3.4$ cm

11. Rectangle
$l = 8$ ft
$w = 7$ ft

12. Rectangle
$l = 3\frac{1}{2}$ yd
$w = 1\frac{1}{2}$ yd

13. Square
$s = 3$ m

14. Square
$s = 8$ cm

15. Square
$s = 6$ mm

16. Square
$s = 2.5$ m

17. Square
$s = 1$ cm

18. Square
$s = 2\frac{1}{2}$ ft

PROBLEM CORNER

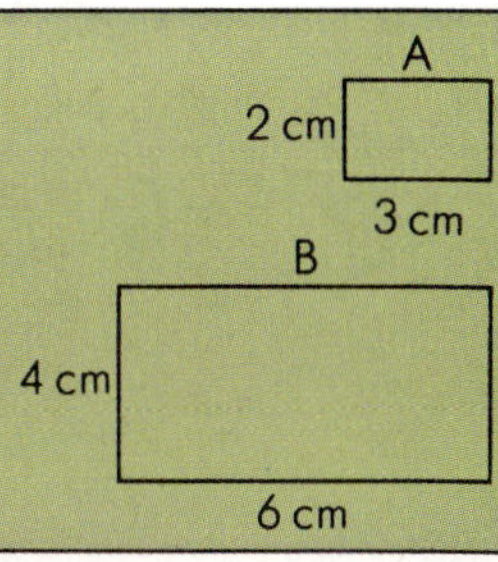

1. Find the perimeter of rectangle *A*.
2. In rectangle *B* each side is double that of rectangle *A*. Find the perimeter of rectangle *B*.
3. Compare the perimeters. Is the perimeter of rectangle *B* double that of rectangle *A*?

More Practice, page 334

CIRCUMFERENCE

The distance around a circle is the **circumference** (C) The distance across a circle through the center is the **diameter** (d) The **radius** (r) is the distance from the center to the outer edge of the circle.

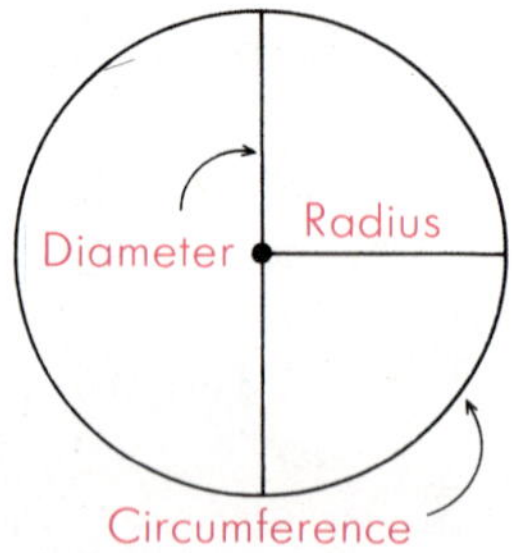

The ratio $\frac{C}{d}$ is the same for all circles. We use pi (π) for this number: $\frac{C}{d} = \pi$, or $C = \pi d$. Since the diameter is twice the radius, we also have the formula $C = 2\pi r$

EXAMPLE 1. Find the circumference.

$C = \pi d$
$C = 3.14 \times 4$
$C = 12.56$

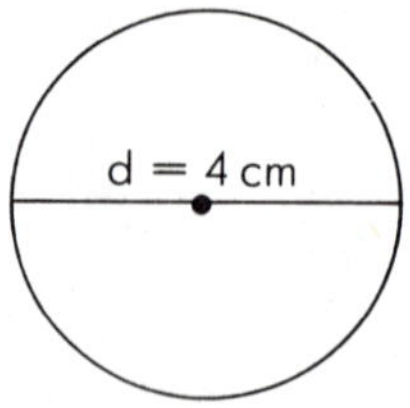

The circumference is approximately 12.56 cm.

TRY THIS

Find the circumference.

1. 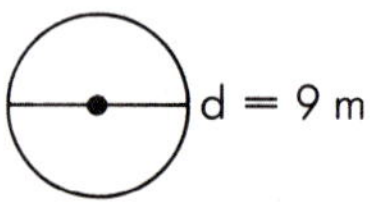

2. 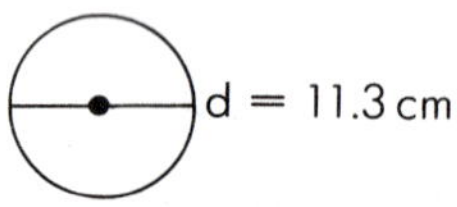

EXAMPLE 2. Find the circumference.

$C = 2\pi r$
$C = 2 \times 3.14 \times 5$
$C = 31.4$

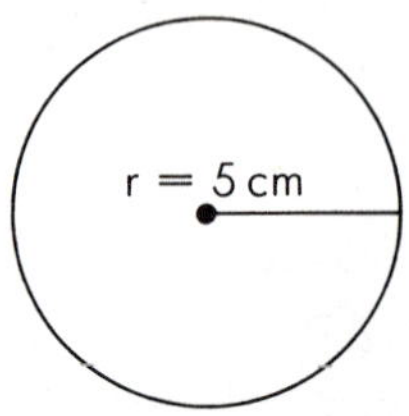

The circumference is approximately 31.4 cm.

TRY THIS

Find the circumference.

3. 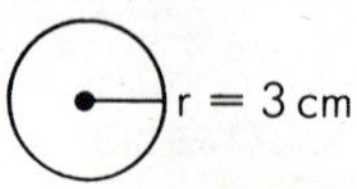

4. 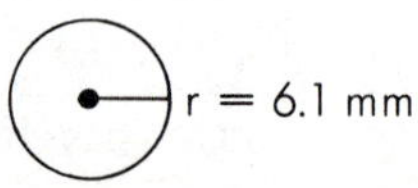

PRACTICE

Find the circumference.

1.

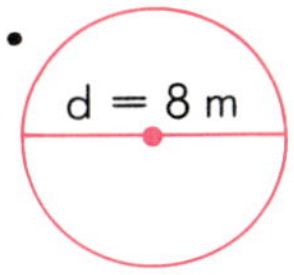

2.

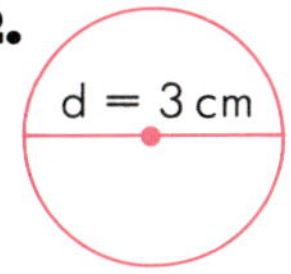

3.

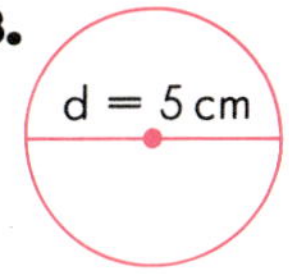

4.

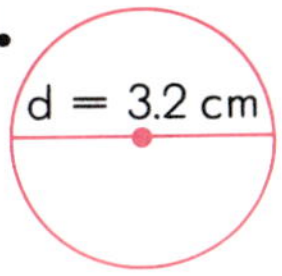

5.

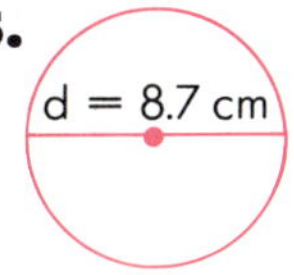

6.

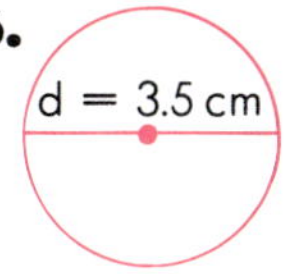

7. Circle
$d = 12$ cm

8. Circle
$d = 6$ in.

9. Circle
$d = 4$ ft

10.

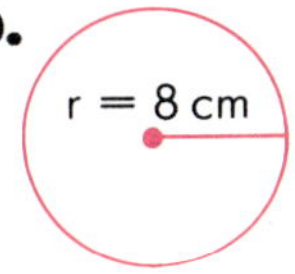

11.

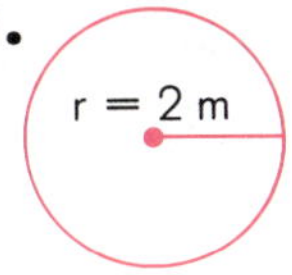

12.

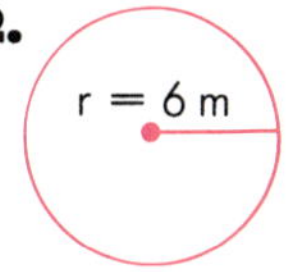

13.

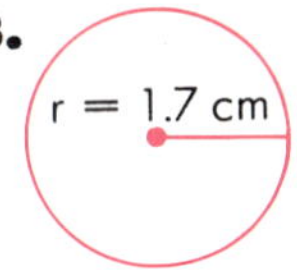

14.

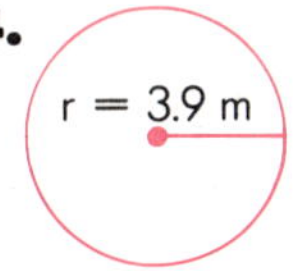

15.

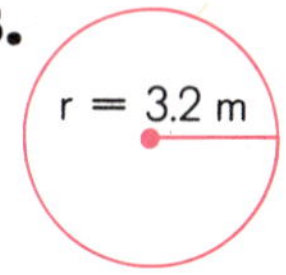

16. Circle
$r = 16$ mm

17. Circle
$r = 12$ in.

18. Circle
$r = 16$ ft

PROBLEM CORNER

1. Find the circumference of circle *A*.

2. In circle *B* the radius is double that of circle *A*. Find the circumference of circle *B*.

3. Compare the circumferences. Is the circumference of circle *B* double that of circle *A*?

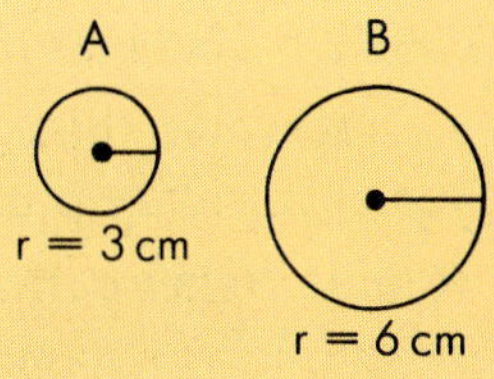

APPLICATION

RENTING AN APARTMENT

Money spent for rent is generally between 20% and 25% of the net income.

EXAMPLE Net monthly income: $ 602.58
The rent budget is between 20% and 25% of about $ 600.
20% of $ 600 = $ 120 25% of 600 = $ 150
The budget for rent is about $ 120 to $ 150 per month.

For each of the following estimate amount that should be budgeted for rent as 20% to 25% of net income.

1. Net monthly income: $ 847.95
2. Net monthly income: $ 514.32
3. Net weekly income: $ 318.96
4. Net weekly income: $ 163.25
5. Vince Valdez is a heavy equipment operator earning $ 62.25 take-home pay per day for a five day week.
6. Jane Childs takes home $ 9640 per year as a teacher.
7. Kris Karman manages The Pizza Place. He takes home $ 8175 yearly in salary plus 10% in tips.
8. Bonnie Baker earns $ 13,119 working in a U.S. Post Office. About 20% is taken out in deductions.

Utility bills must also be considered when renting an apartment.

The manager tells prospective tenants that electricity costs are about 20% of the monthly rent. About how much is the full cost of these apartments?

9. studio
10. 2 bedroom
11. 1 bedroom, full bath
12. 3 bedroom

HAINES MANOR
Furnished Apartments

Pool—Tennis
Covered Parking $15 per mo

Close to schools, shopping and transportation

Studio Efficiency . . . . $130 mo
1 Bdrm, full bath . . . . $160 mo
1 Bdrm, 1½ bath . . . . $195 mo
2 Bdrm, 1½ bath . . . . $235 mo
3 Bdrm, 2 bath . . . . $280 mo

Utilities included except electricity
Laundry Facilities—Health Club
Day Care Center available

Gloria Wren earns $975 net per month. She budgeted 22% of her salary for rent.

13. Can Gloria rent the studio apartment with covered parking?
14. Diane and Chris Wild earn $1875 net monthly. They budget 19% for housing. For how much under their budget can they rent a 2-bedroom apartment at Alberti Apartments?

FURNISHED APTS
All Utilities Included
2 bedroom, 1 bath apts $275
800 sq ft

ALBERTI APARTMENTS
Centerville 453-2146

How much will it cost per month to rent each of the following?

15. a 1-bedroom apartment, 400 sq ft
16. a 3-bedroom apartment, 650 sq ft
17. a 2-bedroom apartment, 540 sq ft

How large are these apartments?

18. $240.50 monthly rent
19. $377.00 monthly rent

5 STAR APARTMENTS
1 BDRM—from $165 up
5 sizes to pick from
Pay by the size you want
All Utilities included
65¢ per square ft/mo
All Electric Cooking-Solar Heat
All Luxury Apartments
Pools-Tennis Courts
Health Clubs-Laundry

ALL INCLUDED IN ONE LOW PRICE

AREA OF RECTANGLES AND SQUARES

Area is measured with square units. This is a square centimeter. It is 1 cm on each side. We write: 1 cm^2. We say: 1 square centimeter or 1 centimeter squared.

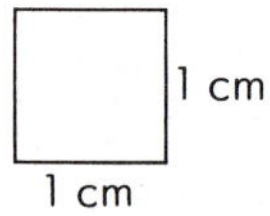

Find the area of the rectangle.
Area = length × width

$A = lw$
$A = 3 \times 2$
$A = 6$

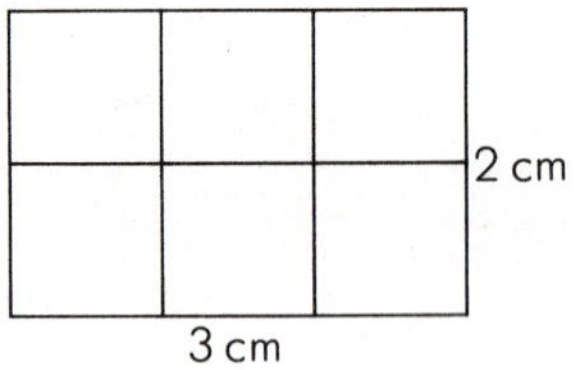

The area is 6 cm^2.

EXAMPLE 1. Find the area.

$A = lw$
$A = 8 \times 4$
$A = 32$

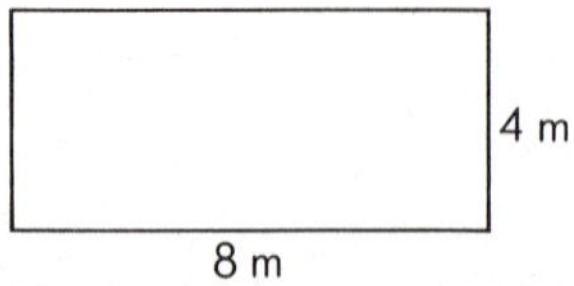

The area is 32 m^2.

A square is a special rectangle. All sides have the same length.
Area = side × side

$A = s \times s$ or s^2 (s^2 is read: side squared)

EXAMPLE 2. Find the area.

$A = s \times s$
$A = 5 \times 5$
$A = 25$

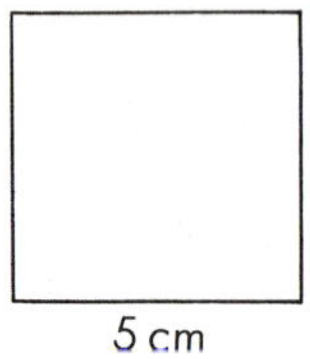

The area is 25 cm^2.

TRY THIS

Find the area.

1.

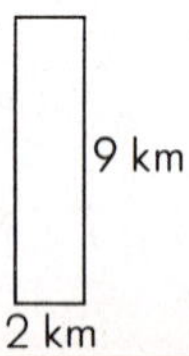

2.

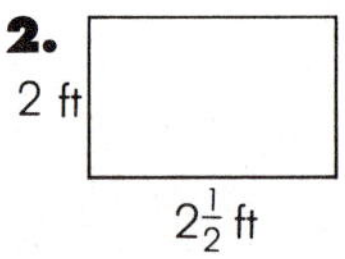

3.

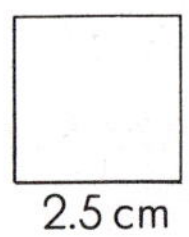

4.

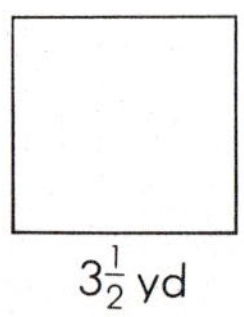

PRACTICE

Find the area.

1.

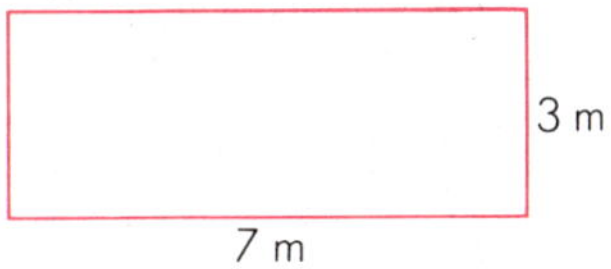

2.

3.

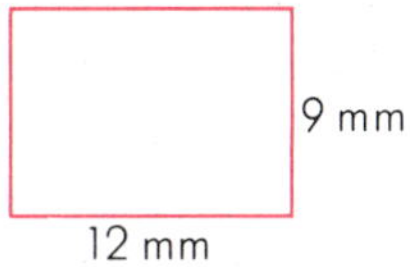

4.

5.

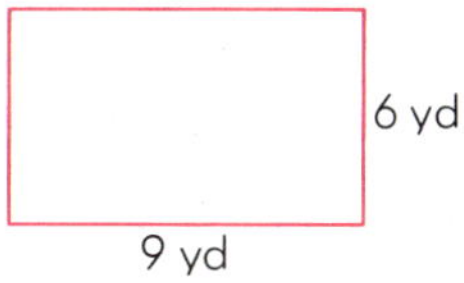

6.

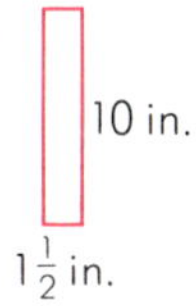

7. Rectangle
$l = 8.4$ cm
$w = 6$ cm

8. Rectangle
$l = 12$ m
$w = 4.6$ m

9. Rectangle
$l = 8.7$ m
$w = 3.7$ m

10.

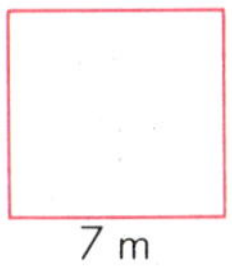

11.

12.

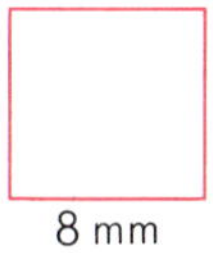

13.

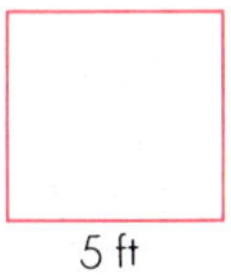

14.

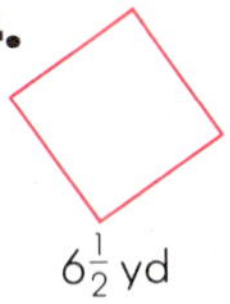

15.

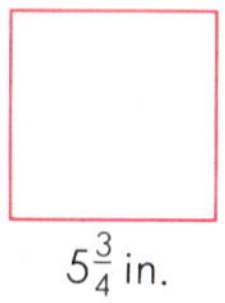

16. Square
$s = 4.6$ m

17. Square
$s = 9.7$ m

18. Square
$s = 3.7$ cm

PROBLEM CORNER

1. Find the perimeters of rectangle *A* and square *B*. Compare the perimeters.

2. Find the areas of rectangle *A* and square *B*. Compare the areas.

3. Which type of figure has more area for a given perimeter?

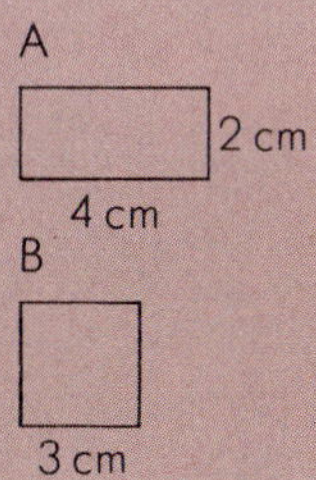

More Practice, page 334

AREA OF PARALLELOGRAMS, TRIANGLES

Parallelogram

Think of sliding the left triangle over to make a rectangle.

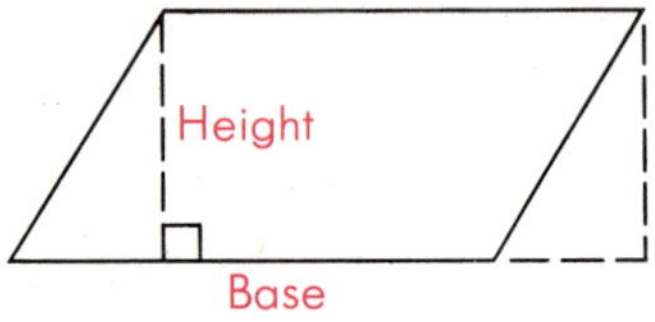

Area = base × height

$A = bh$

Triangle

Think of a triangle as half of a parallelogram.

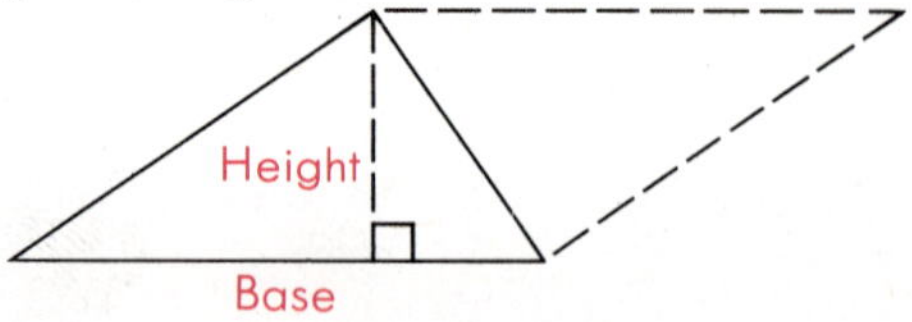

Area = $\frac{1}{2}$ × base × height

$A = \frac{1}{2}bh$

EXAMPLE 1. Find the area.

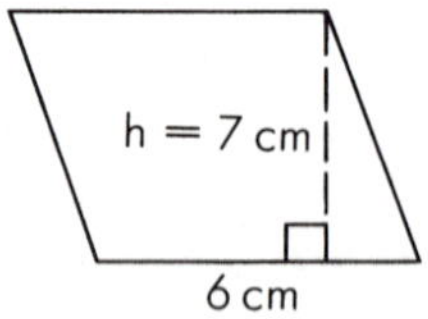

$A = bh$
$A = 6 \times 7$
$A = 42$

The area is 42 cm^2.

TRY THIS

Find the area.

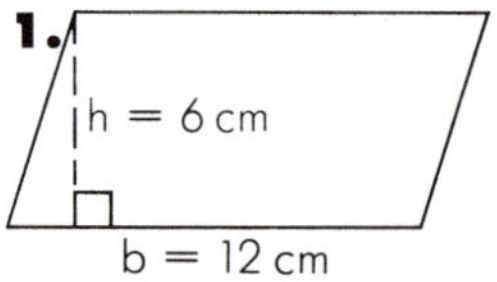

2.

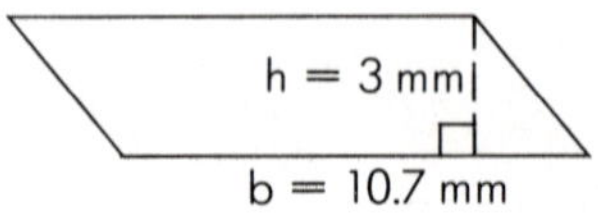

EXAMPLES. Find the area.

2.

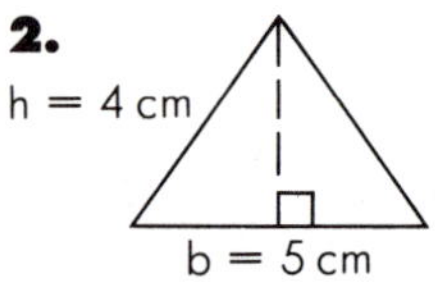

$A = \frac{1}{2}bh$

$A = \frac{1}{2} \times 5 \times 4$

$A = \frac{1}{2} \times 20 = 10$

The area is 10 cm^2.

3.

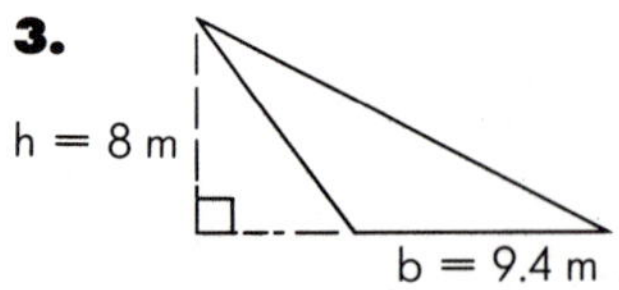

$A = \frac{1}{2}bh$

$A = \frac{1}{2} \times 9.4 \times 8$

$A = \frac{1}{2} \times 75.2 = \frac{75.2}{2} = 37.6$

The area is 37.6 m^2.

TRY THIS

Find the area.

3.

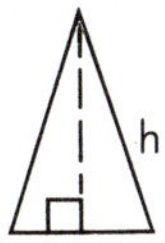

4.

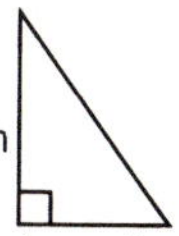

5.

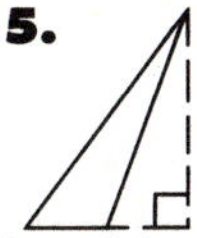

PRACTICE

Find the area.

1.

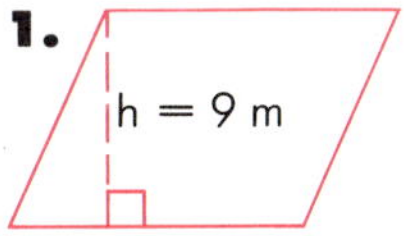

2.

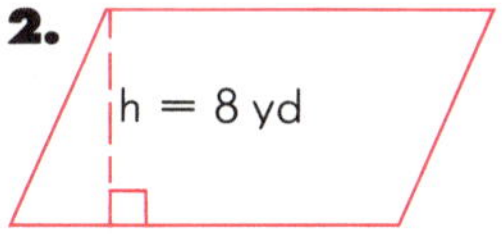

3.

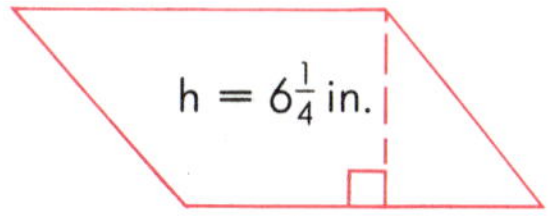

4.

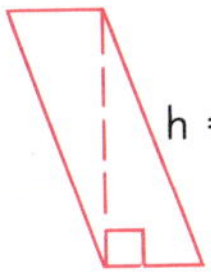

5.

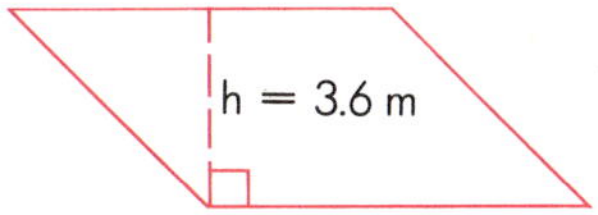

6.

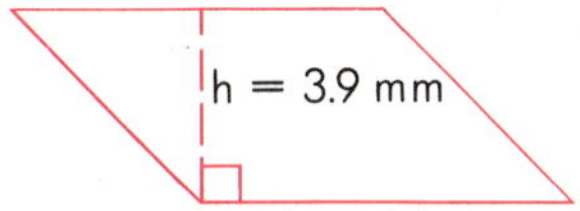

7. Parallelogram
$b = 20$ ft
$h = 8\frac{1}{2}$ ft

8. Parallelogram
$b = 11$ cm
$h = 8$ cm

9. Parallelogram
$b = 30$ mm
$h = 10$ mm

10.

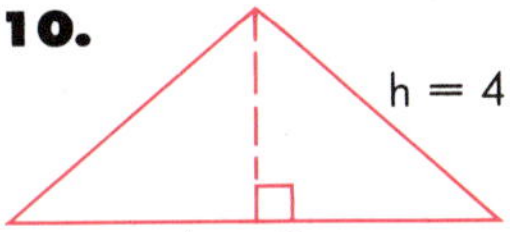

11.

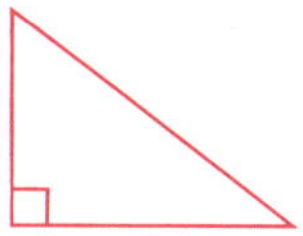

12.

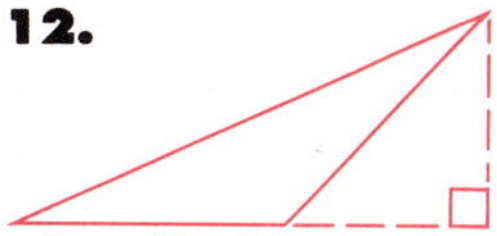

13.

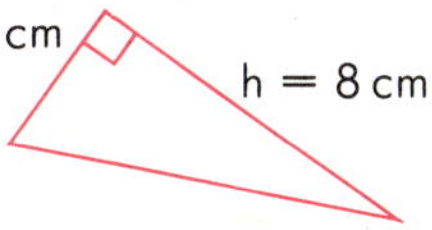

14.

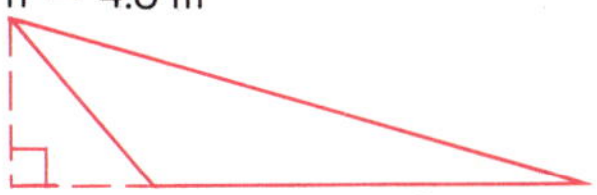

15.

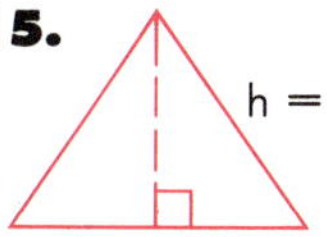

16. Triangle
$b = 7$ ft
$h = 3$ ft

17. Triangle
$b = 21$ cm
$h = 28$ cm

18. Triangle
$b = 40$ mm
$h = 31$ mm

More Practice, page 335

AREA OF TRAPEZOIDS

To find the area of a trapezoid we multiply the average of its bases ($\frac{1}{2}$ of base 1 + $\frac{1}{2}$ of base 2) times the height.

Area = $\frac{1}{2} \times$ (base 1 + base 2) $\times$ height

$A = \frac{1}{2} \times (b_1 + b_2) \times h$

$A = \frac{1}{2}(b_1 + b_2)h$

The formula is usually written $A = \frac{1}{2}h(b_1 + b_2)$.

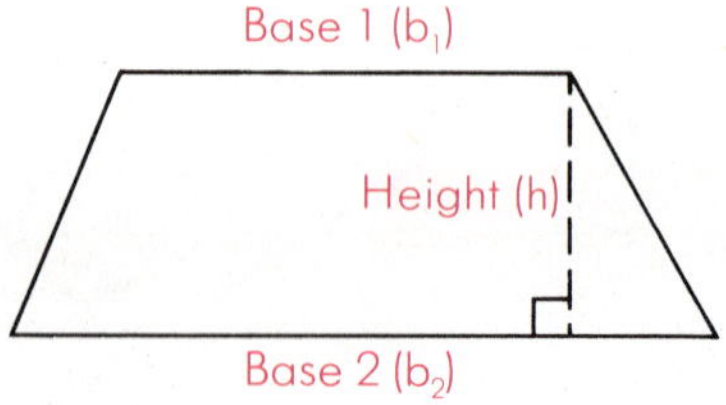

EXAMPLE 1. Find the area.

$A = \frac{1}{2}h(b_1 + b_2)$

$A = \frac{1}{2} \times 5 \times (8 + 10)$

$A = \frac{1}{2} \times 5 \times 18$ *Add.*

$A = 45$ *Then multiply.*

The area is 45 cm².

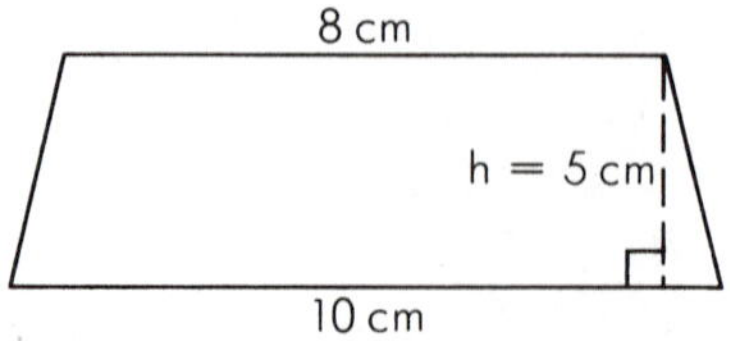

EXAMPLE 2. Find the area.

The height is one of the sides.

$A = \frac{1}{2}h(b_1 + b_2)$

$A = \frac{1}{2} \times 6.8 \times (8 + 12)$

$A = \frac{1}{2} \times 6.8 \times 20$

$A = \frac{1}{2} \times 136 = 68$

The area is 68 mm².

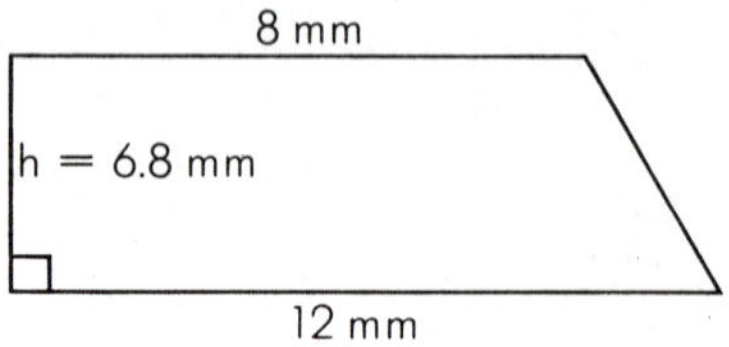

TRY THIS

Find the area.

1.

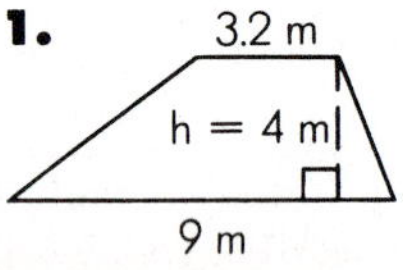

2.

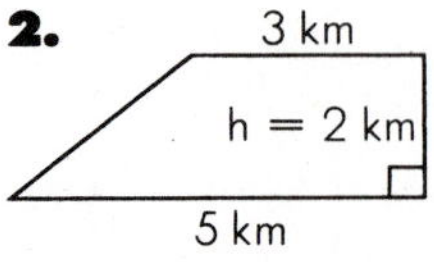

PRACTICE

Find the area.

1.
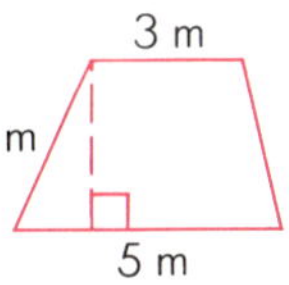

2.
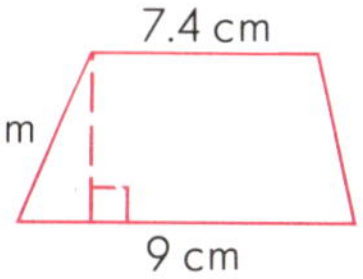

3.
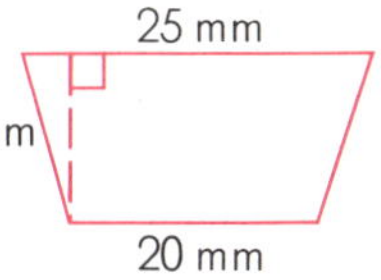

4. Trapezoid
$h = 5$ in.
$b_1 = 6$ in.
$b_2 = 8\frac{1}{2}$ in.

5. Trapezoid
$h = 8$ ft
$b_1 = 20$ ft
$b_2 = 12$ ft

6. Trapezoid
$h = 10$ yd
$b_1 = 30$ yd
$b_2 = 15$ yd

7. Trapezoid
$h = 8$ cm
$b_1 = 9$ cm
$b_2 = 12$ cm

8. Trapezoid
$h = 6$ mm
$b_1 = 8$ mm
$b_2 = 4$ mm

9. Trapezoid
$h = 8$ m
$b_1 = 10$ m
$b_2 = 12.1$ m

10.
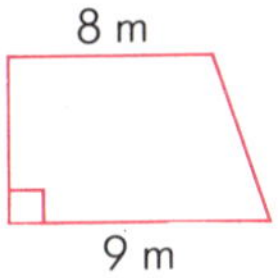

11.
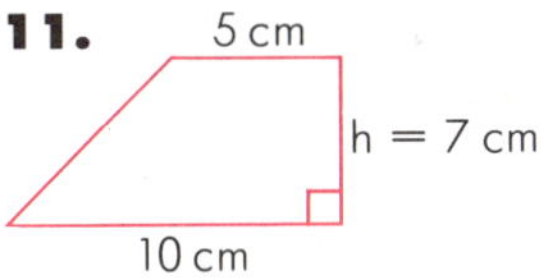

12.

13. Trapezoid
$h = 4$ in.
$b_1 = 15$ in.
$b_2 = 8$ in.

14. Trapezoid
$h = 7\frac{1}{4}$ ft
$b_1 = 6$ ft
$b_2 = 3$ ft

15. Trapezoid
$h = 8$ yd
$b_1 = 16$ yd
$b_2 = 10$ yd

PROBLEM CORNER

1. Find the area of trapezoid *A*.
2. In trapezoid *B* the lengths are tripled. Find the area of trapezoid *B*.
3. Compare the areas. Is the area of trapezoid *B* triple that of trapezoid *A*? What is the ratio of the area of trapezoid *B* to the area of trapezoid *A*?

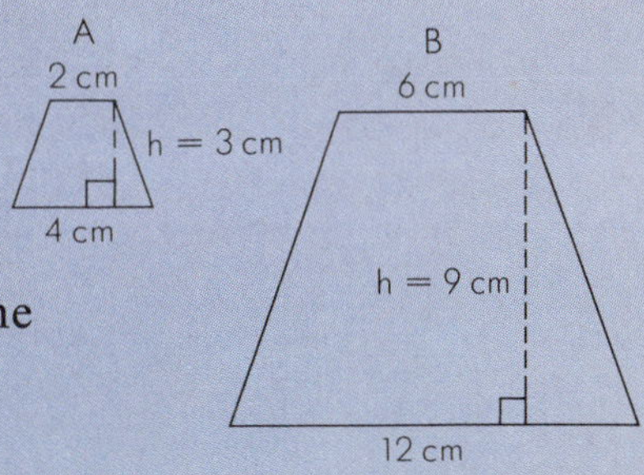

More Practice, page 335

APPLICATION

BUYING A HOUSE

The construction cost for a new house is often expressed as the cost per square foot.

What is the cost of construction of these houses?

1.

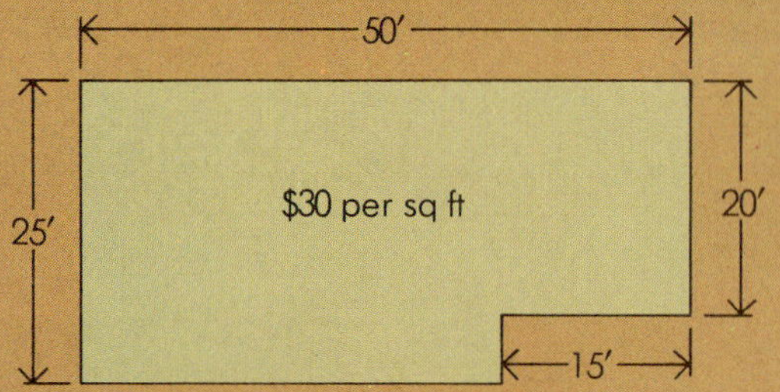

2.

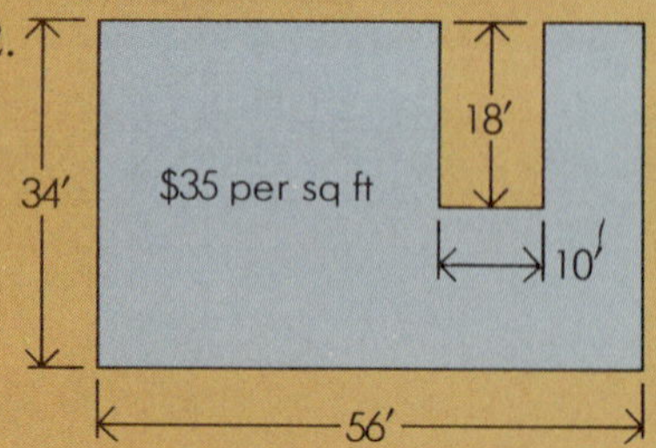

3.

4.

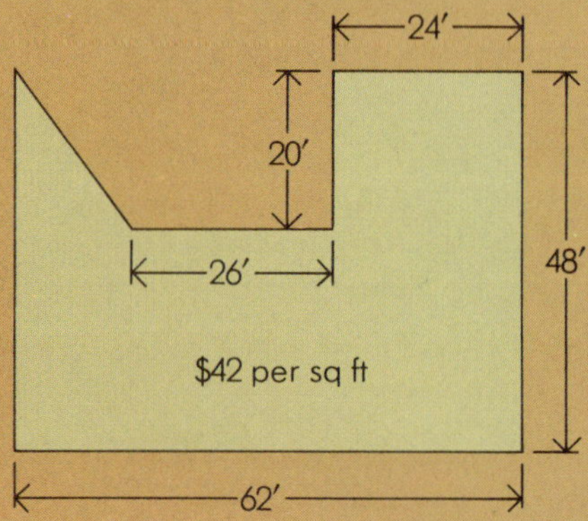

Many people consider buying mobile homes.

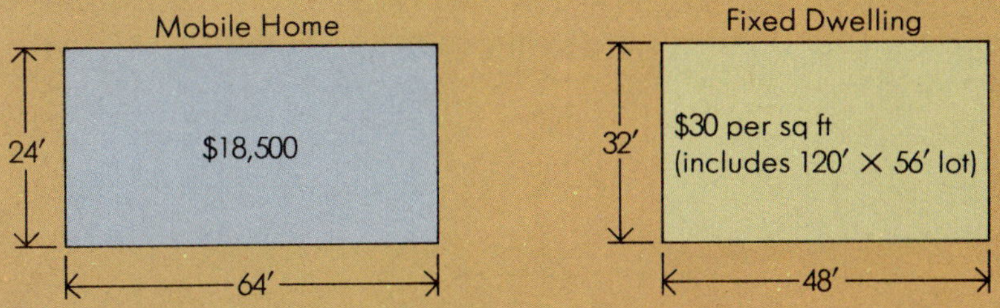

Cost of a 120′ × 56′ lot: $7800

5. What is the total cost of the mobile home?
6. What is the total cost of the fixed dwelling?
7. Compare the total costs. What percent of the total cost of the fixed dwelling is the cost of the mobile home?
8. List some advantages and disadvantages of owning each type of dwelling.

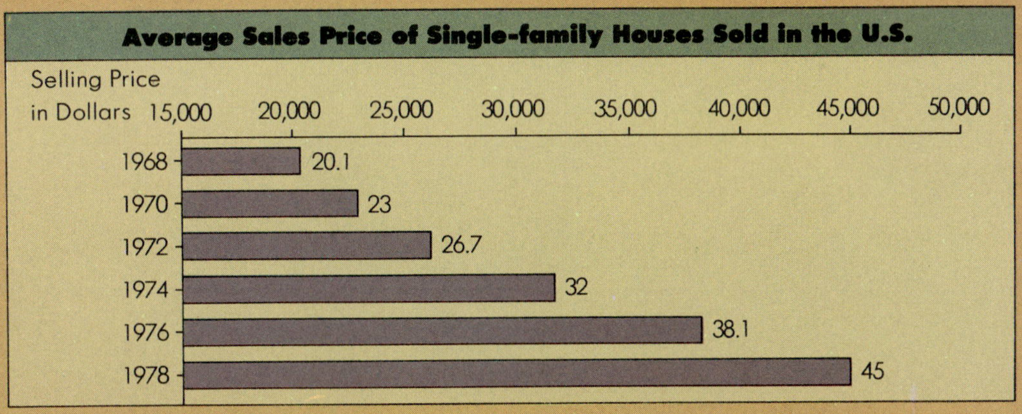

Find the percent of increase between the following years?

9. 1968 and 1970
10. 1970 and 1972
11. 1972 and 1974
12. 1974 and 1976
13. 1976 and 1978

Predict the average selling price in these years.

14. 1980
15. 1990

AREA OF CIRCLES

To find the area of a circle, we multiply $\pi(3.14)$ by the radius squared. $(r^2 = r \times r)$. Area $= \pi \times r \times r$

$$A = \pi r^2$$

EXAMPLE 1. Find the area.

$A = \pi r^2$
$A = 3.14 \times 3^2$ $\quad$ *($3^2 = 3 \times 3 = 9$)*
$A = 3.14 \times 9$
$A = 28.26$

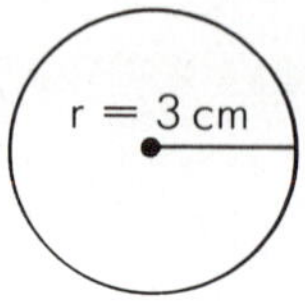

The area is approximately 28.26 cm².

TRY THIS

Find the area.

1. 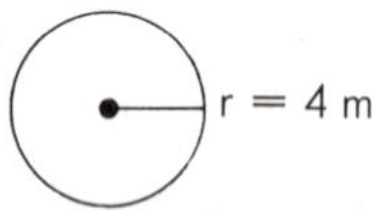

2. 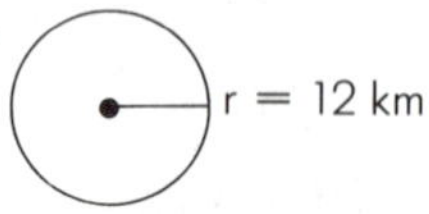

To find the area when the diameter is given we first find the radius.

EXAMPLE 2. Find the area.

$d = 14$
$r = \frac{1}{2} \cdot 14 = 7$
$A = \pi r^2$
$A = 3.14 \times 7^2$
$A = 3.14 \times 49$
$A = 153.86$

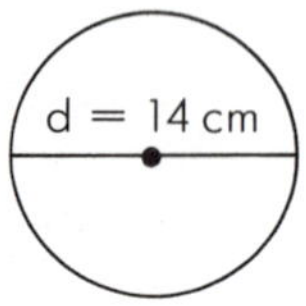

The area is approximately 153.86 cm².

TRY THIS

Find the area.

3. 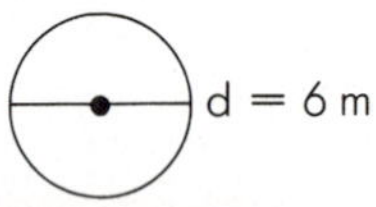

4. 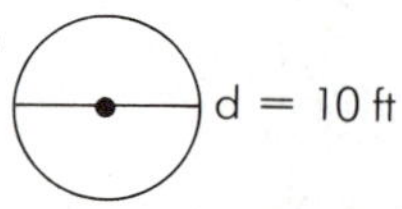

PRACTICE

Find the area.

1.

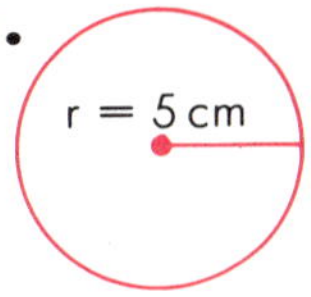

2.

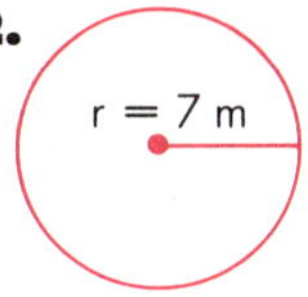

3.

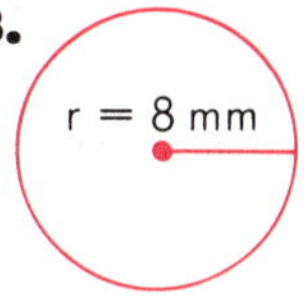

4.

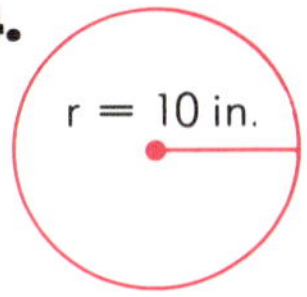

5.

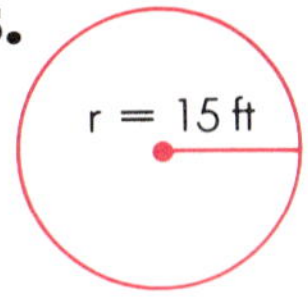

6.

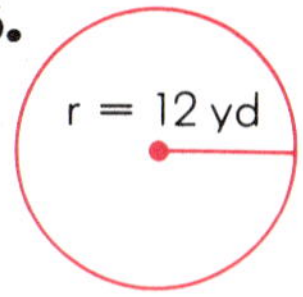

7. Circle
$r = 4$ m

8. Circle
$r = 8$ cm

9. Circle
$r = 3$ mm

10.

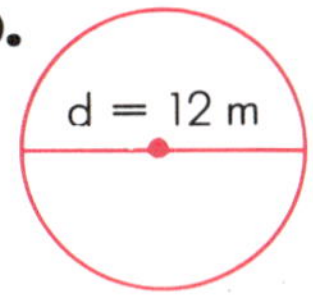

11.

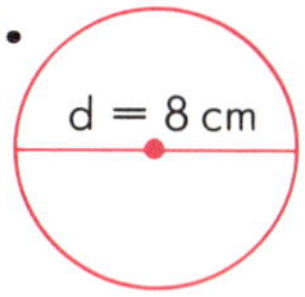

12.

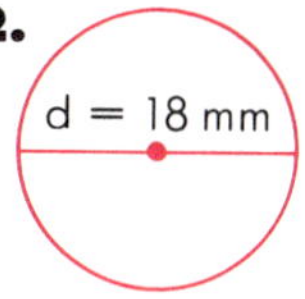

13.

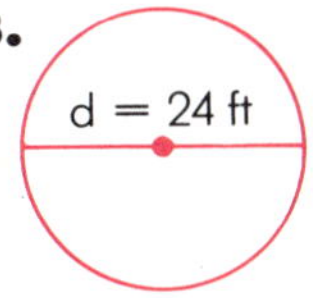

14.

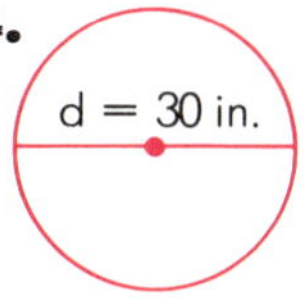

15.

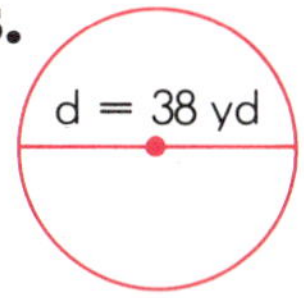

16. Circle
$d = 6$ m

17. Circle
$d = 18$ mm

18. Circle
$d = 10$ cm

PROBLEM CORNER

1. Find the area of circle *A*.

2. The radius of circle *B* is triple that of circle *A*. Find the area of circle *B*.

3. What is the ratio of area *B* to area *A*?

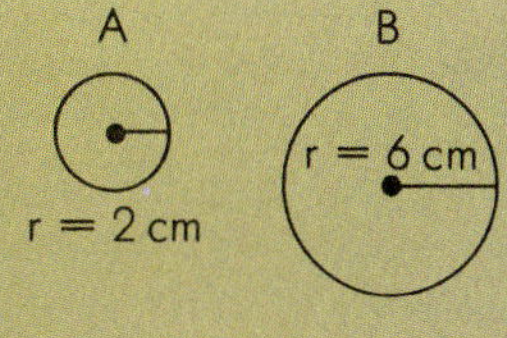

More Practice, page 335

SOLVING PROBLEMS

Drawing a picture can sometimes help with the solution of a problem.

EXAMPLE. The Yingling family needs to build a wood fence around their new swimming pool. The fence will have five sides and a gate. Mr. Yingling drew this picture.

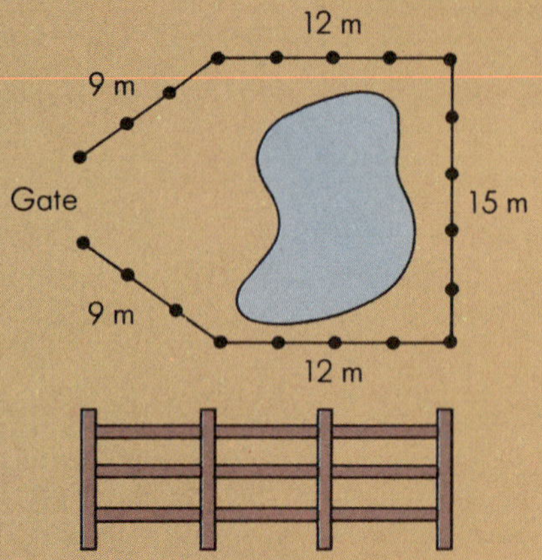

1. What is the approximate distance between fence posts?

2. The cost of posts is $ 4.25 each. How much must be spent for posts?

3. The fence will consist of three rails. What is the total length of rails needed?

4. Rails cost $ 3.25 per linear meter. What will be the cost of the rails?

5. A gate will cost $ 12.50. Find the total cost of materials for the fence.

TRY THIS

Peninsula Garden Service planted rose bushes around two gardens. One garden was rectangular, measuring 12 m by 32 m. The other garden was square, measuring 20 m on a side. The rose bushes were planted 4 meters apart. How many more rosebushes were needed for the rectangular garden than the square garden? Make drawings to help solve the problem.

PRACTICE

1. Joe Gori plans to fence in his garden. The shape is a quadrilateral which has sides of 3 m, 5 m, 5 m, and 6 m. The posts will be 1 meter apart and cost \$ 3.25 each. Wire garden fence will cost \$ 4.95 per meter. There will be no gate. Find the total cost of the fence.

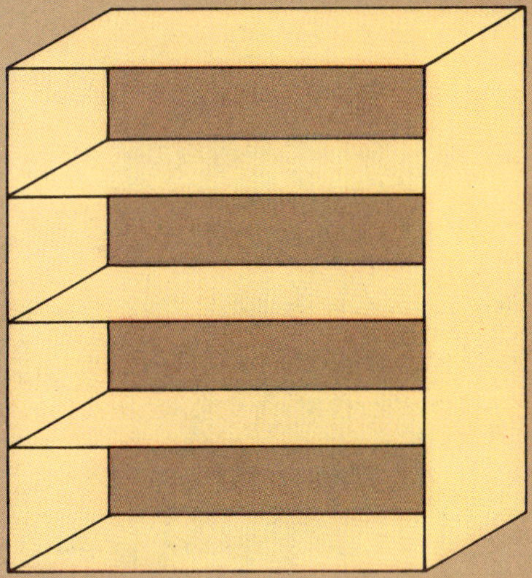

2. A bookcase is 4′ high by 5′ long by 1′ deep. There are four shelves 1′ apart. The backing is one solid piece 4′ × 5′. A quart of paint will cover 44 square feet. How many quarts of paint are needed to paint the bookcase completely?

3. A round pool has a radius of 15 feet. A trapezoidal pool has a width of 16 feet and opposite base sides of 14 feet and 20 feet. Which of these swimming pools has the greater surface area? How much greater is it? Make a drawing to prove (show) your answer.

4. Carpet costs \$ 14.25 per square yard. Follow these steps to find how much it costs to carpet a square room whose perimeter is 48 feet. Draw a square whose perimeter is 48 feet. Use a scale where $\frac{1}{2}$ inch represents 1 foot.

How long is each side on the scale drawing?
How many square feet are there in the room?
How much does the carpet cost? (9 square feet = 1 square yard)

CAREER CAREER CAREER CAREER CAREER

ARCHITECT

Maria Constantine is an architect. Architects design and draw plans for all types of structures.

In a home she is designing, Maria plans to put decorative brick 5 feet high around the base of the house. There is an 8-foot door in the back and a 6-foot door in the front.

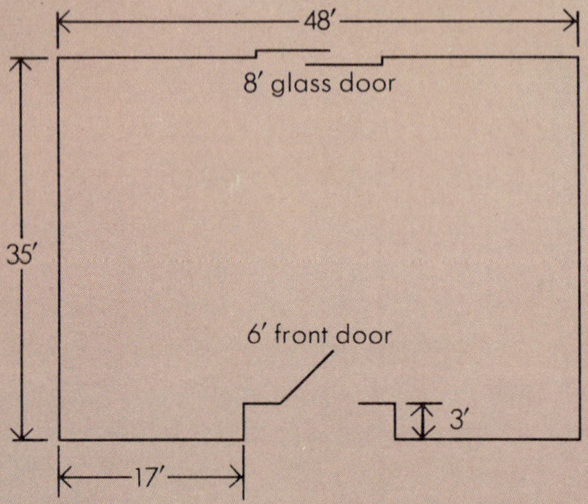

1. How many linear feet of brick will be needed?
2. How many square feet of brick will be needed?
3. Brick costs $ 12 per square foot. Find the total cost of the brick.

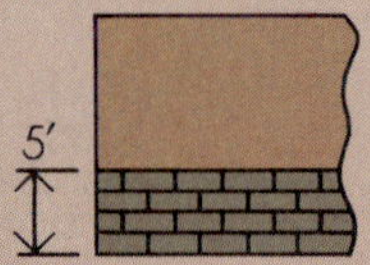

Architect's fees are often based on a percentage of the final construction cost of a house.

4. Maria designs a house with an estimated construction cost of \$55,800. She is to be paid 6% for preparing the working drawings. About how much can she expect to be paid?
5. For preparing working drawings and supervising the construction, Maria was paid $8\frac{1}{2}$% of the construction cost of an \$83,650 home. What was her fee?

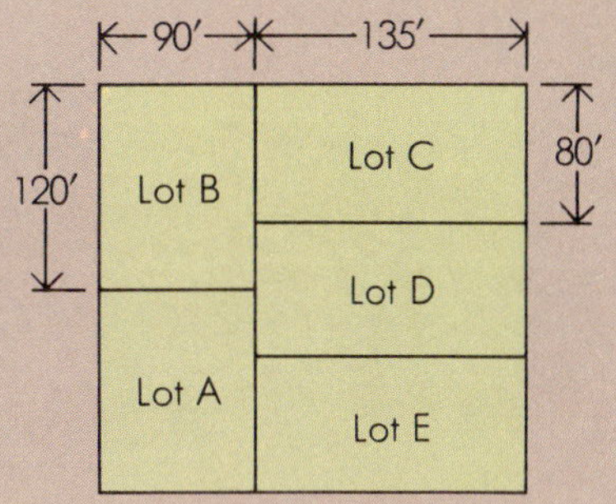

Architects frequently work with developers to plan communities. Maria has a tract to divide into five lots. She made two lots $120' \times 90'$ and three lots $135' \times 80'$.

6. What is the area of lot A?
7. What is the area of lot D?
8. What is the perimeter of lot B?
9. What is the perimeter of lot E?
10. How much fencing will be needed to enclose all five lots completely? (Remember, some parts of the fence will be shared by two lots.)

Fencing costs \$6 per running foot. Lots sharing a fence are assessed an equal part of the cost.

How much is each of the lots assessed for fencing?

11. lot A
12. lot B
13. lot C
14. lot D
15. lot E

CHAPTER 11 REVIEW

Find the perimeter. p. 312

1.

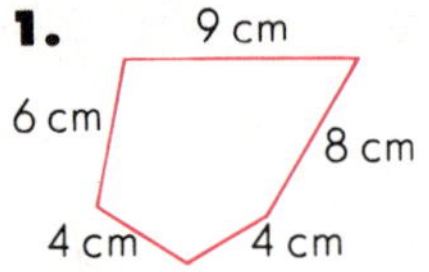

2.

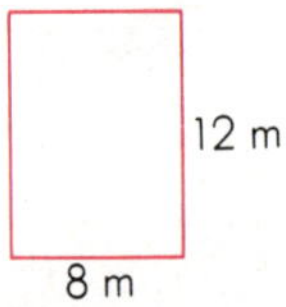

3.

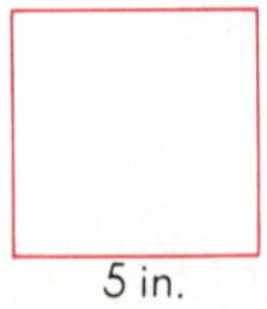

Find the circumference. p. 314

4.

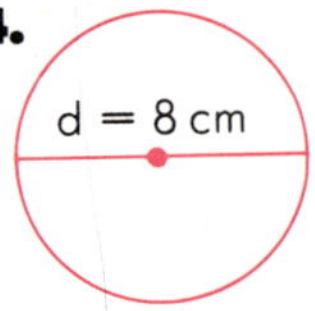

5.

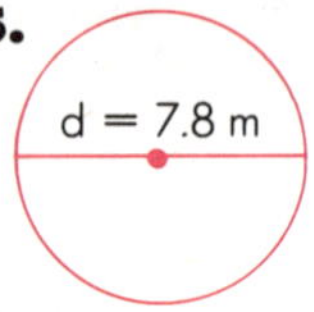

6.

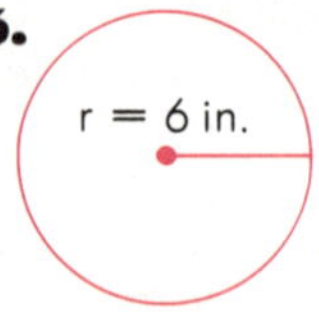

Find the area. pp. 318–326

7.

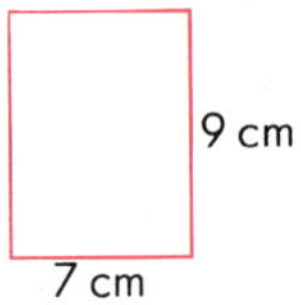

8.

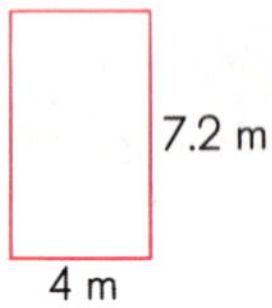

9.

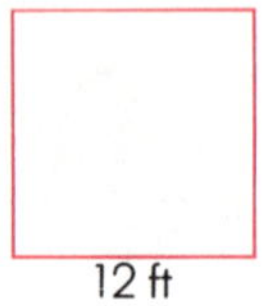

10. Parallelogram
$b = 12$ m
$h = 9$ m

11. Triangle
$b = 10$ cm
$h = 8$ cm

12. Triangle
$b = 8$ in.
$h = 6$ in.

13. Trapezoid
$h = 4$ mm
$b_1 = 15$ mm
$b_2 = 7$ mm

14. Trapezoid
$h = 6$ m
$b_1 = 18$ m
$b_2 = 8$ m

15. Trapezoid
$h = 3$ yd
$b_1 = 10$ yd
$b_2 = 5$ yd

16. Circle
$d = 18$ m

17. Circle
$r = 3$ cm

18. Circle
$r = 7$ ft

Application pp. 316, 324

19. Randy budgets 23% of his net income for housing. His net income is $ 650 per month. How much does Randy budget for housing?

20. In 1970 the average sales price of a house was $ 23,000. In 1974, the price was $ 32,000. What was the percent of increase in the sales price?

CHAPTER 11 TEST

Find the perimeter.

1. 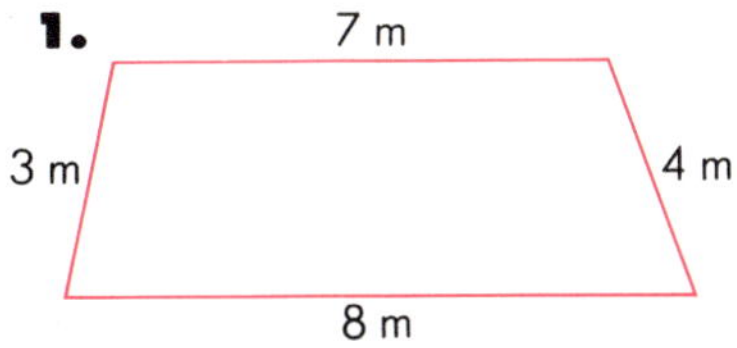

2. 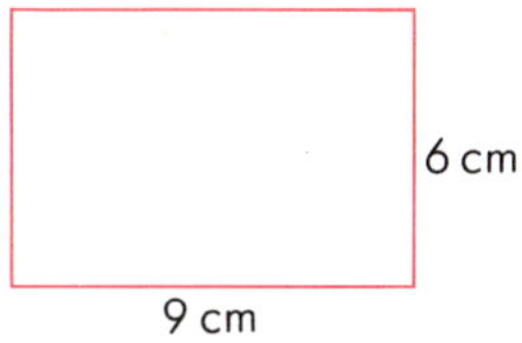

3. 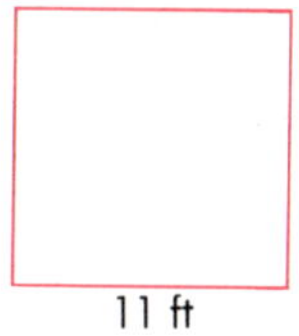

Find the circumference.

4. 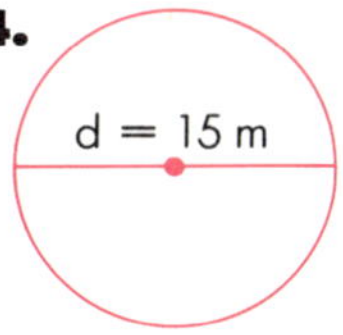

5. 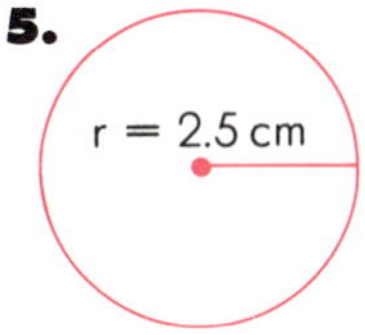

6. 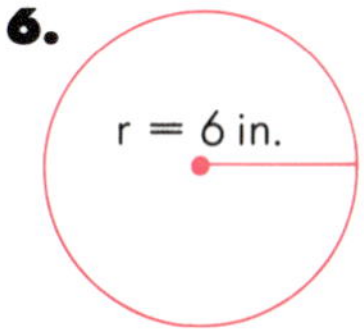

Find the area.

7. 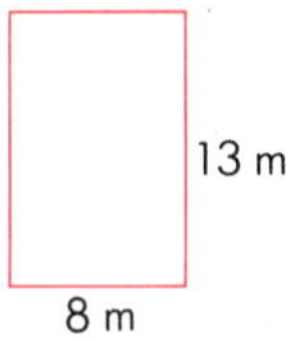

8. 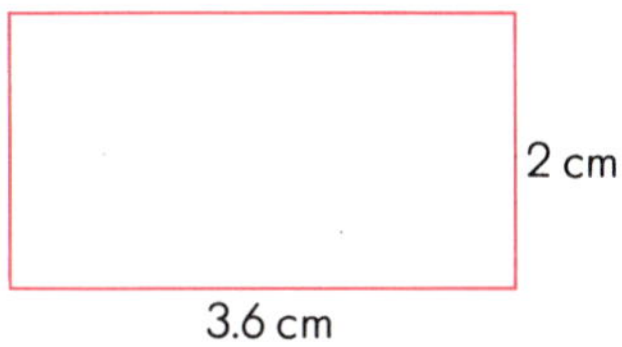

9. 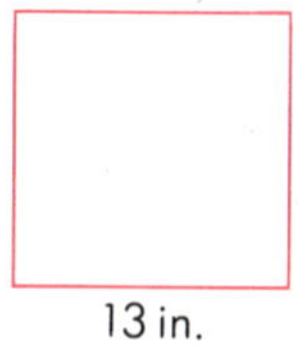

10. Parallelogram
$b = 14$ cm
$h = 8$ cm

11. Triangle
$b = 13$ m
$h = 5$ m

12. Triangle
$b = 24$ yd
$h = 18$ yd

13. Trapezoid
$h = 8$ m
$b_1 = 13$ m
$b_2 = 9$ m

14. Trapezoid
$h = 5$ cm
$b_1 = 10$ cm
$b_2 = 5$ cm

15. Trapezoid
$h = 8$ ft
$b_1 = 12$ ft
$b_2 = 12$ ft

16. Circle
$d = 24$ cm

17. Circle
$r = 9$ m

18. Circle
$r = 8$ in.

Application

19. Jane budgets 22% of her net income for housing. Her net income is \$ 550 per month. How much does Jane budget for housing?

20. A 2-story house is 40 feet long and 35 feet wide. Construction costs \$ 35 per square foot. What is the cost of constructing the house?

MORE PRACTICE

Find the perimeter. Use after page 313.

1.
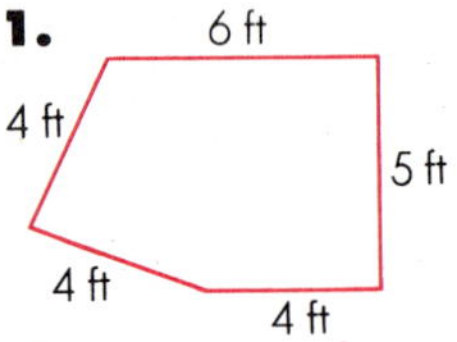

2.
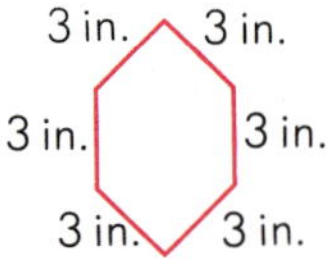

3.
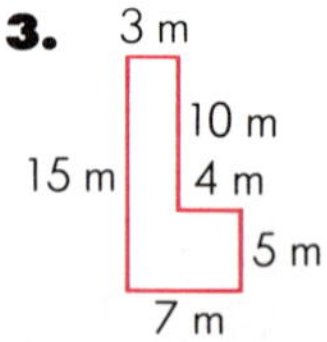

4.
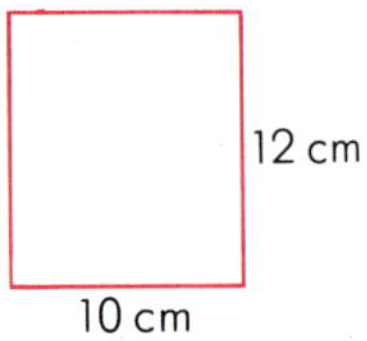

5.
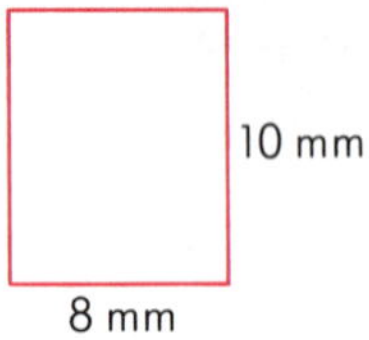

6.
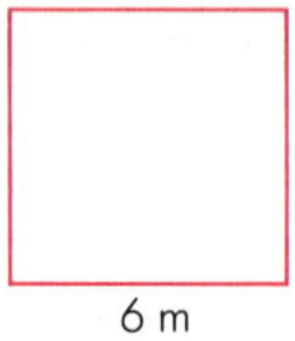

7. Rectangle
$l = 20$ yd
$w = 5$ yd

8. Square
$s = 5$ cm

9. Square
$s = 4\frac{1}{2}$ ft

Find the circumference. Use after page 315.

10.
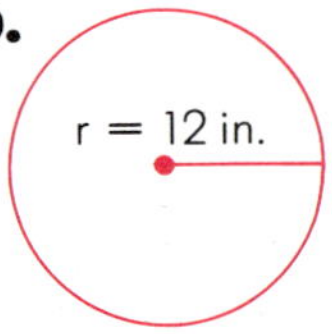

11.
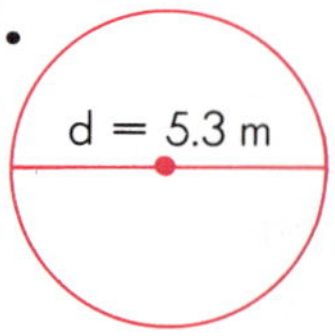

12.
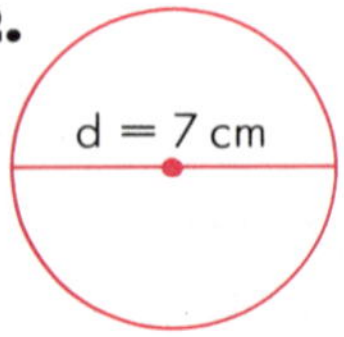

13. Circle
$d = 4$ mm

14. Circle
$r = 10$ ft

15. Circle
$r = 9.3$ cm

Find the area. Use after page 319.

16.

17.
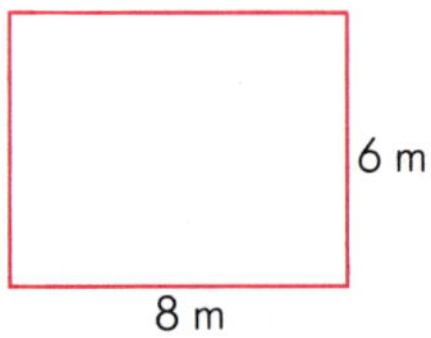

18.
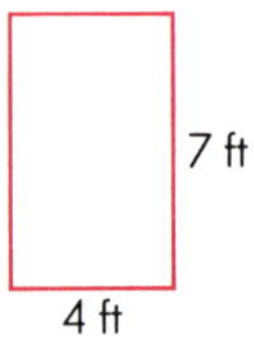

19. Rectangle
$l = 9.8$ cm
$w = 4$ cm

20. Square
$s = 6.3$ m

21. Square
$s = 4$ mm

Find the area. Use after page 321.

1.

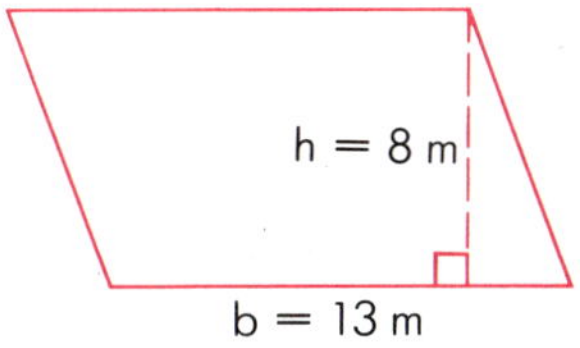

2.

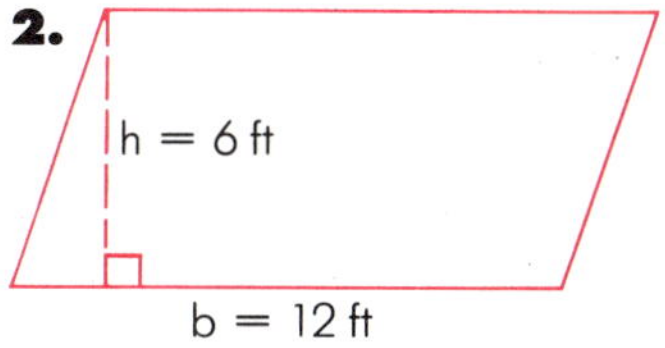

3.

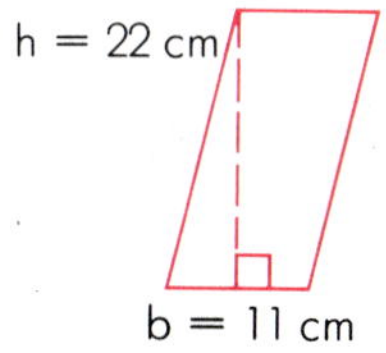

4.

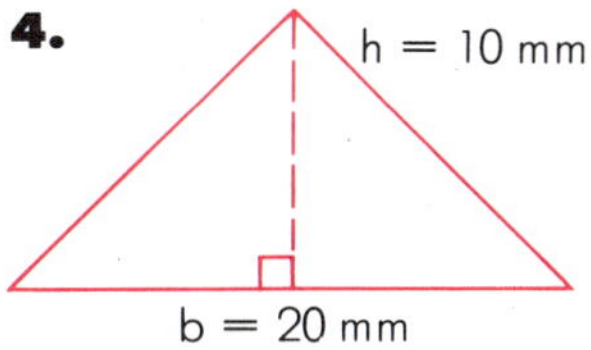

5.

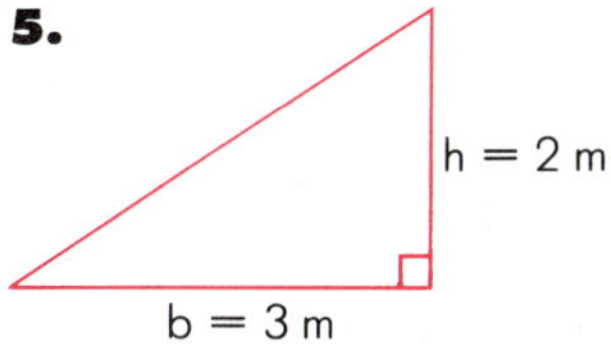

6.

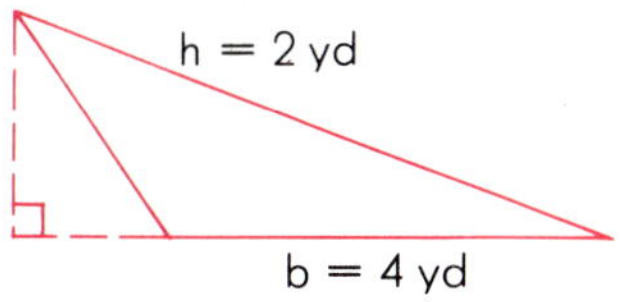

7. Triangle
$b = 12$ cm
$h = 9$ cm

8. Triangle
$b = 8$ in.
$h = 5$ in.

9. Triangle
$b = 4.4$ m
$h = 5.3$ m

Find the area. Use after page 323.

10.

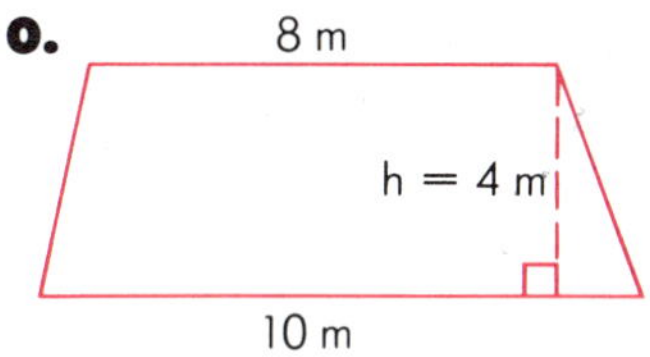

11.

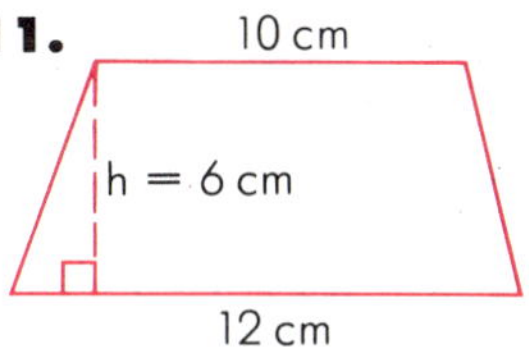

12.

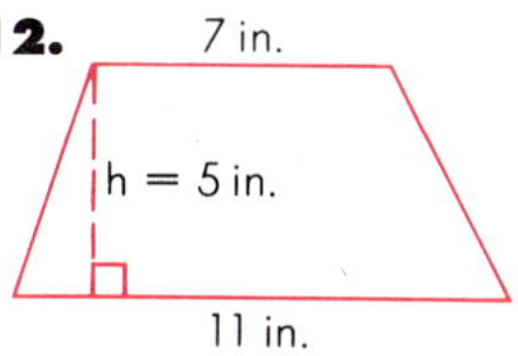

13. Trapezoid
$h = 5$ ft
$b_1 = 4$ ft
$b_2 = 6$ ft

14. Trapezoid
$h = 7$ m
$b_1 = 11$ m
$b_2 = 15$ m

15. Trapezoid
$h = 15$ mm
$b_1 = 10$ mm
$b_2 = 14$ mm

Find the area. Use after page 327.

16.

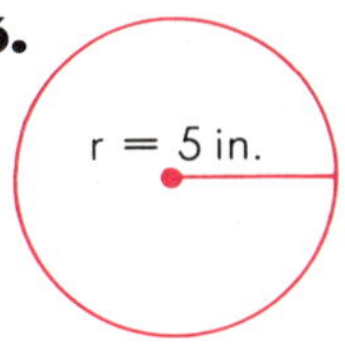

17.

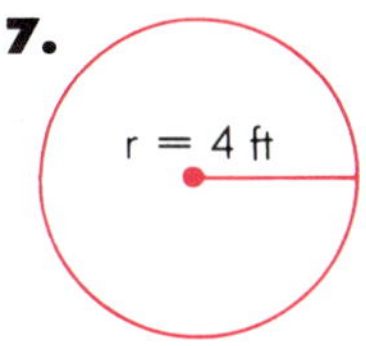

18.

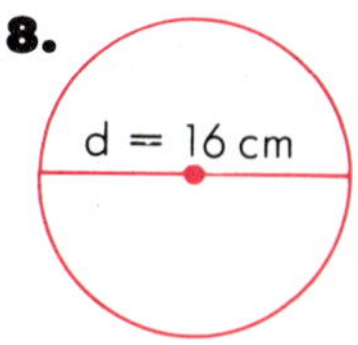

19. Circle
$r = 9$ mm

20. Circle
$d = 12$ cm

21. Circle
$d = 6$ m

CALCULATOR CORNER

THREE OF A KIND

Players one or more
Materials one calculator, paper and pencil

Suppose a calculator has 345.6 on the display.

How many ways can you get a number ▢▢▢. 8 on the display where the three boxes contain the same digit?

Try these with your calculator.

1. Add 321.2 to 345.6
   Result: 666.8
2. Subtract 122.8 from 345.6
   Result: 222.8

Find 4 more ways to get such a number on the display. Write down your steps.

EXERCISES Write down your steps for each of the following.

1. Start with 236.5 on the display. Find 5 ways to get a number ▢▢▢.7 on the display, where the three boxes contain the same digit.
2. Start with 342.16 on the display. Find 5 ways to get a number ▢▢▢.48 on the display, where the three boxes contain the same digit.
3. Start with 423.152 on the display. Find 5 ways to get number 6▢▢.▢66 on the display, where the three boxes contain the same digit.

READY FOR CHAPTER 12

Find the volume.

1\.

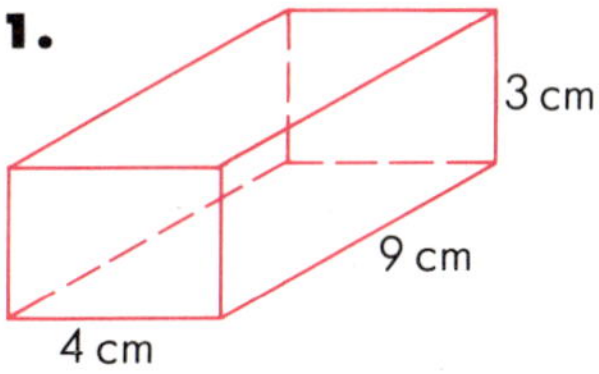

2\.

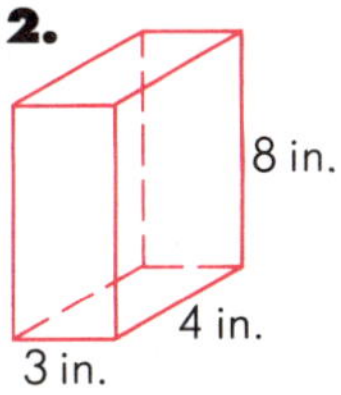

3\.

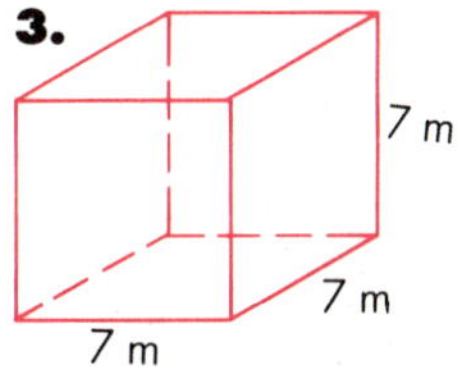

4. Rectangular prism
$l = 12$ m
$w = 6$ m
$h = 5$ m

5. Cube
$s = 3$ cm

6. Cube
$s = 6$ ft

7\.

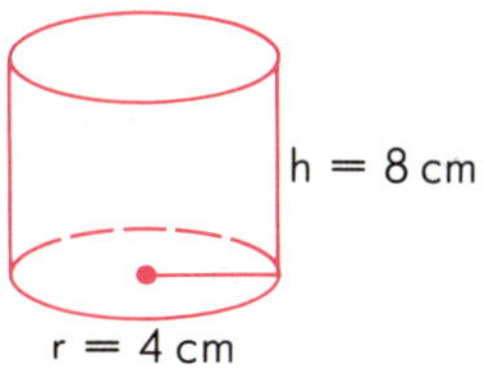

8\.

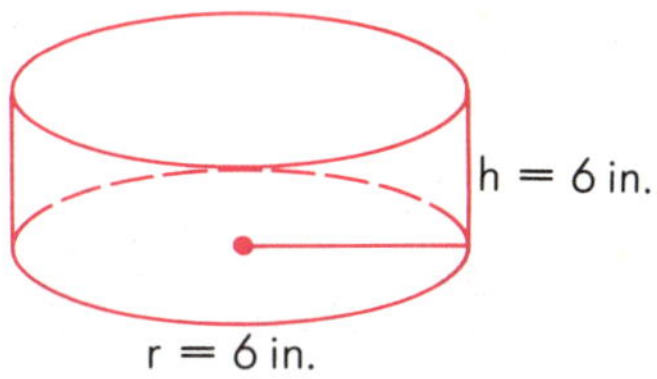

9\.

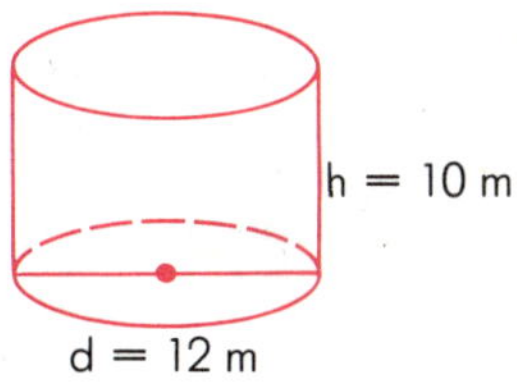

10. Cylinder
$r = 3$ m
$h = 7$ m

11. Cylinder
$d = 18$ cm
$h = 12$ cm

12. Cylinder
$d = 3$ ft
$h = 2$ ft

13\.

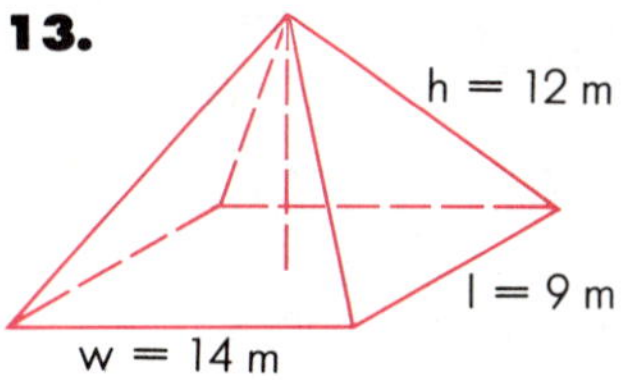

14\.

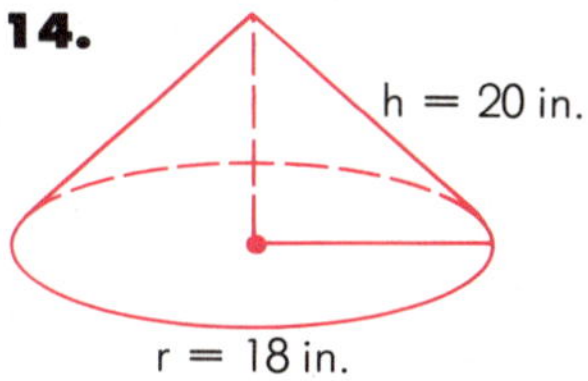

15\.

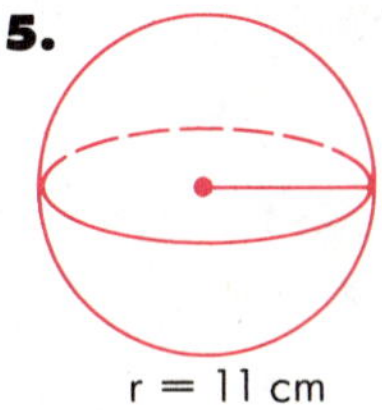

Application

16. A bill for electricity is $98 for 2000 kilowatt-hours of electricity. What is the cost per kilowatt-hour?

17. A barrel of oil generates 500 kilowatt-hours of electricity. How many kilowatt-hours are generated by 100,000 barrels?

12
VOLUME

GETTING STARTED

CUBE COST

If 1 cube costs 5¢, how much would each stack of cubes cost?

1.

2.

3.

4.

5.

6.

7.

8.

RECTANGULAR PRISMS AND CUBES

Volume is measured with cubic units.
This is a cubic centimeter. It is 1 cm on each side.
We write: 1 cm^3.
We say: 1 cubic centimeter or 1 centimeter cubed.

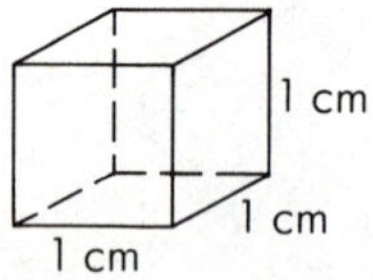

Find the volume of the rectangular prism.

Volume = length × width × height

$V = lwh$
$V = 4 \times 3 \times 2$
$V = 24\ m^3$

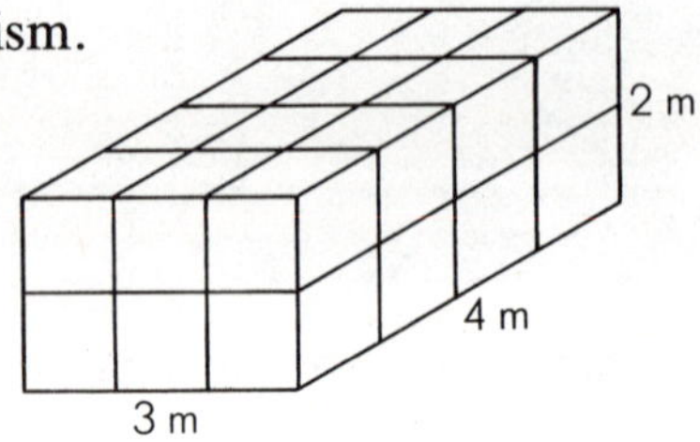

EXAMPLE 1. Find the volume.

$V = lwh$
$V = 3 \times 2 \times 5$
$V = 30\ cm^3$

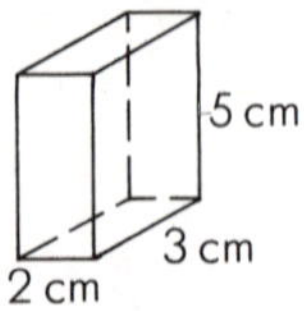

A cube is a special rectangular prism. Its length, width, and height are the same.

EXAMPLE 2. Find the volume.

Volume = side × side × side

$V = s \times s \times s$
$V = 3 \times 3 \times 3$
$V = 27\ cm^3$

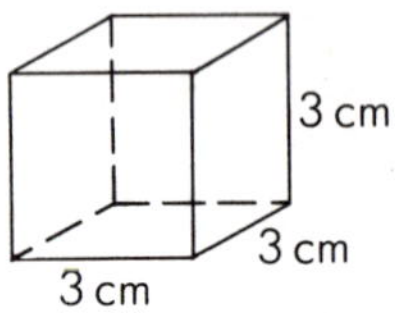

TRY THIS

Find the volume.

1.

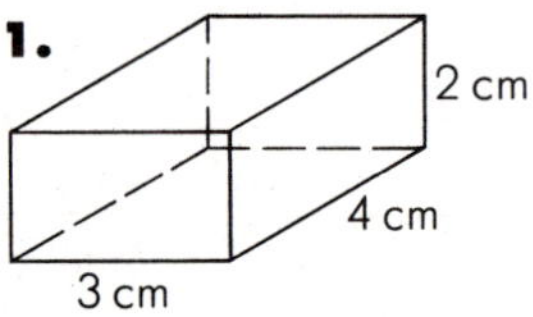

2.

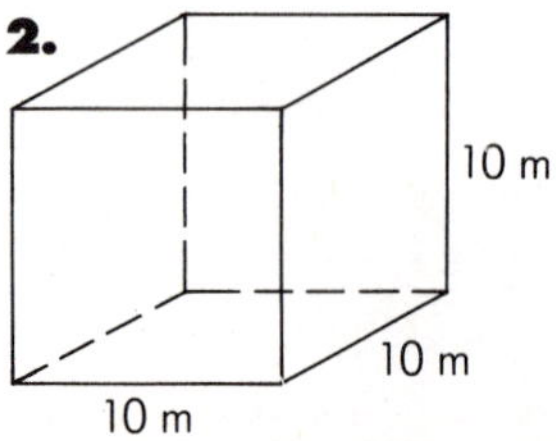

PRACTICE

Find the volume.

1.

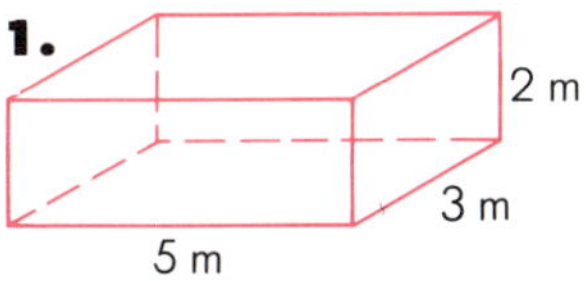

2.

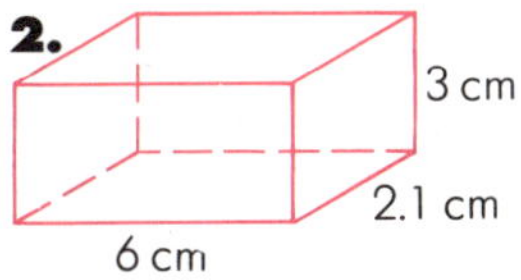

3.

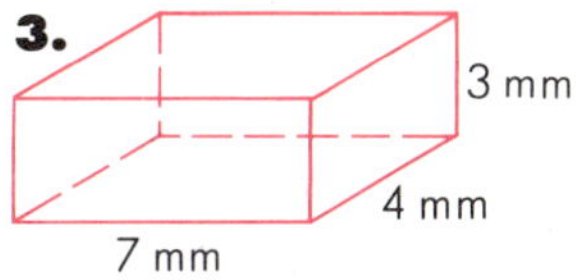

4. Rectangular prism
$l = 3\frac{1}{2}$ in.
$w = 3$ in.
$h = 7$ in.

5. Rectangular prism
$l = 3$ ft
$w = 7$ ft
$h = 5$ ft

6. Rectangular prism
$l = 6$ yd
$w = 2$ yd
$h = 7$ yd

7. Rectangular prism
$l = 8$ mm
$w = 6$ mm
$h = 4$ mm

8. Rectangular prism
$l = 4$ m
$w = 3$ m
$h = 5$ m

9. Rectangular prism
$l = 6$ cm
$w = 2$ cm
$h = 5.6$ cm

10.

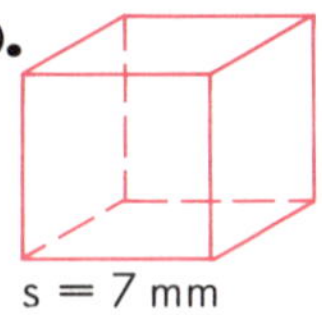

11.

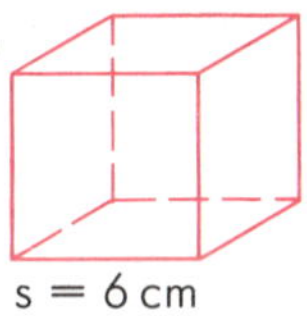

12.

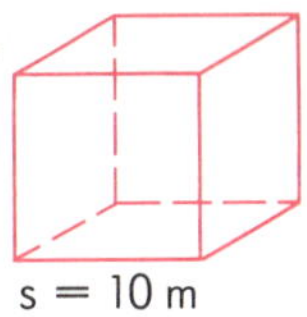

13. Cube
$s = 4\frac{1}{2}$ yd

14. Cube
$s = 8$ ft

15. Cube
$s = 9$ in.

16. Cube
$s = 3$ m

17. Cube
$s = 5$ cm

18. Cube
$s = 12$ mm

PROBLEM CORNER

1. Find the volume of cube *A*.

2. In cube *B* the sides are doubled. Find the volume of cube *B*.

3. Is the volume of cube *B* double that of cube *A*? What is the ratio of the volume of cube *B* to the volume of cube *A*?

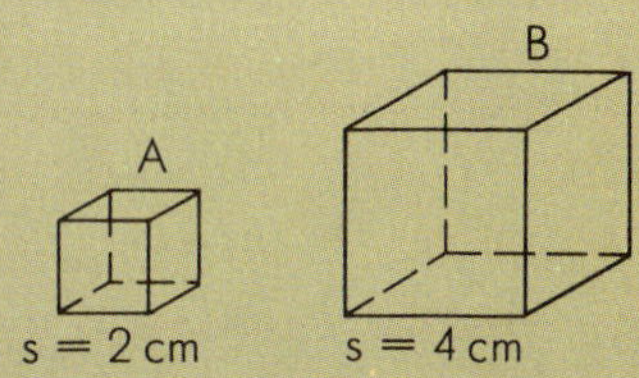

More Practice, page 356

APPLICATION

ENERGY BILLS

The cost of electricity, gas, and water add to the cost of housing. Their cost depends on the number of units used.

Utility	**Unit**
Electricity	Kilowatt-hour (kWh)
Gas	Cubic Foot
Water	Gallon or Cubic Foot

$$\text{UNIT RATE} = \frac{\text{TOTAL COST}}{\text{NUMBER OF UNITS}}$$

What is the cost per kWh for these electric bills?

	kWh	**Cost**
1. San Francisco, California	2160	$ 114.48
2. New York City, New York	1800	156.60
3. Dallas, Texas	1995	95.76
4. Pittsburgh, Pennsylvania	1460	75.92
5. Atlanta, Georgia	978	34.23

6. Find an electric bill from your city or town. What is the cost per kWh?

ONE KILOWATT = 1000 WATTS

ONE KILOWATT × HOURS USED = kWh

Each electric appliance has a watt rating. The watt rating can be used to compute the cost of energy.

$$\frac{\text{WATTAGE}}{1000} \times \text{COST PER KILOWATT-HOUR} = \begin{array}{l}\text{COST OF ELECTRICITY FOR}\\ \text{1 HOUR OF OPERATION}\end{array}$$

5¢ per kilowatt-hour is commonly used to calculate energy costs.

EXAMPLE. An old color TV has a watt rating of 350.

$\frac{350}{1000} \times \$0.05 = \$0.0175$ cents per hour

If you watch TV from 5 PM to 10 PM every day, the cost would be:

5 hours × 365 days × \$0.0175 = \$31.94 per year *Round to the nearest cent.*

Find the cost to operate these electric appliances for 1 year.

7. a 1000 W hair dryer for 1 hour per week
8. a 650 W microwave oven for 3 hours per week
9. a 1400 W toaster for 1 hour per week
10. a 230 W stereo for 3 hours per day
11. a 2 W clock continuously
12. a 700 W vacuum cleaner for 15 minutes per day
13. a 100 W bulb for 6 hours every day
14. a 40 W bulb for 6 hours every day

How much money will you save in a year in each situation?

15. replacing ten 100 W bulbs by 60 W bulbs, used 6 hours per day, each day of the year
16. using a $7\frac{1}{2}$ W night light instead of a 60 W bulb for 8 hours each night of the year

CYLINDERS

The base of a cylinder is a circle.

Volume = area of the base × height
Volume = $\pi r^2 \times h$
$V = \pi r^2 h$

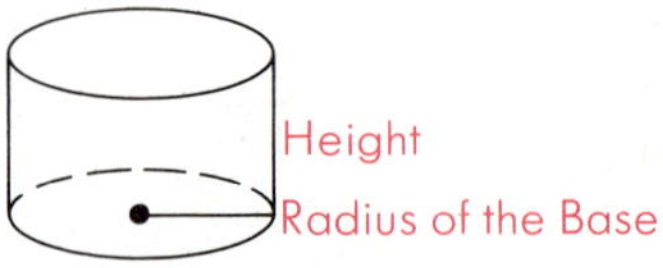

EXAMPLE 1. Find the volume.

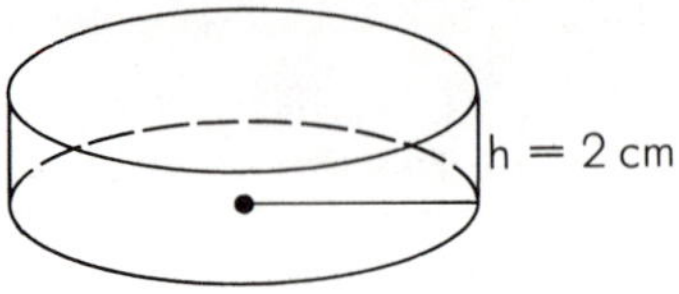

$V = \pi r^2 h$
$V = 3.14 \times 4^2 \times 2$ *Use 3.14 for π.*
$V = 3.14 \times 16 \times 2$ *$4^2 = 4 \times 4 = 16$*
$V = 100.48\ \text{cm}^3$

To find the volume when the diameter is given, we first find the radius.

EXAMPLE 2. Find the volume.

$d = 6$
$r = \frac{1}{2} \cdot 6 = 3$
$V = \pi r^2 h$
$V = 3.14 \times 3^2 \times 6$
$V = 3.14 \times 9 \times 6$
$V = 169.56\ \text{cm}^3$

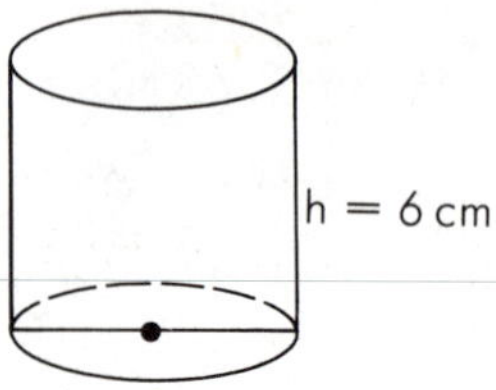

TRY THIS

Find the volume.

1.

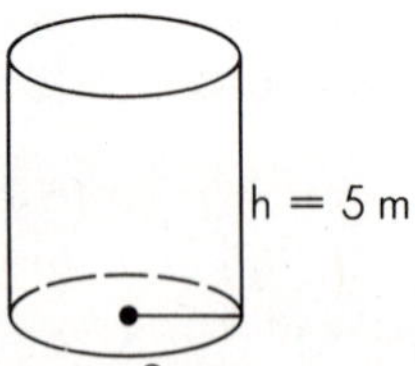

2.

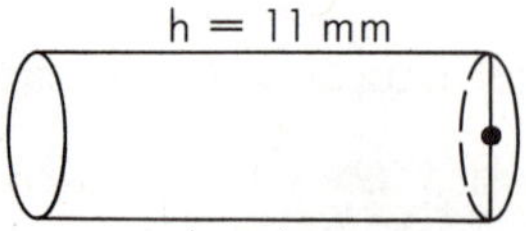

PRACTICE

Find the volume.

1.

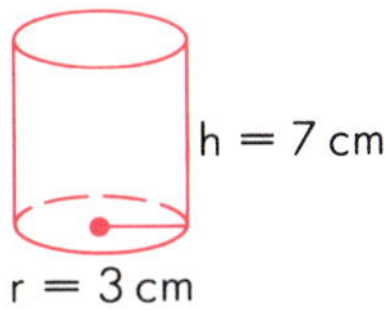

2.

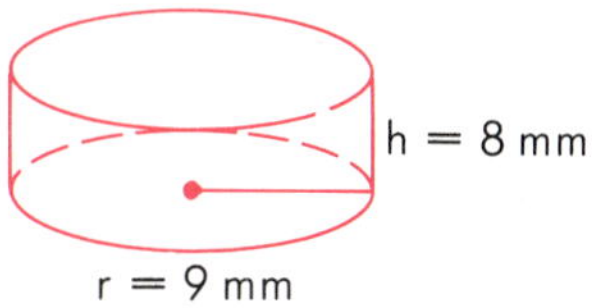

3.

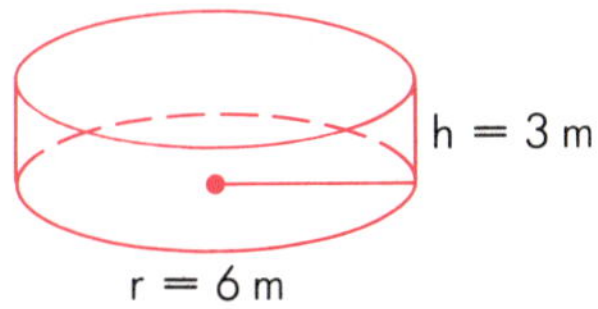

4. Cylinder
$r = 9$ m; $h = 10$ m

5. Cylinder
$r = 8$ cm; $h = 6$ cm

6. Cylinder
$r = 20$ mm; $h = 20$ mm

7. Cylinder
$r = 6$ in.; $h = 5$ in.

8. Cylinder
$r = 3$ ft; $h = 2$ ft

9. Cylinder
$r = 4$ yd; $h = 3$ yd

10.

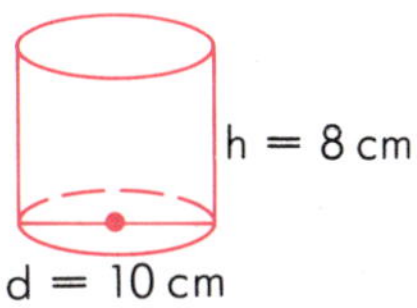

11.

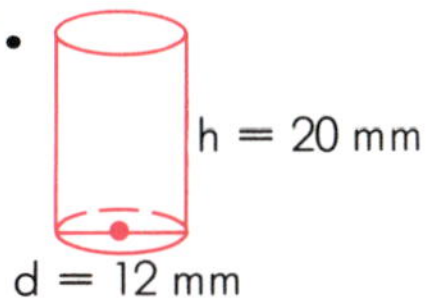

12.

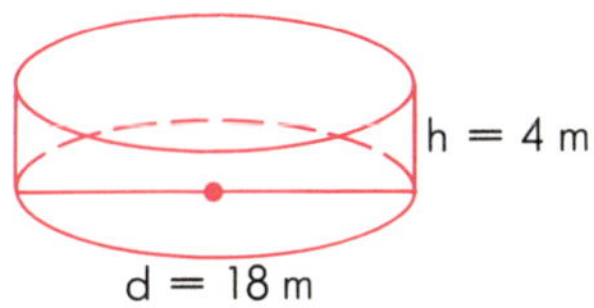

13. Cylinder
$d = 16$ m; $h = 30$ m

14. Cylinder
$d = 14$ cm; $h = 10$ cm

15. Cylinder
$d = 12$ mm; $h = 40$ mm

16. Cylinder
$d = 2$ yd; $h = 3$ yd

17. Cylinder
$d = 8$ ft; $h = 5$ ft

18. Cylinder
$d = 6$ in.; $h = 6$ in.

PROBLEM CORNER

1. Find the volume of cylinder *A*.

2. In cylinder *B* the height is double that of cylinder *A*. The radius is the same. Find the volume of cylinder *B*. What is the ratio of the volume of cylinder *B* to the volume of cylinder *A*?

3. In cylinder *C* the radius is double that of cylinder *A*. The height is the same. Find the volume of cylinder *C*. What is the approximate ratio of the volume of cylinder *C* to the volume of cylinder *A*?

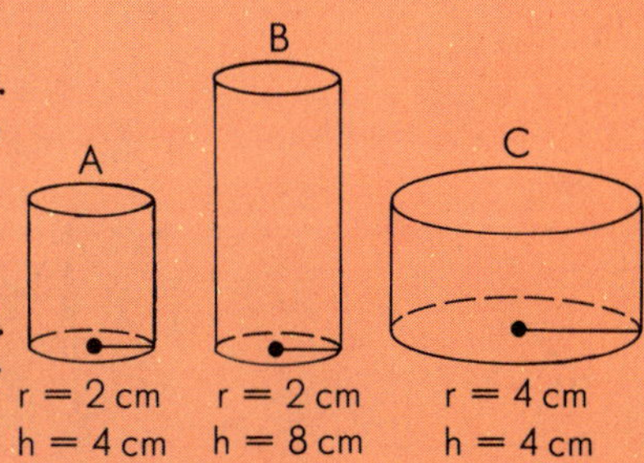

More Practice, page 356

PYRAMIDS AND CONES

Imagine a pyramid full of sand emptied into a prism which has the same base and height as the pyramid.

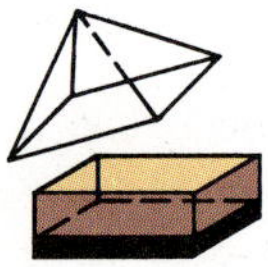

The volume of the pyramid is $\frac{1}{3}$ the volume of the prism.

Prism: $V = lwh$

Pyramid: $V = \frac{1}{3}lwh$

Imagine a cone full of sand emptied into a cylinder which has the same base and height as the cone.

The volume of the cone is $\frac{1}{3}$ the volume of the cylinder.

Cylinder: $V = \pi r^2 h$

Cone: $V = \frac{1}{3}\pi r^2 h$

EXAMPLE 1. Find the volume.

$V = \frac{1}{3}lwh$

$V = \frac{1}{3} \times 5 \times 4 \times 3$

$V = 20\ \text{cm}^3$

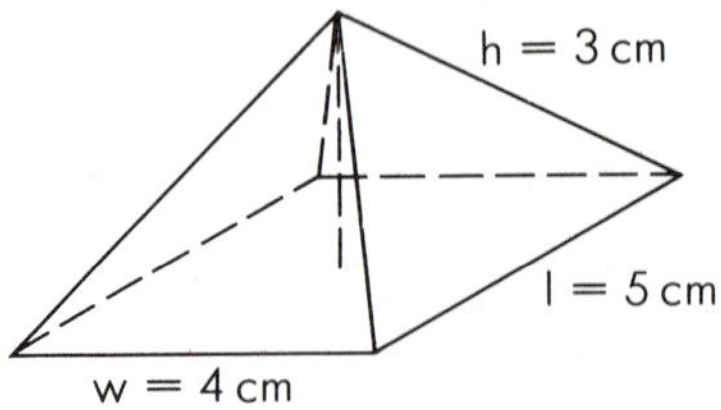

EXAMPLE 2. Find the volume.

$V = \frac{1}{3}\pi r^2 h$

$V = \frac{1}{3} \times 3.14 \times 3^2 \times 5$

$V = \frac{1}{3} \times 3.14 \times 9 \times 5$

$V = \frac{1}{3} \times \frac{3.14}{1} \times \frac{9}{1} \times \frac{5}{1}$

$V = \frac{141.30}{3} = 47.1\ \text{cm}^3$

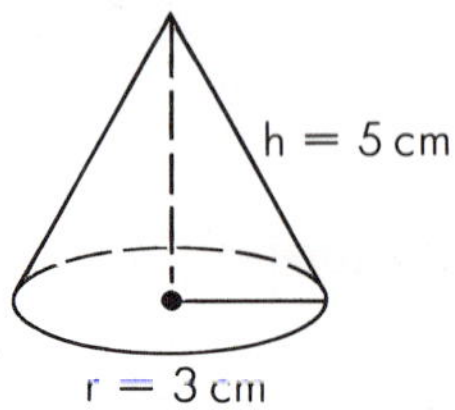

TRY THIS

Find the volume.

1.

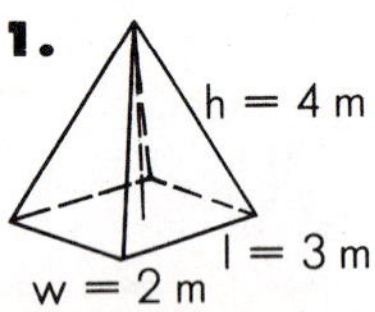

2.

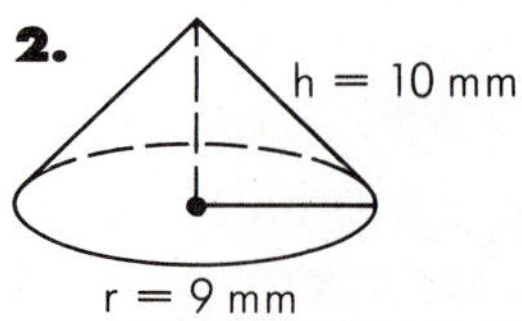

PRACTICE

Find the volume.

1. 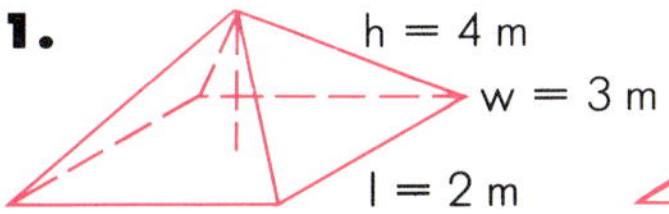

2. 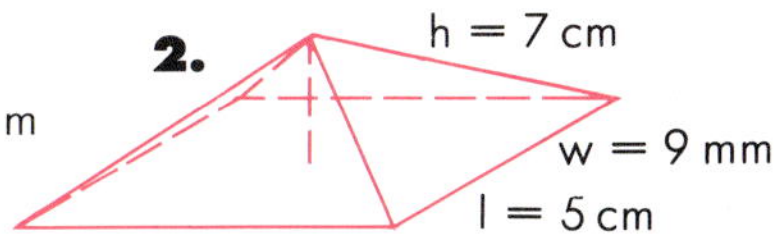

3. 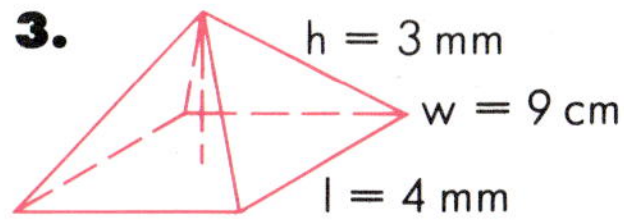

4. Pyramid
l = 9 mm
w = 12 mm
h = 18 mm

5. Pyramid
l = 10 cm
w = 10 cm
h = 10 cm

6. Pyramid
l = 4 m
w = 7 m
h = 20 m

7. Pyramid
l = 7 in.
w = 8 in.
h = $4\frac{1}{2}$ in.

8. Pyramid
l = 9 ft
w = 4 ft
h = 6 ft

9. Pyramid
l = $6\frac{1}{2}$ yd
w = 10 yd
h = 20 yd

10. 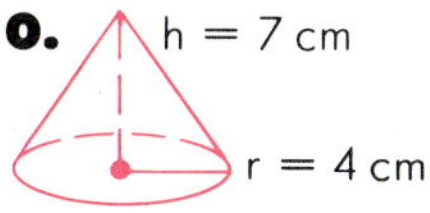

11. 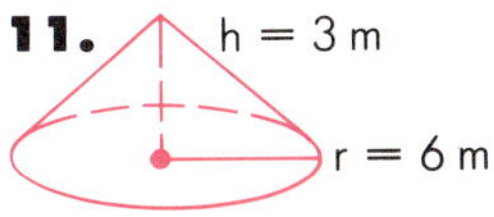

12. 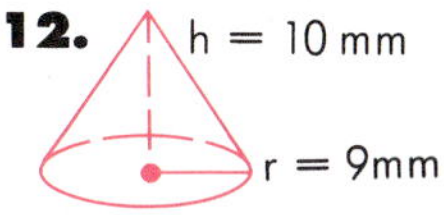

13. Cone
r = 12 mm
h = 15 mm

14. Cone
r = 7 cm
h = 3 cm

15. Cone
r = 3 m
h = 10 m

16. Cone
r = 4 in.
h = 6 in.

17. Cone
r = $2\frac{1}{2}$ ft
h = 3 ft

18. Cone
r = 7 yd
h = 10 yd

PROBLEM CORNER

1. Find the volume of cone *A*.

2. In cone *B* the lengths are doubled. Find the volume of cone *B*.

3. Is the volume of cone *B* double that of cone *A*? What is the ratio of the volume of cone *B* to the volume of cone *A*?

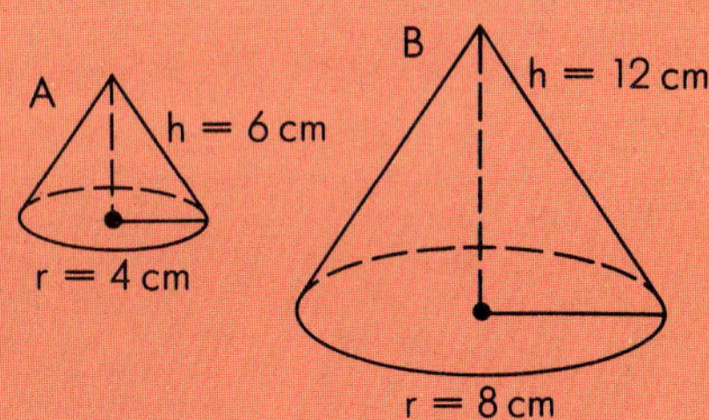

More Practice, page 357

SPHERES

The **radius** of a sphere is the distance from the center to any point on the sphere.

Volume $= \frac{4}{3} \times \pi \times$ radius cubed ($r^3 = r \times r \times r$)

$V = \frac{4}{3}\pi r^3$

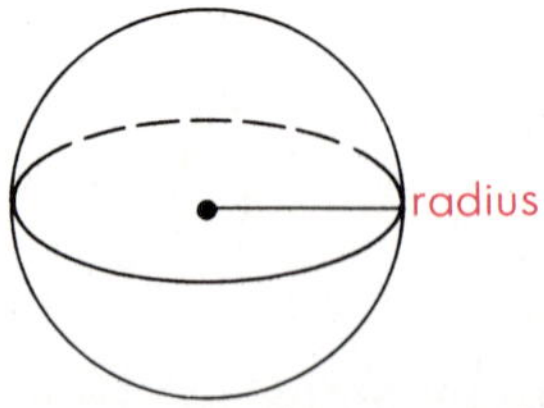

EXAMPLE 1. Find the volume.

$$V = \frac{4}{3}\pi r^3$$
$$V = \frac{4}{3} \times 3.14 \times 2^3$$
$$V = \frac{4}{3} \times 3.14 \times 8$$
$$V = \frac{4}{3} \times \frac{3.14}{1} \times \frac{8}{1}$$
$$V = \frac{100.48}{3} = 33.49 \text{ cm}^3$$

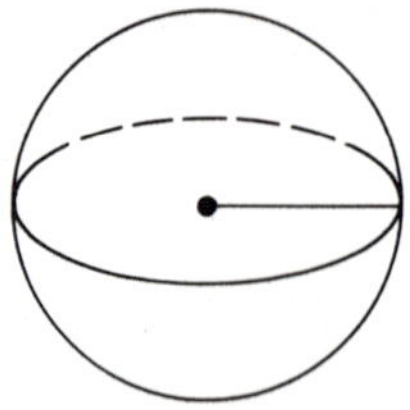

r = 2 cm

To find the volume when the diameter is given, we first find the radius.

EXAMPLE 2. Find the volume.

$$d = 10$$
$$r = \frac{1}{2} \cdot 10 = 5$$
$$V = \frac{4}{3}\pi r^3$$
$$V = \frac{4}{3} \times 3.14 \times 5^3$$
$$V = \frac{4}{3} \times \frac{3.14}{1} \times \frac{125}{1}$$
$$V = \frac{1570}{3} = 523.33 \text{ m}^3$$

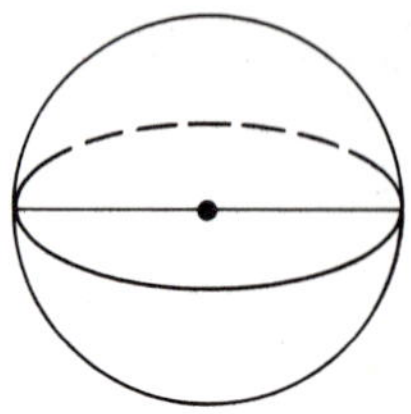

d = 10 m

TRY THIS

Find the volume.

1.

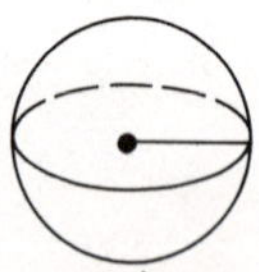

r = 4 cm

2.

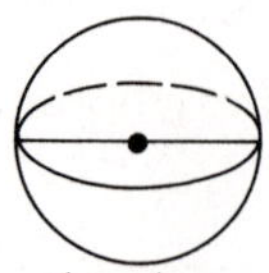

d = 6 m

PRACTICE

1.

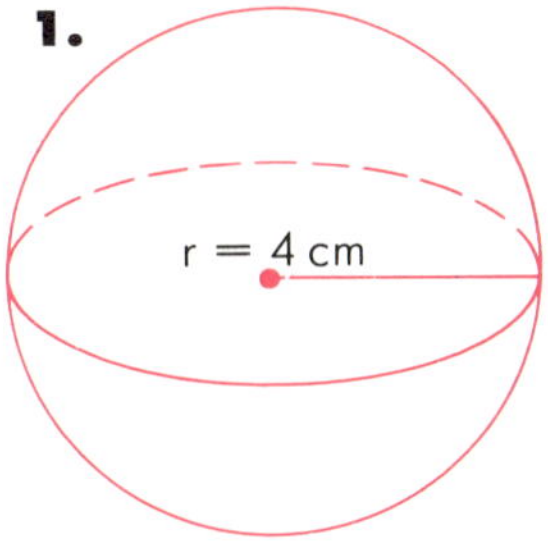

2.

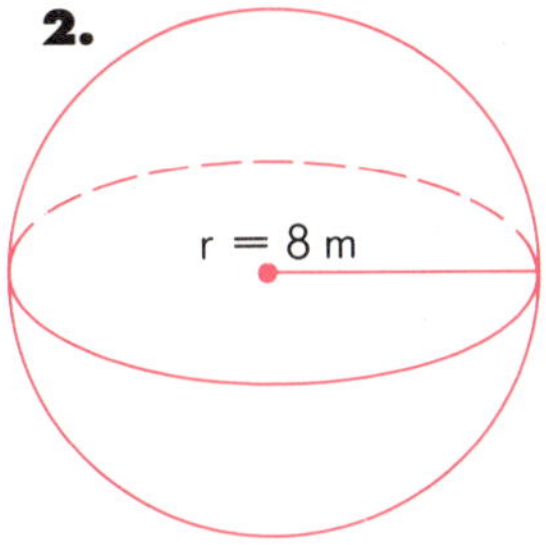

3.

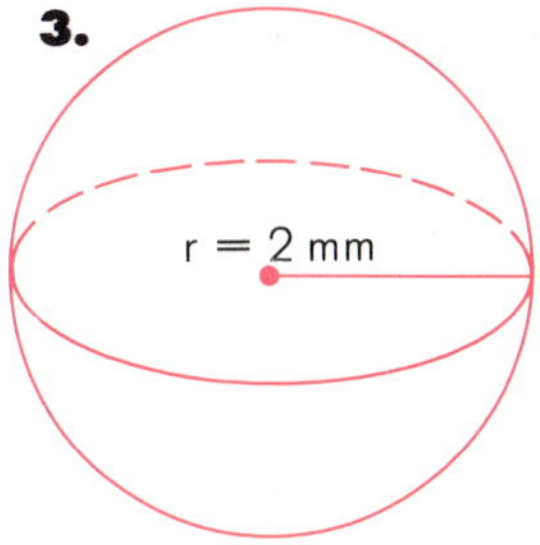

4. Sphere
$r = 6$ in.

5. Sphere
$r = 3$ yd

6. Sphere
$r = 5$ ft

7. Sphere
$r = 9$ mm

8. Sphere
$r = 1$ m

9. Sphere
$r = 7$ cm

10.

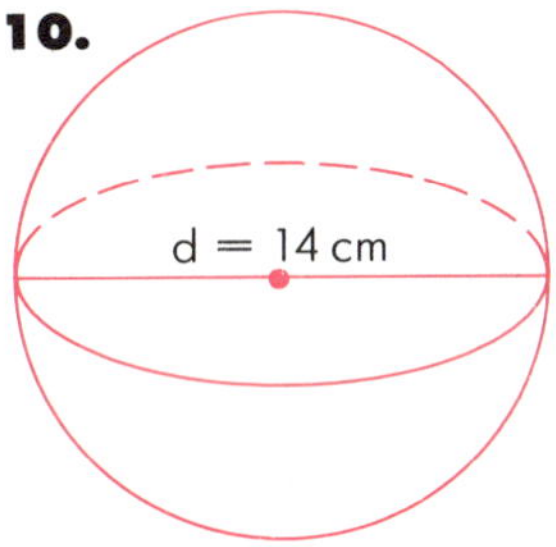

11.

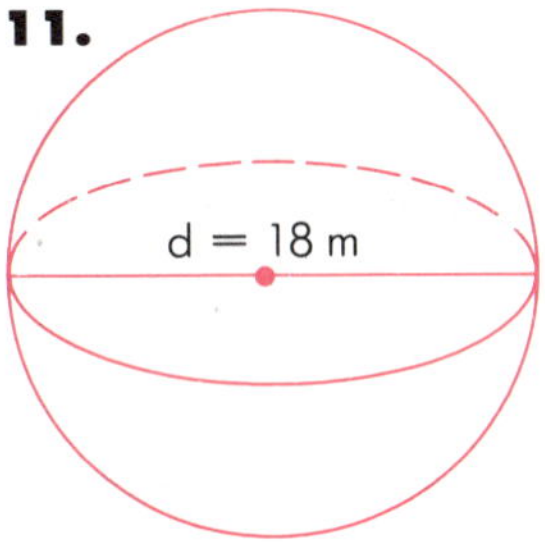

12.

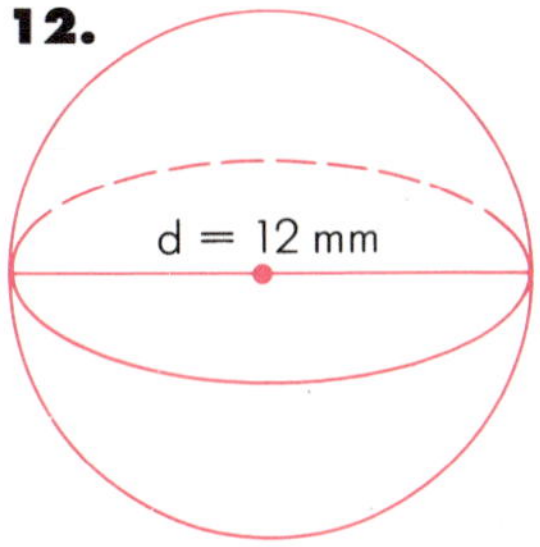

13. Sphere
$d = 10$ in.

14. Sphere
$d = 8$ ft

15. Sphere
$d = 2$ yd

16. Sphere
$d = 100$ mm

17. Sphere
$d = 8$ cm

18. Sphere
$d = 24$ cm

PROBLEM CORNER

1. Find the volume of the hemisphere.
(Hint: the prefix "hemi" means half.)

2. A hemisphere dome on the top of a silo has a diameter of 40 feet. Find the volume of the dome.

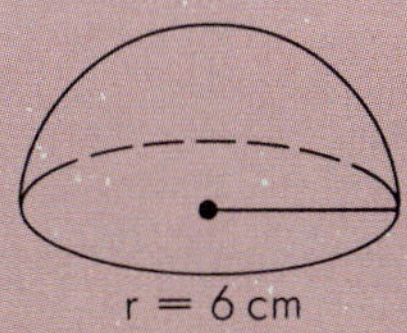

More Practice, page 357

APPLICATION

ENERGY CONSERVATION

There are 42 gallons of oil in a barrel. A barrel of oil generates 500 kWh of electricity.

Suppose thermostats in all homes in the U.S. are lowered an average of 6° in winter. The country would save 570,000 barrels of oil each day.

1. How many gallons of oil would be saved daily?
2. How many gallons would be saved in a 90 day winter?
3. How many kWh would be saved in a 90 day winter?
4. At an average rate of $ 0.05 per kWh, how much money would be saved?
5. The present consumption is about 35,000,000 barrels of oil per day in the U.S. What percent could be saved per day by lowering thermostats 6° in winter?

A large amount of oil can be saved daily by not wasting electricity. Suppose you leave two 100 W lights on when you leave for school each day. How much oil would be wasted?

Remember: $\frac{\text{WATTS}}{1000} \times \text{HOURS} = \text{kWh}$ $\qquad$ $\text{kWh} \times \text{RATE} = \text{COST}$

$$\frac{200 \text{ watts}}{1000} \times \text{8 hours of unneeded lights each day} \times \text{180 days of school} = \text{288 wasted kWh}$$

It takes about one gallon of oil to generate 11.9 kWh of electricity.

$288 \div 11.9 = 24.2$ gallons of oil wasted *(Rounded to tenths)*

For each situation tell:

A. how much money is spent in one year (365 days) at the rate of $0.05 per kWh
B. how much oil is used yearly (1 gallon for 11.9 kWh)
C. how much oil is saved by the conservation way

Wasteful Way	**Conservation Way**
6. Leave two 75 W lights burning at night in bathrooms from 7 PM to 7 AM.	7. Leave a $7\frac{1}{2}$ W night light on from 7 PM to 7 AM.

8. An electric clothes dryer uses 4.8 kilowatts per hour, full or empty. Use the electric dryer for 10 hours weekly with partial loads	9. Use the dryer at the manufacturer's recommended full load for 6 hours weekly. 10. Hang most clothing outdoors. Use the dryer one hour per week for emergency loads or special clothes.
11. Uninsulated hot water tanks and pipes can use 600 kWh monthly for an average family's hot water needs.	12. Properly insulated tanks and pipes save 30% of the electricity that is used to produce hot water.

CAREER CAREER CAREER CAREER CAREER

OCEANOGRAPHER

Chris Tanaka is an oceanographer. She is an expert on the way oceans are formed. She studies the things that live in the ocean.

To go deep in the ocean, Chris uses a bathysphere. A bathysphere is made to withstand great water pressures. Only rounded objects have been found to be strong enough. Water pressure at one mile in depth is 2280 pounds per square inch. The pressure increases at the rate of 2280 pounds per square inch for each mile of depth.

What is the water pressure per square inch at the following depths?

1. 4 miles
2. 6.5 miles
3. 15,840 feet

SURFACE AREA OF A SPHERE $= 4\pi r^2$

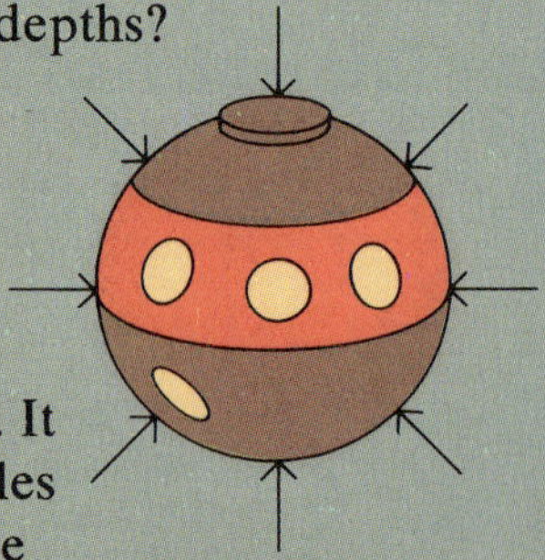

4. A bathysphere is 12 feet in diameter. What is the surface area of the bathysphere?
5. A bathysphere has a surface area of 45,216 square inches. It descends into the Marianas Trench which is about 6.5 miles deep. What is the total weight of water on the bathysphere at that depth?

Chris often takes samples of the ocean floor. To do so, she drills a hole that is the shape of a cylinder. What is the volume of each of the following?

6. a core sample 46 cm in diameter and 56.4 meters in length retrieved from a depth of 3 miles
7. a core 28 cm in diameter, 120 meters long
8. a core 37 cm in diameter, 184 meters long

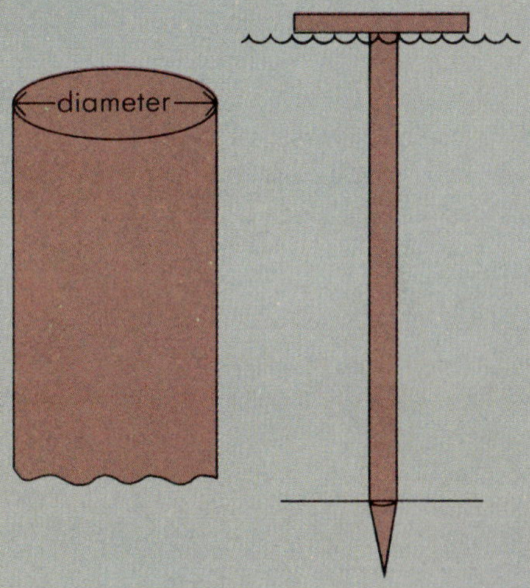

Some oceanographers specialize in the chemistry of the sea. Almost every element known is found in the ocean. Each cubic meter of sea water contains 0.004 mg of gold. How much gold is there in these average seawater samples?

9.

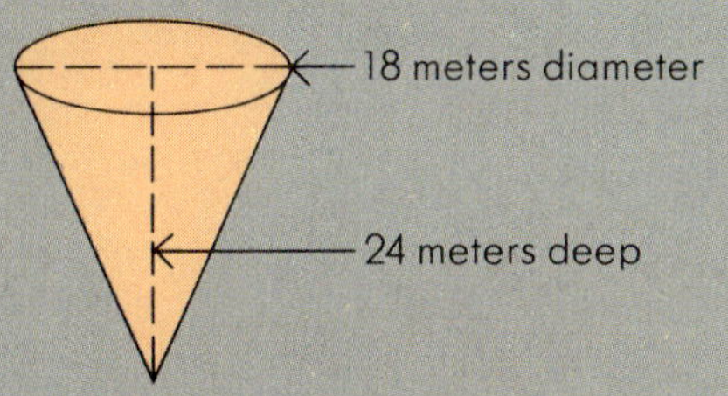

10.

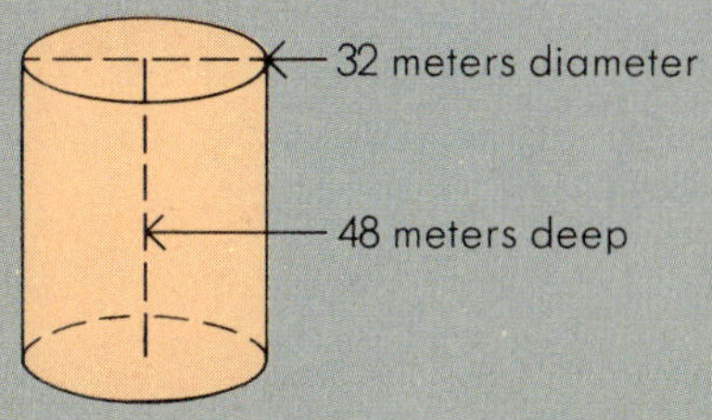

Icebergs are studied by oceanographers from the Arctic to the Antarctic. Some icebergs are so large that early explorers thought they were islands.

11. One iceberg measured 30 kilometers wide by 100 kilometers long and 800 meters thick. What was the volume of this iceberg?
12. Another iceberg measured 25 kilometers wide by 75 kilometers long and 600 meters thick. What was the volume of this iceberg?

CHAPTER 12 REVIEW

Find the volume. pp. 340–348

1.

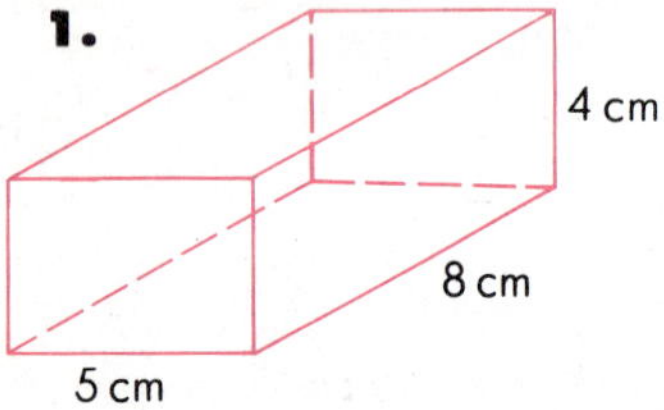

2.

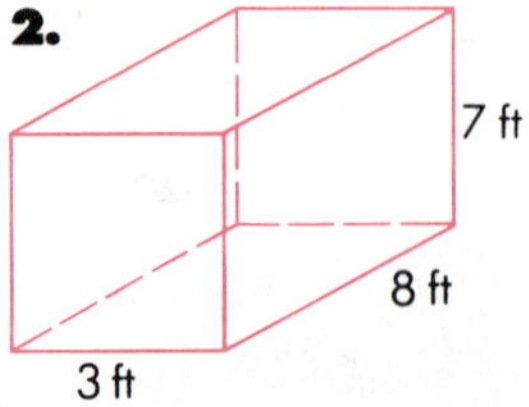

3.

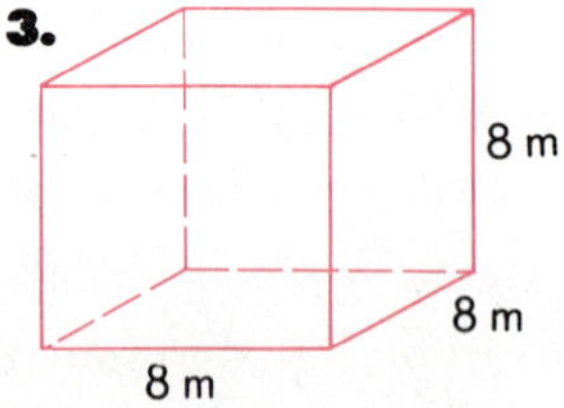

4. Rectangular prism
$l = 8$ m
$w = 5$ m
$h = 6$ m

5. Cube
$s = 4$ cm

6. Cube
$s = 5$ in.

7.

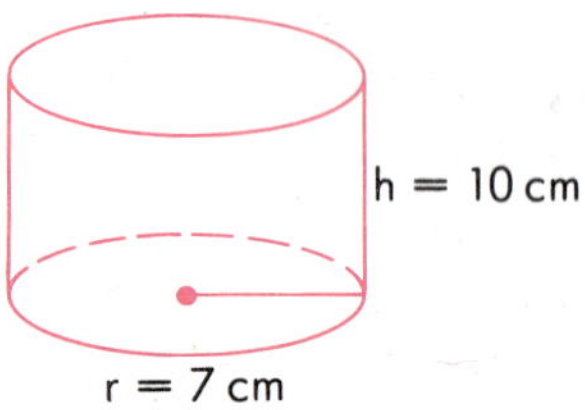

8.

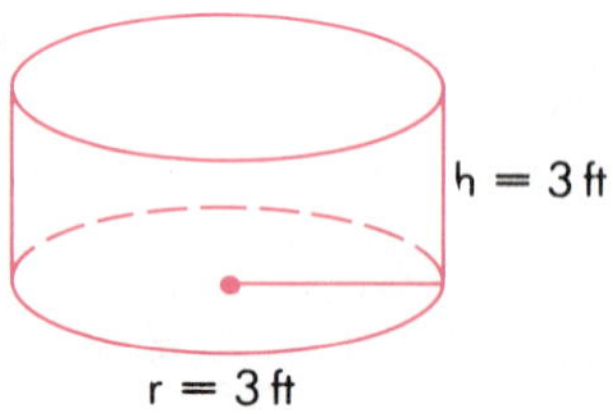

9.

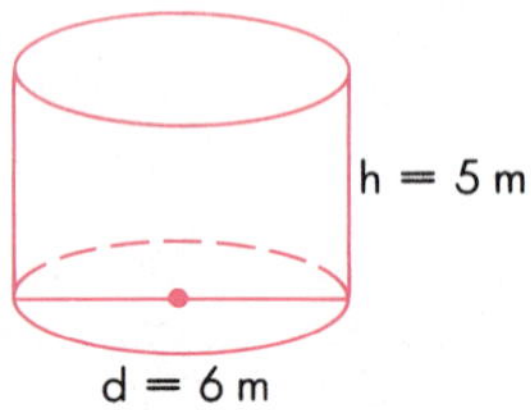

10. Cylinder
$r = 6$ m
$h = 14$ m

11. Cylinder
$d = 9$ cm
$h = 6$ cm

12. Cylinder
$d = 6$ in.
$h = 2$ in.

13.

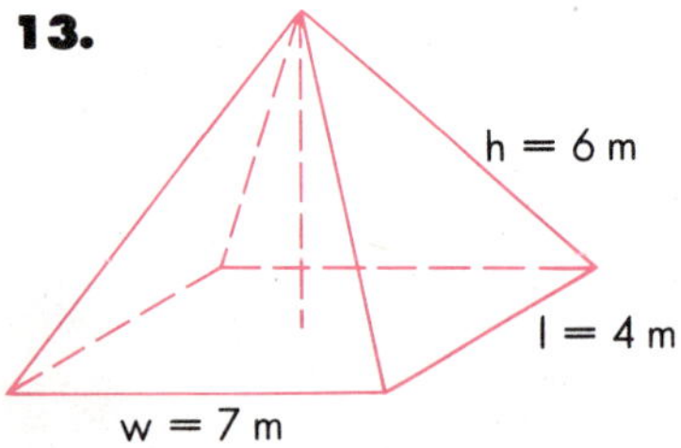

14.

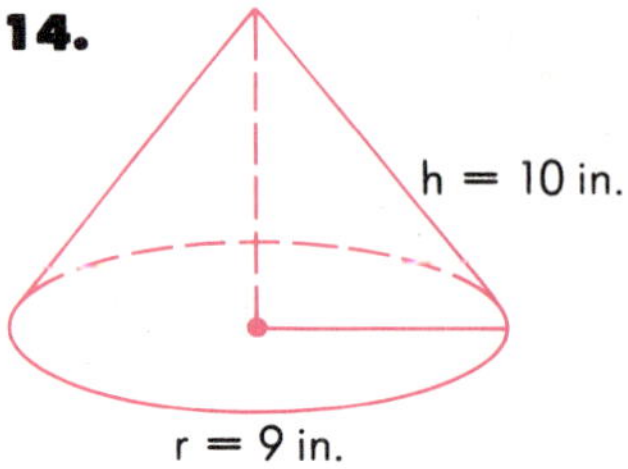

15.

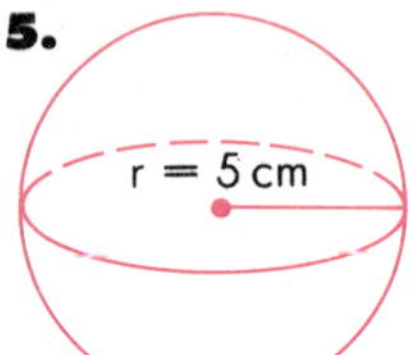

Application pp. 342, 350

16. A bill for electricity was $ 112.50. At an average rate of $ 0.05 per kilowatt-hour, how many kilowatt-hours were used?

17. During the month of January the Duncans left extra lights on in their house. They used 350 more kilowatt-hours than usual. At $ 0.06 per kWh, how much higher was their electricity bill?

CHAPTER 12 TEST

Find the volume.

1.

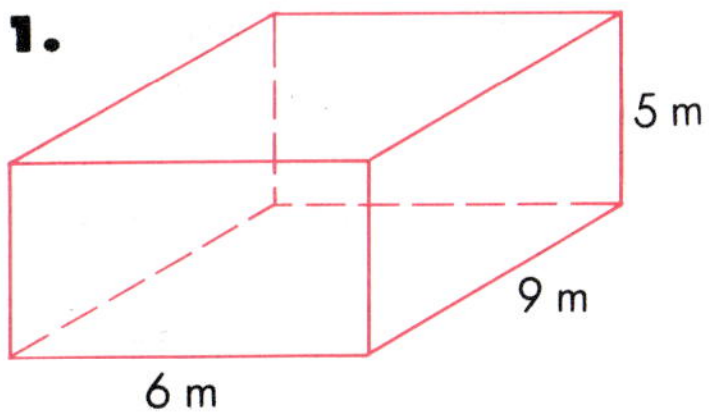

2.

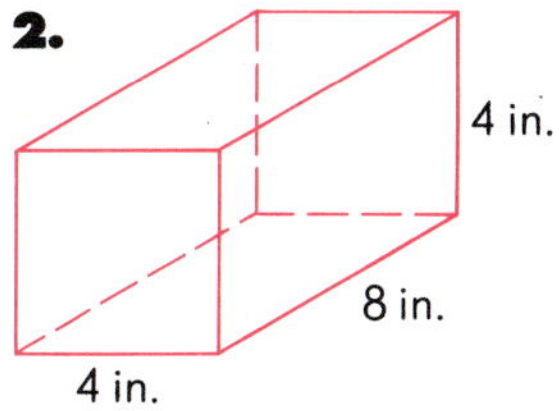

3.

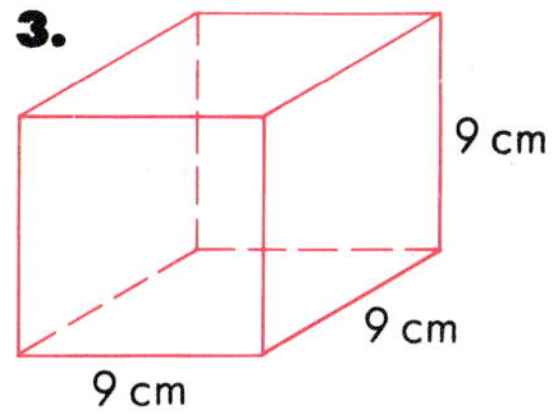

4. Rectangular prism
$l = 9$ cm
$w = 6$ cm
$h = 7$ cm

5. Cube
$s = 2$ m

6. Cube
$s = 6$ ft

7.

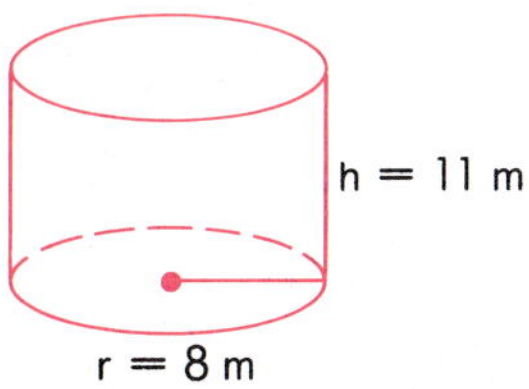

8.

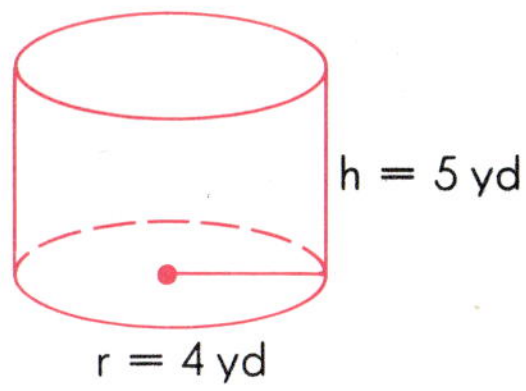

9.

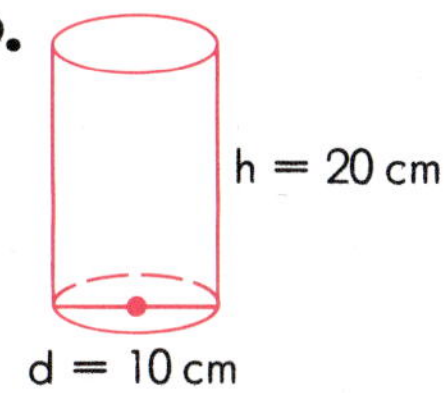

10. Cylinder
$r = 7$ cm
$h = 15$ cm

11. Cylinder
$d = 8$ m
$h = 6$ m

12. Cylinder
$d = 3$ ft
$h = 4$ ft

13.

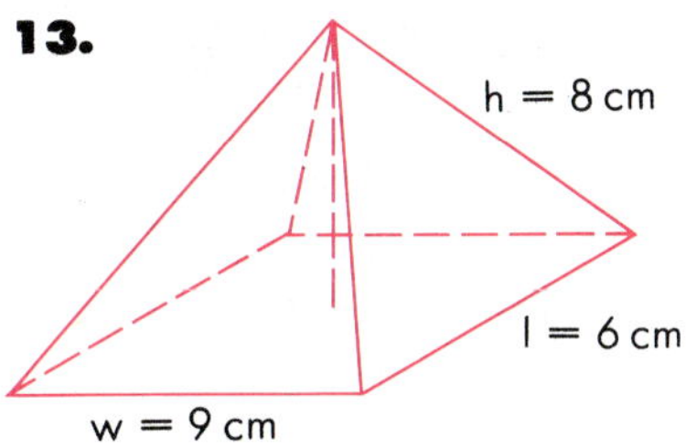

14.

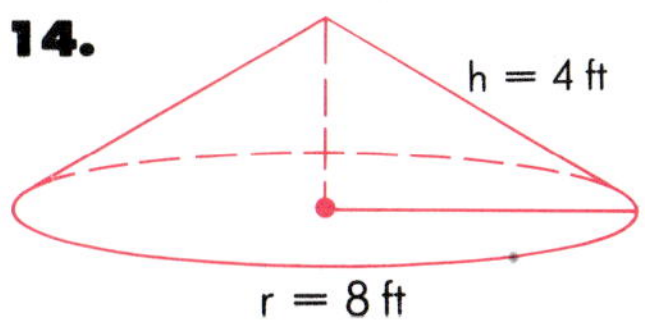

15.

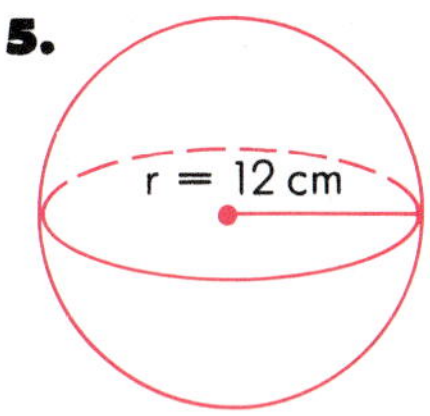

Application

16. A bill for electricity is \$ 73.47 for 1800 kilowatt-hours of electricity. What is the cost per kilowatt-hour?

17. The Perrys saved 1542 kilowatt-hours of electricity in a year. At an average rate of \$ 0.05 per kWh, how much money was saved?

MORE PRACTICE

Find the volume. Use after page 341.

1.

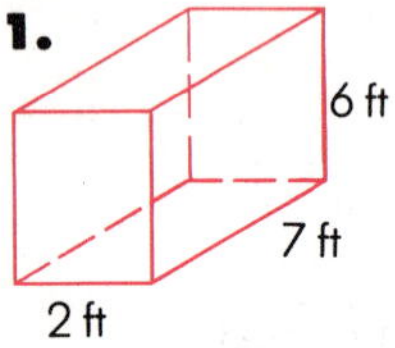

2.

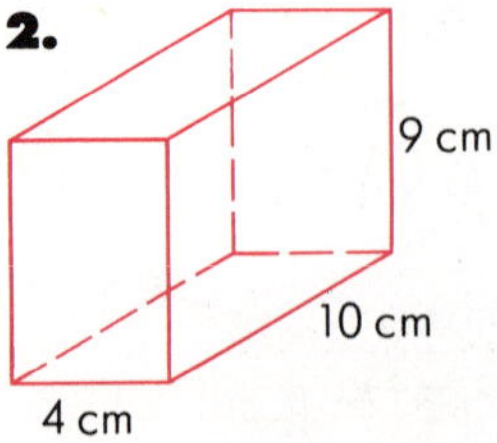

3.

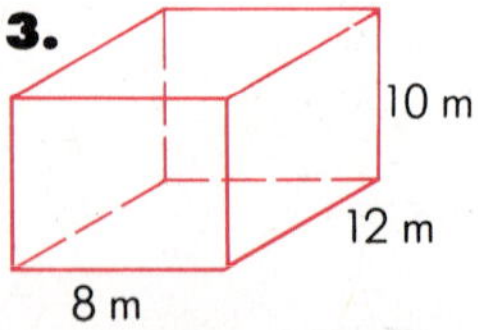

4. Rectangular prism
$l = 7$ in.
$w = 7$ in.
$h = 3$ in.

5. Rectangular prism
$l = 9$ m
$w = 10$ m
$h = 5$ m

6. Rectangular prism
$l = 8$ cm
$w = 6$ cm
$h = 4$ cm

7. Cube
$s = 6$ in.

8. Cube
$s = 10$ m

9. Cube
$s = 11$ cm

Find the volume. Use after page 345.

10.

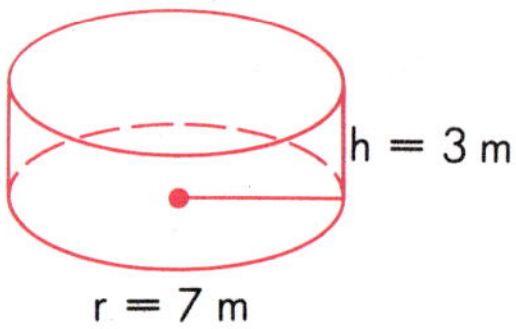

11.

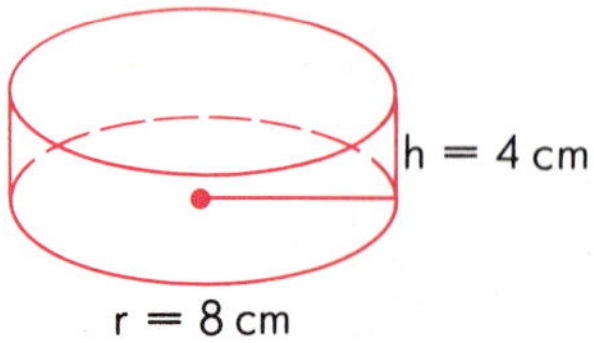

12.

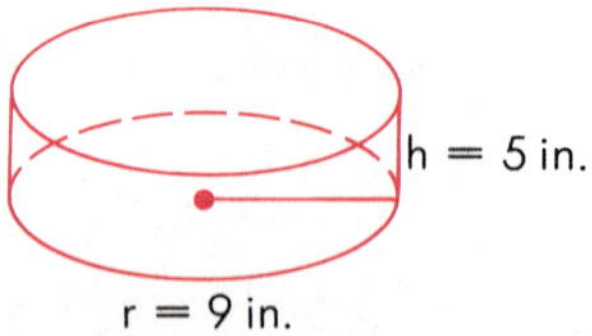

13. Cylinder
$r = 2$ ft
$h = 3$ ft

14. Cylinder
$r = 10$ mm
$h = 6$ mm

15. Cylinder
$r = 5$ cm
$h = 5$ cm

16. Cylinder
$d = 20$ cm
$h = 4$ cm

17. Cylinder
$d = 4$ in.
$h = 20$ in.

18. Cylinder
$d = 6$ cm
$h = 6$ cm

19. Cylinder
$d = 8$ cm
$h = 10$ cm

20. Cylinder
$d = 4$ ft
$h = 2$ ft

21. Cylinder
$d = 6$ m
$h = 5$ m

Find the volume. Use after page 347.

1.

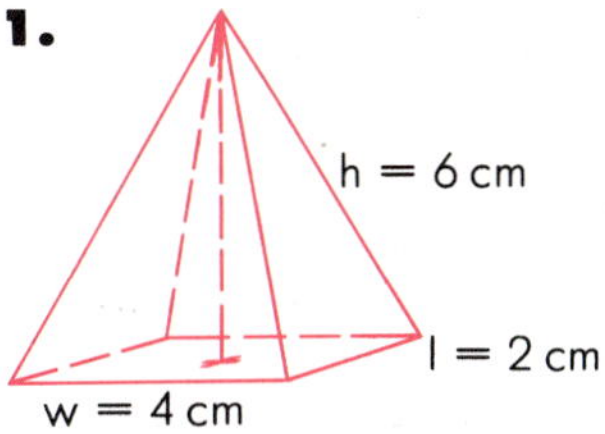

2.

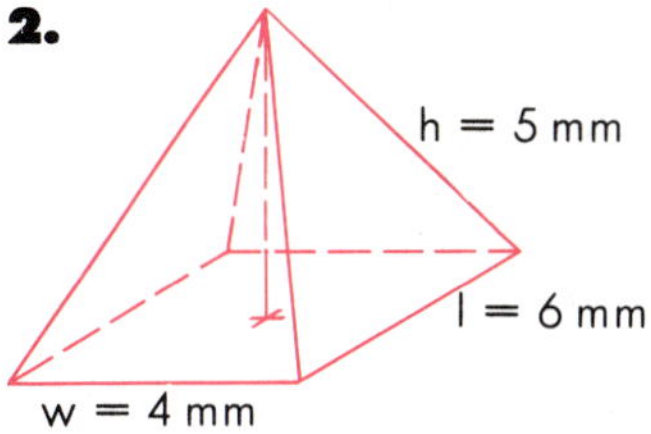

3.

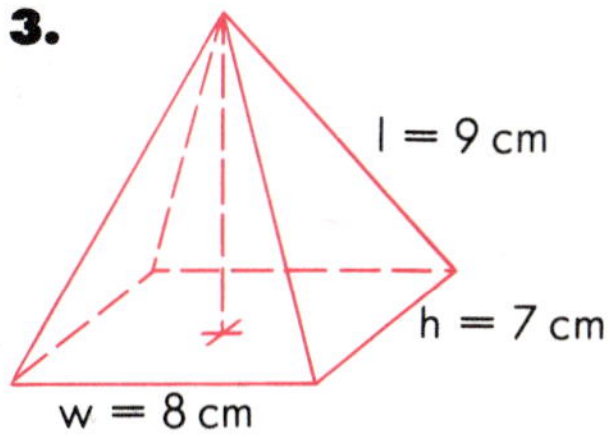

4. Pyramid
$l = 10$ in.
$w = 2$ in.
$h = 6$ in.

5. Pyramid
$l = 4$ cm
$w = 9$ cm
$h = 4$ cm

6. Pyramid
$l = 5$ m
$w = 5$ m
$h = 5$ m

7.

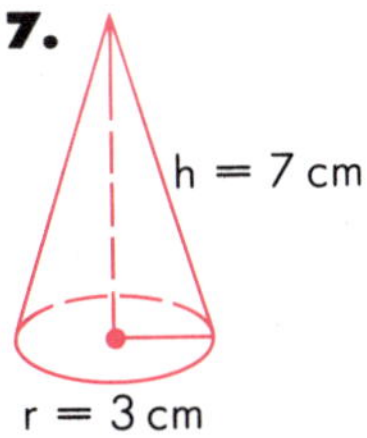

8.

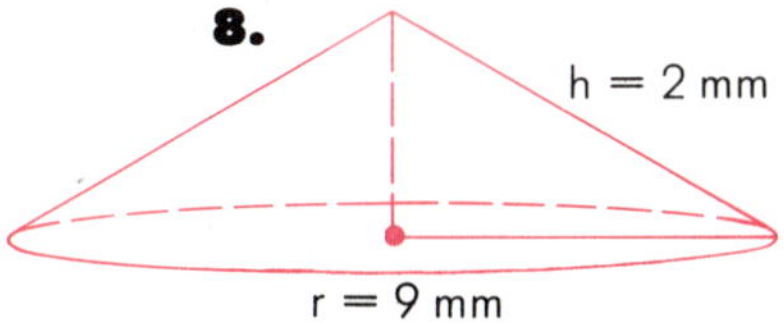

9.

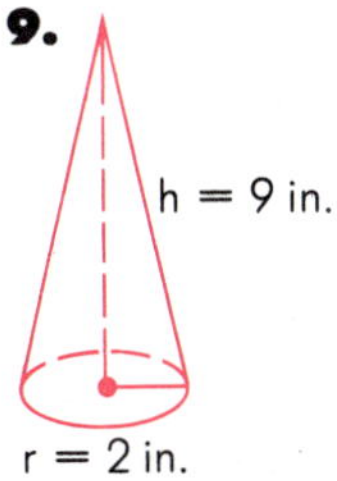

10. Cone
$r = 7$ cm
$h = 6$ cm

11. Cone
$r = 8$ m
$h = 10$ m

12. Cone
$r = 4$ cm
$h = 8$ cm

Find the volume. Use after page 349.

13.

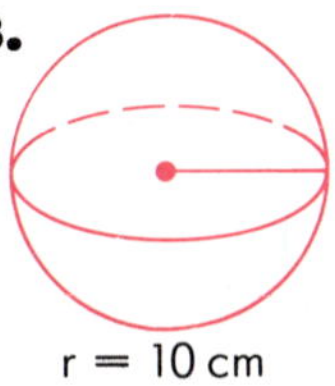

14.

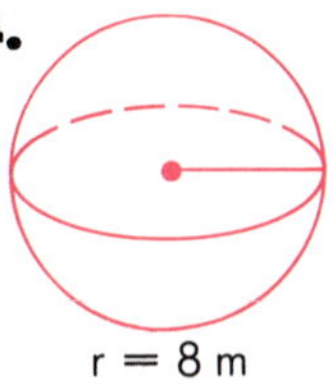

15.

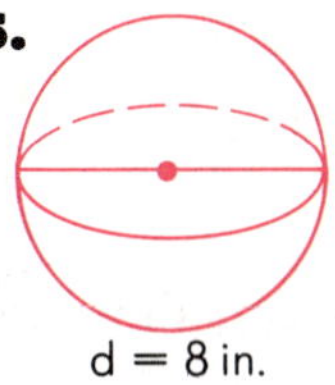

16. Sphere
$d = 12$ cm

17. Sphere
$d = 10$ m

18. Sphere
$d = 6$ ft

19. Sphere
$r = 4$ ft

20. Sphere
$r = 9$ m

21. Sphere
$r = 6$ cm

SKILLS REVIEW

Multiply and simplify. Refer to pages 168–177.

1. $\frac{3}{4} \cdot \frac{3}{4}$ **2.** $\frac{7}{8} \times \frac{4}{5}$ **3.** $\frac{3}{8} \times \frac{4}{9}$ **4.** $\frac{4}{5} \cdot \frac{5}{6} \cdot \frac{2}{5}$

5. $\frac{1}{2} \times 12$ **6.** $9 \cdot \frac{2}{3}$ **7.** $5 \times \frac{4}{5}$ **8.** $\frac{2}{3} \times 8$

9. $8 \cdot 2\frac{2}{3}$ **10.** $1\frac{2}{3} \times \frac{4}{9}$ **11.** $7\frac{1}{2} \cdot 3\frac{2}{3}$ **12.** $6\frac{1}{4} \cdot 2\frac{2}{5}$

Divide and simplify. Refer to pages 180–183.

13. $6 \div \frac{1}{3}$ **14.** $7 \div \frac{4}{7}$ **15.** $\frac{4}{5} \div 4$ **16.** $10 \div \frac{1}{2}$

17. $\frac{2}{3} \div \frac{7}{3}$ **18.** $\frac{3}{4} \div \frac{1}{2}$ **19.** $\frac{7}{10} \div \frac{3}{5}$ **20.** $\frac{4}{5} \div \frac{4}{10}$

21. $1\frac{1}{2} \div 1\frac{1}{2}$ **22.** $6\frac{2}{3} \div 1\frac{1}{4}$ **23.** $7\frac{3}{4} \div 2\frac{1}{2}$ **24.** $6\frac{4}{5} \div 5\frac{3}{10}$

Add and simplify. Refer to pages 196–203.

25. $\frac{2}{3} + \frac{2}{3}$ **26.** $\frac{7}{10} + \frac{3}{10}$ **27.** $\frac{9}{12} + \frac{6}{12}$ **28.** $\frac{4}{5} + \frac{2}{5}$

29. $\frac{3}{4} + \frac{1}{2}$ **30.** $\frac{2}{3} + \frac{3}{4}$ **31.** $\frac{7}{8} + \frac{1}{4} + \frac{1}{2}$ **32.** $\frac{3}{5} + \frac{7}{10} + \frac{1}{2}$

33. $2\frac{1}{2} + 3\frac{1}{4}$ **34.** $3\frac{5}{8} + 2\frac{3}{4}$ **35.** $4\frac{3}{10} + 1\frac{4}{5}$ **36.** $3\frac{1}{2} + 2\frac{3}{4} + 3\frac{7}{8}$

Subtract and simplify. Refer to pages 206–211.

37. $\frac{9}{10} - \frac{4}{10}$ **38.** $\frac{7}{8} - \frac{1}{8}$ **39.** $\frac{11}{12} - \frac{3}{4}$ **40.** $\frac{7}{9} - \frac{2}{3}$

41. $8\frac{7}{10} - 4\frac{3}{10}$ **42.** $9\frac{3}{4} - 6\frac{1}{4}$ **43.** $4\frac{9}{10} - 1\frac{2}{5}$ **44.** $8\frac{3}{4} - 7\frac{1}{2}$

45. $6 - 3\frac{3}{4}$ **46.** $10 - 3\frac{1}{2}$ **47.** $12\frac{1}{3} - 6\frac{2}{3}$ **48.** $5\frac{1}{5} - 4\frac{7}{10}$

49. $17\frac{15}{16} - 9$ **50.** $12\frac{2}{5} - 6\frac{2}{3}$ **51.** $22\frac{1}{4} - 8\frac{4}{5}$ **52.** $10\frac{3}{8} - 8\frac{5}{6}$

Solve. Refer to page 230.

1. $\frac{3}{4} = \frac{x}{20}$ **2.** $\frac{4}{5} = \frac{20}{x}$ **3.** $\frac{2}{3} = \frac{7}{x}$ **4.** $\frac{7}{10} = \frac{x}{3}$

5. $\frac{x}{5} = \frac{11}{4}$ **6.** $\frac{4}{x} = \frac{9}{8}$ **7.** $\frac{x}{8} = \frac{7}{10}$ **8.** $\frac{9}{x} = \frac{2}{5}$

Solve. Refer to page 232.

9. If you drive 110 miles in 2 hours, how far will you drive in 3 hours?

10. It takes Carie 2 hours to drive 110 miles. How many hours does it take her to drive 330 miles?

11. If you receive $ 17.00 for 4 hours of work, how much will you receive for 6 hours of work?

12. Bo receives $ 17.00 for 4 hours of work. How many hours must he work to receive $ 25.50?

13. If 3 tapes cost $ 9, how many tapes can you buy for $ 15?

14. If 5 pounds of mushrooms cost $ 7, how much do 2 pounds cost?

15. A team won 4 out of 7 games. At this rate, how many will it win in 21 games?

16. If you save $ 50 every 3 months, how much will you save in 12 months?

Solve. Refer to pages 256–263.

17. 60% of 10 is what number?

18. What is 70% of 100?

19. What percent of 10 is 6?

20. 12 is what percent of 24?

21. 75 is 50% of what number?

22. 9 is $12\frac{1}{2}\%$ of what number?

23. 75% of 20 is what number?

24. What is $33\frac{1}{3}\%$ of 27?

25. What percent of 50 is 5?

26. 100 is what percent of 200?

27. 12 is 50% of what number?

28. 10 is $66\frac{2}{3}\%$ of what number?

29. 90% of 50 is what number?

30. What is $87\frac{1}{2}\%$ of 8?

31. What percent of 300 is 200?

32. 13 is what percent of 39?

Make these changes. Refer to page 284.

1. 6 km = ______ m　**2.** 4378 m = ______ km　**3.** 12 m = ______ cm

4. 76 cm = ______ m　**5.** 814 mm = ______ cm　**6.** 16 cm = ______ mm

7. 79 mm = ______ cm　**8.** 4.1 cm = ______ mm　**9.** 4.8 km = ______ m

10. 8 m = ______ cm　**11.** 9.21 km = ______ m　**12.** 6501 m = ______ km

13. 6.2 km = ______ m　**14.** 371 m = ______ km　**15.** 89 mm = ______ cm

Make these changes. Refer to page 290.

16. 4 L = ______ mL　**17.** 7.4 L = ______ mL　**18.** 6.5 L = ______ mL

19. 6247 mL = ______ L　**20.** 9000 mL = ______ L　**21.** 4367 mL = ______ L

22. 0.91 L = ______ mL　**23.** 438 mL = ______ L　**24.** 0.45 L = ______ mL

25. 0.6L = ______ mL　**26.** 73 mL = ______ L　**27.** 142 mL = ______ L

28. 17,000 mL = ______ L　**29.** 750 mL = ______ L　**30.** 57 mL = ______ L

Make these changes. Refer to page 292.

31. 8605 g = ______ kg　**32.** 4 g = ______ mg　**33.** 4763 mg = ______ g

34. 7.3 t = ______ kg　**35.** 2476 kg = ______ t　**36.** 1.5 kg = ______ g

37. 814 mg = ______ g　**38.** 8.8 kg = ______ g　**39.** 23 g = ______ mg

40. 9 t = ______ kg　**41.** 704 g = ______ kg　**42.** 598 kg = ______ t

43. 7141 mg = ______ g　**44.** 12 kg = ______ g　**45.** 2.1 g = ______ mg

Make these changes. Refer to page 298.

46. 7 ft = ______ in.　**47.** 8 pt = ______ cups　**48.** 2 lb = ______ oz

49. 5 ft = ______ yd　**50.** $3\frac{1}{2}$ qt = ______ pt　**51.** 2 mi = ______ yd

52. $2\frac{1}{2}$ yd = ______ in.　**53.** 96 oz = ______ lb　**54.** $1\frac{1}{2}$ gal = ______ qt

55. 3 mi = ______ ft　**56.** 48 in. = ______ ft　**57.** 4 yd = ______ ft

58. 40 fl oz = ______ cups　**59.** 3000 lb = ______ ton　**60.** 6 cups = ______ pt

Find the perimeter. Refer to page 312.

1.
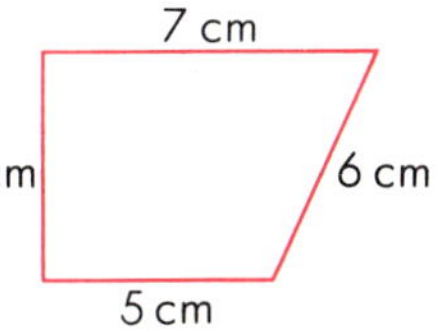

2.
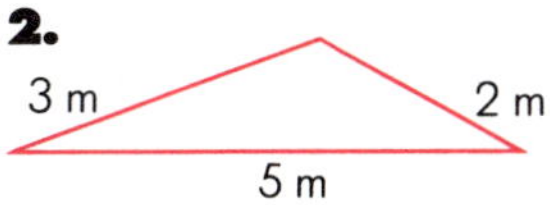

3.
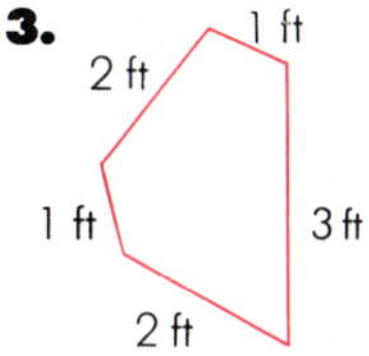

4.
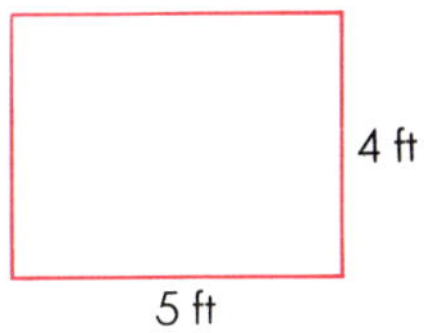

5.
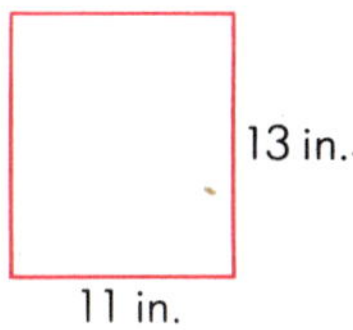

6.
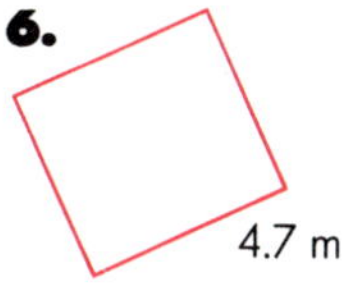

7.
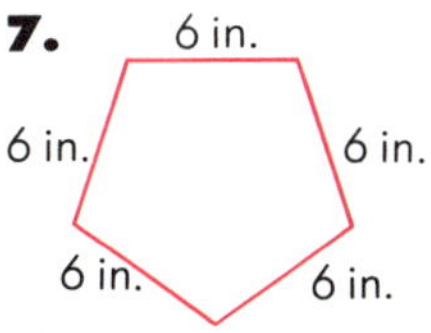

8.
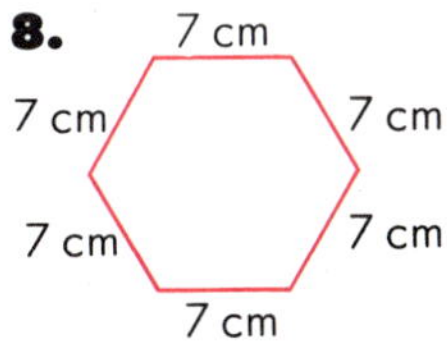

9.
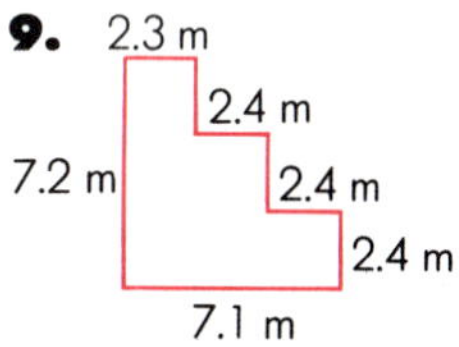

10. Square
$s = 5$ cm

11. Square
$s = 12$ mm

12. Rectangle
$l = 4\frac{1}{2}$ yd
$w = 2\frac{1}{2}$ yd

Find the circumference. Refer to page 314.

13.
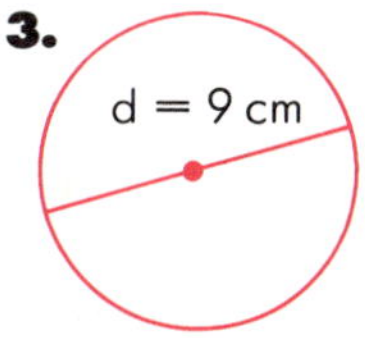

14.
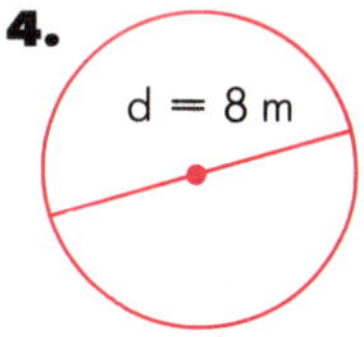

15.
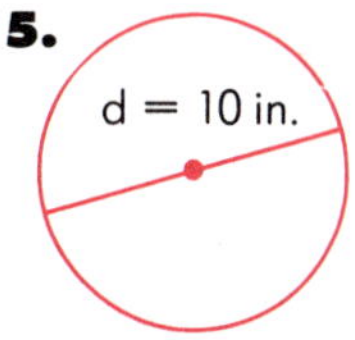

16. Circle
$d = 12$ mm

17. Circle
$d = 3.2$ m

18. Circle
$d = 10.3$ cm

19.
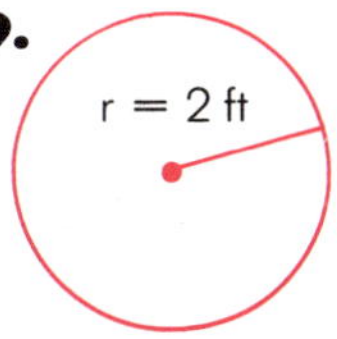

20.
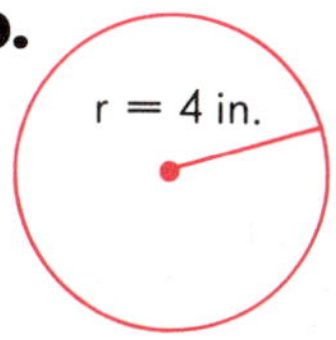

21.
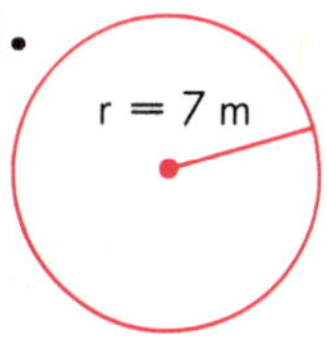

22. Circle
$r = 6$ cm

23. Circle
$r = 1.3$ m

24. Circle
$r = 2.2$ cm

Find the area. Refer to page 318.

1.

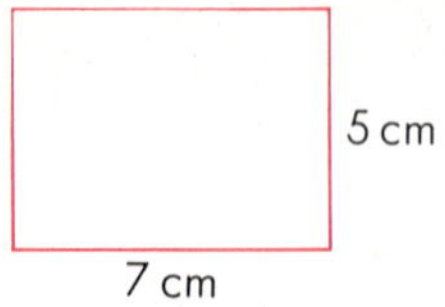

2.

3.

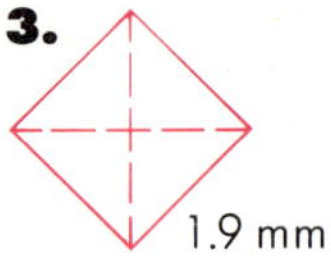

4. Square
$s = 7$ in.

5. Square
$s = 3.7$ m

6. Rectangle
$l = 8.1$ m
$w = 2.7$ m

Find the area. Refer to page 320.

7.

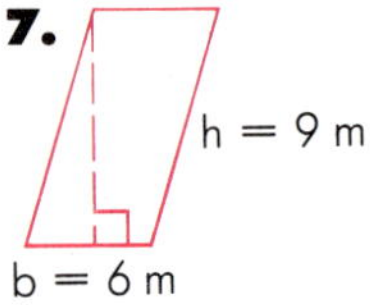

8.

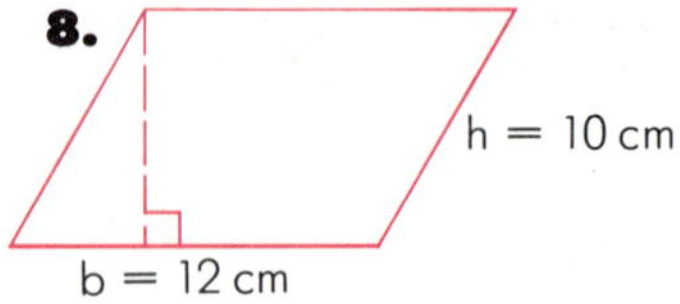

9.

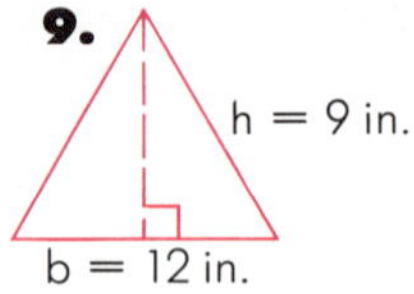

10. Triangle
$b = 10$ cm
$h = 6$ cm

11. Triangle
$b = 4$ m
$h = 3.6$ m

12. Triangle
$b = 5.2$ m
$h = 4.4$ m

Find the area. Refer to page 322.

13.

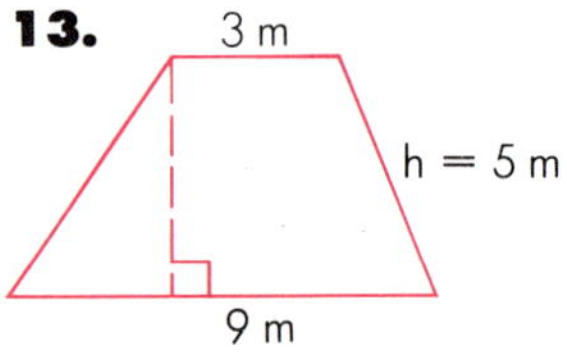

14.

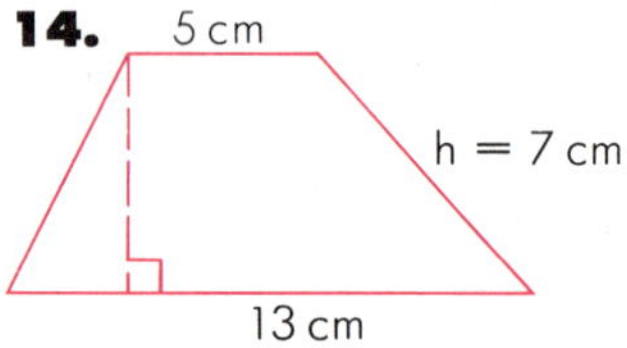

15.

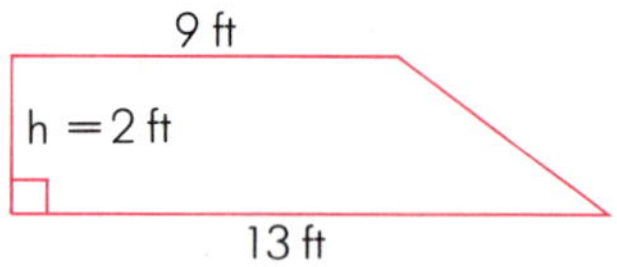

16. Trapezoid
$h = 20$ mm
$b_1 = 12$ mm
$b_2 = 16$ mm

17. Trapezoid
$h = 10$ m
$b_1 = 10$ m
$b_2 = 12$ m

18. Trapezoid
$h = 4$ in.
$b_1 = 8$ in.
$b_2 = 10$ in.

Find the area. Refer to page 326.

19.

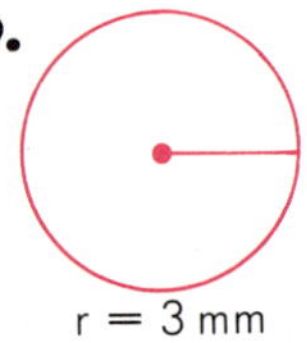

20.

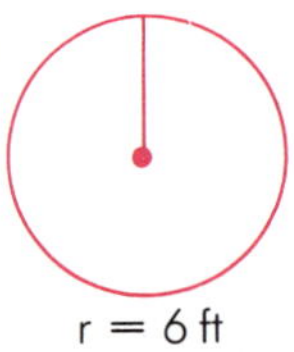

21.

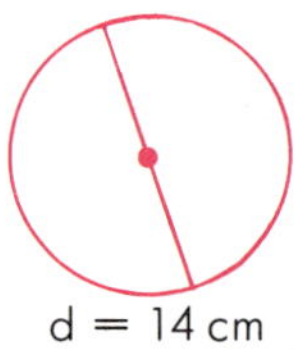

22. Circle
$d = 10$ m

23. Circle
$d = 8$ cm

24. Circle
$r = 4$ in.

Find the volume. Refer to page 340.

1.

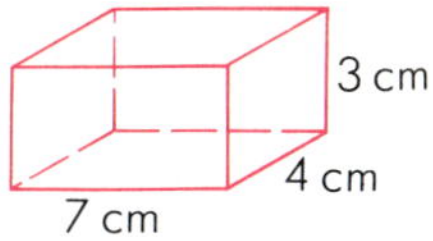

2.

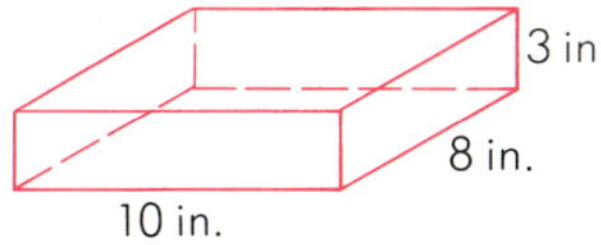

3.

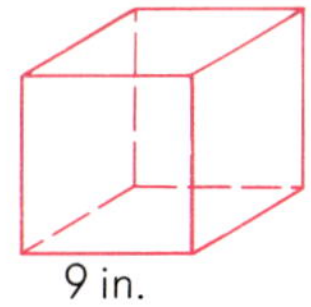

4. Cube
$s = 3$ m

5. Cube
$s = 8$ cm

6. Rectangular prism
$l = 3$ ft
$w = 2$ ft
$h = 4$ ft

Find the volume. Refer to page 344.

7.

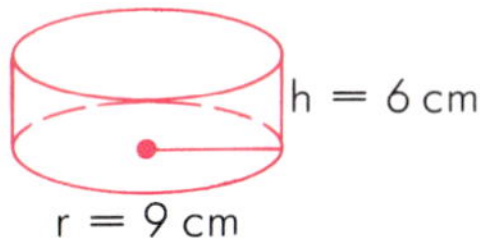

8.

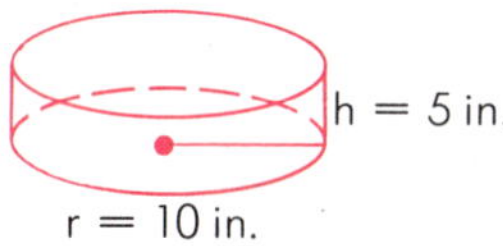

9.

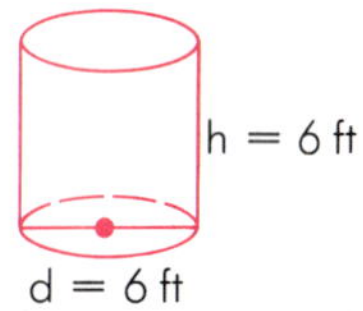

10. Cylinder
$d = 12$ mm
$h = 12$ mm

11. Cylinder
$d = 20$ cm
$h = 5$ cm

12. Cylinder
$r = 5$ ft
$h = 10$ ft

Find the volume. Refer to page 346.

13.

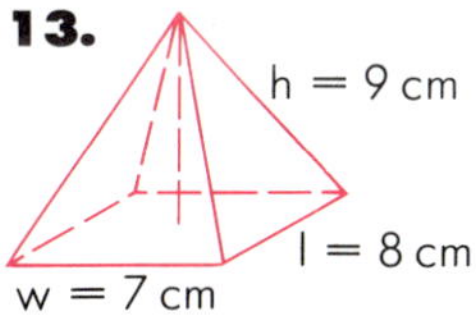

14.

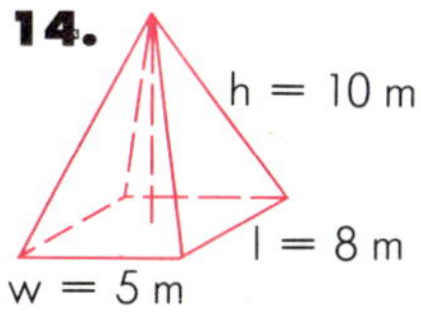

15.

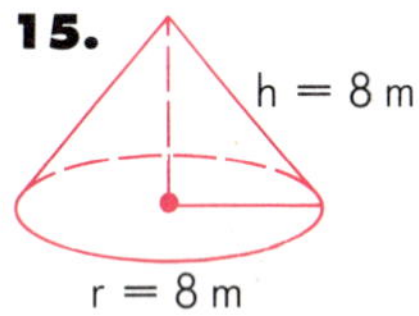

16. Cone
$r = 6$ in.
$h = 4$ in.

17. Cone
$r = 4$ cm
$h = 6$ cm

18. Pyramid
$l = 12$ cm
$w = 3$ cm
$h = 10$ cm

Find the volume. Refer to page 348.

19.

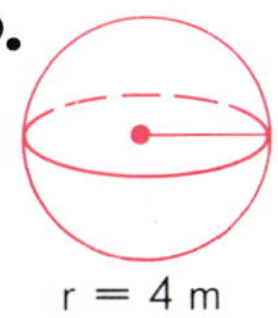

20.

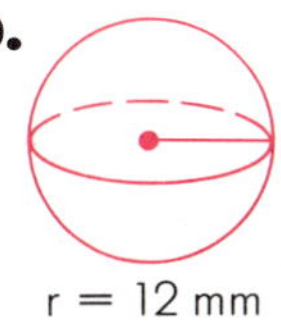

21.

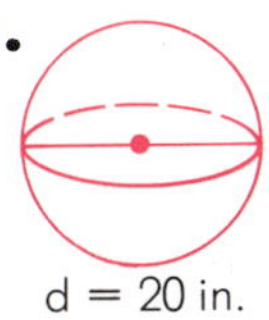

22. Sphere
$d = 16$ ft

23. Sphere
$d = 10$ m

24. Sphere
$r = 9$ in.

SOLVING PROBLEMS

Sometimes problems can be solved best by doing a quick mental calculation.

SITUATION 1 You are out to lunch with 3 friends. The bill is \$ 16.56. You decide to divide the bill four ways.

To mentally divide by 4, take half and then take half again. Divide the bill by 4 mentally. How much should each person pay?

SITUATION 2 You are making a fertilizer mixture which calls for $1\frac{1}{2}$ tablespoons of fertilizer per gallon of water. You wish to make 8 gallons.

To mentally multiply by 8, double the number, double again and finally double a third time. Multiply $1\frac{1}{2}$ by 8 mentally. How much fertilizer should you use?

SITUATION 3 Your dinner bill is \$ 14.80. You wish to leave a 15% tip.

To mentally take 15%, take 10% and then add half of your result. Mentally take 15% of \$ 14.80. How much tip should you leave?

SITUATION 4 You are chairman of a committee to plan a party for 25 people. The caterer says that the cost will be \$ 8.50 per person.

To mentally multiply by 25, multiply by 100 and then divide by 4. Mentally multiply \$ 8.50 by 25. How much will the party cost?

READY FOR CHAPTER 13

1. Graph the ordered pair (4,5) on a grid.

2. Find an ordered pair for the point A on the grid.

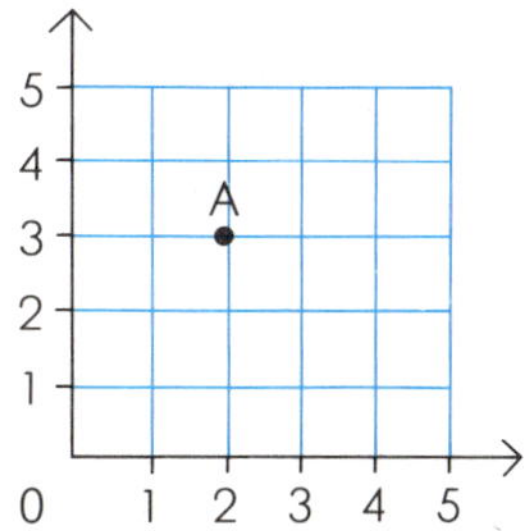

3. Make a line graph for the data in the table.

Height (in thousands of feet)	View (in miles)
5	87
10	122
15	149
20	172

4. Make a bar graph for the data in the table.

Causes of Death	Number of Deaths (in thousands)
Heart Trouble	230
Cancer	175
Accidents	80

5. Find the angle for each percent and make a circle graph.

House Color	Percent
White	48%
Gray	34%
Yellow	16%
Other	2%

Find these statistics.

6. the mean: 14, 28, 16, 10, 11
7. the median: 78, 75, 73, 77, 80
8. the mode: 96, 89, 96, 97, 99, 103

Application

9. Rona went on a weekend ski trip and spent \$ 38.60 for transportation, \$ 18.25 for food, \$ 14 for lift tickets, and \$ 41.63 for a hotel room. How much did she spend in all on her trip?

10. A football team won 13 out of 16 games. Find the percentage of games won.

13
GRAPHING AND STATISTICS

GETTING STARTED

COORDINATE MATCH

WHAT'S NEEDED A 5 × 5 grid drawn on the blackboard

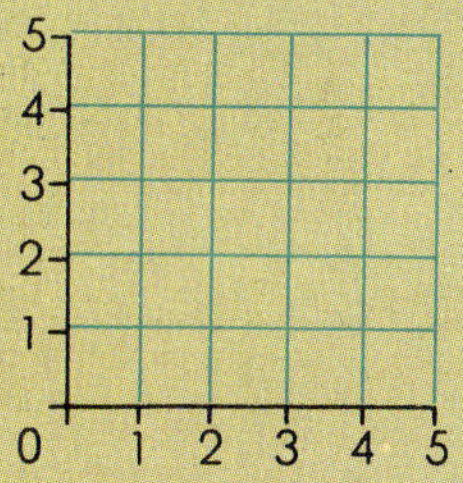

THE RULES

1. On a sheet of paper each player makes a scorecard. Then each player writes down five ordered pairs.
2. The leader marks ten points on the grid.
3. The players compare their ordered pairs with the points on the grid and check the ones that match. Each player records on the scorecard the number of points that match.
4. Play three rounds and total the scores. The player with the highest total wins.

SAMPLE

Round 1

(3,3)
(0,4)
(1,2)
(2,1)
(5,0)

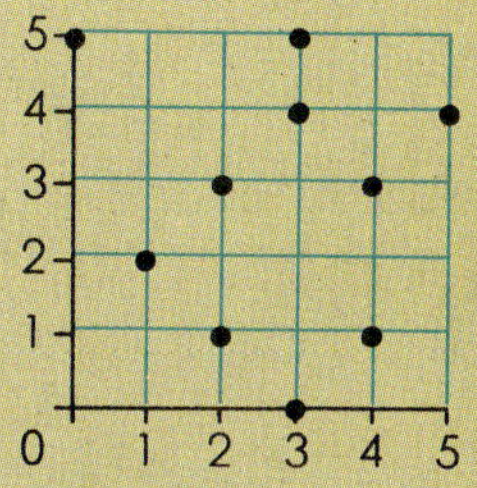

Round 2

(4,1)
(5,5)
(3,2)
(0,4)
(1,5)

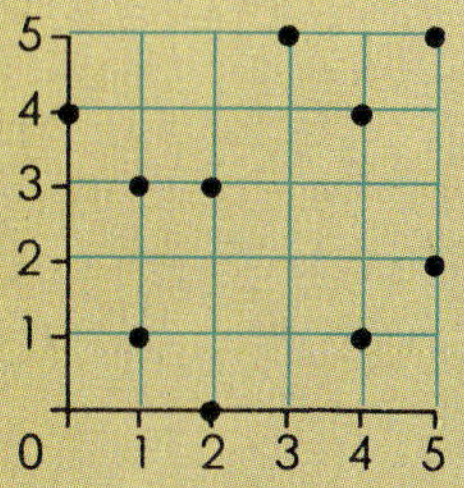

Round 3

(0,0)
(2,4)
(3,5)
(4,0)
(1,1)

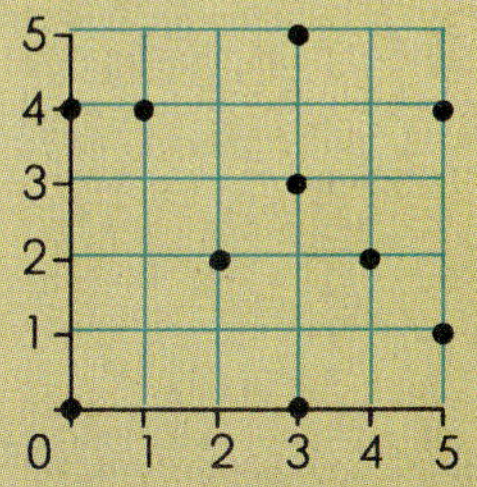

Scorecard

1. 2
2. 3
3. 2
Total 7

GRAPHING ORDERED PAIRS

The first number in an ordered pair is the number of units to the right of (0,0). The second number is the number of units up.

EXAMPLE 1. Graph this ordered pair on a grid. (3,4).

(3,4)
Go 3 units to the right of (0,0).
Go 4 units up.
Mark and label the point.

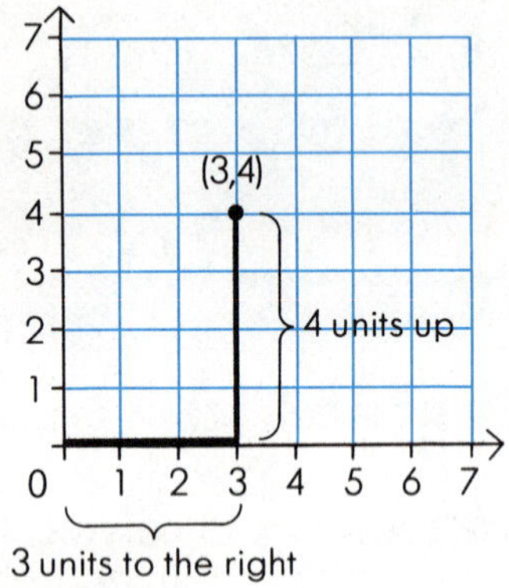

TRY THIS

Graph each ordered pair on a grid.

1. (4,5) **2.** (1,2) **3.** (0,3) **4.** (2,0)

EXAMPLE 2. Find an ordered pair for point A on the grid.

Point A is 7 units to the right of (0,0). → (,)

Point A is 5 units up. →

The ordered pair is (7,5).

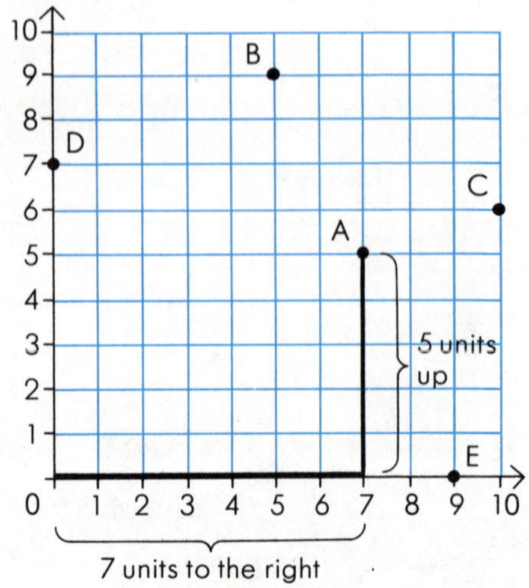

TRY THIS

Find an ordered pair for each point on the grid in Example 2.

5. B **6.** C **7.** D **8.** E

PRACTICE

Graph each ordered pair on a grid.

1. (2,1) **2.** (3.7) **3.** (2,5) **4.** (8,4)

5. (0,6) **6.** (0,8) **7.** (0,5) **8.** (0,10)

9. (5,0) **10.** (3,0) **11.** (7,0) **12.** (1,0)

13. (4,4) **14.** (9,9) **15.** (10,10) **16.** (2,2)

Find an ordered pair for each point on the grid.

17. A **18.** B **19.** C

20. D **21.** E **22.** F

23. G **24.** H **25.** I

26. J **27.** K **28.** L

29. M **30.** N **31.** P

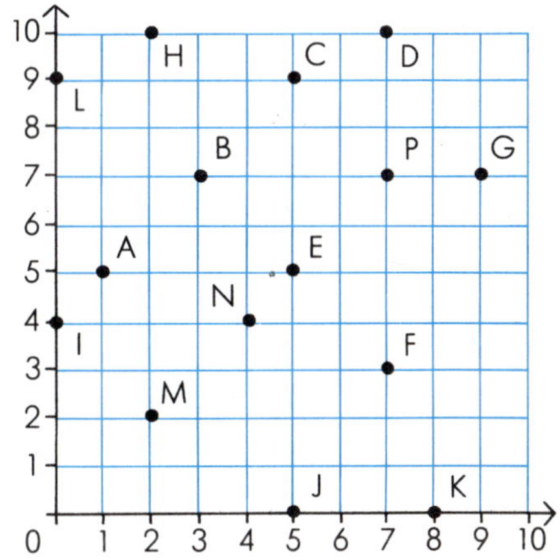

PROBLEM CORNER

A bottle washed up on shore contains the following message.

Clues to Buried Treasures

Treasure A: (2,3)
Treasure B: (7,9)
Treasure C: (10,5)
Treasure D: (5,0)

Make a graph to show the location of the treasures.

MAKING LINE GRAPHS

A line graph shows the change in data.

EXAMPLE 1. Make a line graph for the data in the table.

The table shows the decreasing temperature in a glass of water as it cools.

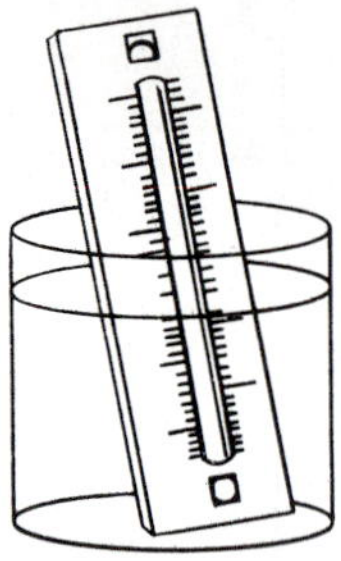

Time (in minutes)	Temperature (in Fahrenheit)
0	160°
30	120°
60	90°
90	70°
120	70°

To make a line graph follow these steps.

1. Draw and label a vertical axis. It does not have to start at 0. We start at some number less than the smallest number in the table. The numbers increase along the axis by the same amount.
2. Draw and label a horizontal axis. This time it is convenient to start with 0.
3. Graph ordered pairs from the table. For example, the table gives us the ordered pair (30,120°).
4. Connect each pair of points with a line.

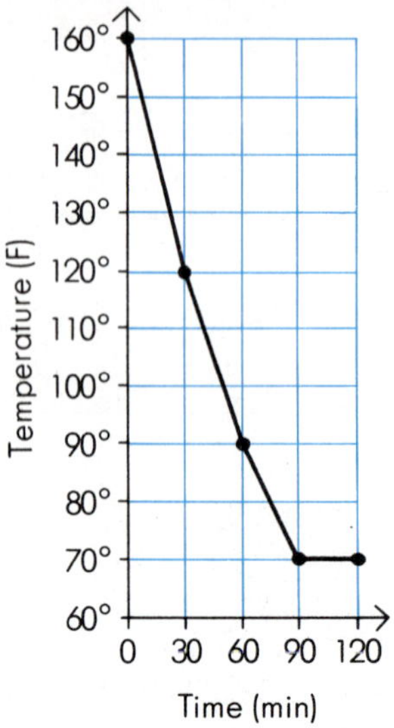

TRY THIS

Make a line graph for the data in the table.

Year	Value of a One-Carat Diamond
1940	$900
1950	$1500
1960	$2500
1970	$6300

PRACTICE

Make a line graph.

1.

Time	Temperature (C)
9:00	18°
10:00	20°
11:00	21°
12:00	23°
1:00	22°

2.

Year	Earnings
1940	$ 5500
1950	$ 6000
1960	$ 8000
1970	$14,000
1980	$12,000

3.

Year	Rainfall (Inches)
1940	28
1950	25
1960	30
1970	27
1980	30

4.

Speed (mph)	Gas Mileage (miles per gallon)
20	17
30	18
40	19
50	18
60	16

PROBLEM CORNER

How many TV's were sold each year?

1. 1978 **2.** 1979

3. 1980 **4.** 1981

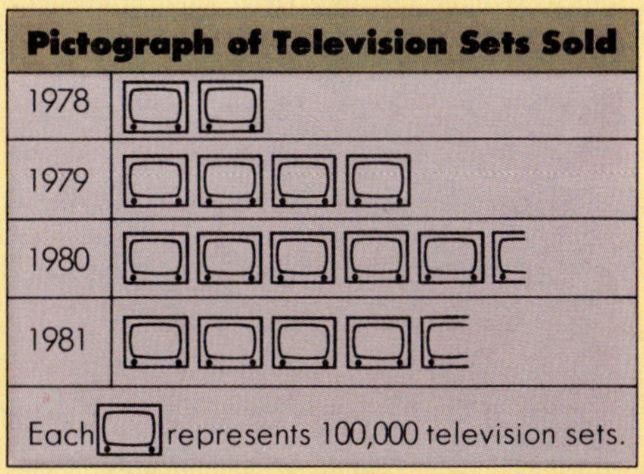

MAKING BAR GRAPHS

A bar graph shows the relationship of data.

EXAMPLE 1. Make a bar graph for the data in the table.

To make a bar graph follow these steps.

1. Draw and label a vertical axis. The last number should be large enough so that every number in the table can be graphed.
2. Draw and label a horizontal axis. Space the names of the nations evenly.
3. Draw vertical bars to show the percents.

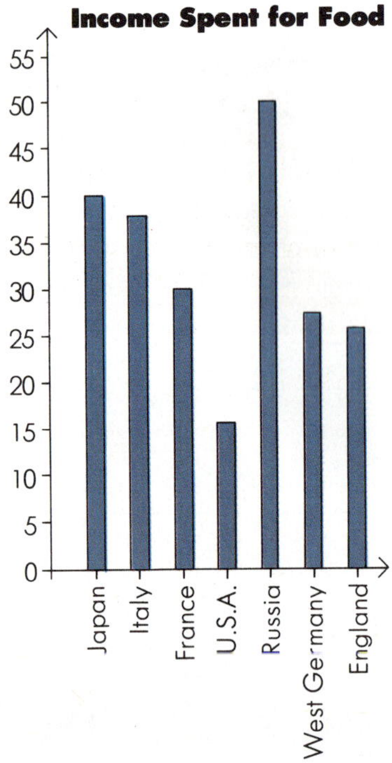

Nation	Percent of Income Spent for Food
Japan	40
Italy	38
France	31
U.S.A.	17
Russia	50
West Germany	28
England	26

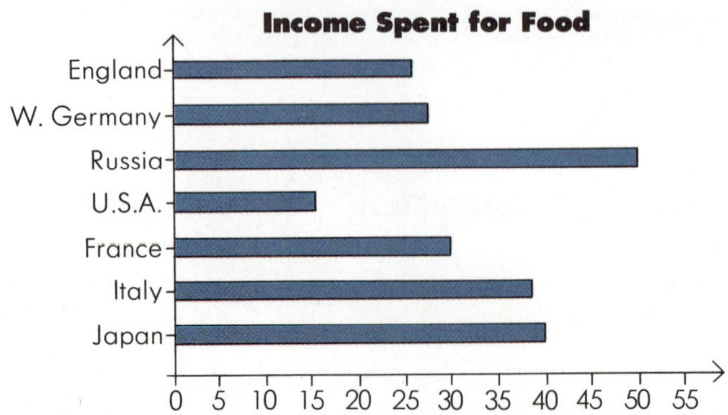

TRY THIS

1. Make a table and a bar graph for the following data. The number of calories in a one-scoop ice cream cone: Chocolate Fudge, 230; French Vanilla, 220; Strawberry, 170; Peach, 170; Butter Pecan, 200; Chocolate Mint, 190.

PRACTICE

Make a bar graph.

1.

Activity	Calories Burned Per Hour by Person Weighing 152 lbs
Tennis	420
Jogging	650
Hiking	590
Office Work	180
Sleeping	70

2.

Speed (mph)	Gas Mileage (miles per gallon)
20	17
30	18
40	19
50	18
60	16

3.

Education (years)	Income
8	$ 8000
12	$12,000
16	$18,000

4. Make a table and a bar graph for the following data. The number of passengers in one year (in millions) to fly out of a given airport: Chicago (O'Hare), 41; Atlanta, 27; Los Angeles, 26; London (Heathrow), 24; New York (Kennedy), 21; San Francisco, 18.

PROBLEM CORNER

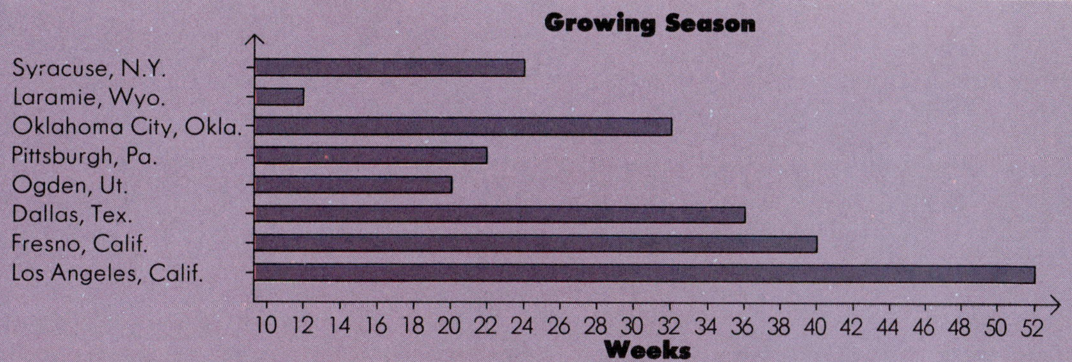

1. Which city has the longest growing season?

2. Which city has the shortest growing season?

3. How many times longer is the growing season in Fresno, California than in Ogden, Utah? in Dallas, Texas than in Laramie, Wyoming?

More Practice, page 388

MAKING CIRCLE GRAPHS

A circle graph shows data as percents.

EXAMPLE 1. Find the angle for each percent in the table and make a circle graph.

1. To find the angle convert each percent to a decimal and multiply by 360, the number of degrees in a circle.

Hair Color	Percent
Brown	36%
Black	28%
Blonde	24%
Red	12%

$0.36 \times 360° \approx 130°$
$0.28 \times 360° \approx 101°$
$0.24 \times 360° \approx 86°$
$0.12 \times 360° \approx 43°$

2. Draw a circle and a radius. Draw the angles given by the degree measures from Step 1. Label each angle as it is drawn.

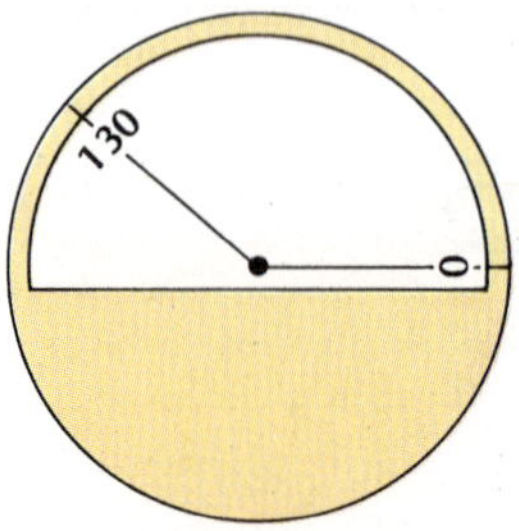

Hair Color

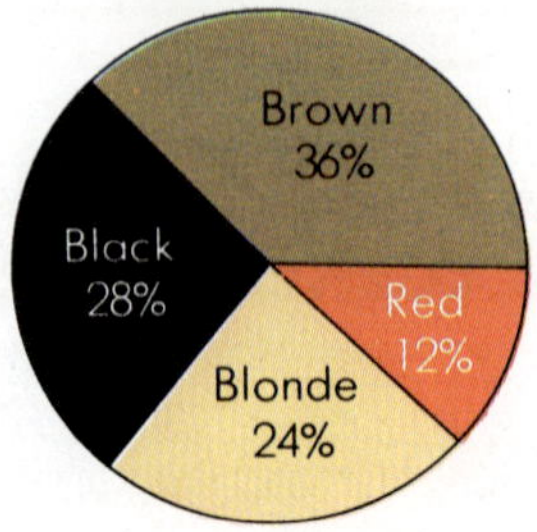

TRY THIS

Find the angle for each percent and make a circle graph.

1.

Family Expenses	Percent
Food	20%
Auto	14%
Clothing	10%
Housing	26%
Other	30%

PRACTICE

Find the angle for each percent in the table and make a circle graph.

1.

Study Time	Percent
Math	24%
English	38%
Social Studies	20%
Science	18%

2.

Sales	Percent
Shoes	10%
Suits	50%
Shirts	15%
Ties	5%
Pants	20%

3.

Favorite TV Shows	Percent
Movies	12%
Sports	20%
News	4%
Drama	16%
Comedy	20%
Music	28%

4.

Personal Income in 1980	Percent
$25,000 and over	13%
$15,000–$25,000	33%
$10,000–$15,000	23%
$7000–$10,000	12%
Below $7000	19%

PROBLEM CORNER

Compare the graphs. Does the percent increase or decrease from 1965 to 1975 for each situation?

1. Households with no TV sets

2. Households with 1 TV set

3. Households with 2 or more TV sets

What percent of U.S. households had at least 1 TV set in the following years?

4. 1965 **5.** 1975

Percent of Households in U.S. With TV Sets

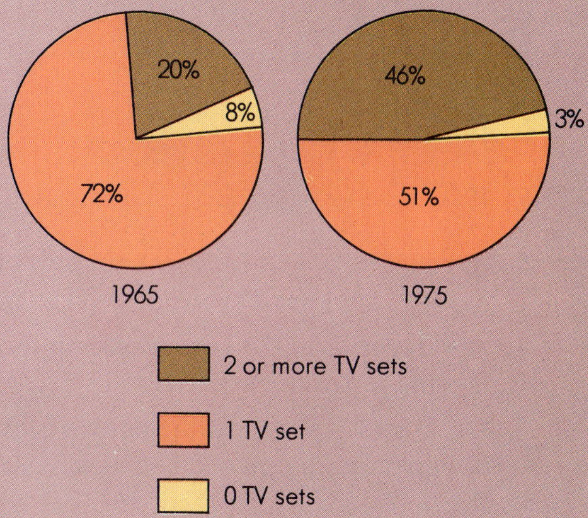

APPLICATION

TRAVEL COSTS

Each year many more people travel for business and for pleasure. The amount of money they spend is also increasing.

Find how much the average American traveler spent abroad in each of the following years.

1. 1960
2. 1965
3. 1970
4. 1975
5. 1980
6. In which year was the most amount of money spent per traveler? the least amount?
7. Are travelers to the U.S. spending more or less money each year?
8. Draw a graph to show the comparison of dollars spent by travelers to and from the U.S.

Overseas Travel from the U.S.

Year	Number of Travelers	Millions of Dollars Spent
1960	1,634,000	2623
1965	2,623,000	3768
1970	5,260,000	6180
1975	6,354,000	10,143
1980	7,100,000	10,700

Travel to the U.S.

Year	Millions of Dollars Spent
1960	1025
1965	1545
1970	2708
1975	5606
1980	7200

Business travelers must often report their travel expenses. Tax deductible expenses must be shown with receipts. It is necessary to keep accurate records of travel expenses.

Betty Riggs is a buyer for a department store. On a business trip last month she kept a list of her expenses.

Travel Expenses	Amount	Paid With
Airfare	$ 115.98	Credit card
Hotel	62.30	Credit card
Meals	22.89	Cash
Tips	4.00	Cash
Cab fare	18.25	Cash
Convention	25.00	Check
Phone	8.50	Cash

9. How much did Betty charge on her credit card?
10. What was the total amount of cash spent?
11. How much did Betty spend in all?

Betty could have saved 20% on the airfare by traveling at night.

12. How much money could she have saved?
13. What was the price of the night airfare?
14. James went skiing last weekend. He spent $ 29.75 for transportation, $ 18.34 for food, $ 12 for lift tickets, and $ 35 for a hotel room. How much did he spend in all?
15. On her vacation Tanja spent $ 153.50 for airfare, $ 75.80 for hotel rooms, $ 42.37 for food, and $ 23.40 for souvenirs. How much did she spend in all?

The Clark twins compared their vacation expenses. Jerry spent $ 83 for transportation. Jill spent $ 28 for food.

16. Jerry spent 25% more than Jill for food. How much did he spend for food?
17. Jill spent 30% more than Jerry for transportation. How much did she spend for transportation?
18. Make a chart to show their individual and total expenses.

FINDING THE MEAN

The mean of a set of numbers is the average of the numbers.

Find the mean. 78, 81, 82, 79

Find the sum of the numbers.	→	How many numbers were added?	→	Divide by that number.

$$\begin{array}{r} \overset{2}{7}8 \\ 81 \\ 82 \\ +79 \\ \hline 320 \end{array} \qquad \left.\begin{array}{r} 78 \\ 81 \\ 82 \\ +79 \end{array}\right\} 4 \qquad \text{Mean} = \frac{320}{4} = 80$$
$$\begin{array}{r} \hline 320 \end{array}$$

EXAMPLE 1. Find the mean. 75, 36.8, 95.7, 12.1, 47.6

Find the sum of the numbers.	*How many numbers were added?*	*Divide by that number.*
$\begin{array}{r} \overset{2\,2}{75} \\ 36.8 \\ 95.7 \\ 12.1 \\ +47.6 \\ \hline 267.2 \end{array}$	$\left.\begin{array}{r} 75 \\ 36.8 \\ 95.7 \\ 12.1 \\ 47.6 \end{array}\right\} 5$ $\begin{array}{r}\hline 267.2\end{array}$	$\frac{267.2}{5} = 53.44$ Mean $= 53.44$

TRY THIS

Find the mean.

1. 86, 91, 68, 76
2. 81, 59, 32, 67, 45
3. 105.2, 108.9, 110, 200, 240
4. 162.3, 247.3, 99.9, 158.7

Sometimes the sum is given and all we need to do is divide.

EXAMPLE 2. A car went 432 miles on 24 gallons of gasoline. What was the average number of miles per gallon?

$$\frac{432 \text{ miles}}{24 \text{ gallons}} = 18 \text{ miles per gallon}$$

We divide 432 miles by 24 gallons.

TRY THIS **5.** A car went 522 miles on 18 gallons of gasoline. What was the average number of miles per gallon?

PRACTICE

Find the mean.

1. 8, 7, 15, 15, 15, 12

2. 72, 83, 85, 88, 92

3. 5, 10, 15, 20, 25, 30, 35

4. 13, 13, 25, 27, 32

5. 1.2, 4.3, 5.7, 7.4, 7.4

6. 13.4, 13.4, 12.6, 42.9

7. Sandra Palmer scored 78, 74, 71, and 72 to win the U.S. Open Golf Tournament. What was her average score?

8. The following temperatures were recorded for seven days: 43°, 40°, 23°, 38°, 54°, 35°, 47° What was the average temperature?

9. The following prices per pound of steak were found at five supermarkets: $ 3.79, $ 3.59, $ 3.69, $ 3.79, $ 3.89. What was the average price per pound?

10. The following prices per pound of hamburger were found at five supermarkets: $ 1.39, $ 1.29, $ 1.49, $ 1.09, $ 0.99. What was the average price per pound?

11. The attendance at five basketball games was 30,050, 28,740, 35,170, 29,080, and 32,610. What was the average attendance?

12. O. J. Simpson set an NFL rushing record gaining 2003 yards in a 14-game season. What was the average number of yards gained per game?

PROBLEM CORNER

1. What was each person's average score?

2. What was the team total and average in Game 1? Game 2? Game 3?

3. What was the team average for the 3 games?

The Oakwood Bowling Team Scores

	Sally	Randy	Chip	Leanne
Game 1	115	196	144	162
Game 2	132	182	138	166
Game 3	124	156	126	172

More Practice, page 389

FINDING THE MEDIAN AND MODE

In studying data it is often required to find the median and mode of a set of numbers.

Find the median. 78, 81, 82, 76, 84

Arrange the numbers in order. → The middle number is the median.

76, 78, 81, 82, 84 Median = 81

If there are two middle numbers, the median is the number halfway between.

EXAMPLE 1. Find the median. 23, 24, 25, 26

23, 24, 25, 26 *The median is halfway between 24 and 25.*

Median = 24.5 $\frac{24 + 25}{2} = \frac{49}{2} = 24.5$

TRY THIS

Find the median.

1. 20, 13, 17, 14, 18
2. 37, 57, 27, 47, 27
3. 18, 14, 16, 13, 19, 20
4. 8, 7, 8, 6, 9, 10

Find the mode. 14, 23, 25, 23, 27, 27, 29, 27

Arrange the numbers in order. → The number which appears the greatest number of times is the mode.

14, 23, 23, 25, 27, 27, 27, 29 Mode = 27

EXAMPLE 2. Find the mode. 69, 60, 60, 61, 62, 63, 63, 59

59, 60, 60, 61, 62, 63, 63, 69 *Both 60 and 63 appear twice. This set of numbers has 2 modes, 60 and 63.*

(60, 60: Mode) (63, 63: Mode)

TRY THIS

Find the mode.

5. 78, 66, 78, 81, 81, 78, 78
6. 14, 16, 15, 15, 14, 16, 16
7. 103, 101, 101, 101, 102, 103
8. 9, 8, 4, 5, 4, 2, 3, 7

PRACTICE

Find the median.

1. 72, 83, 85, 92, 88
2. 80, 81, 79, 78, 76
3. 8, 7, 15, 15, 12, 13, 14
4. 11, 9, 17, 13, 27, 29, 28
5. 36, 37, 39, 40
6. 41, 42, 44, 45
7. 36, 37, 38, 39
8. 41, 42, 43, 44
9. 8, 7, 15, 15, 15, 12
10. 13, 32, 25, 27, 13

Find the mode.

11. 1, 3, 3, 3, 4
12. 5, 5, 6, 7, 8
13. 14, 14, 15, 15, 15, 12, 14
14. 19, 17, 19, 17, 19, 20, 20, 17
15. 9, 9, 9, 9, 9, 9, 9, 7
16. 20, 20, 20, 20, 20, 18, 20
17. 34, 34, 35, 35, 36, 34, 34
18. 102, 101, 102, 103, 102, 102, 101
19. 38, 38, 37, 38, 39, 39, 40, 39
20. 97, 98, 99, 97, 98, 99, 97, 98, 100

PROBLEM CORNER

Marcy received the following scores on five tests:

1. What was the average?
2. What was the median?
3. What was the mode?

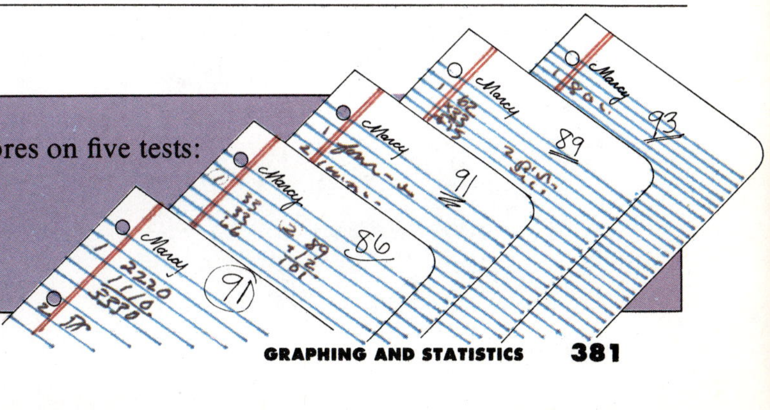

APPLICATION

SPORTS STATISTICS

In many different sports an important figure is the percentage of games won.

$$\text{PERCENTAGE OF GAMES WON} = \frac{\text{NUMBER OF GAMES WON}}{\text{NUMBER OF GAMES PLAYED}}$$

The Boston Celtics won the Eastern Conference Championship of the NBA 14 times from 1957 to 1976. They won each year from 1957 to 1965 and each year from 1972 to 1976. Here are their records from 1957 to 1965.

1. How many games were played each year?
2. What was the percentage of games won each year?
3. In 1976, the Celtics played 82 games. Their percentage of games won was 0.659. How many games did they win? lose?
4. In 1975, the Chicago Bulls played 82 games. Their percentage of games won was 0.573. How many games did they win? lose?

Year	Games Won	Games Lost
1957	44	28
1958	49	23
1959	52	20
1960	59	16
1961	57	22
1962	60	20
1963	58	22
1964	57	21
1965	62	18

A basketball game can be won or lost from the free throw line.

$$\text{FREE THROW PERCENTAGE} = \frac{\text{NUMBER OF COMPLETIONS}}{\text{NUMBER OF ATTEMPTS}}$$

What is the free throw percentage of each player?

5. Manny
6. Philip
7. Sam
8. Ron
9. Bob
10. What is the team free throw percentage?

Player	Attempts	Completions
Manny	62	24
Philip	57	19
Sam	41	16
Ron	71	32
Bob	45	17

We often use batting averages to compare baseball players' performances.

$$\text{BATTING AVERAGE} = \frac{\text{NUMBER OF HITS}}{\text{NUMBER OF TIMES AT BAT}}$$

In one season Melanie had 25 hits out of 58 times at bat. What was her batting average?

$$\text{BATTING AVERAGE} = \frac{25}{58} = 0.431$$

Player	Hits	Times at Bat
Jay	20	60
Rita	12	62
Ken	21	55
Connie	26	63
Paul	16	61
Julie	15	65

Find the batting average for each player.

11. Jay
12. Rita
13. Ken
14. Connie
15. Paul
16. Julie

The 100-meter women's butterfly stroke has been an event in the summer Olympics since 1956. The graph shows the winning times.

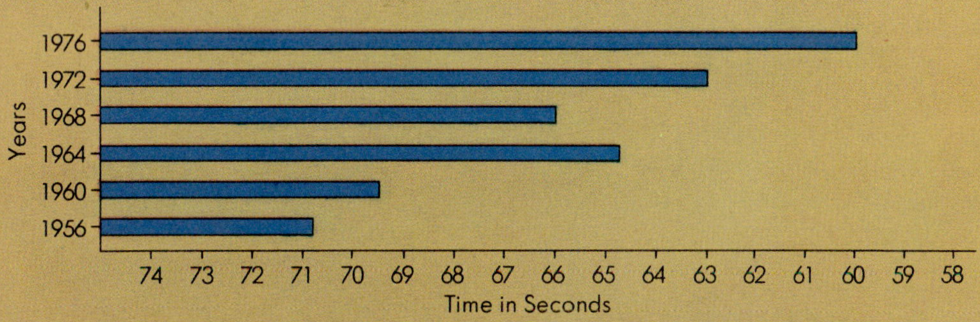

17. Did the winning times get faster or slower from 1956 to 1976?
18. Which year does not follow the pattern?
19. How many seconds difference is there between the winning time in 1956 and 1976?

If this race is held in 1988, would you expect the winning time to be faster than each of these times?

20. 68 seconds
21. 63 seconds
22. 59 seconds

ADVERTISING LAYOUT ARTIST

Valerie Edwards is an advertising layout artist. She helps plan advertising campaigns. She works with the research staff. Her job is to present information about products in an attractive way. Simple graphs can be made more appealing with a creative design.

Valerie designed an ad for salad dressing. It shows the growth in sales from 1979 to 1982.

1. What type of graph is this?
2. Does it show exactly how many bottles were sold each year?
3. In which year was the increase in growth the greatest? least?
4. About how many bottles must be sold in 1983 to increase the 1982 sales by $\frac{1}{3}$?

Look How We've Grown in Just 4 Years

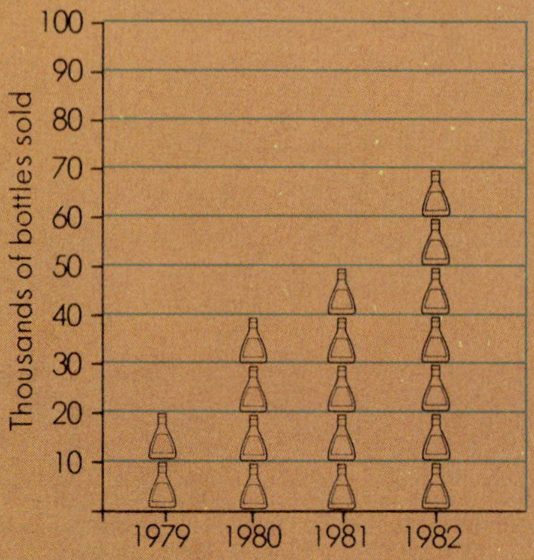

Valerie must be creative in her advertising designs. She needs to design several graphs and labels from the information in the chart. Can you help her out?

5. Design an attractive label to show the information in the table. The label is to be 3 in. by 3 in.
6. Design a pictograph to advertise the vitamins.

Vital Vitamins

One DNG capsule provides the adult RDA

Vitamin A – 125%
B_1 – 250%
B_2 – 225%
C – 125%
D – 100%

Valerie knows she can design graphs which give either a positive or negative impression. She designed Graphs A and B using the same data.

Graph A

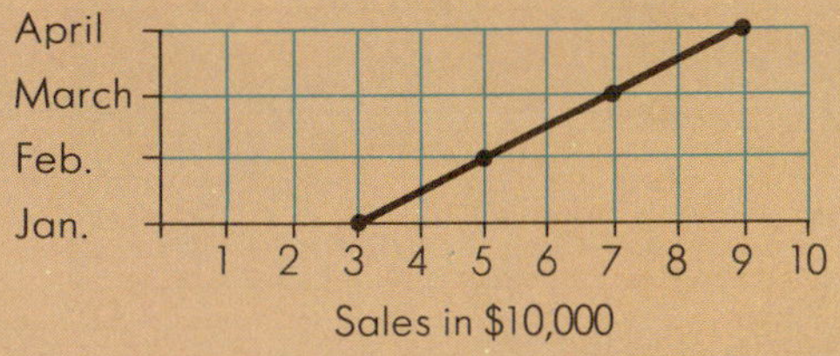

Graph B

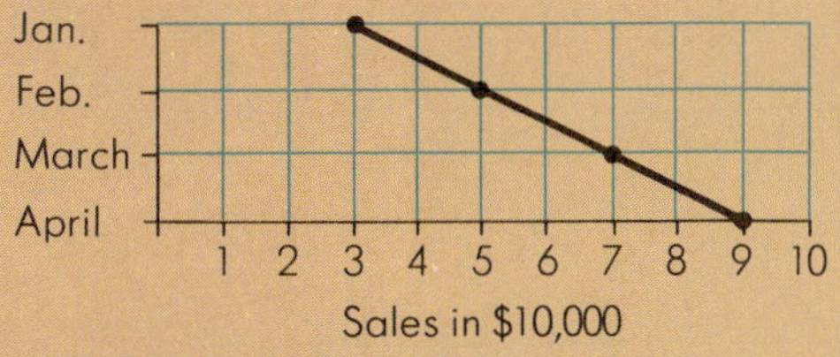

7. Which graph gives a positive impression of the company?
8. Use the information in Graph A. Draw a line graph which would give a more positive impression than Graph A.
9. Draw a bar graph to show the information in Graph A.
10. Draw a bar graph to show the information in Graph B.

For her monthly report, Valerie must keep a record of the time she spends each week on various projects.

Week of June 15	Time
TV ads	10 hours
Magazine ads	12 hours
Company promotions	8 hours
Annual reports	5 hours
Client conferences	5 hours

11. How many hours in all did Valerie work during the week of June 15?
12. What fractional part of the total time did she spend on each project?
13. What percent of the total time did she spend on each project?
14. Make a circle graph to show the information in the table.

CHAPTER 13 REVIEW

1. Graph the ordered pair (5,4) on a grid. p. 368

2. Find an ordered pair for the point A on the grid. p. 368

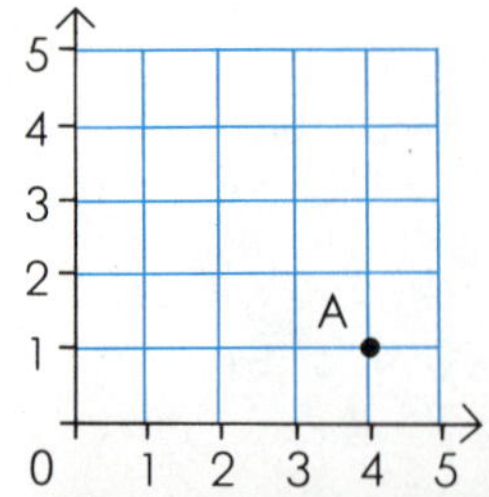

3. Make a line graph for the data in the table. p. 370

Distance (in miles)	Cost of 3 Min Evening Direct-Dial Phone Call
20	37¢
30	44¢
40	50¢
50	59¢

4. Make a bar graph for the data in the table. p. 372

Baseball Player	Lifetime Batting Average
Babe Ruth	0.342
Roger Maris	0.263
Rogers Hornsby	0.358

5. Find the angle for each percent and make a circle graph. p. 374

Sales	Percent
Bats	18%
Balls	21%
Rackets	13%
Shoes	14%
Other	34%

Find these statistics. pp. 378-381

6. the mean: 8, 10, 11, 7, 6, 5

7. the median: 8, 10, 11, 7, 6, 5

8. the mode: 24, 23, 27, 25, 27, 31, 27

Application pp. 376, 382

9. On vacation Jason spent $ 114 for plane fare, $ 53.75 for hotel rooms, $ 37.80 for food, and $ 18.27 for souvenirs. How much did he spend in all on his trip?

10. A basketball team won 23 out of 25 games. Find the percentage of games won.

CHAPTER 13 TEST

1. Graph the ordered pair (1,4) on a grid.

2. Find an ordered pair for the point A on the grid.

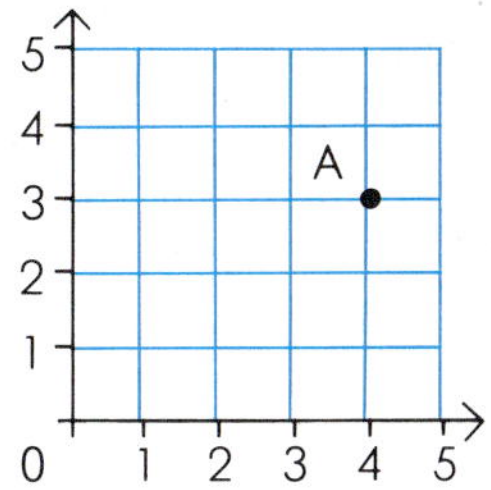

3. Make a line graph for the data in the table.

Distance (in miles)	Cost of 3 Min Weekday Direct-Dial Phone Call
20	55¢
30	67¢
40	77¢
50	89¢

4. Make a bar graph for the data in the table.

Player	Earnings in Women's Billiards Championship
Jean Balukas	$3000
Mieko Harada	$1500
Gerry Titcomb	$800
Becky Brown	$800
Palmer Byrd	$600

5. Find the angle for each percent and make a circle graph.

Batters	Percent
Right-Handed	78%
Left-Handed	20%
Both	2%

Find these statistics.

6. the mean: 88, 86, 76, 20, 100, 20
7. the median: 88, 86, 76, 20, 100, 20
8. the mode: 88, 86, 76, 20, 100, 20

Application

9. Marguerite went to a convention and spent $ 89 for transportation, $ 3.50 for tips, $ 63.47 for a hotel room, and $ 22.75 for food. How much did she spend in all on her trip?
10. A football team won 12 out of 16 games. Find the percentage of games won.

MORE PRACTICE

Graph each ordered pair on a grid. Use after page 369.

1. (2,1) **2.** (3,8) **3.** (3,5)

4. (7,4) **5.** (0,5) **6.** (0,9)

7. (4,0) **8.** (0,1) **9.** (4,4)

10. (5,0) **11.** (1,9) **12.** (9,1)

Find an ordered pair for each point on the grid. Use after page 369.

13. A **14.** B **15.** C

16. D **17.** E **18.** F

19. G **20.** H **21.** I

22. J **23.** K **24.** L

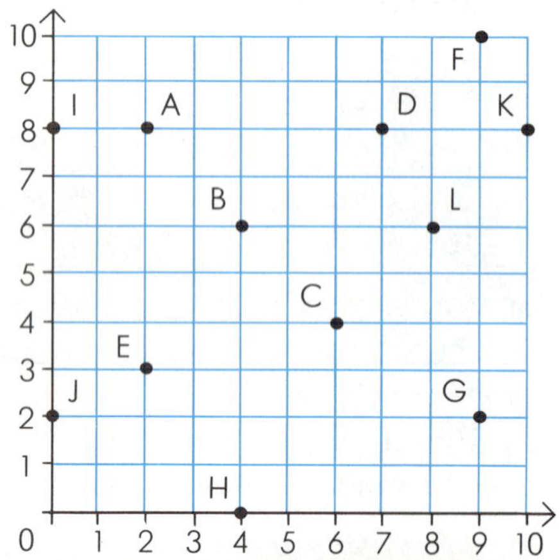

Make a line graph for the data in each table. Use after page 371.

25.

Year	Demand for Natural Gas (in quadrillion BTU)
1950	19
1960	21
1970	22
1980	27

26.

Olympic Runner	Time in 100 Meter Dash (in seconds)
Betty Cuthbert	11.5
Wilma Rudolph	11.0
Wyomia Tyus	11.4
Renate Stecher	11.07

Make a bar graph for the data in the table. Use after page 373.

27.

Cricket Chirps per Minute	Temperature (F°)
60	55°
76	59°
88	62°
100	65°

Make a table and a bar graph for the following data. Use after page 373.

1. The population of the U.S. (in millions) for a given year: 1976, 216; 1977, 219; 1978, 222; 1979, 226; 1980, 229.

Find the angle for each percent and make a circle graph. Use after page 375.

2.

Investment	Percent
Savings Account	48%
Stock	12%
Mutual Funds	16%
Retirement Fund	24%

3.

Family Expenses	Percent
Food	22%
Auto	18%
Clothing	11%
Housing	25%
Recreation	10%
Other	14%

Find the mean. Use after page 379.

4. 18, 17, 22, 48, 50

5. 102, 103, 110, 98, 120, 100

6. $ 1.79, $ 1.89, $ 1.99, $ 1.95

7. 32°, 27°, 45°, 50°, 10°, 20°, 12°

8. 87, 63, 72, 80

9. 59, 95, 83, 74, 69

Find the median. Use after page 381.

10. 18, 17, 48, 22, 50

11. 102, 103, 110, 98, 120, 100, 110

12. 102, 103, 110, 98, 120, 100

13. 78, 77, 79, 80

14. 78, 90, 62, 73, 89

15. 65, 45, 81, 57, 77, 63

Find the mode. Use after page 381.

16. 17, 16, 17, 19, 20, 14

17. 23, 44, 23, 45, 44, 22, 28

18. 96, 97, 95, 95, 96, 99, 100

19. 10, 8, 11, 8, 8, 7, 6, 8

20. 82, 59, 73, 59, 91

21. 79, 74, 62, 74, 88, 82, 62

CALCULATOR CORNER

TUG OF WAR

Players two
Materials two calculators, paper and pencils

Sample Try this with your calculators.

Jay starts with a two-digit number, such as 15, on the display. Gina starts with a four-digit number, such as 1000, on the display.

Jay must always add a whole number to the number that shows on his calculator. His answer must always be larger than his preceding number.

Gina must always subtract a whole number from the number that shows on her calculator. Her answer must always be smaller than her preceding number.

If Jay allows his number to become equal to or greater than Gina's number, he loses.

If Gina allows her number to become less than or equal to Jay's number, she loses.

	Jay		Gina
	15		1000
Play #1	+ 10	**Play #2**	− 1
	25		999
Play #3	+ 500	**Play #4**	− 10
	525		989
Play #5	+ 200	**Play #6**	− 9
	725		980
Play #7	+ 100	**Play #8**	− 80
	825		900
Play #9	+ 74	**Play #10**	− 1
	899		899

Gina loses, since her number is equal to Jay's.

Play the game on your own.

READY FOR CHAPTER 14

Give a positive or negative number for each situation.

1. a loss of \$ 148 **2.** 14,254 ft above sea level **3.** 4 hours ago

Compare. Use $<$ or $>$.

4. -3 and 7 **5.** -8 and -7 **6.** 0 and -2

Add.

7. $5 + (-3)$ **8.** $-9 + 6$ **9.** $-7 + (-4)$

10. $-19 + (-16)$ **11.** $24 + (-11)$ **12.** $-18 + 10$

Subtract.

13. $2 - 7$ **14.** $-3 - 6$ **15.** $-4 - (-5)$

16. $8 - (-8)$ **17.** $7 - 10$ **18.** $-3 - (-9)$

Multiply.

19. -4×5 **20.** $-8 \times (-6)$ **21.** $10 \times (-9)$

Divide.

22. $54 \div (-6)$ **23.** $-63 \div 7$ **24.** $-75 \div (-25)$

Application

25. Was it colder in Ft. Wayne or Indianapolis?

26. What was the difference between the high and low temperatures in each city?

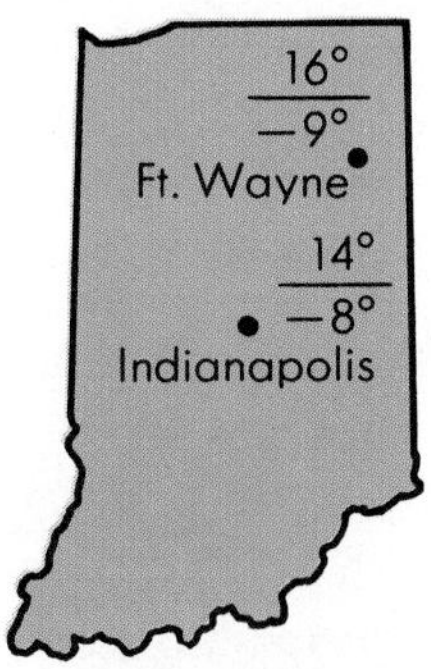

14 POSITIVE AND NEGATIVE NUMBERS

GETTING STARTED

THE GREATEST GREEN

WHAT'S NEEDED 20 squares of paper or cardboard about this size

Write the digits 0 through 9 in green on 10 squares, one digit on each square. Write the digits 0 through 9 in blue on 10 squares, one digit on each square. Put them in a box and mix well.

THE RULES

1. On a sheet of paper each player makes a scorecard. A leader draws two digits from the box. If both numbers are the same color, the players add. The sum is labeled the same color as the digits. If one number of each color is drawn, the players subtract. The difference is labeled the color of the larger number.
2. The leader continues to draw digits, one at a time. Each number is added or subtracted from the previous total, depending on the colors of the digits.
3. The game continues until five digits have been drawn. A player may stop after any number. The goal is to get the highest green number.
4. Play three games and find the total score. The player with the highest green total wins.

SAMPLE

Game 1

5 Green
<u>9</u> Green
14 Green
<u>7</u> Blue
7 Green
<u>9</u> Blue
2 Blue
<u>3</u> Blue
5 Blue

Scorecard

1.	5 Blue
2.	9 Green
Subtotal	4 Green
3.	5 Green
Total	9 Green

POSITIVE AND NEGATIVE NUMBERS

We use positive and negative numbers to describe many situations.

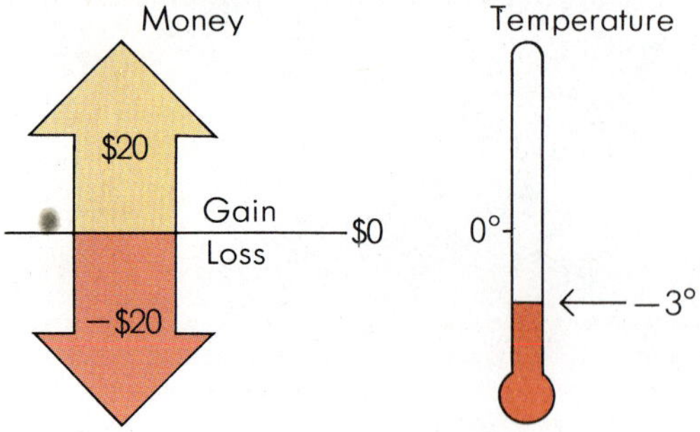

A loss of \$20 is −\$20. 3° below zero is −3°.
A gain of \$20 is \$20. 70° above zero is 70°.

EXAMPLES. Give a positive or negative number for each situation.

1. 280 ft below sea level −280
2. 4 hours from now 4
3. 2 hours before departure −2

TRY THIS

Give a positive or negative number for each situation.

1. a loss of 8 yards **2.** 6 hours ago
3. a gain of 6 pounds **4.** 10 seconds before liftoff

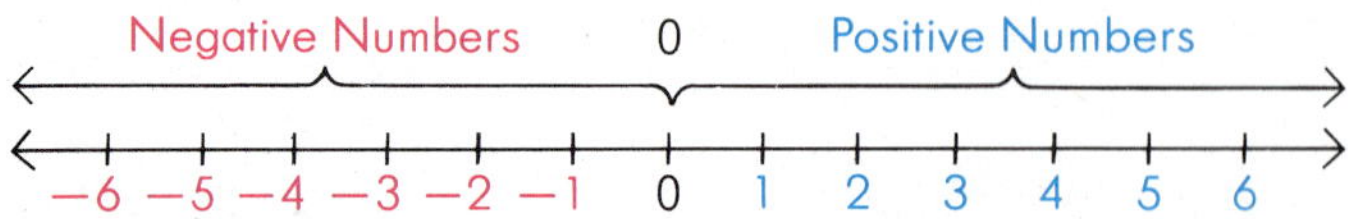

One number is less than another if it is to the left of the other on the number line. One number is greater than another if it is to the right of the other on the number line.

EXAMPLES. Compare. Use $<$ or $>$.

4. −3 and 4 $-3 < 4$ *−3 is to the left of 4*
5. −2 and −1 $-2 < -1$ *−2 is to the left of −1*
6. −4 and −6 $-4 > -6$ *−4 is to the right of −6*

TRY THIS

Compare. Use > or <.

5. −5 and 2 **6.** 0 and −3 **7.** −5 and −2

8. 0 and 3 **9.** 5 and −5 **10.** −3 and 3

PRACTICE

Give a positive or negative number for each situation.

1. a loss of $ 65
2. a loss of $ 74
3. a loss of $ 51
4. a gain of $ 34
5. a gain of $ 17
6. a gain of $23
7. 3 hours ago
8. 4 hours ago
9. 6 hours ago
10. 3 hours from now
11. 5 hours from now
12. 8 hours from now
13. gain 5 lbs
14. lose 4 lbs
15. lose 8 lbs
16. 230 ft below sea level
17. 240 ft above sea level
18. 30,000 ft above sea level
19. 15° below zero
20. 17° below zero
21. 1° below zero
22. 10 seconds before rocket liftoff
23. 10 seconds after rocket liftoff
24. 30 minutes before departure

Compare. Use > or <.

25. −2 and 4
26. −1 and 5
27. −4 and 8
28. 4 and −2
29. 5 and −1
30. 8 and −4
31. −4 and −5
32. −6 and −8
33. −7 and −9
34. −3 and −2
35. −4 and −1
36. −9 and −7
37. 5 and 0
38. −9 and 0
39. −9 and 5
40. −6 and 6
41. −7 and 7
42. −8 and −5
43. −5 and −11
44. −3 and −4
45. −6 and −5

More Practice, page 416

ADDING ON A NUMBER LINE

Add, using a number line. $2 + (-5)$

Find the first addend on a number line. → If the second addend is negative, move to the left. → The arrow stops at the answer.

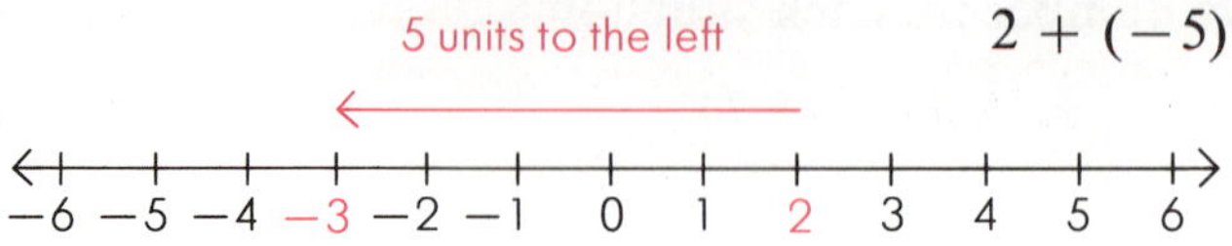

$2 + (-5) = -3$

EXAMPLE 1. Add, using a number line. $-2 + (-4)$

$-2 + (-4) = -6$ *The sum of two negative numbers is always negative.*

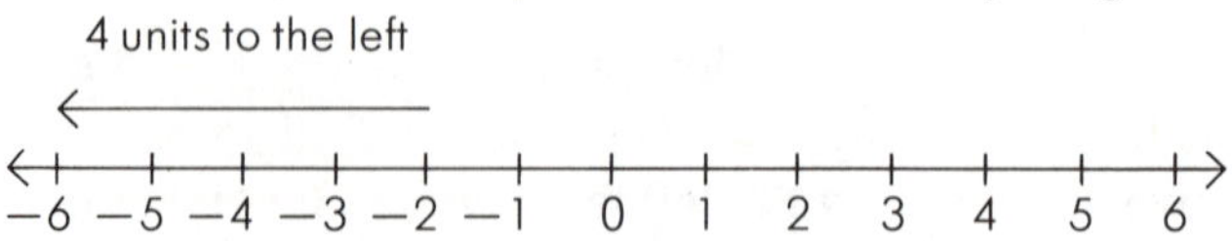

TRY THIS

Add, using a number line.

1. $3 + (-5)$ **2.** $0 + (-3)$
3. $-1 + (-3)$ **4.** $6 + (-4)$
5. $4 + (-4)$ **6.** $-3 + (-2)$

Add, using a number line. $-6 + 4$

Find the first addend on a number line. → If the second addend is positive, move to the right. → The arrow stops at the answer.

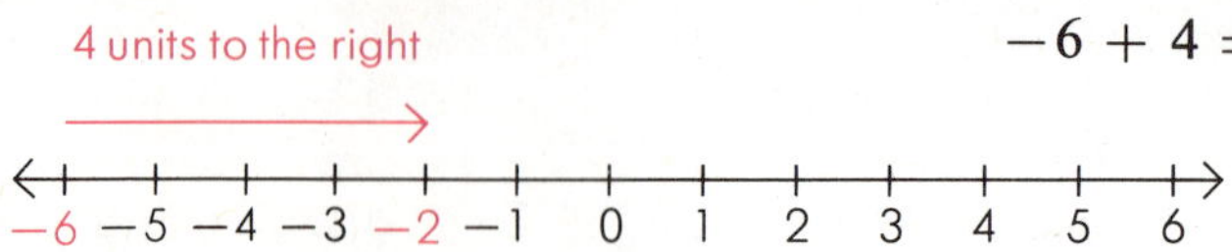

$-6 + 4 = -2$

EXAMPLE 2. Add, using a number line. $-3 + 3$

3 units to the right $-3 + 3 = 0$

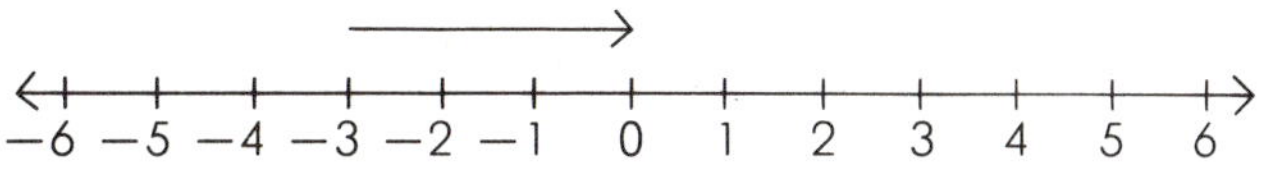

TRY THIS

Add, using a number line.

7. $-2 + 2$ **8.** $-3 + 5$ **9.** $-6 + 2$ **10.** $-6 + 4$

PRACTICE

Add, using a number line.

1. $8 + (-5)$ **2.** $7 + (-5)$ **3.** $9 + (-3)$ **4.** $6 + (-5)$

5. $7 + (-8)$ **6.** $8 + (-10)$ **7.** $6 + (-8)$ **8.** $4 + (-7)$

9. $-7 + (-5)$ **10.** $-4 + (-2)$ **11.** $-8 + (-7)$ **12.** $-6 + (-3)$

13. $-2 + 6$ **14.** $-4 + 7$ **15.** $-6 + 9$ **16.** $-5 + 6$

17. $-4 + 4$ **18.** $-2 + 2$ **19.** $-10 + 10$ **20.** $-9 + 9$

21. $5 + (-5)$ **22.** $6 + (-6)$ **23.** $8 + (-8)$ **24.** $7 + (-7)$

PROBLEM CORNER

1. The temperature one day was 5°C and it dropped 8°. What was the new temperature?

2. The temperature one day was −3°F and it rose 12°. What was the new temperature?

3. The temperature one day was −15°C and it dropped 9°. What was the new temperature?

ADDING

The sum of two positive numbers is positive.

The sum of two negative numbers is negative.

Add. $-4 + (-5)$

Add as with whole numbers. → Make the answer negative.

$4 + 5 = 9$ $\qquad$ $-4 + (-5) = -9$

EXAMPLES. Add.

1. $-7 + (-8) = -15$ $\qquad$ **2.** $-12 + (-14) = -26$

TRY THIS Add.

1. $-3 + (-8)$ $\qquad$ **2.** $-11 + (-18)$ $\qquad$ **3.** $-24 + (-13)$

The sum of a positive number and a negative number has the same sign as the addend further from 0.

Add. $-14 + 2$

Subtract as with whole numbers. → Decide which addend is further from 0. → Make the answer the same sign as that addend.

$14 - 2 = 12$

$-14 + 2$
-14 is further from 0.

$-14 + 2 = -12$

EXAMPLES. Add.

3. $-5 + 12 = 7$ $\qquad$ **4.** $-9 + 23 = 14$

5. $45 + (-18) = 27$ $\qquad$ **6.** $-27 + 15 = -12$

TRY THIS

Add.

4. $-5 + 8$ **5.** $-9 + 4$ **6.** $9 + (-2)$ **7.** $4 + (-10)$

PRACTICE

Add.

1. $-5 + 6$	**2.** $-3 + 9$	**3.** $-5 + 7$
4. $-8 + 7$	**5.** $-10 + 8$	**6.** $-8 + 2$
7. $-3 + (-6)$	**8.** $-7 + (-5)$	**9.** $-2 + (-4)$
10. $6 + (-2)$	**11.** $9 + (-7)$	**12.** $7 + (-3)$
13. $18 + (-7)$	**14.** $29 + (-8)$	**15.** $17 + (-12)$
16. $-19 + (-14)$	**17.** $-18 + (-17)$	**18.** $-26 + (-17)$
19. $-32 + 17$	**20.** $18 + (-45)$	**21.** $29 + (-41)$
22. $63 + (-18)$	**23.** $78 + (-29)$	**24.** $100 + (-42)$
25. $-65 + 47$	**26.** $-36 + 19$	**27.** $-28 + (-53)$
28. $-57 + (-72)$	**29.** $-103 + (-75)$	**30.** $-96 + 84$
31. $-35 + 250$	**32.** $-43 + 230$	**33.** $-73 + (-104)$

PROBLEM CORNER

Use positive and negative numbers to solve each problem.

1. Sal spent \$ 15 for a shirt and \$ 7 for a record. Find how much he spent in all.

2. Lorraine owed \$ 28 for a tennis racquet. She made a \$ 15 payment. Find how much she still owes.

More Practice, page 416

APPLICATION

READING CHARTS

Wind speed does not affect the actual temperature, but it can make a person feel colder. The wind chill temperature is what the temperature would have to be without wind to give the same chilling effect.

Wind Chill Temperature in °F

Wind Speed	Actual Thermometer Reading in °F								
Calm	50	40	30	20	10	0	−10	−20	−30
5 MPH	48	37	27	16	6	−5	−15	−26	−36
10 MPH	40	28	16	4	−9	−21	−33	−46	−58
15 MPH	36	22	9	−5	−18	−36	−45	−58	−72
20 MPH	32	18	4	−10	−25	−39	−53	−67	−82
25 MPH	30	16	0	−15	−29	−44	−59	−74	−88
30 MPH	28	13	−2	−18	−33	−48	−63	−79	−94
35 MPH	27	11	−4	−20	−35	−49	−67	−82	−98
40 MPH	26	10	−6	−21	−37	−53	−69	−85	−100

(Wind speeds greater than 40 mph have little additional effect.)

What is the wind chill temperature when the actual temperature is 20° F and the wind speed is 30 mph?

Locate 30 mph on the left side of the chart. Go across until you are under 20° F. The wind chill temperature is −18° F.

Find the wind chill temperatures for the following actual temperatures and wind speeds.

1. 20°; 35 mph
2. −10°; 20 mph
3. 0°; 25 mph
4. 10°; 10 mph
5. −30°; 40 mph
6. 30°; 25 mph
7. The actual temperature is 10° F and the wind speed is 20 mph. The actual temperature drops 20°. How much does the wind chill temperature drop?
8. The actual temperature is 20° F. The wind speed changes from 20 mph to 40 mph. How much does the wind chill temperature drop?

	Chicago, Ill	Denver, Colo.	Houston, Tex.	Las Vegas, Nev.	Mexico, D.F.	Miami, Fla.	Montreal, Que.	Nashville, Tenn.	New York, N.Y.	San Francisco, Calif.	Seattle, Wash.	Toronto, Ont.	Vancouver, B.C.	Winnipeg, Man.
Chicago, Ill.		996	1067	1772	2045	1329	828	446	802	2142	2013	465	2268	860
Denver, Colo.	996		1019	777	1746	2037	1815	1156	1771	1235	1307	1479	1453	1039
Houston, Tex.	1067	1019		1417	979	1190	1827	769	1608	1912	2274	1491	2420	1515
Las Vegas, Nev.	1772	777	1417		1834	2521	2591	1801	2548	564	1152	2255	1298	1704
Mexico, D.F.	2045	1746	979	1834		2169	2805	1747	2587	2291	2852	2469	2998	2421
Miami, Fla.	1329	2037	1190	2521	2169		1654	897	1308	3053	3273	1509	3419	2154
Montreal, Que.	828	1815	1827	2591	2805	1654		1074	378	2961	2685	341	2940	1450
Nashville, Tenn.	446	1156	769	1801	1747	897	1074		892	2333	2376	754	2522	1257
New York, N.Y.	802	1771	1608	2548	2587	1308	378	892		2934	2815	477	3070	1662
San Francisco, Calif.	2142	1235	1912	564	2291	3053	2961	2333	2934		808	2625	954	1919
Seattle, Wash.	2013	1307	2274	1152	2852	3273	2685	2376	2815	808		2600	146	1444
Toronto, Ont.	465	1479	1491	2255	2469	1509	341	754	477	2625	2600		2800	1310
Vancouver, B.C.	2268	1453	2420	1298	2998	3419	2940	2522	3070	954	146	2800		1490
Winnipeg, Man.	860	1039	1515	1704	2421	2154	1450	1257	1662	1919	1444	1310	1490	

MILEAGE CHART: CANADIAN, MEXICAN, AND U.S. CITIES

The time needed to drive between cities depends not only on distance, but on traffic conditions, types of roads, and speed limits.

To the nearest hour, find how long it takes to drive from city to city. Assume a maximum speed limit of 55 mph.

9. Miami to Montreal
10. Winnipeg to Houston
11. Seattle to Mexico
12. Chicago to Toronto
13. Vancouver to Denver
14. New York to San Francisco
15. Nashville to Las Vegas
16. Mexico to Winnipeg
17. Is Houston or Nashville further from Seattle? How much further?
18. Is Winnipeg or Montreal closer to Chicago? How much closer?

Jason and Sheila drove from Vancouver to New York in 6 days.

19. How many miles did they drive?
20. How many miles did they average each day?
21. Jason and Sheila used 135 gallons of gasoline. To the nearest mile, how many miles per gallon did they average?
22. If they paid an average of $ 0.95 per gallon, how much did Jason and Sheila spend for gas?

SUBTRACTING ON A NUMBER LINE

Subtract, using a number line. $6 - 8$

Start at the number being subtracted. → Move to the other number. Count the number of units moved. → If the move is to the left, the answer is negative.

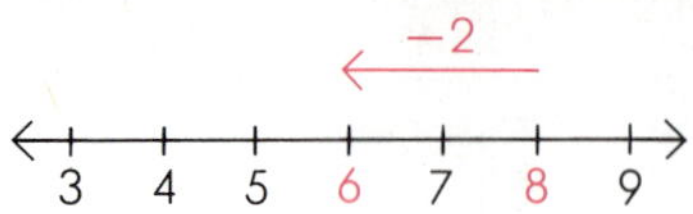

$6 - 8 = -2$

EXAMPLE 1. Subtract, using a number line. $-1 - 5$

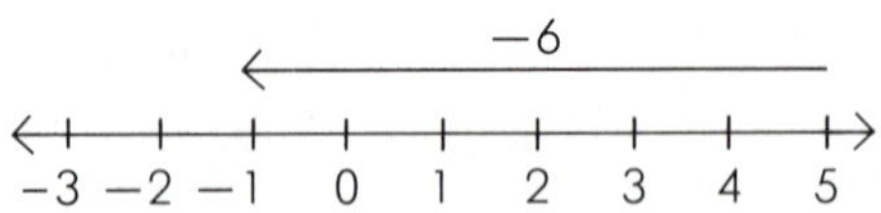

$-1 - 5 = -6$

TRY THIS

Subtract, using a number line.

1. $4 - 7$ **2.** $-2 - 6$ **3.** $-6 - 3$ **4.** $6 - 8$

Subtract, using a number line. $-3 - (-7)$

Start at the number being subtracted. → Move to the other number. Count the number of units moved. → If the move is to the right, the answer is positive.

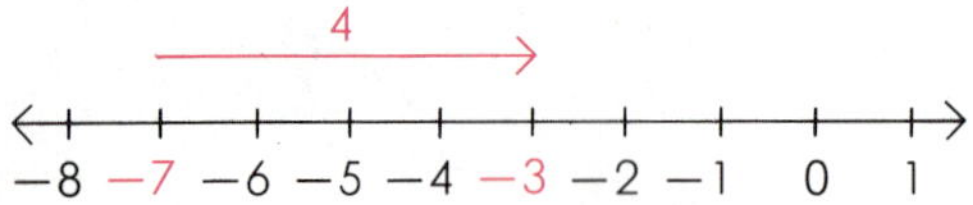

$-3 - (-7) = 4$

EXAMPLE 2. Subtract, using a number line. $4 - (-4)$

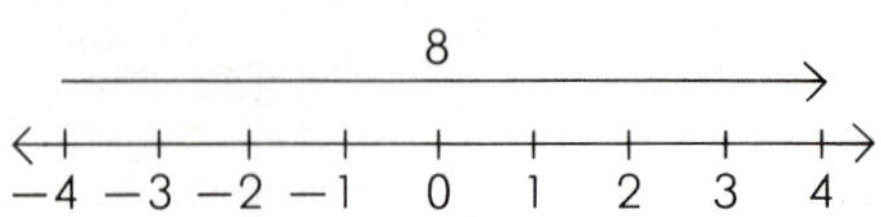

$4 - (-4) = 8$

TRY THIS Subtract, using a number line.

5. $-2 - (-3)$ **6.** $7 - (-3)$ **7.** $5 - (-6)$ **8.** $-5 - (-2)$

PRACTICE

Subtract, using a number line.

1. $2 - 5$	**2.** $1 - 5$	**3.** $3 - 5$	**4.** $5 - 9$
5. $-5 - 4$	**6.** $-3 - 5$	**7.** $-2 - 4$	**8.** $-1 - 3$
9. $-7 - (-2)$	**10.** $-4 - (-9)$	**11.** $-5 - (-3)$	**12.** $-6 - (-4)$
13. $-5 - 5$	**14.** $-2 - 2$	**15.** $-3 - 3$	**16.** $-8 - 8$
17. $8 - (-8)$	**18.** $6 - (-6)$	**19.** $10 - (-10)$	**20.** $9 - (-9)$
21. $8 - (-3)$	**22.** $7 - (-2)$	**23.** $6 - (-7)$	**24.** $5 - (-4)$
25. $0 - 5$	**26.** $0 - 4$	**27.** $0 - 10$	**28.** $0 - 8$
29. $8 - 12$	**30.** $7 - 10$	**31.** $6 - 14$	**32.** $8 - 9$
33. $-8 - 1$	**34.** $-9 - 4$	**35.** $-7 - 6$	**36.** $-9 - 2$
37. $-5 - (-5)$	**38.** $-8 - (-8)$	**39.** $-11 - (-11)$	**40.** $-12 - (-12)$

PROBLEM CORNER

1. The temperature one day was $-7°F$ and it dropped 4°. What was the new temperature?

2. The temperature one day was 9°C and it dropped 12°. What was the new temperature?

3. The temperature in St. Louis was $-3°F$. In Minneapolis it was $-7°F$. What was the difference between the temperatures of the two cities?

SUBTRACTING

Numbers like 4 and -4 are called opposites. Their sum is 0.

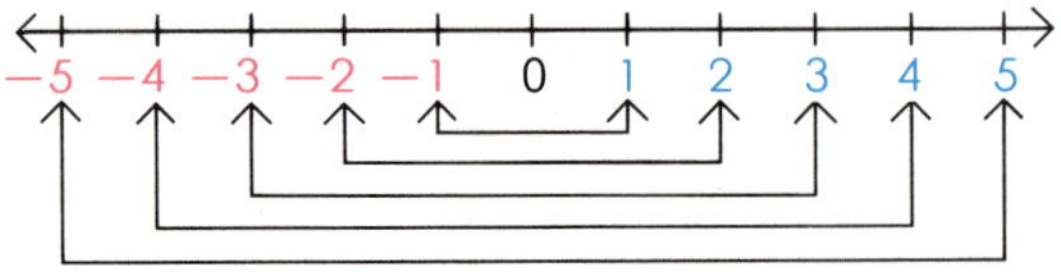

$2 + (-2) = 0$
$-1 + 1 = 0$
$5 + (-5) = 0$

EXAMPLES. Find the opposite.

1. -3 The opposite of -3 is 3.
2. 4 The opposite of 4 is -4
3. 0 The opposite of 0 is 0.

TRY THIS Find the opposite.

1. -5 **2.** -6 **3.** 6 **4.** 12 **5.** 0 **6.** 200

Subtract. $3 - 8$

Find the opposite of the number being subtracted. → Then add.

The opposite of 8 is -8.

$$3 - 8 = 3 + (-8)$$
$$= -5$$

EXAMPLES. Subtract.

4. $-5 - 4 = -5 + (-4) = -9$

5. $2 - 6 = 2 + (-6) = -4$

6. $5 - (-9) = 5 + 9 = 14$

7. $-7 - (-2) = -7 + 2 = -5$

TRY THIS Subtract.

7. $-6 - 3$ **8.** $5 - 9$ **9.** $-4 - 8$
10. $-5 - (-9)$ **11.** $-2 - (-2)$ **12.** $-6 - (-4)$

PRACTICE

1. $3 - 4$ **2.** $2 - 6$ **3.** $7 - 9$ **4.** $1 - 6$

5. $-3 - 5$ **6.** $-5 - 4$ **7.** $-2 - 7$ **8.** $-2 - 3$

9. $-8 - (-3)$ **10.** $-2 - (-7)$ **11.** $-6 - (-4)$ **12.** $-9 - (-1)$

13. $-8 - 8$ **14.** $-9 - 9$ **15.** $-12 - 12$ **16.** $-13 - 13$

17. $9 - (-9)$ **18.** $7 - (-7)$ **19.** $10 - (-10)$ **20.** $16 - (-16)$

21. $6 - 9$ **22.** $9 - 16$ **23.** $7 - 10$ **24.** $3 - 8$

25. $0 - 12$ **26.** $0 - 8$ **27.** $0 - 14$ **28.** $0 - 9$

29. $7 - (-2)$ **30.** $8 - (-3)$ **31.** $5 - (-7)$ **32.** $9 - (-4)$

33. $-7 - 6$ **34.** $-9 - 2$ **35.** $-8 - 5$ **36.** $-2 - 8$

37. $-9 - (-9)$ **38.** $-5 - (-5)$ **39.** $-7 - (-7)$

40. $28 - 53$ **41.** $65 - 79$ **42.** $15 - (-84)$

PROBLEM CORNER

Use positive and negative numbers to solve each problem.

1. Rocco has \$ 75 in his savings account and owes \$ 88 on his credit card account. What is the difference between the balances of these accounts?

2. Mae owes \$ 34 on a layaway account. She also owes \$ 17 on her gas credit card. What is the difference between the amounts Mae owes?

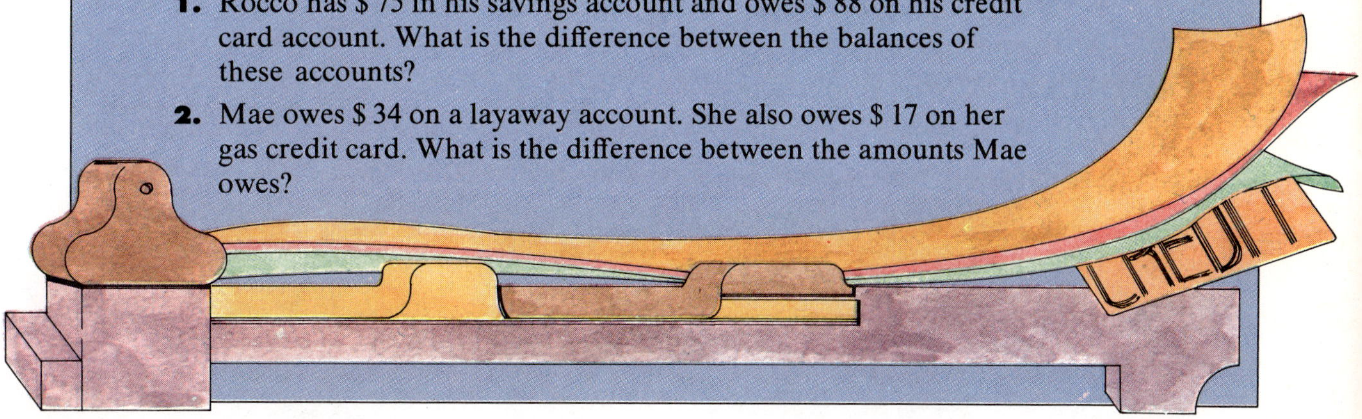

More Practice, page 417

MULTIPLYING

The product of two positive numbers is positive.

The product of two negative numbers is positive.

Multiply. $-6 \times (-9)$

Multiply as with whole numbers. → Make the answer positive.

$6 \times 9 = 54$ $\quad$ $-6 \times (-9) = 54$

EXAMPLES. Multiply.

1. $-4 \times (-5) = 20$ **2.** $-7 \times (-11) = 77$
3. $-18 \times (-6) = 108$

TRY THIS Multiply.

1. $-3 \times (-8)$ **2.** $-9 \times (-14)$ **3.** $-12 \times (-4)$

The product of a positive and a negative number is negative.

Multiply. -4×5

Multiply as with whole numbers. → Make the answer negative.

$4 \times 5 = 20$ $\quad$ $-4 \times 5 = -20$

EXAMPLES. Multiply.

4. $-3 \times 5 = -15$ **5.** $7 \times (-8) = -56$
6. $7 \times (-10) = -70$

TRY THIS Multiply.

4. -6×7 **5.** -3×6 **6.** $20 \times (-5)$

PRACTICE

Multiply.

1. $-8 \times (-2)$
2. $-2 \times (-5)$
3. $-7 \times (-6)$
4. $-9 \times (-2)$
5. $-9 \times (-5)$
6. $-8 \times (-3)$
7. $-9 \times (-8)$
8. $-10 \times (-3)$
9. $-6 \times (-15)$
10. $-8 \times (-22)$
11. $-25 \times (-8)$
12. $-35 \times (-4)$
13. $8 \times (-3)$
14. $9 \times (-5)$
15. $8 \times (-9)$
16. $3 \times (-10)$
17. -2×5
18. -8×2
19. -7×6
20. -9×2
21. -8×5
22. 9×-7
23. -10×4
24. $12 \times (-3)$
25. $15 \times (-8)$
26. -10×12
27. $-25 \times (-40)$
28. -200×5
29. $125 \times (-8)$
30. $-420 \times (-9)$
31. -42×8
32. $56 \times (-9)$
33. -86×43
34. $-95 \times (-74)$
35. $-32 \times (-41)$
36. -63×27
37. $23 \times (-35)$
38. $56 \times (-15)$
39. -24×24
40. $-34 \times (-89)$
41. $-32 \times (-49)$
42. $-58 \times (-125)$

PROBLEM CORNER

1. Rita spent \$38 each month for food during the first 6 months of the year. Find the total amount spent, using positive and negative numbers.
2. Victor spent \$29 each month for gas for 3 months. Use positive and negative numbers to find the total amount spent.

More Practice, page 417

DIVIDING

The quotient of two positive numbers is positive.

The quotient of two negative numbers is positive.

Divide. $-24 \div (-8)$

Divide as with whole numbers. → Make the answer positive.

$24 \div 8 = 3$ $\quad$ $-24 \div (-8) = 3$

EXAMPLES. Divide.

1. $-56 \div (-7) = 8$ **2.** $-34 \div (-2) = 17$

3. $-42 \div (-6) = 7$

TRY THIS Divide.

1. $-21 \div (-3)$ **2.** $-28 \div (-2)$ **3.** $-36 \div (-9)$

The quotient of a negative number and a positive number is negative.

Divide. $-48 \div 8$

Divide as with whole numbers. → Make the answer negative.

$48 \div 8 = 6$ $\quad$ $-48 \div 8 = -6$

EXAMPLES. Divide.

4. $-56 \div 7 = -8$ **5.** $28 \div (-2) = -14$

6. $-100 \div 25 = -4$

TRY THIS Divide.

4. $-24 \div 4$ **5.** $-48 \div 16$ **6.** $75 \div (-15)$

PRACTICE

Divide

1. $-8 \div (-2)$	**2.** $-10 \div (-5)$	**3.** $-15 \div (-3)$
4. $-24 \div 6$	**5.** $-30 \div 5$	**6.** $-32 \div 4$
7. $30 \div (-5)$	**8.** $36 \div (-6)$	**9.** $26 \div (-2)$
10. $-16 \div (-8)$	**11.** $-22 \div (-2)$	**12.** $-63 \div (-9)$
13. $-72 \div 9$	**14.** $-84 \div 7$	**15.** $-64 \div 2$
16. $49 \div (-7)$	**17.** $64 \div (-8)$	**18.** $25 \div (-5)$
19. $-100 \div (-50)$	**20.** $-200 \div (-40)$	**21.** $-400 \div (-80)$
22. $-300 \div 75$	**23.** $-300 \div 25$	**24.** $200 \div (-8)$
25. $90 \div (-18)$	**26.** $-98 \div 14$	**27.** $-84 \div 12$
28. $-92 \div (-23)$	**29.** $133 \div (-19)$	**30.** $-128 \div (-4)$
31. $55 \div (-11)$	**32.** $-72 \div (-12)$	**33.** $-65 \div (-13)$
34. $432 \div (-18)$	**35.** $-442 \div 17$	**36.** $-756 \div (-28)$

PROBLEM CORNER

1. The 10 members of the Bee-See Club sponsored a dance. They took in only \$ 16, but their expenses were \$ 38. If the club members decide to share the cost equally, how much does each one owe?

2. The *A Capella* Trio owes \$ 57 for uniforms. If the 3 singers divide the cost evenly, how much does each one owe?

More Practice, page 417

APPLICATION

READING MAPS

The scale on a map is used to find the actual distance between two places. First, measure the distance on the map. Then compare this measurement to the scale on the map.

Approximate the actual distances between the following cities.

1. Detroit and Montreal
2. St. Louis and Washington, D.C.
3. Vancouver and Minneapolis
4. Chihuahua and Phoenix
5. Pittsburgh and Atlanta
6. Salt Lake City and Winnipeg
7. Portland and Dallas
8. Vancouver and Chihuahua

A map may show the elevations of many landmarks and cities. What is the difference in elevation between these landmarks?

9. Lake Ontario, Mt. Cadillac
10. Mt. Robson, Mt. Rushmore
11. Pikes Peak, Grand Canyon
12. Mammoth Cave, Lake Okeechobee
13. Sierra de la Madera, Salton Sea
14. Mammoth Cave, Salton Sea
15. Both the highest and lowest points in the continental U.S. are in California. Mt. Whitney has an elevation of 4418 meters. Bad Water, California in Death Valley has an elevation of -86 meters. What is the difference in elevation between these two landmarks?
16. The highest and lowest points on the earth's surface are in Asia. Mt. Everest in Nepal has an elevation of 8848 meters. The Dead Sea in Israel and Jordan has an elevation of -397 meters. What is the difference in elevation between these two places?

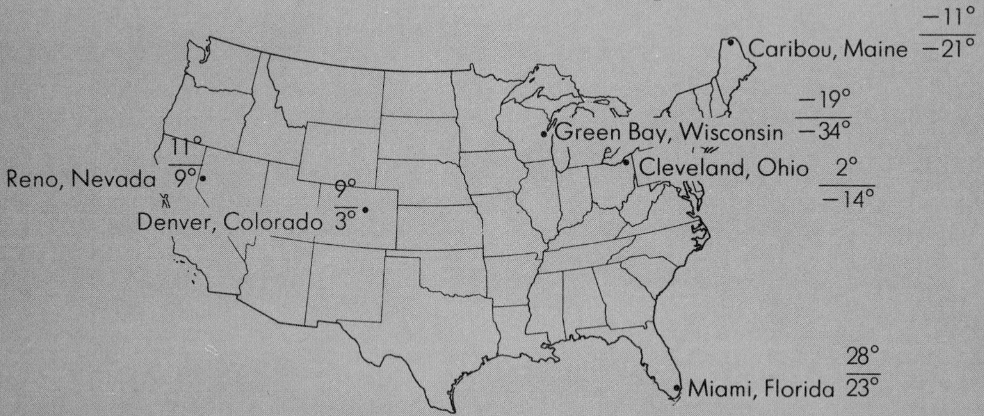

A weather map often gives high and low temperatures of an area. In this weather map each temperature is given in degrees Celcius.

$\frac{29°}{16°}$ ← High temperature
← Low temperature

For each city, what is the difference between the high and low temperatures?

17. Green Bay, Wisconsin
18. Reno, Nevada
19. Cleveland, Ohio
20. How much warmer was it in Reno, Nevada than in Denver, Colorado?
21. Was it colder in Caribou, Maine or Green Bay, Wisconsin? How much colder?
22. How much colder was it in Cleveland, Ohio than in Miami, Florida?
23. In which cities did the temperature not rise above 0° C?

CAREER CAREER CAREER CAREER CAREER

WELL DRILLER

Oil wells are drilled through land and ocean. Neal Morris works for World Petroleum Corporation. He builds oil-drilling derricks on land and from platforms on the ocean.

How far are these layers of material from the top of the oil rig?

1. 1st layer of rock
2. 2nd layer of rock
3. limestone and methane layer
4. petroleum layer

How far apart are these layers?

5. gravel and clay from petroleum
6. 1st rock layer from the 2nd
7. limestone and methane gas from the 2nd rock layer

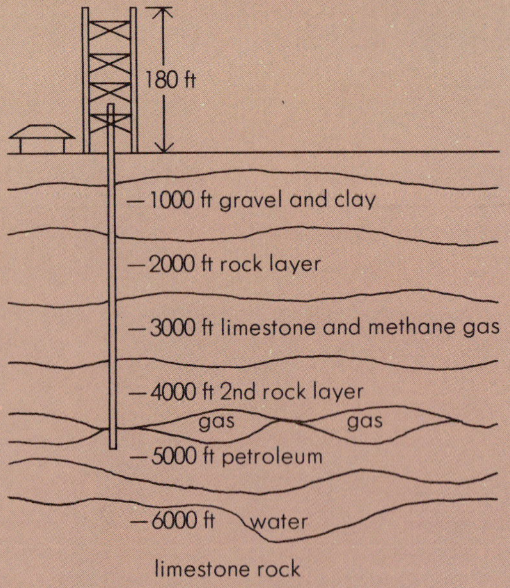

Well depth is measured from ground level. It is common to have wells drilled in excess of 15,000 ft.

Neal is working on a well with a target depth of 24,087 ft. The well is in Tulsa, Oklahoma which is 668 ft above sea level.

8. How far below sea level is the target depth?
9. After one month, the drill reaches 1784 ft. How far does Neal have to drill to get to the target depth?
10. Neal expects the drilling to take 14 months. How many feet per month is this?

The drill bit needs to be cooled, since it heats up from friction and the heat in the ground. Neal cools the drill bit by pumping mud down the drill rod. At a depth of 12,500 ft the well temperature was 130° F. What is the difference between the well temperature and each of the following surface temperatures?

11. −12° F 12. 32° F (Freezing) 13. 50° F 14. 78° F

Neal travels to many states and countries checking on drilling sites. To what elevation was each well drilled?

	Location	Well Depth	Elevation of the Field
15.	Tierra del Fuego, Argentina	17,000 ft	6500 ft
16.	Prudhoe Bay, Alaska	15,000 ft	650 ft
17.	Galveston, Texas	20,000 ft	300 ft
18.	Ghawar, Saudi Arabia	6920 ft	650 ft
19.	Gippsland Shelf, Australia	7200 ft	−600 ft
20.	Swan Hills, Alberta, Canada	8100 ft	6000 ft

CHAPTER 14 REVIEW

Give a positive or negative number for each situation. p. 394

1. 500 ft below sea level. **2.** 5 hours from now **3.** 9° below zero

Compare. Use $<$ or $>$. p. 394

4. -6 and -5 **5.** 8 and -4 **6.** -4 and 0

Add. p. 398

7. $-3 + (-4)$ **8.** $6 + (-2)$ **9.** $-7 + 1$

10. $19 + (-14)$ **11.** $-23 + 10$ **12.** $25 + (-17)$

Subtract. p. 404

13. $3 - 8$ **14.** $-5 - (-3)$ **15.** $-4 - 5$

16. $5 - (-7)$ **17.** $-4 - 8$ **18.** $-10 - (-12)$

Multiply. p. 406

19. $-6 \times (-7)$ **20.** $8 \times (-3)$ **21.** -5×4

Divide. p. 408

22. $-28 \div (-7)$ **23.** $36 \div (-9)$ **24.** $-80 \div 10$

Application p. 410

25. Was it colder in Juneau or Anchorage?

26. What was the difference between the high and low temperatures in each city?

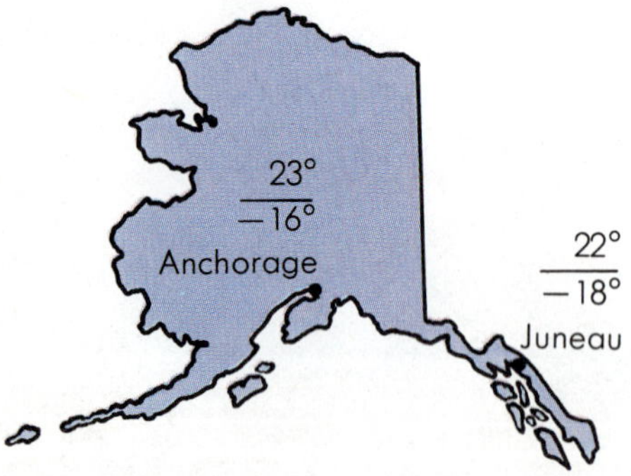

CHAPTER 14 TEST

Give a positive or negative number for each situation.

1. a gain of $ 678 **2.** 135 ft below sea level **3.** 7 hours ago

Compare. Use < or >.

4. -9 and -10 **5.** -7 and 0 **6.** 10 and -3

Add.

7. $-2 + 5$ **8.** $-4 + (-6)$ **9.** $-8 + 3$

10. $-18 + 23$ **11.** $14 + (-18)$ **12.** $-17 + (-21)$

Subtract.

13. $-7 - 5$ **14.** $5 - 6$ **15.** $-2 - (-3)$

16. $5 - (-8)$ **17.** $-3 - 11$ **18.** $-12 - (-8)$

Multiply.

19. $7 \times (-9)$ **20.** $-5 \times (-10)$ **21.** -8×7

Divide.

22. $72 \div (-9)$ **23.** $-64 \div 8$ **24.** $-56 \div (-7)$

Application

25. Was it colder in Syracuse or Buffalo?

26. What was the difference between the high and low temperatures in each city?

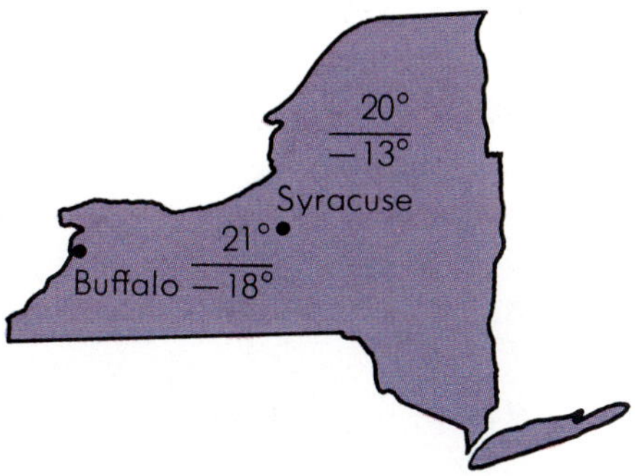

MORE PRACTICE

Give a positive or negative number for each situation. Use after page 395.

1. a loss of \$45   **2.** a gain of \$68

3. 5 hours ago   **4.** 6 hours from now

5. 180 ft below sea level   **6.** 10,500 ft above sea level

7. 45° above zero   **8.** 16° below zero

9. 8 seconds before liftoff   **10.** 10 seconds after liftoff

11. lose 3 pounds   **12.** gain 7 yards

Compare. Use $<$ or $>$. Use after page 395.

13. -3 and 7   **14.** 7 and -3 **15.** -8 and -10

16. 0 and -3 **17.** 0 and 6   **18.** -7 and 0

19. -4 and 4   **20.** 5 and -5 **21.** -4 and -6

22. -10 and -9 **23.** -12 and -1 **24.** 8 and -3

25. -35 and -33 **26.** 14 and -1 **27.** -6 and 16

28. 0 and -1 **29.** -20 and -22 **30.** -7 and 7

Add. Use after page 399.

31. $-8 + 6$ **32.** $-2 + (-4)$ **33.** $-2 + 10$

34. $8 + (-8)$ **35.** $-9 + 2$ **36.** $-3 + 7$

37. $-6 + 7$ **38.** $-11 + 9$ **39.** $-3 + (-5)$

40. $-14 + (-8)$ **41.** $7 + (-5)$ **42.** $-20 + (-15)$

43. $100 + (-53)$ **44.** $-80 + (-64)$ **45.** $-97 + 34$

46. $-58 + (-22)$ **47.** $-72 + (-109)$ **48.** $-47 + (-207)$

Subtract. Use after page 405.

1. $-8 - 6$ **2.** $-2 - (-4)$ **3.** $-2 - 10$

4. $-5 - 9$ **5.** $-3 - 7$ **6.** $-3 - (-9)$

7. $-12 - (-5)$ **8.** $-13 - 7$ **9.** $-6 - (-8)$

10. $7 - (-5)$ **11.** $-20 - (-15)$ **12.** $17 - 35$

13. $1 - 100$ **14.** $10 - (-20)$ **15.** $-1 - 100$

16. $-17 - 19$ **17.** $14 - 75$ **18.** $-23 - (-19)$

Multiply. Use after page 407.

19. $-5 \times (-8)$ **20.** -13×5 **21.** $-6 \times (-8)$

22. $-6 \times (-6)$ **23.** $-5 \times (-5)$ **24.** $6 \times (-22)$

25. -10×14 **26.** $9 \times (-9)$ **27.** $8 \times (-8)$

28. $-120 \times (-14)$ **29.** $-35 \times (-45)$ **30.** -44×85

31. -75×19 **32.** $46 \times (-18)$ **33.** -25×25

34. -24×6 **35.** $-15 \times (-21)$ **36.** $12 \times (-8)$

37. $-140 \times (-7)$ **38.** -13×195 **39.** $-39 \times (-56)$

Divide. Use after page 409.

40. $-48 \div 12$ **41.** $-72 \div (-6)$ **42.** $-54 \div 6$

43. $48 \div (-16)$ **44.** $49 \div (-7)$ **45.** $-64 \div 8$

46. $-98 \div 49$ **47.** $-104 \div 4$ **48.** $-125 \div (-25)$

49. $400 \div (-10)$ **50.** $-300 \div 75$ **51.** $-58 \div (-29)$

52. $-456 \div 12$ **53.** $-232 \div 4$ **54.** $306 \div (-17)$

55. $255 \div (-17)$ **56.** $-196 \div (-7)$ **57.** $-308 \div 28$

CALCULATOR CORNER

MYSTERY NUMBERS

Mystery Number A

Clue 1 If you enter 14 on your calculator and press + 10 = over and over, Mystery Number A will appear.

Clue 2 If you start with 0 on your calculator and press + 4 = over and over, Mystery Number A will appear.

List some numbers that Mystery Number A could be.

Clue 3 Mystery Number A is the largest such number less than 1000.

What is Mystery Number A?

Mystery Number B

Clue 1 If you start with 0 on your calculator and press + 4 = over and over, Mystery Number B will appear.

Clue 2 If you enter 13 on your calculator and press + 5 = over and over, Mystery Number B will appear.

List some numbers that Mystery Number B could be.

Clue 3 Mystery Number B is the largest such number less than 30,000.

What is Mystery Number B?

Use your calculator to make up several mystery number problems.

READY FOR CHAPTER 15

Evaluate each expression when $x = -2$ and $y = 3$.

1. $2x + y$ **2.** $x - 3y$ **3.** $-4xy$

Solve.

4. $x + 8 = 5$ **5.** $t - 7 = 15$ **6.** $m + \frac{2}{3} = \frac{3}{4}$

Solve.

7. $8x = -48$ **8.** $-7x = -63$ **9.** $\frac{t}{8} = -9$

Solve.

10. $5x + 4 = 39$ **11.** $-2x + 40 = 28$ **12.** $1.3x - 7 = 0.28$

Solve.

13. Eight plus what number is -48?

14. The length of a rectangle is 18 meters and the area is 252 square meters. What is the width?

15. A bicycle is on sale for \$ 60. This is 80% of the original price. What was the original price?

16. An antique clock was sold for \$ 450. This was 5 times the original price. What was the original price?

Graph the equation.

17. $y = 2 + x$

Application

18. Ben keeps \$ 1000 in a savings account for 4 months at 6%. Find the simple interest.

19. Lisa still owes \$ 800 on her credit card account. Interest is $1\frac{1}{2}\%$ per month. How much interest will she owe next month? How much will she owe in all?

15 EXPRESSIONS AND EQUATIONS

GETTING STARTED

EXPRESS IT

WHAT'S NEEDED 31 squares of paper or cardboard. Write the numbers −15 through 15 on the squares, one number on each square. Put them in a box and mix well.

THE RULES

1. On a sheet of paper each player makes a scorecard. The players each draw a number from the box.
2. The leader writes an expression from the list on the blackboard.
3. Each player evaluates the expression, using his or her number. The players record their answers on their scorecards.
4. Each round the leader writes a new expression on the blackboard.

   Play three rounds and total the scores. The player with the highest total wins.

EXPRESSIONS

$-2x - 3$

$5 + 4x$

$3 - 8x$

$7x - 5$

$-6x + 9$

$\frac{x}{4} - 3$

$5 - \frac{x}{3}$

$9x - 1$

$0.3x - 1.5$

$-4x + 10$

$7 - 0.6x$

$4 - \frac{x}{6}$

$5x - 9$

$8 - \frac{x}{5}$

$8x + 3$

$0.4x + 1$

SAMPLE Les has the number −3.

Round 1	Round 2	Round 3
$-2x - 3$	$5 - \frac{x}{3}$	$0.4x + 1$
$(-2)(-3) - 3$	$5 - \frac{-3}{3}$	$0.4(-3) + 1$
$6 - 3 = 3$	$5 - (-1)$	$-1.2 + 1 = -0.2$
	$5 + 1 = 6$	

Scorecard

1.	3
2.	6
3.	−0.2
Total	8.8

EVALUATING EXPRESSIONS

Evaluate the expression, $a + b - 8$, when $a = 2$ and $b = -3$.

Replace each letter with its value. → Simplify.

$a + b - 8$ $\quad$ $2 + (-3) - 8$

$2 + (-3) - 8$ $\quad$ $-1 - 8 = -9$

EXAMPLE 1. Evaluate $2x - y - 54$ when $x = 4$ and $y = -3$.

$$\begin{aligned} 2x - y - 54 &= 2 \times 4 - (-3) - 54 \\ &= 8 - (-3) - 54 \\ &= 8 + 3 - 54 \\ &= 11 - 54 \\ &= -43 \end{aligned}$$

Replace x with 4 and y with −3. Simplify.

TRY THIS

Evaluate each expression when $a = -4$ and $b = 9$.

1. $a - b + 29$ **2.** $3b - a - 21$ **3.** $2b - a - 55$

We write ab as an abbreviation for $a \times b$ or $a \cdot b$.

EXAMPLE 2. Evaluate $-9ab$ when $a = 3$ and $b = -6$.

$$\begin{aligned} -9ab &= -9 \cdot 3 \cdot (-6) \\ &= (-27) \cdot (-6) \\ &= 162 \end{aligned}$$

Replace a with 3 and b with −6. Multiply.

EXAMPLE 3. Evaluate $-4ab + 9$ when $a = -8$ and $b = -7$.

$$\begin{aligned} -4ab + 9 &= (-4) \cdot (-8) \cdot (-7) + 9 \\ &= 32 \cdot (-7) + 9 \\ &= -224 + 9 = -215 \end{aligned}$$

Replace a with −8 and b with −7. Multiply first. Then add.

TRY THIS

Evaluate each expression when $x = -6$ and $y = 10$.

4. $-5xy$ **5.** $7xy - 4$ **6.** $-2xy - 150$

PRACTICE

Evaluate each expression when $x = 4$.

1. $x + 8$ **2.** $5x$ **3.** $9 - x$ **4.** $16 \div x$

Evaluate each expression when $y = -3$.

5. $15 \div y$ **6.** $19 - y$ **7.** $-8y$ **8.** $2y + 10$

Evaluate each expression when $a = 5$ and $b = -6$.

9. ab **10.** $5ab$ **11.** $\frac{1}{2}ab$ **12.** $-3ab$

13. $a + b$ **14.** $a - b$ **15.** $b - a$ **16.** $b + a$

17. $a - b - 27$ **18.** $a + b + 45$ **19.** $89 - a + b$ **20.** $100 + a - b$

Evaluate each expression when $r = 8$ and $d = 16$.

21. $3.14d$ **22.** $6.28r$ **23.** $3.14 \times r \times r$ **24.** $0.785 \times d \times d$

Evaluate each expression when $b = 14$ and $h = 15$.

25. bh **26.** $\frac{1}{2}bh$ **27.** $\frac{1}{2} \times (b + h)$ **28.** $\frac{1}{2}b + h$

Evaluate each expression when $x = -3$ and $y = -4$.

29. $2xy - 16$ **30.** $51 + x + y$ **31.** $-2 - 3xy$

32. $-5xy - 3y$ **33.** $2x + y - x$ **34.** $5y - y + 4y$

35. $0.3x + 0.4y$ **36.** $1.2x - 3.2y$ **37.** $5.01x + y$

PROBLEM CORNER

1. Regina traveled for 6 hours at 88 km per hour. How far did she travel? (Hint: $d = r \cdot t$)

2. Mario traveled for 7 hours at 78 km per hour. How far did he travel?

APPLICATION

INTEREST

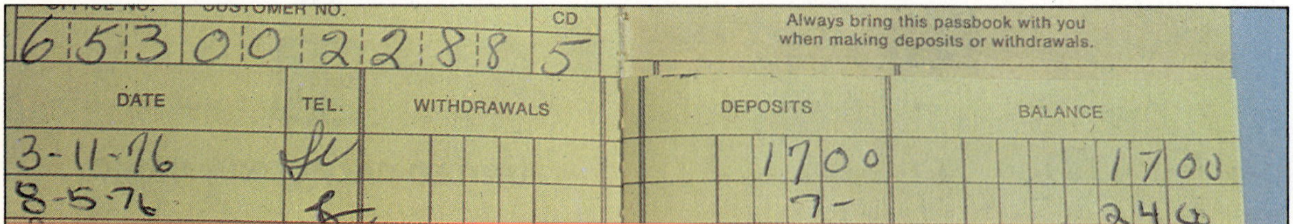

Interest is paid to the lender for the use of money. When you borrow money, you must pay interest for the use of the money. If you put money in a savings account, the bank pays you interest for the use of the money.

$I = Prt$ is the formula for simple interest.
I = interest
P = the principal, the amount loaned or borrowed
r = interest rate
t = length of time, in years

Julio put \$ 200 in a savings account for 6 months at 8% interest. Find the simple interest.

$$\begin{aligned} I &= Prt \\ &= \$200 \times 8\% \times \tfrac{6}{12} \qquad 6 \text{ months} = \tfrac{6}{12}\text{yr} = \tfrac{1}{2}\text{yr} \\ &= \$200 \times 0.08 \times \tfrac{1}{2} \\ &= \$200 \times 0.04 \\ &= \$80 \end{aligned}$$

Find the simple interest.

1. Sara kept \$ 100 in a savings account at $7\frac{1}{2}\%$ for 1 year.
2. Bernie bought a \$ 500 savings certificate for 6% that matures in 6 months.
3. Martha borrowed \$ 350 for 4 months at 8%.
4. Larry borrowed \$ 500 for 2 months at 9%.
5. Vera deposited \$ 1000 in a savings account for 1 year at 5.5%.
6. Fred put \$ 150 in a savings account for 2 years at $6\frac{1}{4}\%$.

Compound interest is money paid on the principal and the interest already earned. Interest compounded semiannually means that there are 2 payment periods per year. Interest compounded quarterly means that interest is paid 4 times per year.

Sumi has \$ 100 in a savings account at 6% interest compounded semiannually. Find how much is in her account after 1 year. Find the compound interest.

1st payment period
$$
\begin{aligned}
I &= Prt \\
&= \$\,100 \times 6\% \times \tfrac{1}{2} \\
&= \$\,100 \times 0.06 \times \tfrac{1}{2} \\
&= \$\,100 \times 0.03 = \$\,3
\end{aligned}
$$

2nd payment period
$$
\begin{aligned}
P &= \$\,100 + \$\,3 = \$\,103 \\
I &= Prt \\
&= \$\,103 \times 6\% \times \tfrac{1}{2} \\
&= \$\,103 \times 0.06 \times \tfrac{1}{2} \\
&= \$\,103 \times 0.03 = \$\,3.09
\end{aligned}
$$

After 1 year $P = \$\,103 + \$\,3.09 = \$\,106.09$

The amount in the account is \$ 106.09

$$
\begin{aligned}
\text{Compound Interest} &= \text{Compound Amount} - \text{Original Principal} \\
&= \$\,106.09 - \$\,100 = \$\,6.09
\end{aligned}
$$

The compound interest is \$ 6.09

For the given amount and interest rate, compounded semiannually, find how much is in each account after 1 year. Find the compound interest.

7. \$ 400; 8%
8. \$ 1000; 7%
9. \$ 800; 6%
10. \$ 1500; 5%

Find the compound amount and compound interest. The interest is compounded annually.

11. \$ 500 at 7% for 3 years
12. \$ 700 at 8% for 3 years
13. \$ 200 at 6.5% for 2 years
14. \$ 1200 at $7\frac{1}{2}\%$ for 2 years
15. Suppose in #11–14 the interest was not compounded. Find the simple interest for each problem. How much more interest was earned by compounding the interest?

THE ADDITION PRINCIPLE

The number 2 is a solution of the equation $x + 5 = 7$.
When we replace x with 2, we get a true equation: $2 + 5 = 7$

We use the following principle to solve equations.

THE ADDITION PRINCIPLE

If an equation $a = b$ is true, then $a + c = b + c$ is true for any number c.

EXAMPLE 1. Solve. $x + 9 = -2$

$$x + 9 = -2$$

$$x + 9 + (-9) = -2 + (-9)$$ *Add −9 on both sides to get rid of the 9.*

$$x + 0 = -2 + (-9)$$

$$x = -11$$ *The solution is −11.*

Check: $x + 9 = -2$

$-11 + 9$	-2
-2	

Replace x with the solution, −11.

TRY THIS Solve.

1. $x + 8 = -4$ **2.** $x + 14 = 25$ **3.** $x + 11 = 8$

EXAMPLE 2. Solve. $y - 7 = 15$

$$y - 7 = 15$$

$$y - 7 + 7 = 15 + 7$$ *Add 7 on both sides to get rid of the −7.*

$$y + 0 = 22$$

$$y = 22$$ *The solution is 22.*

Check: $y - 7 = 15$

$22 - 7$	15
15	

Replace y with the solution, 22.

TRY THIS Solve.

4. $y - 8 = 27$ **5.** $m - 5 = -4$ **6.** $p - 6 = -9$

PRACTICE

Solve.

1. $x + 2 = 6$	**2.** $x + 3 = 9$	**3.** $x + 4 = 10$
4. $y + 9 = 4$	**5.** $m + 8 = 2$	**6.** $t + 10 = 5$
7. $t + 12 = -7$	**8.** $y + 11 = -4$	**9.** $x + 14 = -8$
10. $x + 39 = 58$	**11.** $y + 59 = 86$	**12.** $m + 47 = 83$
13. $x - 8 = 3$	**14.** $x - 9 = 15$	**15.** $x - 7 = 14$
16. $p - 8 = -3$	**17.** $q - 9 = -15$	**18.** $t - 7 = -14$
19. $x - 2.3 = 7.4$	**20.** $x - 3.7 = 8.4$	**21.** $x - 2.6 = 8.3$
22. $19 + x = 11$	**23.** $17 + y = 26$	**24.** $22 + p = 14$
25. $234 + t = 709$	**26.** $567 + y = 1123$	**27.** $t - 899 = 1045$
28. $x - \frac{5}{6} = \frac{7}{8}$	**29.** $x - \frac{1}{3} = \frac{1}{7}$	**30.** $r + \frac{2}{3} = 1$

PROBLEM CORNER

1. Julio can finish his experiment by solving the equation $x - 16 = 9$, where x represents the number of milliliters of water. How much water does he need?

2. To find the amount of salt solution, Rosa uses the formula $27 + y = 33$. In the equation y represents the number of liters of solution. How much salt solution is needed?

THE MULTIPLICATION PRINCIPLE

Here is another principle for solving equations.

THE MULTIPLICATION PRINCIPLE

If an equation $a = b$ is true,
then $a \cdot c = b \cdot c$ is true for any number c.

EXAMPLE 1. Solve. $-8x = 32$

$$-8x = 32$$

$$-\frac{1}{8} \cdot (-8x) = -\frac{1}{8} \cdot 32$$ *Multiply by $-\frac{1}{8}$ on both sides to get rid of the -8.*

$$1x = -4$$

$$x = -4$$ *The solution is -4.*

Check: $-8x = 32$

$(-8) \cdot (-4)$	32
32	

Replace x with the solution, -4.

TRY THIS

Solve.

1. $-6x = 48$ **2.** $12x = -72$ **3.** $5x = -35$

EXAMPLE 2. Solve. $\frac{m}{7} = 38$

$$\frac{m}{7} = 38$$

$$\frac{1}{7}m = 38$$ $\frac{m}{7} = \frac{1}{7}m$

$$7 \cdot \frac{1}{7}m = 7 \cdot 38$$ *Multiply by 7 on both sides to get rid of the $\frac{1}{7}$.*

$$1m = 266$$

$$m = 266$$ *The solution is 266.*

Check: $\frac{m}{7} = 38$

$\frac{266}{7}$	38
38	

Replace m with the solution, 266.

TRY THIS

Solve.

4. $\frac{m}{9} = 45$ **5.** $\frac{t}{8} = -13$ **6.** $\frac{r}{6} = -14$

PRACTICE

Solve.

1. $6x = 36$	**2.** $3x = 39$	**3.** $4x = 28$
4. $9x = -36$	**5.** $7x = -49$	**6.** $9x = -63$
7. $-8m = 64$	**8.** $-9y = 81$	**9.** $-7t = 70$
10. $-12x = -72$	**11.** $-15x = -105$	**12.** $-16x = -80$
13. $\frac{t}{2} = 15$	**14.** $\frac{y}{4} = 9$	**15.** $\frac{n}{7} = 8$
16. $\frac{x}{14} = 12$	**17.** $\frac{m}{13} = 15$	**18.** $\frac{y}{18} = 15$
19. $\frac{x}{9} = 8$	**20.** $\frac{y}{3} = 8$	**21.** $\frac{t}{6} = 9$
22. $\frac{x}{10} = -18$	**23.** $\frac{t}{12} = -16$	**24.** $\frac{m}{14} = -15$
25. $3.3y = 66$	**26.** $2.8t = 56$	**27.** $7.4p = 74$
28. $6.3y = 44.1$	**29.** $38.7x = 309.6$	**30.** $29.4m = 235.2$

PROBLEM CORNER

Beto finds the total amount he owes for his stereo by solving the equation $\frac{m}{12} = 15$. He lets m represent the total amount of money. How much does he owe for the stereo?

USING THE PRINCIPLES TOGETHER

When we need to use both the Addition and Multiplication Principles, we add first and then multiply.

EXAMPLE 1. Solve. $3x + 4 = 13$

$$\begin{aligned} 3x + 4 &= 13 \\ 3x + 4 + (-4) &= 13 + (-4) && \textit{Add } -4 \textit{ on both sides to get } 3x \textit{ alone.} \\ 3x + 0 &= 9 \\ 3x &= 9 \\ \frac{1}{3} \cdot 3x &= \frac{1}{3} \cdot 9 && \textit{Multiply by } \tfrac{1}{3} \textit{ on both sides.} \\ x &= 3 && \textit{The solution is 3.} \end{aligned}$$

Check: $3x + 4 = 13$

$3 \cdot 3 + 4$	13
$9 + 4$	
13	

Replace x with the solution, 3.

TRY THIS

Solve.

1. $9x + 6 = 51$ **2.** $8y + 7 = 79$ **3.** $3t + 12 = 45$

EXAMPLE 2. Solve. $-5x - 6 = 19$

$$\begin{aligned} -5x - 6 &= 19 \\ -5x - 6 + 6 &= 19 + 6 && \textit{Add 6 on both sides to get } -5x \textit{ alone.} \\ -5x + 0 &= 25 \\ -5x &= 25 \\ -\frac{1}{5} \cdot (-5x) &= -\frac{1}{5} \cdot 25 && \textit{Multiply by } -\tfrac{1}{5} \textit{ on both sides.} \\ x &= -5 && \textit{The solution is } -5. \end{aligned}$$

Check: $-5x - 6 = 19$

$-5 \cdot (-5) - 6$	19
$25 - 6$	
19	

Replace x with the solution, −5.

TRY THIS

Solve.

4. $-8x - 4 = 28$ **5.** $3x - 4 = 8$ **6.** $-6x - 14 = 34$

PRACTICE

Solve.

1. $5x + 6 = 31$	**2.** $3x + 6 = 30$	**3.** $4x + 7 = 27$
4. $4x - 6 = 34$	**5.** $6x - 3 = 15$	**6.** $8x - 7 = 41$
7. $-6t + 5 = 59$	**8.** $-2m + 5 = 55$	**9.** $-3n + 9 = 72$
10. $7k + 2 = -54$	**11.** $5y + 4 = -41$	**12.** $9t + 8 = -64$
13. $-10x - 2 = 78$	**14.** $-9x - 5 = 58$	**15.** $-7x - 8 = 76$
16. $12y - 6 = 30$	**17.** $15t - 9 = 66$	**18.** $14m - 4 = 66$
19. $3.2x - 5 = 7$	**20.** $1.6y + 7 = 11$	**21.** $6.4t + 5 = 13$
22. $\frac{x}{7} - 3 = 9$	**23.** $\frac{t}{4} + 8 = 3$	**24.** $\frac{m}{9} - 2 = 10$
25. $0.5x + 0.8 = 1.6$	**26.** $0.4x - 0.9 = 0.3$	**27.** $0.2x - 0.6 = 1.4$
28. $\frac{y}{6} - 19 = 2.3$	**29.** $\frac{x}{4} - 42 = 5.6$	**30.** $\frac{t}{5} + 39 = -42$

PROBLEM CORNER

What you could wish for if you had one wish: 10 . . .

To answer the riddle, solve each equation.
Then cross out the false answers in the chart.

$2x - 12 = 8$
$2y + 4 = 18$
$6w - 9 = 9$
$5t + 6 = 26$
$-3m + 10 = -5$
$-3a + 9 = 6$

$x = 10$	W
$x = 2$	A
$y = 7$	I
$y = 1$	P
$w = 3$	S
$w = 5$	T
$t = 4$	H
$t = 10$	B
$m = 5$	E
$a = 4$	T
$a = 1$	S

SOLVING PROBLEMS WITH EQUATIONS

To solve problems we can translate to an equation and solve.

EXAMPLE 1. 4 times what number plus 7 is 27?

$$4 \times n + 7 = 27$$ *Translate.*

$$\begin{aligned} 4n + 7 &= 27 && \textit{Write the equation.} \\ 4n + 7 + (-7) &= 27 + (-7) && \textit{Add } -7 \textit{ on both sides} \\ 4n + 0 &= 20 && \textit{to get 4n alone.} \\ 4n &= 20 \\ \frac{1}{4} \cdot 4n &= \frac{1}{4} \cdot 20 && \textit{Multiply by } \tfrac{1}{4} \textit{ on both sides.} \\ 1n &= 5 \\ n &= 5 && \textit{The solution is 5.} \end{aligned}$$

4 times 5 plus 7 is 27.

EXAMPLE 2. Marcel received the same number of points on 6 tests. On another test he received 86 points. His total number of points was 518. How many points did Marcel receive on each of the 6 tests?

6 times number of points per test plus 86 is total points

$$6 \times n + 86 = 518$$

$$\begin{aligned} 6n + 86 &= 518 && \textit{Write the equation.} \\ 6n + 86 + (-86) &= 518 + (-86) && \textit{Add } -86 \textit{ on both sides} \\ 6n &= 432 && \textit{to get 6n alone.} \\ \frac{1}{6} \cdot 6n &= \frac{1}{6} \cdot 432 && \textit{Multiply by } \tfrac{1}{6} \textit{ on both sides.} \\ 1n &= 72 \\ n &= 72 && \textit{The solution is 72.} \end{aligned}$$

Marcel received 72 points on each test.

TRY THIS

Solve.

1. -6 times what number minus 8 is 46?
2. Cici earned the same amount of money for 4 days. Another day she earned \$ 18. She earned \$ 114 in all. How much did she earn on each of the 4 days?

PRACTICE

Solve.

1. What number plus 9 is 13?
2. What number plus 7 is 22?
3. What number plus 13 is 8?
4. What number plus 19 is 11?
5. What number minus 7 is 11?
6. What number minus 8 is 19?
7. 5 times what number is 80?
8. 8 times what number is 96?
9. 9 times what number is -54?
10. 7 times what number is -63?
11. 9 times what number plus 6 is 51?
12. 8 times what number plus 7 is 79?
13. -3 times what number minus 9 is -72?
14. -7 times what number minus 8 is 76?
15. When 18 is subtracted from six times a certain number, the result is 96. What is the number?
16. When 28 is added to five times a certain number, the result is 232. What is the number?
17. The width of a rectangle is 7 cm and the area is 84 square centimeters. What is the length?
18. The length of a rectangle is 27 ft and the area is 108 square ft. What is the width?
19. A shirt is on sale for \$ 9.60. This is 60% of the original price. What was the original price?
20. A blouse is on sale for \$ 9.24. This is 66% of the original price. What was the original price?
21. Debbie earned the same amount of money for 3 days. Another day she earned \$ 24. She earned \$ 84 in all. How much did she earn on each of the 3 days?
22. Marcos earned the same amount of money for 5 days. Another day he earned \$ 32. He earned \$ 122 in all. How much did he earn on each of the 5 days?
23. Kirk sold his painting for \$ 42. This was 6 times what he paid for the materials. What did he pay for the materials?
24. Jamie sold her ivy plant for \$ 18. This was 3 times what she paid for it. What did she pay for the plant?

More Practice, page 443

APPLICATION

CREDIT COSTS

Over 2000 banks and credit card companies offer credit cards. In addition thousands of individual stores have their own credit cards. Credit card purchases are paid for by monthly payments. The interest rate is often $1\frac{1}{2}\%$ per month, which sounds small, but amounts to 18% per year.

After one month Trent has a balance of \$400 in his credit card account. How much interest does he owe? How much does he owe in all?

$I = Prt$

$I = \$400 \times 18\% \times \frac{1}{12}$

$I = \$400 \times 1\frac{1}{2}\%$ $\qquad P = \$400 + \$6$

$I = \$400 \times 0.015 = \$6$ $\qquad P = \$406$

Trent owes the credit card company \$406.

For each balance, compute the interest and the amount owed after one month.

1. \$800 2. \$1000 3. \$400 4. \$700

A color TV sells for $ 400 cash. It can be purchased on credit for $ 90 down and $ 40 per month for 9 months.

5. What is the cost of buying the TV in installments? The cost is the total amount paid.
6. How much interest is paid?
7. What percent of the $ 400 is the interest?

Sal bought a CB radio for $ 150. The sales tax was 3.5%. Sal made a down payment of $ 25 and charged the rest on his charge card. The bill each month was $ 30 plus $1\frac{1}{2}$% interest on the unpaid balance.

8. What was the purchase price including tax?
9. How much was charged to the credit card?
10. What was the first monthly payment?
11. What was the second monthly payment?
12. What was the third monthly payment?

How much was the interest for each of the following months?

13. the first month
14. the second month
15. the third month
16. Estimate how many months it took Sal to pay for the radio.

Rita bought new auto tires for $ 150 and charged them on her charge card. She paid $ 14.75 per month for 12 months.

17. How much did Rita pay in all for the tires?
18. How much did she pay in interest?

Some credit card companies also make money by charging the merchant. Nadine takes a group of people to The Spaghetti Shop for dinner. The bill is $ 57.85. Nadine charges it on her charge card. The Spaghetti Shop has to pay the charge card company 6% of the bill.

19. How much does The Spaghetti Shop have to pay the charge card company?
20. How much does The Spaghetti Shop actually get from the dinner bill?

GRAPHING EQUATIONS

The graphs of many equations are straight lines.

EXAMPLE 1. Graph $y = 2x - 1$.

1. Make a table. Choose a value for x. In the equation replace x with that value. Solve for y. Repeat for other values of x.

$$y = 2x - 1$$
$$y = 2 \cdot (-2) - 1$$
$$y = -4 - 1$$
$$y = -5$$

$$y = 2x - 1$$
$$y = 2 \cdot (-1) - 1$$
$$y = -2 - 1$$
$$y = -3$$

$$y = 2x - 1$$
$$y = 2 \cdot 0 - 1$$
$$y = 0 - 1$$
$$y = -1$$

$$y = 2x - 1$$
$$y = 2 \cdot 1 - 1$$
$$y = 2 - 1$$
$$y = 1$$

$$y = 2x - 1$$
$$y = 2 \cdot 2 - 1$$
$$y = 4 - 1$$
$$y = 3$$

x	y
−2	−5
−1	−3
0	−1
1	1
2	3

2. Graph and label the ordered pairs in the table.

To graph an ordered pair like (−2,−5), start at (0,0). Go to the left 2 units, then down 5 units.

3. Connect the points.

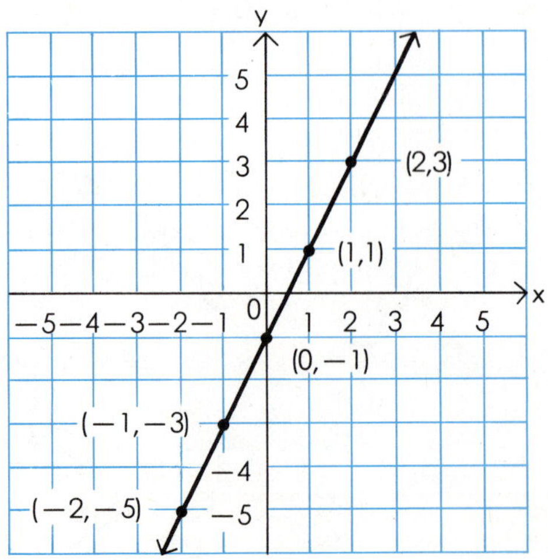

TRY THIS

Graph each equation.

1. $y = x + 1$ **2.** $y = 1 - 2x$

PRACTICE

Graph each equation.

1. $y = x + 4$	**2.** $y = x + 2$	**3.** $y = x + 3$
4. $y = x - 2$	**5.** $y = x - 4$	**6.** $y = x - 5$
7. $y = 2x$	**8.** $y = x$	**9.** $y = -1 \cdot x$
10. $y = -3x$	**11.** $y = -2x$	**12.** $y = -4x$
13. $y = 2x + 1$	**14.** $y = 3x + 1$	**15.** $y = 2x + 2$
16. $y = 4 - x$	**17.** $y = 3 - x$	**18.** $y = 5 - x$
19. $y = 10 - 3x$	**20.** $y = 9 - 2x$	**21.** $y = 8 - 3x$
22. $y = 3x - 5$	**23.** $y = 4x - 3$	**24.** $y = 5x - 4$
25. $y = -2x + 1$	**26.** $y = -2x + 2$	**27.** $y = -2x - 1$

PROBLEM CORNER

1. Luis is using the equation $y = x^2$ in his experiment. He chose these values for x: $-3, -2, -1, 0, 1, 2, 3$. Make a table and find the values for y.

2. Graph $y = x^2$.

More Practice, page 443

CAREER CAREER CAREER CAREER CAREER

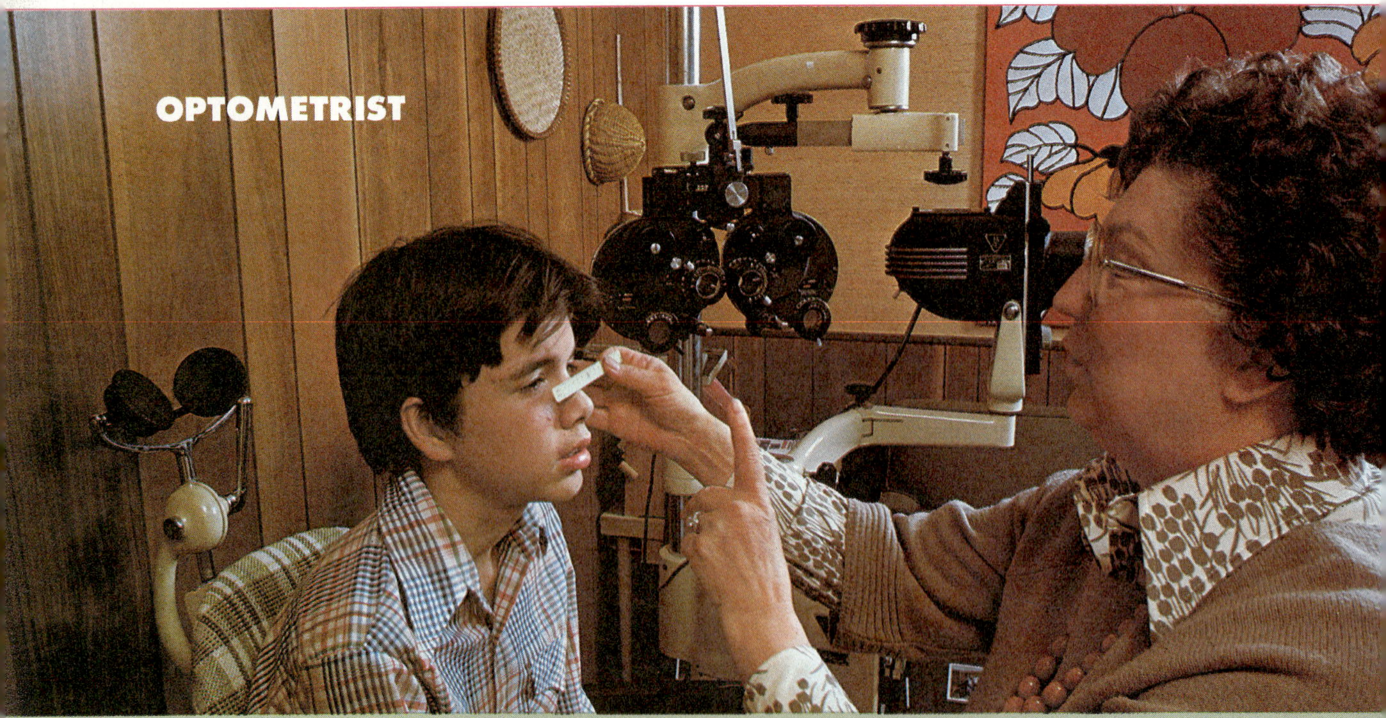

OPTOMETRIST

Dr. Mary Chaney is an optometrist. She relies on precise measurements and calculations to determine eyeglass prescriptions.

One important measurement is the pupillary distance, PD. This is the distance, in millimeters, between the pupils of the eyes. This distance is widest when a person looks into the distance. It is narrowest when a person looks at objects close by.

Patient	Far PD	Near PD
Elaine Tyler	65	58
Doug Wong	70	65
Angela Perez	61	53
Mike Barnes	59	51
Sammy Cole	69	64
Julia Spinola	63	56

1. For each patient find the difference of the measures of the far and near PDs.
2. For each patient find the average of the near and far PDs.
3. For the six patients what is the largest PD?

4. What is the smallest PD?
5. What is the difference between the largest and smallest PD?
6. What is the average near PD?
7. What is the average far PD?

Objects are not always the same size as they appear to be. Dr. Chaney uses this fact in fitting lenses to make objects appear larger or smaller.

The angular size of an object is the size that it appears to be. It is the ratio of the height of the object, h to its distance, d, from the eye.

$$\text{Angular size} = \frac{h}{d}$$

NOTE: h and d must be measured in the same unit.

Find the angular size of each of the following.

8. A $\frac{3}{4}$-inch penny held 24 inches from the eye
9. A 9-inch plate held 24 feet from the eye
10. A full moon, 2400 miles in diameter, 240,000 miles from the eye
11. How do the angular sizes of the penny and the plate compare?
12. The angular size of the penny is about how many times as large as that of the moon?
13. Which of these has the larger angular size?

    a 6-ft baseball player seen from the bleachers 95 yards away or a 2-inch image of the player seen on a TV screen, 6 ft from the eye?

We can use the angular size to find other lengths.

14. A pen, 13 centimeters long, has an angular size of 0.5. How far is it from the eye?
15. A pencil is held 76 centimeters from the eye and has an angular size of 0.25. How long is the pencil?
16. You are standing 100 feet from a tree. The tree has an angular size of 0.12. How tall is the tree?

CHAPTER 15 REVIEW

Evaluate each expression when $x = -2$ and $y = -4$. p. 422

1. $18 + x - y$ **2.** $5x + 4y$ **3.** $8xy$

Solve. p. 426

4. $x + 9 = -4$ **5.** $y - 3.8 = 9.4$ **6.** $t - \frac{5}{6} = \frac{2}{3}$

Solve. p. 428

7. $\frac{t}{-3} = 8$ **8.** $1.4x = 56$ **9.** $-9x = -81$

Solve. p. 430

10. $5x - 6 = 34$ **11.** $-8n + 9 = 81$ **12.** $\frac{m}{8} - 3 = 10$

Solve. p. 432

13. -4 times what number plus 7 is 61?

14. The sales tax on a record is \$ 0.40. This is 5% of the purchase price. What is the purchase price?

15. The length of a rectangle is 19 meters and the area is 323 square meters. What is the width?

16. A painting was sold for \$ 140. This was 7 times the cost of the materials. What was the cost of the materials?

Graph the equation. p. 436

17. $y = -2x$

Application pp. 424, 434

18. Ellen keeps \$ 900 in a savings account for 8 months at 5%. Find the simple interest.

19. Tom still owes \$ 500 on his credit card account. Interest is $1\frac{1}{2}$ % per month. How much interest will he owe next month? How much will he owe in all?

CHAPTER 15 TEST

Evaluate each expression when $x = 4$ and $y = -5$.

1. $-3xy$ **2.** $x - 3y$ **3.** $5x + 2y - 40$

Solve.

4. $t - 5 = -12$ **5.** $x + \frac{1}{3} = \frac{5}{6}$ **6.** $y - 8.4 = 2.9$

Solve.

7. $3.2x = 64$ **8.** $-8t = 96$ **9.** $\frac{n}{4} = -13$

Solve.

10. $3x + 7 = 40$ **11.** $-4m - 9 = -48$ **12.** $-2.5y - 7 = 23$

Solve.

13. -8 times what number plus 9 is -47?

14. The width of a rectangle is 4.8 feet and the area is 72 square feet. What is the length?

15. A stereo is on sale for \$ 240. This is 75% of the original price. What was the original price?

16. A horse was sold for \$ 1200. This was 3 times the price of the horse as a two-year-old. What was the price of the horse as a two-year-old?

Graph the equation.

17. $y = 3 - x$

Application

18. Darren keeps \$ 700 in a savings account for 6 months at 8%. Find the simple interest.

19. Samantha still owes \$ 600 on her credit card account. Interest is $1\frac{1}{2}$ % per month. How much interest will she owe next month? How much will she owe in all?

MORE PRACTICE

Evaluate each expression when $x = -6$. Use after page 423.

1. $32 \div x$ **2.** $4x$ **3.** $-3x + 10$ **4.** $20 - x$

Evaluate each expression when $a = -2$ and $b = -6$. Use after page 423.

5. $a - b + 40$ **6.** $-7ab$ **7.** $8ab$ **8.** $2a + b$

9. $5a + 7b$ **10.** $5a - 4b$ **11.** $5 \times a \times a$ **12.** $-2a - 10b$

Solve. Use after page 427.

13. $x + 6 = 11$ **14.** $z + 4 = 1$ **15.** $x + 5 = -2$

16. $h + 27 = 63$ **17.** $m + 28 = 19$ **18.** $t + 17 = -10$

19. $x - 9 = 4$ **20.** $r - 11 = -7$ **21.** $g - 3.4 = 9.7$

22. $t - 7 = -23$ **23.** $v - 23 = -48$ **24.** $24 + y = 56$

25. $x + 2348 = 5667$ **26.** $y + 103 = 5467$ **27.** $x + \frac{1}{4} = \frac{2}{3}$

28. $x + \frac{7}{8} = 1$ **29.** $t - \frac{3}{4} = \frac{5}{6}$ **30.** $m - 1 = \frac{2}{5}$

Solve. Use after page 429.

31. $4x = 20$ **32.** $9x = 72$ **33.** $-6x = 54$

34. $2y = -18$ **35.** $6z = -18$ **36.** $-9t = -45$

37. $\frac{t}{3} = 17$ **38.** $\frac{x}{7} = -9$ **39.** $\frac{y}{-5} = 13$

40. $\frac{y}{10} = -13$ **41.** $\frac{z}{-45} = -81$ **42.** $10x = 7$

43. $1.8y = 90$ **44.** $4.4z = 88$ **45.** $5.3d = 121.9$

46. $-25t = 75$ **47.** $-35y = 2100$ **48.** $14h = 35$

49. $2.8a = -126$ **50.** $5.6b = -151.2$ **51.** $3.4x = 255$

Solve. Use after page 431.

1. $2x + 5 = 11$ **2.** $3x + 8 = 20$ **3.** $5t + 6 = -34$

4. $4y - 7 = 25$ **5.** $-3m - 25 = 7$ **6.** $8t - 10 = -82$

7. $-5t - 11 = 69$ **8.** $1.8y - 12 = 24$ **9.** $6.4m + 2 = 10$

10. $\frac{x}{-3} + 8 = 29$ **11.** $\frac{x}{6} - 5 = 11$ **12.** $\frac{y}{4} - 4 = 13$

13. $3.2t - 4 = 8$ **14.** $0.2x - 0.7 = 1.3$ **15.** $0.8x + 0.4 = 2.9$

16. $0.5y + 0.9 = 1.7$ **17.** $45t + 13 = 1067$ **18.** $12x - 24 = 672$

Solve. Use after page 433.

19. What number plus 9 is 23?

20. Six times what number is -36?

21. -5 times what number minus 12 is 68?

22. Nine times what number plus 9 is 81?

23. When 37 is subtracted from five times a certain number, the result is 243. What is the number?

24. When 16 is added to 3 times a certain number, the result is 64. What is the number?

25. The length of a rectangle is 14 cm and the area is 378 cm^2. What is the width?

26. The width of a rectangle is 19 m and the area is 494 m^2. What is the length?

27. A book is on sale for \$ 11.96. This is 80% of the original price. What was the original price?

28. The sales tax on a plant is \$ 0.39. This is 6% of the purchase price. What is the purchase price?

29. Leslie sold her bookcase for \$ 21. This was 3 times what she paid for the wood. What did she pay for the wood?

30. Trevor sold his coin collection for \$ 340. This was 4 times what he paid for the coins. What did he pay for the coins?

Graph each equation. Use after page 437.

31. $y = x - 1$ **32.** $y = x + 1$ **33.** $y = 3x$ **34.** $y = -2x$

35. $y = 2x - 1$ **36.** $y = -2x + 2$ **37.** $y = 6 - 3x$ **38.** $y = -3x - 4$

SKILLS REVIEW

Make these changes. Refer to pages 286–299

1. 9 km = _____ m **2.** 7 L = _____ mL **3.** 8268 m = _____ km

4. 5 kg = _____ g **5.** 748 g = _____ kg **6.** 4678 mL = _____ L

7. 6 t = _____ kg **8.** 48 kg = _____ t **9.** 98 mm = _____ cm

10. 7 cm = _____ mm **11.** 3 g = _____ mg **12.** 4286 mg = _____ g

13. 4 ft = _____ in. **14.** 18 in. = _____ ft **15.** 20 yd = _____ ft

16. 10 ft = _____ yd **17.** 4 gal = _____ qt **18.** 24 oz = _____ lb

19. 3 mi = _____ yd **20.** 6 qt = _____ gal **21.** 5 ton = _____ lb

22. 2 mi = _____ ft **23.** 5 lb = _____ oz **24.** 2500 lb = _____ ton

Find the perimeter. Refer to page 312.

25.

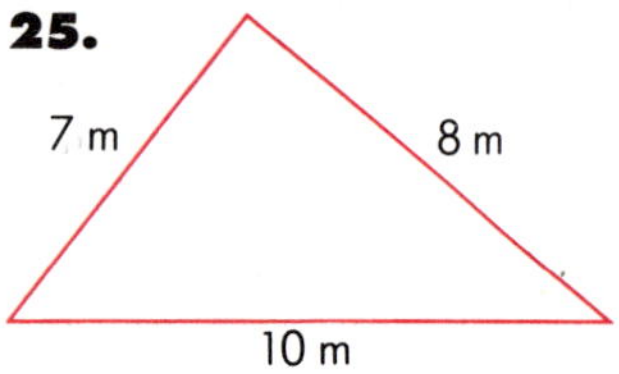

26.

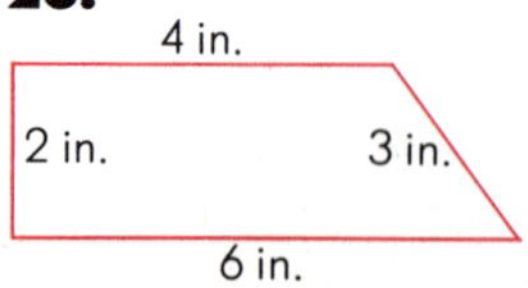

27.

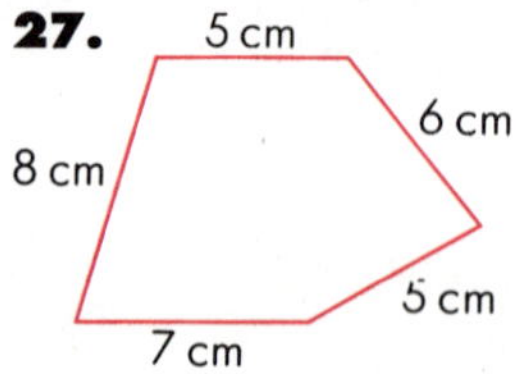

28. Rectangle
$l = 3$ ft
$w = 2$ ft

29. Rectangle
$l = 23.7$ cm
$w = 12.6$ cm

30. Square
$s = 3.4$ m

Find the circumference. Refer to page 314.

31.

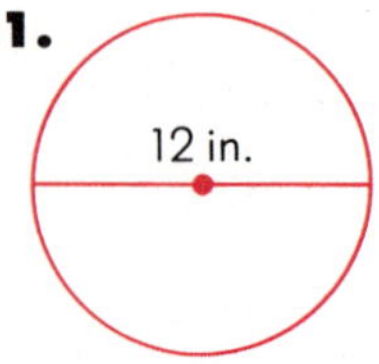

32.

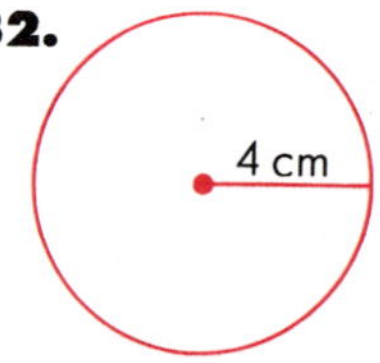

33.

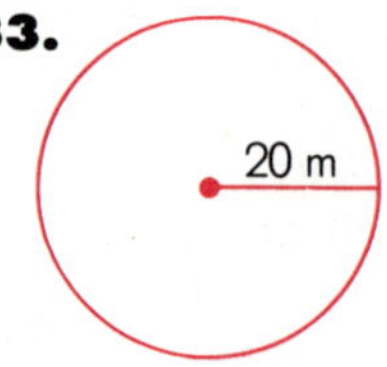

34. Circle
$r = 7$ ft

35. Circle
$d = 25$ mm

36. Circle
$r = 3$ yd

Find the area. Refer to pages 318–327.

1.

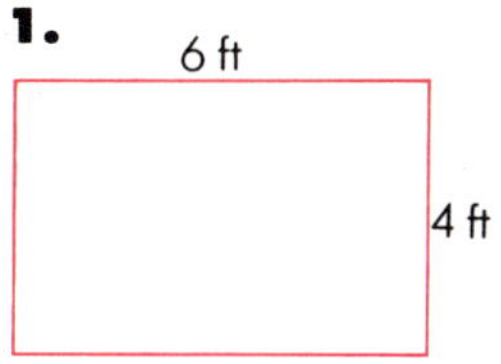

2.

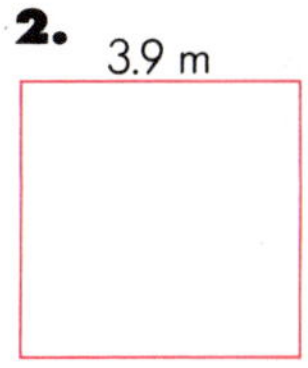

3.

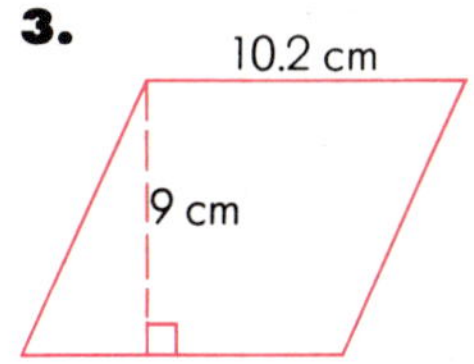

4.

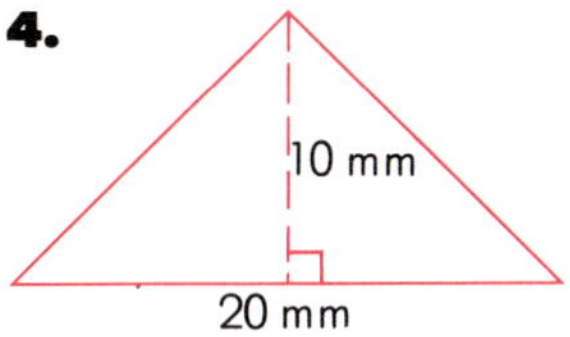

5.

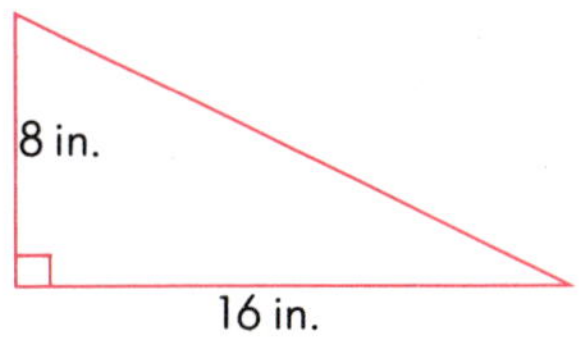

6.

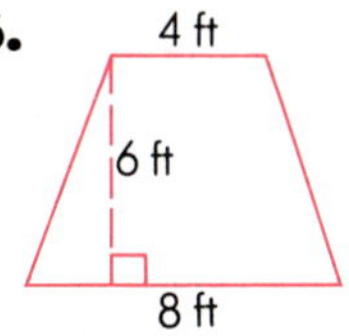

7. Trapezoid
$b_1 = 7$ m
$b_2 = 9$ m
$h = 4.3$ m

8. Circle
$r = 18$ cm

9. Circle
$d = 10$ yd

Find the volume. Refer to pages 340–349.

10.

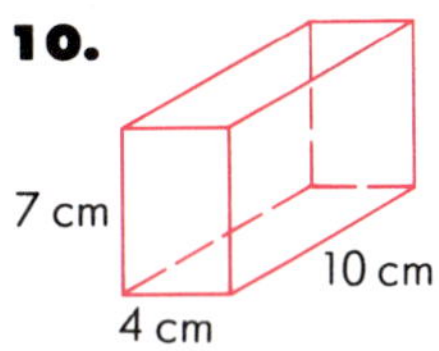

11.

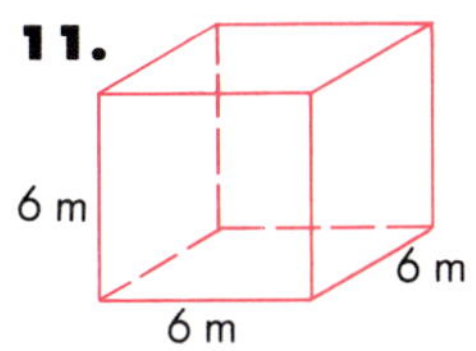

12.

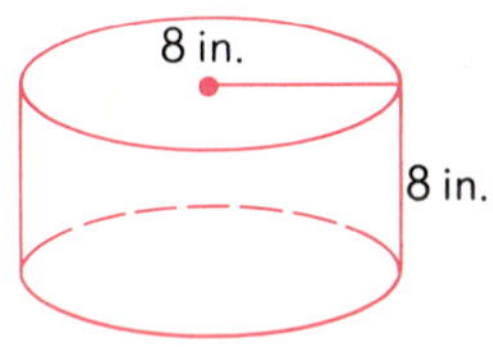

13.

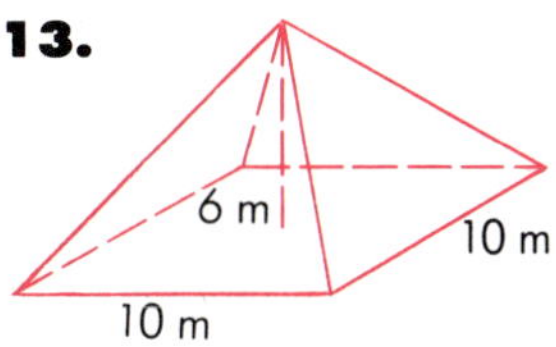

14.

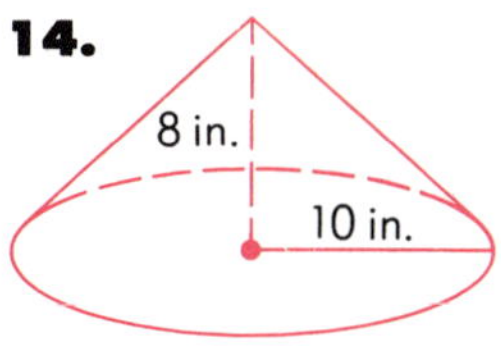

15.

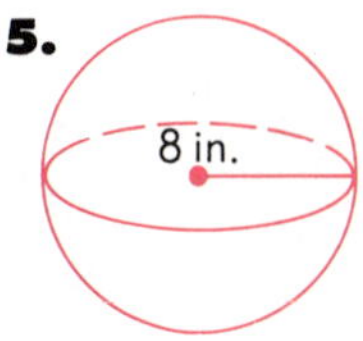

16. Rectangular prism
$l = 4$ m
$w = 1$ m
$h = 8$ m

17. Cube
$s = 12$ in.

18. Cylinder
$d = 12$ cm
$h = 7$ cm

19. Pyramid
$l = 12$ yd
$w = 7$ yd
$h = 9$ yd

20. Cone
$r = 12$ cm
$h = 12$ cm

21. Sphere
$d = 18$ cm

1. Make a line graph for the data in the table. Refer to page 370.

National Longevity Records			
Country	Years	Country	Years
United States	113	U.S.S.R.	110
Canada	113	France	109
Japan	112	Germany	108
South Africa	111	Finland	107
Australia	110	Spain	106

2. Make a bar graph for the data in the table Refer to page 372.

Moons of the Sun's Planets			
Planet	Number of Moons	Planet	Number of Moons
Mercury	0	Saturn	9
Venus	0	Uranus	4
Earth	1	Neptune	1
Mars	2	Pluto	0
Jupiter	9		

3. Find the angle for each percent in the table and make a circle graph. Refer to page 374.

Time Spent in One Week	
Activity	Percent
Sleeping	40%
School	25%
Recreation	25%
Eating	4%
Other	6%

Find the mean. Refer to page 378.

4. 2, 4, 6, 0 **5.** 9, 9, 12, 15, 19 **6.** 37, 48, 65, 27

Find the median. Refer to page 380.

7. 15, 7, 9, 10, 12 **8.** 24, 498, 76, 98, 376, 105, 381

9. 75, 86, 94, 88, 87 **10.** 6, 6, 8, 8, 7, 7, 9

11. 3, 7, 9, 10 **12.** 12, 4, 6, 11, 10, 9

Find the mode. Refer to page 380.

1. 4, 7, 4, 8, 4, 9

2. 7, 6, 5, 6, 4, 3, 2

3. 21, 48, 74, 21, 76, 90

4. 4, 3, 2, 4, 3, 2, 4

5. 3, 9, 4, 9, 3, 7

6. 6, 6, 7, 8, 12, 7, 11

Give a positive or negative number for each situation. Refer to page 394.

7. a loss of $ 76

8. a gain of $ 10

9. 5 hours ago

10. 17° below 0

11. 25° above 0

12. 90 ft above sea level

13. 130 ft below sea level

14. 12 sec after liftoff

15. 3 min before liftoff

Compare. Use $>$ or $<$. Refer to page 394.

16. 5 and -5

17. 0 and -4

18. -6 and 7

19. 4 and -5

20. -6 and -5

21. 9 and -12

22. -4 and -10

23. 0 and 7

24. 0 and -7

25. -7 and 7

26. -9 and -10

27. -4 and -1

Add. Refer to page 398.

28. $-6 + 7$

29. $-11 + 7$

30. $-6 + 8$

31. $-8 + 6$

32. $-3 + (-7)$

33. $-8 + (-6)$

34. $-3 + (-5)$

35. $-4 + (-6)$

36. $16 + (-3)$

37. $9 + (-9)$

38. $7 + (-12)$

39. $6 + (-8)$

Subtract. Refer to page 404.

40. $4 - 5$

41. $-4 - 5$

42. $-9 - (-4)$

43. $-7 - 7$

44. $8 - (-8)$

45. $7 - 10$

46. $0 - 7$

47. $8 - (-3)$

48. $-8 - 15$

49. $-9 - (-9)$

50. $37 - 43$

51. $-20 - 10$

Multiply. Refer to page 406.

52. 9×6

53. $9 \times (-4)$

54. -3×6

55. $-8 \times (-3)$

56. $-8 \times (-6)$

57. $-7 \times (-15)$

58. $12 \times (-7)$

59. -100×7

Divide. Refer to page 408.

1. $-8 \div 4$ **2.** $-8 \div (-4)$ **3.** $-24 \div 3$

4. $-16 \div (-2)$ **5.** $72 \div (-9)$ **6.** $42 \div (-6)$

7. $-30 \div 6$ **8.** $35 \div (-5)$ **9.** $-9 \div (-3)$

10. $28 \div (-4)$ **11.** $-60 \div 12$ **12.** $-72 \div (-6)$

13. $-300 \div 50$ **14.** $-480 \div (-60)$ **15.** $400 \div (-25)$

Evaluate each expression when $a = 5$ and $b = -4$. Refer to page 422.

16. $a + b$ **17.** $a - b$ **18.** $b - a$

19. ba **20.** $a \div b$ **21.** $b \div a$

22. $2a + b$ **23.** $2(a + b)$ **24.** $2a - b$

25. $\frac{1}{2}b + a$ **26.** $\frac{1}{2}(b + a)$ **27.** $a + b + 17$

28. $3a + 3b$ **29.** $3a - 3b$ **30.** $2a + 3b + 10$

Solve. Refer to page 426.

31. $x + 3 = 6$ **32.** $y + 8 = 4$ **33.** $t + 13 = -7$

34. $x - 9 = 4$ **35.** $p - 9 = -3$ **36.** $x + 3 = 8$

37. $y + 12 = -4$ **38.** $x + 4 = 11$ **39.** $t + 10 = 11$

40. $p - 9 = -4$ **41.** $x - 3.1 = 4.4$ **42.** $y + 4.5 = 9$

Solve. Refer to page 428.

43. $7x = 35$ **44.** $6x = -36$ **45.** $-8x = 24$

46. $7a = -35$ **47.** $-8y = 64$ **48.** $7x = 28$

49. $-6t = 60$ **50.** $6x = 66$ **51.** $7x = -56$

52. $\frac{t}{3} = 15$ **53.** $\frac{a}{4} = 8$ **54.** $\frac{b}{10} = -2$

Solve. Refer to page 430.

1. $3x + 3 = 12$ **2.** $4x + 7 = 31$ **3.** $9x - 6 = 30$

4. $-4t + 7 = -1$ **5.** $-6a + 9 = 45$ **6.** $8b + 3 = -5$

7. $-4x - 3 = 17$ **8.** $-5a + 7 = 42$ **9.** $-6x - 5 = 25$

10. $1.2x - 3 = 1.8$ **11.** $3.5b + 1 = -2.5$ **12.** $7.6a + 3 = 25.8$

13. $\frac{a}{3} - 1 = 8$ **14.** $\frac{b}{2} + 3 = -6$ **15.** $\frac{x}{5} + 5 = 6$

Solve. Refer to page 432.

16. What number plus 8 is 14?

17. What number plus 10 is -1?

18. What number plus 12 is 7?

19. What number plus 6 is 0?

20. What number minus 7 is 13?

21. What number minus 10 is 10?

22. What number minus 8 is -1?

23. What number minus -4 is 11?

24. Six times what number is 30?

25. Seven times what number is 56?

26. Five times what number is -15?

27. -4 times what number is -8?

28. -6 times what number is 42?

29. What number divided by 5 is 5?

30. What number divided by 7 is -1?

31. What number divided by -3 is 8?

32. Six times what number plus 7 is 49?

33. -8 times what number minus 7 is 57?

34. When 15 is subtracted from 7 times a certain number, the result is 27. What is the number?

35. The length of a rectangle is 9 feet. The area is 72 square feet. What is the width of the rectangle?

Graph each equation. Refer to page 436.

36. $y = x + 5$ **37.** $y = x + 10$ **38.** $y = x - 7$

39. $y = x - 11$ **40.** $y = 3x$ **41.** $y = -2x$

42. $y = 4x + 1$ **43.** $y = 4x - 1$ **44.** $y = 7 - x$

APPENDIX

MEASUREMENT TABLES AND FORMULAS

METRIC MEASURES

Length

1 kilometer (km) = 1000 meters (m)
1 meter (m) = 100 centimeters (cm)
1 centimeter (cm) = 10 millimeters (mm)

Capacity

1 liter (L) = 1000 milliliters (mL)

Mass

1 metric ton (t) = 1000 kilograms (kg)
1 kilogram (kg) = 1000 grams (g)
1 gram (g) = 1000 milligrams (mg)

Temperature

0°C = freezing point of water
37°C = normal body temperature
100°C = boiling point of water

CUSTOMARY MEASURES

Length

12 inches (in.) = 1 foot (ft)
3 feet = 1 yard (yd)
36 inches = 1 yard
5280 feet = 1 mile (mi)
1760 yards = 1 mile

Capacity

8 fluid ounces (fl oz) = 1 cup (c)
2 cups = 1 pint (pt)
2 pints = 1 quart (qt)
4 quarts = 1 gallon (gal)

Weight

16 ounces (oz) = 1 pound (lb)
2000 pounds = 1 ton

Temperature

32°F = freezing point of water
98.6°F = normal body temperature
212°F = boiling point of water

FORMULAS

Perimeter of a rectangle	$P = 2l + 2w$
Perimeter of a square	$P = 4s$
Circumference of a circle	$C = \pi d$ or $C = 2\pi r$
Area of a rectangle	$A = lw$
Area of a square	$A = s \times s$
Area of a parallelogram	$A = bh$
Area of a triangle	$A = \frac{1}{2}bh$
Area of a trapezoid	$A = \frac{1}{2}h(b_1 + b_2)$
Area of a circle	$A = \pi r^2$
Volume of a rectangular prism	$V = lwh$
Volume of a cube	$V = s \times s \times s$
Volume of a cylinder	$V = \pi r^2 h$
Volume of a pyramid	$V = \frac{1}{3}lwh$
Volume of a cone	$V = \frac{1}{3}\pi r^2 h$
Volume of a sphere	$V = \frac{4}{3}\pi r^3$
Interest	$I = Prt$

SKILLS TEST

Add.

1. $58 + 94$

2. $\begin{array}{r} \$17.95 \\ 6.49 \\ \underline{8.75} \end{array}$

3. $\begin{array}{r} 430{,}532 \\ 769{,}298 \\ \underline{97{,}305} \end{array}$

4. $9.78 + 4.056$

Subtract.

5. $\begin{array}{r} 90 \\ \underline{-\ 28} \end{array}$

6. $\begin{array}{r} 37{,}329 \\ \underline{-\ 19{,}293} \end{array}$

7. $\begin{array}{r} \$90.50 \\ \underline{-\ 38.76} \end{array}$

8. $19.3 - 8.567$

Multiply.

9. $\begin{array}{r} \$9.58 \\ \underline{\times\ \ 74} \end{array}$

10. $\begin{array}{r} 897 \\ \underline{\times 535} \end{array}$

11. $\begin{array}{r} 96.2 \\ \underline{\times 0.55} \end{array}$

12. 1000×5.13

Divide.

13. $7365 \div 3$

14. $9.8 \div 7$

15. $6.9 \div 100$

16. $0.16\overline{)9.6}$

Estimate.

17. $\begin{array}{r} 7341 \\ \underline{+\ 8956} \end{array}$

18. $\begin{array}{r} 5.298 \\ \underline{-\ 2.99} \end{array}$

19. $\begin{array}{r} 71 \\ \underline{\times 39} \end{array}$

20. $8.37\overline{)31.796}$

What part is shaded?

21. 

22. 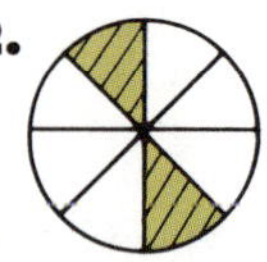

23. 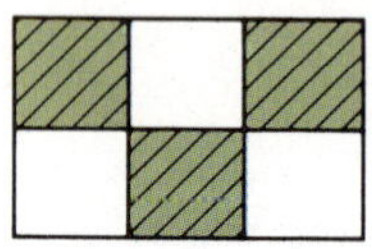

Compare. Use $<$ or $>$.

24. $\frac{5}{8}$ and $\frac{3}{5}$

25. $\frac{5}{6}$ and $\frac{8}{9}$

Convert to a fraction.

26. $1\frac{2}{3}$

27. $4\frac{3}{5}$

Convert to a mixed number.

28. $\frac{19}{5}$

29. $\frac{49}{10}$

Convert to a decimal.

30. $9\frac{7}{100}$

31. $18\frac{4}{5}$

Multiply and simplify.

32. $\frac{3}{4} \cdot \frac{2}{5}$

33. $\frac{5}{8} \cdot \frac{7}{10}$

34. $10 \times 3\frac{2}{5}$

35. $2\frac{2}{3} \times 2\frac{1}{4}$

Divide and simplify.

36. $18 \div \frac{1}{3}$ **37.** $\frac{6}{5} \div \frac{2}{3}$ **38.** $4\frac{1}{5} \div \frac{3}{10}$ **39.** $4\frac{1}{4} \div 1\frac{1}{8}$

Add and simplify.

40. $1\frac{3}{8} + \frac{7}{8}$ **41.** $\frac{5}{6} + \frac{3}{4}$ **42.** $6\frac{3}{5} + 8\frac{9}{10}$ **43.** $\frac{3}{4} + \frac{2}{5} + \frac{3}{10}$

Subtract and simplify.

44. $\frac{7}{8} - \frac{3}{8}$ **45.** $14\frac{5}{6} - 9\frac{5}{8}$ **46.** $9 - 6\frac{3}{5}$ **47.** $8\frac{1}{3} - 1\frac{5}{6}$

Write the ratio. Simplify, if possible.

48. 240 kilometers in 3 hours

49. 20 sandwiches for 12 people

Find the unit rate.

50. $ 18.40 is received for 5 hours of work. What is the amount per hour?

51. If 3 tennis balls cost $3.27, how much does 1 tennis ball cost?

Solve.

52. $\frac{2}{3} = \frac{4}{x}$ **53.** $\frac{x}{7} = \frac{9}{8}$ **54.** $\frac{3}{5} = \frac{x}{16}$

55. A car travels 800 kilometers in 3 days. How far would it travel in 15 days?

56. If 7 tickets cost $ 45.50, how much do 17 tickets cost?

Convert to a decimal.

57. 89% **58.** 6.7%

Convert to a fraction.

59. 84% **60.** $33\frac{1}{3}\%$

Convert to a percent.

61. 0.75 **62.** 1.08 **63.** $\frac{3}{8}$ **64.** $\frac{19}{20}$

Solve.

65. What is 11% of 39?

66. What percent of 50 is 16?

67. 45 is 20% of what number?

68. 180 is what percent of 400?

Find the percent of increase or decrease.

69. From 4 to 5 **70.** From 15 to 12 **71.** From 30 to 10

Make these changes.

72. 8 km = ______ m

73. 9.3 mm = ______ cm

74. 105 cm = ______ mm

75. 5307 m = ______ km

76. 1.32 L ______ mL

77. 72.16 mL = ______ L

78. 45.7 g = ______ kg

79. 931.2 kg = ______ g

80. 8 t = ______ kg

81. 998.4 kg = ______ t

82. 4 ft = ______ in.

83. 60 in. = ______ ft

84. 7 ft = ______ yd

85. $8\frac{1}{2}$ qt = ______ pt

86. 6 mi = ______ ft

87. 80 oz = ______ lb

Find the perimeter.

88.

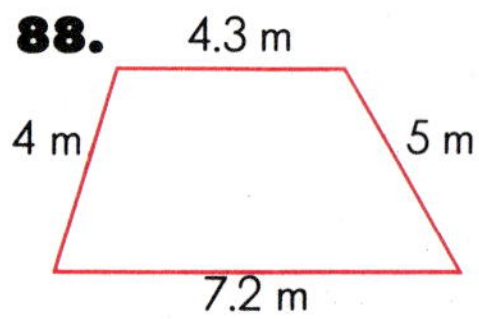

89.

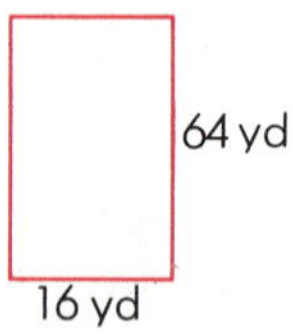

Find the circumference.

90.

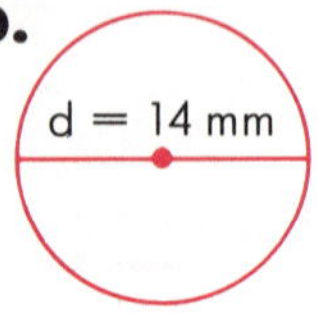

Find the area.

91. Rectangle
$l = 8$ km
$w = 3.4$ km

92. Square
$s = 2\frac{1}{2}$ in.

93. Parallelogram
$b = 8.5$ mm
$h = 3.4$ mm

94. Triangle
$b = 12$ ft
$h = 8$ ft

95. Trapezoid
$h = 40$ m
$b_1 = 14$ m
$b_2 = 18$ m

96. Circle
$r = 10$ cm

Find the volume.

97.

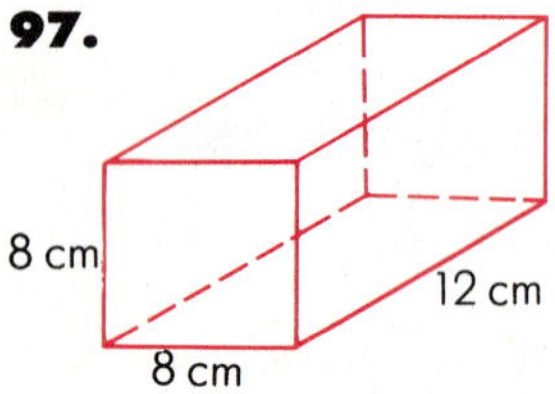

98.

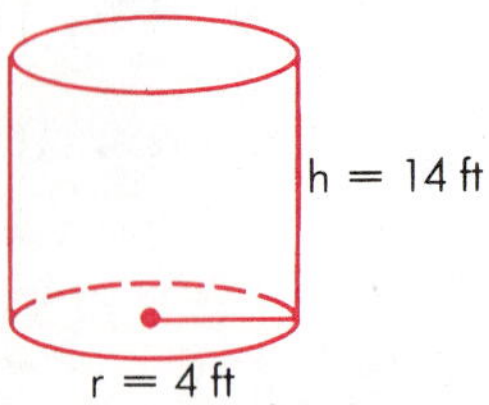

99.

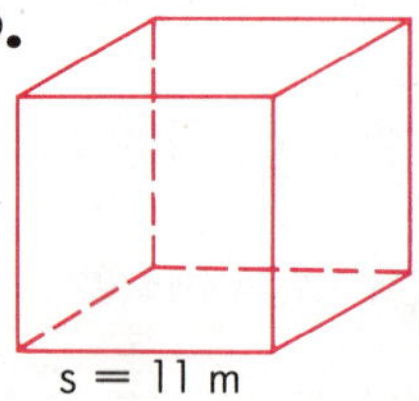

100. Make a bar graph.

Animal	Longest Known Lifetime (years)
Cat	34
Dog	27
Human	114
Horse	50
Chicken	30
Rattlesnake	20

101. Make a line graph.

Continent	Highest Point of Elevation (feet)
Asia	29,028
South America	22,834
North America	20,320
Africa	19,340
Europe	18,510
Antarctica	16,860
Australia	7,310

Find the mean.

102. 17, 20, 15, 22, 16, 14

103. 92, 69, 80, 75

Find the median.

104. 17, 20, 15, 22, 16, 14

Find the mode.

105. 77, 52, 46, 52, 90, 83

Add.

106. $-12 + 8$ **107.** $-14 + (-18)$ **108.** $23 + (-9)$

Subtract.

109. $3 - 8$ **110.** $9 - (-5)$ **111.** $-4 - (-7)$

Multiply.

112. $-8 \times (-7)$ **113.** $9 \times (-6)$ **114.** -14×8

Divide.

115. $-56 \div (-14)$ **116.** $-64 \div 8$ **117.** $81 \div (-9)$

Evaluate each expression when $a = -3$ and $b = -5$.

118. $-6ab$ **119.** $2(a - b)$ **120.** $7b + a$ **121.** $-3a + 8b$

Solve.

122. $x + 9 = -1$ **123.** $-7\text{p} = 42$ **124.** $\frac{t}{8} = -11$

125. $-8x - 83 = 13$ **126.** -3 times what number minus 4 is -34?

Graph each equation.

127. $y = x - 2$ **128.** $y = -1 \cdot x$

APPLICATIONS SKILLS TEST

1. Write a check from Juanita Perez to City Gas and Electric Company in the amount of $ 98.76. Date it August 9, 1981.

2. Write a check from Joe Wong to Silver Art Supply in the amount of $ 19.57. Date it March 27, 1982.

3. Find the net deposit.

		Dollars	Cents
Cash	Currency	186	00
	Coin	48	75
Checks	12-29	38	76
	11-104	448	46
Total			
Less Cash Received			-0-
Net Deposit			

4. Find the balance carried forward.

Date Aug. 6, 1982 $98.76 325
To City Gas and Electric
For Gas Bill

	Dollars	Cents
Balance Forward	268	74
Amount of Deposit	198	70
Total		
Amount of Check		
Balance Carried Forward		

5. Complete the list to give change.
Charge: $ 12.72 Paid with: $ 20
Machine computes $ 7.28 change.

Clerk gives	Clerk says

6. Is the change counted correctly?
Charge: $ 6.39 Paid with: $ 10

Clerk gives	Clerk says
Item	$ 6.39
1 penny	6.50
1 quarter	6.75
1 quarter	7.00
1 dollar	8.00
1 dollar	9.00
1 dollar	10.00

Period Ending	Hours	Rate	Earnings: Amount Earned Regular Rate	Earnings: Overtime and Other	Gross Earnings	Deductions: FICA	Deductions: Federal Withholding Tax	Deductions: State Withholding Tax	Deductions: Others	Net Pay
7-8-81	30	4.50		38.00		10.80	24.50	3.90	48.80	

Statement of Earnings and Deductions for Employee's Record—Detach Before Cashing Check

7. Find the gross earnings.

8. Find the net pay.

9. Wilma spent 2 hr, 40 min studying and 3 hr, 35 min playing the piano. How much time did she spend in all?

10. Fred ran a marathon in 4 hr, 3 min. Larry ran the marathon in 3 hr, 25 min. How much longer did Fred run than Larry?

11. When it is 1 AM in Chicago, Illinois, what time is it in Los Angeles, California?

12. When it is 2 PM in Albuquerque, New Mexico, what time is it in Baltimore, Maryland?

13. Rhonda cuts $3\frac{2}{5}$ feet off of a rope which is $8\frac{1}{4}$ feet long. How much rope is left?

14. Mike can save $\frac{5}{8}$ of his gasoline budget of \$ 1120 by taking the bus to work. How much can he save?

15. Five pounds of potatoes cost \$ 1.48. Ten pounds cost \$ 2.90. Which is the more economical buy?

16. Three shirts cost \$ 41.85. What is the unit price?

17. Tom earns a 10% commission on his monthly sales. How much commission does he earn on sales of \$ 7987?

18. Stacey and Steven go out for lunch. The bill is \$ 9.20. If they tip 15%, how much should they leave for the tip?

19. A camera which regularly costs \$ 48.00 is on sale for 10% off. What is the sale price?

20. Blouses which regularly cost \$ 20 are on sale for \$ 15. Find the percent of decrease in the price.

21. Sales tax is 6%. What is the total price of a record album which costs \$ 7.98?

22. Marlene budgets 23% of her income for housing. Her net income is \$ 780 per month. How much does Marlene budget for housing?

23. What was the difference between the high and low temperatures in Minneapolis? in Duluth?

24. Was it colder in Minneapolis or Duluth? How much colder?

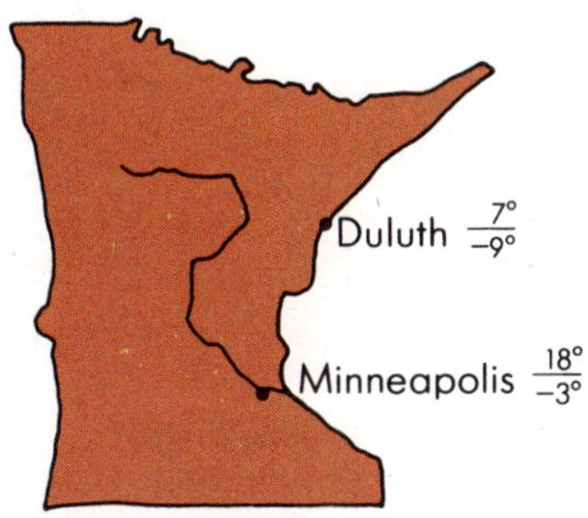

25. Louis still owes \$ 550 on his credit card account. Interest is $1\frac{1}{2}$% per month. How much interest will he owe next month? How much will he owe in all?

26. Bella keeps \$ 2000 in a savings account for 8 months at 7% interest. Find the simple interest.

INDEX

TRY THIS ANSWERS

CHAPTER 1

Pages 4–5

1. 10-21-36, 36-10-21, 21-10-36 **2.** 12-25-81, 81-12-25, 25-12-81 **3.** Social Security Number **4.** Zip Code **5.** Telephone Number **6.** 28:09 **7.** 7:16 **8.** 11:15 **9.** 9:09:09

Pages 6–7

1. tens, 40 **2.** thousands, 5000 **3.** ones, 5 **4.** ninety-eight **5.** three hundred seventy-six **6.** five thousand, two hundred eighty-seven **7.** six thousand, six. **8.** 3651 **9.** 5088 **10.** 3003

Pages 8–9

1. one cent **2.** seventeen dollars and forty-six cents **3.** two hundred thirty-six dollars **4.** 30¢ or $0.30 **5.** $30 or $30.00 **6.** $1.07 **7.** nine and $\frac{89}{100}$ dollars **8.** fifty-six and $\frac{04}{100}$ dollars **9.** one hundred seventy-four and $\frac{00}{100}$ dollars

Pages 12–13

1. ten-thousands, 70,000 **2.** hundred-millions, 700,000,000 **3.** hundred-billions, 100,000,000,000 **4.** 496 million, 287 thousand, 104 **5.** 380 billion, 87 million, 6 thousand, 100 **6.** 986,280,186 **7.** 4,000,000,000 **8.** 20,000,020

Page 14

1. 489 **2.** 1135 **3.** 1605 **4.** 371 **5.** $5.87 **6.** $1.72 **7.** $11.56 **8.** $16.61

Page 16

1. 22,376 **2.** 1,222,115 **3.** $1039.83 **4.** 1,399,339 **5.** $1510.24 **6.** 1,892,592

Pages 20–21

1. tenths, 0.4 **2.** hundredths, 0.09 **3.** thousandths, 0.006 **4.** 7 and 3 tenths **5.** 27 and 386 thousandths **6.** 7 and 492 ten-thousandths **7.** 0.06 **8.** 47.058

Page 22

1. 38.66 **2.** 10.681 **3.** 202.045 **4.** 1.655 **5.** 26.596 **6.** 10.656 **7.** 25.636 **8.** 2.134

Page 25

1. $168

Pages 26–27

1. 500 **2.** 6700 **3.** 14,400 **4.** 7.37 **5.** 9.49 **6.** 9 **7.** 342 **8.** $4 **9.** $41

Page 28

1. 140 **2.** 5600 **3.** 53 **4.** $49

CHAPTER 2

Page 40

1. 54 **2.** 831 **3.** 39 **4.** 288 **5.** $6.36 **6.** $0.21 **7.** $5.75

Page 42

1. 2,494 **2.** 18,899 **3.** 92,457 **4.** $20.86 **5.** $1128.19 **6.** $6,436.86

Page 44

1. 177 **2.** 7629 **3.** $20.56 **4.** 20,726 **5.** 50,759 **6.** $307.29

Page 48

1. 0.20 **2.** 0.70 **3.** 0.90 **4.** $<$ **5.** $>$ **6.** $=$

Pages 50–51

1. 1.55 **2.** 7.687 **3.** 4.46 **4.** 4.48 **5.** 5.3 **6.** 7.79

Page 54

1. 50 **2.** 3700 **3.** 590 **4.** 5 **5.** $9 **6.** 5

CHAPTER 3

Page 68

1. 1286 **2.** 156 **3.** 72,135 **4.** $5.68 **5.** $30.32 **6.** $1646.33

Pages 70–71

1. 62,440 **2.** 2646 **3.** 57,168 **4.** $8.76 **5.** $212.50 **6.** $3180.24

Pages 72-73

1. 87,400 **2.** 80,730 **3.** $1524.75 **4.** 272,035 **5.** 291,150 **6.** $5944.86

Page 76

1. 63,000 **2.** 700 **3.** 400,000 **4.** 400 **5.** 56,000 **6.** 2,000,000

Pages 78–79

1. $24 **2.** $480 **3.** $150 **4.** $3200 **5.** 350 **6.** 20 7. 2000

Page 82

1. 22.8 **2.** 6.966 **3.** 89.76 **4.** 0.024 **5.** 0.049 **6.** 0.00009

Pages 84–85

1. 630 **2.** 9678 **3.** 86.4 **4.** 0.043 **5.** 0.087 **6.** 0.00734

Page 86

1. $30.35

CHAPTER 4

Page 99

1. 6 r 1 **2.** 6 **3.** 6 r 6 **4.** 46 r 1 **5.** 38 r 3 **6.** 80 r 8

Page 100–101

1. 157 r 5 **2.** 920 **3.** 4030 r 3 **4.** $43.73 **5.** $93.59 r 1¢ **6.** $6.04

Page 103

1. 3 r 12 **2.** 2 r 26 **3.** 38 **4.** 20 r 67 **5.** 872 **6.** $1.34

Page 104

1. 4 r 11 **2.** 4 **3.** 8 r 640 **4.** 11 r 382 **5.** 86 r 148 **6.** 30

Pages 108–109

1. 4.3 **2.** 2.3 **3.** 3.34 **4.** 0.5 **5.** 0.02 **6.** 0.002 **7.** 2.4 **8.** 0.65 **9.** 0.375

Page 110

1. 6 **2.** 30 **3.** 5.2 **4.** 6.874 **5.** 0.06 **6.** 0.0007

Pages 112–113

1. 30 **2.** 600 **3.** $7 **4.** 4 **5.** 50 **6.** 5

Pages 116–117

1. 24.8 **2.** 41 **3.** 4321 **4.** 80 **5.** 600 **6.** 500

Page 119

1. 0.8 **2.** 23.5 **3.** 6.67 **4.** 50.71 **5.** 1.143 **6.** 13.286

Page 120

1. $76,000

CHAPTER 5

Page 136

1. $\frac{3}{7}$ **2.** $\frac{2}{7}$ **3.** $\frac{5}{6}$ **4.** $\frac{1}{4}$

Pages 138–139

1. 1 **2.** 1 **3.** 0 **4.** 0 **5.** 9 **6.** 36 **7.** 0 **8.** Not possible **9.** 0 **10.** Not possible

Page 140

1. yes **2.** no **3.** yes

Page 143

1. $>$ **2.** $<$ 3. $<$

Page 144

1. $\frac{679}{1000}$ **2.** $\frac{37}{10}$ **3.** $\frac{1795}{100}$ **4.** $\frac{47,332}{10,000}$ **5.** $\frac{32}{100}$ **6.** $\frac{105}{100}$

Pages 146–147

1. 0.375 **2.** 0.063 **3.** 0.583 **4.** 4.917 **5.** 0.57 **6.** 1.7 **7.** 0.0987

Pages 150–151

1. $8\frac{3}{4}$ **2.** $12\frac{2}{3}$ **3.** $\frac{22}{5}$ **4.** $\frac{61}{6}$ **5.** $\frac{93}{10}$

Page 152

1. $2\frac{1}{3}$ **2.** $1\frac{1}{10}$ **3.** $18\frac{2}{6}$ or $18\frac{1}{3}$

Page 154

1. $4\frac{17}{100}$ **2.** $13\frac{18}{1000}$ **3.** $1\frac{9}{10}$ **4.** 5.6 **5.** 22.8 **6.** 2.033

CHAPTER 6

Page 168

1. $\frac{8}{15}$ **2.** $\frac{4}{6}$ **3.** $\frac{1}{10}$ **4.** $5\frac{5}{6}$ **5.** $6\frac{2}{3}$ **6.** $6\frac{2}{5}$ **7.** $\frac{5}{8}$

Pages 170–171

1. $\frac{4}{6}, \frac{6}{9}, \frac{10}{15}$ **2.** $\frac{14}{16}, \frac{21}{24}, \frac{28}{32}$ **3.** $\frac{6}{9}$ **4.** $\frac{18}{20}$ **5.** $\frac{8}{32}$

Pages 172–173

1. $\frac{3}{5}$ **2.** $\frac{3}{8}$ **3.** $\frac{5}{6}$ **4.** $\frac{8}{5}$ **5.** 9

Page 174

1. $\frac{7}{12}$ **2.** $\frac{1}{3}$ **3.** $2\frac{2}{3}$ **4.** 8 **5.** $10\frac{1}{2}$

Page 176

1. 39 **2.** $12\frac{4}{5}$ **3.** $2\frac{3}{4}$ **4.** $1\frac{5}{6}$ **5.** $8\frac{1}{3}$ **6.** $10\frac{5}{6}$

Page 180

1. $\frac{5}{2}$ **2.** $\frac{2}{5}$ **3.** 5 **4.** $\frac{1}{24}$ **5.** $1\frac{1}{7}$ **6.** $2\frac{2}{3}$ **7.** $\frac{1}{10}$

Page 182

1. $3\frac{1}{9}$ **2.** $\frac{9}{34}$ **3.** $\frac{1}{10}$ **4.** $\frac{12}{25}$

CHAPTER 7

Page 197

1. 1 **2.** $\frac{7}{12}$ **3.** $\frac{7}{8}$

Pages 198–199

1. 18 **2.** 12 **3.** 24 **4.** 20

Page 200

1. $1\frac{5}{24}$ **2.** $\frac{5}{9}$ **3.** $1\frac{1}{2}$

Pages 202–203

1. $11\frac{4}{5}$ **2.** $23\frac{2}{3}$ **3.** $21\frac{7}{8}$ **4.** $13\frac{7}{12}$ **5.** $16\frac{1}{20}$

Pages 206–207

1. $\frac{1}{4}$ **2.** $\frac{1}{2}$ **3.** $\frac{1}{2}$ **4.** $\frac{13}{18}$

Page 209

1. $5\frac{1}{4}$ **2.** $12\frac{7}{8}$ **3.** $3\frac{1}{6}$

Pages 210–211

1. $6\frac{5}{6}$ **2.** $6\frac{3}{8}$ **3.** $2\frac{23}{24}$ **4.** $9\frac{1}{3}$

CHAPTER 8

Pages 224–225

1. $\frac{2}{3}$ **2.** $\frac{3}{2}$ **3.** $\frac{3}{5}$ **4.** $\frac{12}{5}$ **5.** $\frac{72}{60} = \frac{6}{5}$ **6.** $\frac{5}{1}$ **7.** $\frac{1}{4}$

Page 227

1. 19.6 miles/gal **2.** 12¢ **3.** $5.35 per hour

Pages 230–231

1. No **2.** Yes **3.** 20 **4.** $2\frac{1}{2}$ **5.** $4\frac{2}{3}$

Page 233

1. 5 oranges **2.** 424 miles

Page 234

1. 15 yds **2.** 10.5 m

CHAPTER 9

Pages 248–249

1. 0.25 **2.** 1.00 **3.** 0.04 **4.** 0.045 **5.** 72% **6.** 700% **7.** 720% **8.** $66\frac{2}{3}$%

Pages 250–251

1. $\frac{31}{100}$ **2.** $\frac{3}{4}$ **3.** 4 **4.** $1\frac{1}{2}$ **5.** $\frac{1}{200}$ **6.** $\frac{1}{6}$ **7.** $\frac{1}{8}$ **8.** $\frac{103}{1000}$ **9.** $\frac{1}{250}$ **10.** $\frac{1}{2500}$

Page 252

1. 25% **2.** 30% **3.** 140% **4.** 230% **5.** $66\frac{2}{3}$% **6.** 62.5% **7.** $183\frac{1}{3}$% **8.** 287.5%

Pages 254–255

1. $x = 60\% \times 70$ **2.** $76 = 19\% \times n$ **3.** $18 = b \times 36$ **4.** $x = 0.5 \times 48$ **5.** $\frac{5}{8} \times a = 32$

Pages 256–257

1. 12.72 **2.** 0.185 **3.** $8\frac{1}{3}$ **4.** 18

Page 259

1. 50% **2.** $166\frac{2}{3}$% **3.** $77\frac{7}{9}$%

Page 263

1. 200 **2.** 18

Page 265

1. $66\frac{2}{3}$% **2.** 5% **3.** 20% **4.** 40%

CHAPTER 10

Pages 284–285

1. km **2.** mm **3.** cm **4.** m **5.** 30 cm **6.** 40 m **7.** 53 mm, 5.3 cm

Page 286

1. 0.187 **2.** 962 **3.** 0.296 **4.** 870

Page 290

1. 10 mL **2.** 400 L **3.** 3200 **4.** 500 **5.** 0.25

Page 292

1. 50 g **2.** 70 kg **3.** 2 t **4.** 250 mg **5.** 3.786 **6.** 7860 **7.** 1.286 **8.** 0.73

Page 296

1. F **2.** C **3.** 100°F **4.** 40°C

Page 298

1. 15 **2.** 2250 **3.** 10 **4.** $1\frac{1}{9}$

Page 300

1. right **2.** larger **3.** smaller **4.** 31° **5.** 145°

CHAPTER 11

Page 312

1. 21 cm **2.** 28 cm **3.** 32 ft

Page 314

1. 28.26 m **2.** 35.482 cm **3.** 18.84 cm **4.** 38.308 mm

Page 318

1. 18 km^2 **2.** 5 sq ft **3.** 6.25 cm^2 **4.** 12.25 sq yd

Pages 320–321

1. 72 cm^2 **2.** 32.1 mm^2 **3.** $7\frac{1}{2}$ sq ft **4.** 21.5 m^2 **5.** 7 km^2

Page 322

1. 24.4 m^2 **2.** 8 km^2

Page 326

1. 50.24 m^2 **2.** 452.16 km^2 **3.** 28.26 m^2 **4.** 78.5 sq ft

Page 328

1. 2

CHAPTER 12

Page 340

1. 24 cm^3 **2.** 1000 m^3

Page 344

1. 62.8 m^3 **2.** 138.16 mm^3

Page 346

1. 8 m^3 **2.** 847.8 mm^3

Page 348

1. 267.95 cm^3 **2.** 113.04 m^3

CHAPTER 13

Pages 368–369

1.–4.

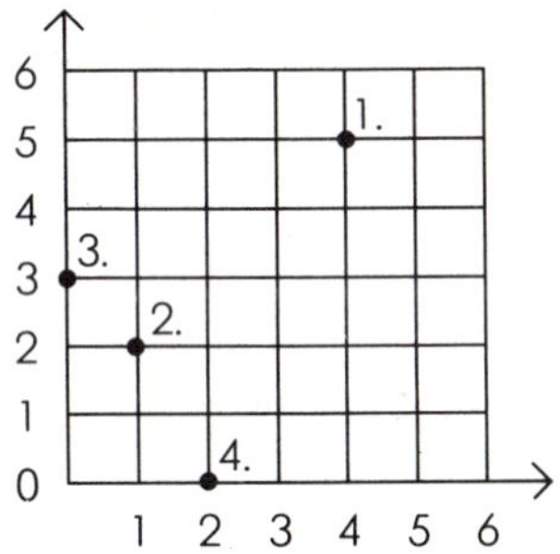

5. (5,9) **6.** (10,6) **7.** (0,7) **8.** (9,0)

Page 370

1.

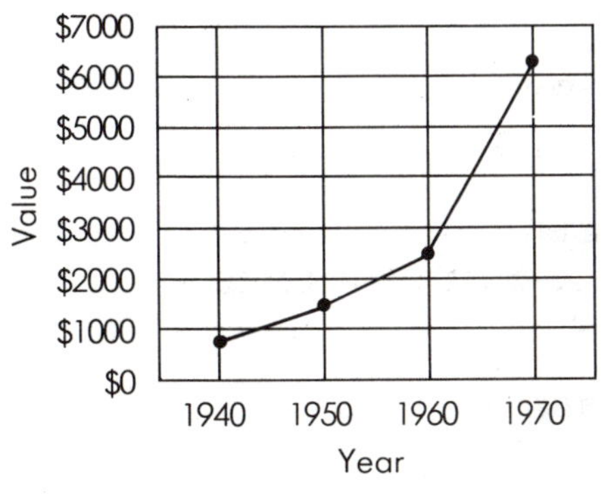

Page 372

1.

Ice Cream Flavor	Calories in a One-Scoop Cone
Chocolate Fudge	230
French Vanilla	220
Strawberry	170
Peach	170
Butter Pecan	200
Chocolate Mint	190

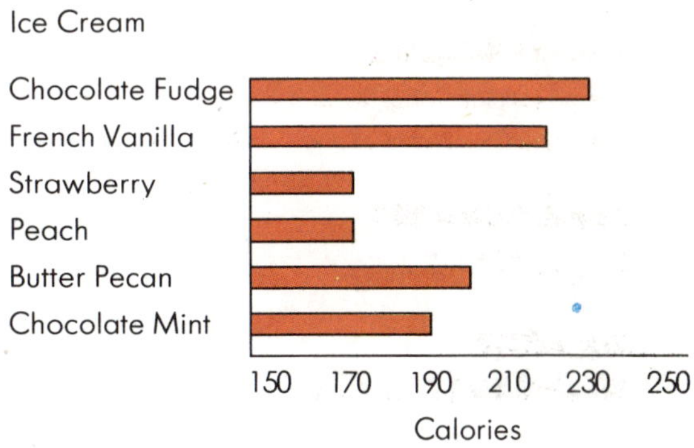

Page 374

1. 20%: 72°; 14%: 50°; 10%: 36°; 26%: 94°; 30%: 108°

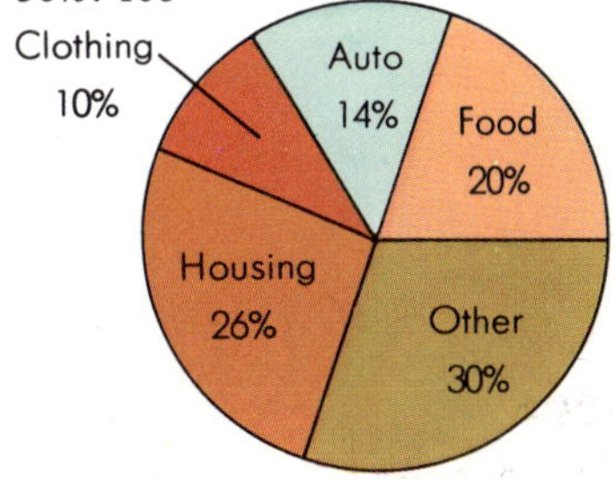

Page 378

1. 80.25 **2.** 152.82 **3.** 29 mi per gal

Page 380–381

1. 17 **2.** 37 **3.** 17 **4.** 8 **5.** 78 **6.** 101, 103

CHAPTER 14

Page 394–395

1. -8 **2.** -6 **3.** 6 **4.** -10 **5.** $<$ **6.** $>$ **7.** $<$ **8.** $<$ **9.** $>$ **10.** $<$

Pages 396–397

1. -2 **2.** -4 **3.** 0 **4.** 0 **5.** 2 **6.** -4

Page 398

1. -11 **2.** -29 **3.** 3 **4.** -5 **5.** 7

Pages 402–403

1. -3 **2.** -8 **3.** -9 **4.** 1 **5.** 10 **6.** 11

Pages 404–405

1. 5 **2.** 6 **3.** -6 **4.** -12 **5.** 0 **6.** -9 **7.** -4 **8.** 4 **9.** 0

Page 406

1. 24 **2.** 126 **3.** 48 **4.** -42 **5.** -18 **6.** -100

Page 408

1. 7. **2.** 14 **3.** 4 **4.** -6 **5.** -3 **6.** -5

CHAPTER 15

Page 422

1. 16 **2.** 10 **3.** 300 **4.** -424

Page 426

1. -12 **2.** 11 **3.** 35 **4.** 1

Pages 428–429

1. -8 **2.** -6 **3.** 405 **4.** -104

Page 430

1. 5 **2.** 9 **3.** -4 **4.** 4

Page 432

1. -9 **2.** \$24

Page 437

1–2.

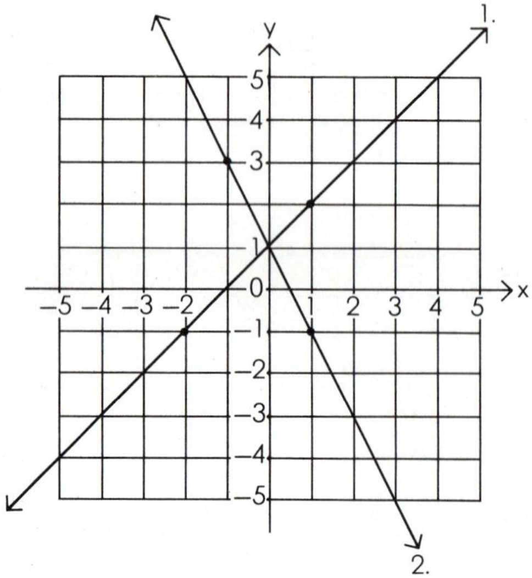